To Mr. Glasnapp

With complements

Reda

Juli 2001

TELECOMMUNICATION NETWORK INTELLIGENCE

IFIP - The International Federation for Information Processing

IFIP was founded in 1960 under the auspices of UNESCO, following the First World Computer Congress held in Paris the previous year. An umbrella organization for societies working in information processing, IFIP's aim is two-fold: to support information processing within its member countries and to encourage technology transfer to developing nations. As its mission statement clearly states,

IFIP's mission is to be the leading, truly international, apolitical organization which encourages and assists in the development, exploitation and application of information technology for the benefit of all people.

IFIP is a non-profitmaking organization, run almost solely by 2500 volunteers. It operates through a number of technical committees, which organize events and publications. IFIP's events range from an international congress to local seminars, but the most important are:

- The IFIP World Computer Congress, held every second year;
- open conferences;
- working conferences.

The flagship event is the IFIP World Computer Congress, at which both invited and contributed papers are presented. Contributed papers are rigorously refereed and the rejection rate is high.

As with the Congress, participation in the open conferences is open to all and papers may be invited or submitted. Again, submitted papers are stringently refereed.

The working conferences are structured differently. They are usually run by a working group and attendance is small and by invitation only. Their purpose is to create an atmosphere conducive to innovation and development. Refereeing is less rigorous and papers are subjected to extensive group discussion.

Publications arising from IFIP events vary. The papers presented at the IFIP World Computer Congress and at open conferences are published as conference proceedings, while the results of the working conferences are often published as collections of selected and edited papers.

Any national society whose primary activity is in information may apply to become a full member of IFIP, although full membership is restricted to one society per country. Full members are entitled to vote at the annual General Assembly, National societies preferring a less committed involvement may apply for associate or corresponding membership. Associate members enjoy the same benefits as full members, but without voting rights. Corresponding members are not represented in IFIP bodies. Affiliated membership is open to non-national societies, and individual and honorary membership schemes are also offered.

TELECOMMUNICATION NETWORK INTELLIGENCE

IFIP TC6/WG6.7 Sixth International Conference on Intelligence in Networks (SmartNet 2000), September 18–22, 2000, Vienna, Austria

Edited by

Harmen R. van As
Institute of Communication Networks
Vienna University of Technology
Austria

KLUWER ACADEMIC PUBLISHERS
BOSTON / DORDRECHT / LONDON

Distributors for North, Central and South America:
Kluwer Academic Publishers
101 Philip Drive
Assinippi Park
Norwell, Massachusetts 02061 USA
Telephone (781) 871-6600
Fax (781) 871-6528
E-Mail <kluwer@wkap.com>

Distributors for all other countries:
Kluwer Academic Publishers Group
Distribution Centre
Post Office Box 322
3300 AH Dordrecht, THE NETHERLANDS
Telephone 31 78 6392 392
Fax 31 78 6546 474
E-Mail <services@wkap.nl>

Electronic Services <http://www.wkap.nl>

Library of Congress Cataloging-in-Publication Data

A C.I.P. Catalogue record for this book is available from the Library of Congress.

Printed on acid-free paper.

Printed in the United States of America.

Contents

Preface

SmartNet'2000 is the sixth in a series of conferences on Intelligence in Networks started by IFIP TC-6 Working Group 6.2 (Broadband Communications), the events being a preceding Workshop on IN, Lappeenranta (August 1994), the Working Conference on IN, Copenhagen (August 1995), the Workshop on Broadband Network Intelligence (WBNI'96), Espoo (November 1996), and the Conference on Intelligent Networks and Network Intelligence (2IN97), Paris, (September 1997). The series now belongs to the area of Working Group 6.7 (Smart Networks) with previous conferences held in Moscow (Smartnet'98, February 1999, postponed from 1998) and Bangkok (Smartnet'99, November 1999). The conference series has been established to be a forum for discussion of issues related to the development of intelligent capabilities in telecommunication networks. These topics generally include the creation, distribution, and management of telecommunication services in narrowband networks, broadband networks as well as mobile networks.

In the world of telecommunications, the ultimate goal is to provide network customers with a variety of versatile services. Today flexibility is a key design issue for emerging network architectures and service platforms in order to be able to adapt instantly to changing customer service demands. In contrast to the traditional way of implementing intelligent network services in centralized service nodes which control the switching nodes of the transport network using a dedicated signaling network, the current trends are towards implementing distributed network intelligence, eventually supported by mobile software agents and active networks. The basic idea of these last two directions is the movement of service code directly to network switching nodes and network control nodes as well as between all kind of applications in the end systems. In addition, this service code movement should be possible in a highly dynamic manner, thus allowing automated, flexible, and customized provision of services in a highly distributed manner. This would enable better service performance as well as optimized control and management of the transport network capabilities.

Two enabling technologies are crucial in open, active and programmable networks: programmable network nodes providing flexibility in the design of connectivity control applications, and platforms for mobile software agents allowing dynamic downloading and movement of service code to specific network nodes. The intelligence inside the service agents permits distributed configuration and provision of service intelligence inside the network, thereby relaxing the load of service control and management systems as well as that of the signaling traffic.

Research in networking and flexible service platforms is exploring ways in which network nodes and end systems may be dynamically programmed by network operators, their customers, and third parties. The advantages of customized routing and communication protocols, dynamic allocation and reallocation of resources among service classes, flow-specific information processing in network nodes, and fast and adaptive service creation are extremely attractive, but the basic conditions of security, reliability, and performance still require considerable efforts.

The proceedings of SmartNet'2000 reflect research activities all around the world for achieving these goals. The discussed topics in the conference are on telecommunication service architectures, flexible service creation, service distribution and management, security services, advanced intelligence in networks, application of agent technology for routing and network control, performance monitoring, network management, QoS management, mobility management, interactive multimedia as well as dynamic switching and active networks.

Many persons have contributed to make this conference a success. I would like to thank the members of the organizing and technical committees, the authors and reviewers, the Vienna University of Technology, the sponsors of the conference, and the staff of the Institute of Communication Networks. Particular thanks are due to Johanna Pfeifer of my institute and Dr. Reda Reda of Siemens Austria for their very successful and highly appreciated organizational support.

Harmen R. van As   Conference Chair   Vienna, September 2000

International Program Committee

Harmen R. van As (Chair)
Finn Arve Aagesen, Norway
Ian F. Akyildiz, USA
Joseph Bannister, USA
Torsten Braun, Switzerland
Augusto Casaca, Portugal
Stefan Covaci, Germany
Andre Danthine, Belgium
Jörg Eberspächer, Germany
Serge Fdida, France
Luigi Fratta, Italy
Maurice Gagnaire, France
Dominique Gaïti, France
Nicolas D. Georganas, Canada
Mario Gerla, USA
David Hutchison, UK
Marjory Johnson, USA
Guenter Karjoth, Switzerland
Ralf Lehnert, Germany
Helmut Leopold, Austria
Thomas Magedanz, Germany
Kurt Maly, USA
Marco Ajmone Marsan, Italy
Olli Martikainen, Finland
Peter Martini, Germany
Ignas Niemegeers, Netherlands
Radu Popescu-Zeletin, Germany
Ramon Puigjaner, Spain
Guy Pujolle, France
Kurt Rothermel, Germany
Jon Schuringa, Austria
Plamen Simeonov, Germany
Otto Spaniol, Germany
Ralf Steinmetz, Germany
Burkhard Stiller, Switzerland
Ahmed Tantawy, USA
Samir Tohme, France
Johnny Wong, Canada
Adam Wolisz, Germany
Kenichi Yoshida, Japan
Martina Zitterbart, Germany

Local Organizing Committee

Reda Reda, Siemens Austria (Chair)
Harmen R. van As, Vienna University of Technology, Austria
Helmut Leopold, Telekom Austria
Georg Mittenecker, Vienna University of Technology, Austria
Rudolf Pailer, Ericsson Austria
Johanna Pfeifer, Vienna University of Technology, Austria
Heinz Weiskirchner, Alcatel Austria

Reviewers

Markus Albrecht
Joe Bannister
Joachim Baumann
Torsten Braun
Augusto Casaca
Sarolta Dibuz
Jörg Eberspächer
Matthias Frank
Maurice Gagnaire
Dominique Gaïti
Nicolas D. Georganas
Mario Gerla
David Hutchison
Ilias Iliadis
Guenter Karjoth
Michael Koester
Ralf Lehnert
Alexander Leonhardi
Helmut Leopold
Kurt Maly
Olli Martikainen
Georg Mittenecker
Ignas Niemegeers
Vladimir Popovic
Aiko Pras
Ramon Puigjaner
Kurt Rothermel
Christoph Ruland
Jochen Schiller
Jon Schuringa
Mathias Schweigel
Christian Schwingenschloegl
Otto Spaniol
Johannes Stadler
Roland Steiner
Ralf Steinmetz
Burkhard Stiller
Samir Tohme
Harmen R. van As
Els Van Herreweghen
Hans-Jörg Vögel
Adam Wolisz
Johnny Wong
Kenichi Yoshida
Martina Zitterbart

Part One

QoS and Security Services

INTERNET SERVICE DELIVERY CONTROL WITH MOBILE CODE

Manuel Günter and Torsten Braun

Institute of Computer Science and Applied Mathematics

University of Berne

Neubrückstrasse 10, CH-3012 Bern, Switzerland

[mguenter,braun]@iam.unibe.ch

Abstract The trend towards value-added Internet services causes network providers to deploy new network based quality-of-service and security services. Today, however, the customer has only limited means of controlling the service delivery. For example the network security guaranteed by virtual private network providers cannot be checked with a traditional static approach. This paper presents a novel approach for controlling new IP services using mobile code, and motivates the approach with two examples of new IP services proposed by the Internet engineering task force (IETF).

Keywords: Mobile agents, network monitoring, value-added Internet services, virtual private networks, differentiated services.

1. INTRODUCTION

The rapid growth of the transport capacity of the Internet and the global trend towards liberalisation of the telecommunication market forces the Internet service providers (ISP) to look for new revenues beyond pure connectivity offerings. Therefore, ISPs that control their own network try to introduce new Internet services including quality features such as premium transport or traffic privacy. Since ISPs have control over a (albeit small) portion of the Internet, they can deploy new technologies such as Differentiated Services (DiffServ) or the Internet Protocol Security architecture (IPSec) to enhance their network service. The deployment of such services brings some challenges, for example: How can enhanced services be set up dynamically? How to provide services that demand for collaboration between providers? In earlier work (Günter et al., 1999) we propose a generic architecture to cope with these problems. However, there are two problems to solve that go beyond the deployment of the service, namely:

1 How to convince the user of the presence of the new quality. For example the privacy of a communication is not easily demonstrated. From the user's point-of-view this is equivalent to the question how can (s)he control if (s)he gets the quality the provider promised.

2 For services involving several providers the question is how to find out who is responsible when the service quality is less than guaranteed to the customer.

It is in the interests of the customers and of honest providers that the customer is able to verify the steady quality of a service and to locate problems when they occur. We refer to this process as *service delivery control* (SDC).

For today's Internet services, there is only very limited support for service delivery control. If a customer happens to detect a problem (which is usually when the customer needs that service badly and does not get it), phone-calls between administrators, local measurements, and manual browsing of log-files will eventually lead to the identification of the problem source. Unfortunately, it is also not uncommon that the involved parties will suspect each other and repudiate any guilt. Note, that this problem not only concerns the relation between customer and provider but also between providers themselves. It is to be expected that the problem becomes worse when new and more expensive network services are deployed that require provider collaboration.

This paper proposes a generic service delivery control architecture based on mobile code. Mobile code allows us to test the service where it is delivered. Thus misbehaviour can be located, even if the provider that causes the problem tries to hide the tracks. Furthermore, mobile code offers a flexible way to deploy new tests for new network services.

The next section will position our work in the field of mobile code research and motivate our approach. In section 3 we will present the architecture of the infrastructure necessary to perform service delivery control with mobile code. Then, we will back-up our arguments by studying two emerging network services proposed by the Internet Engineering Task Force: section 4 describes a virtual private network service and how the privacy that it provides can be controlled. Section 5 discusses service control for differentiated services across multiple domains. Section 6 describes related work in the area of network measurements and section 7 concludes this paper.

2. MOBILITY AND SERVICE DELIVERY CONTROL

2.1. TERMINOLOGY

Today, there are many different trends in the research of mobile code. At the application level the most prominent instance of mobile code is called *mobile agents* (White, 1994, Chess et al., 1997). These are program instances that are able to move self-directed through a network to locally perform a task in behalf of their sender. Different mobile agent platforms have been proposed e.g. for the programming languages Java (Lange and Oshima, 1998, Vitek and Bryce, 1999, Fünfrocken, 1998) and Tcl (Gray, 1998). The term agent is also occupied by other research communities, namely the artificial intelligence research (intelligent agents) (Murch and Johnson, 1999) and the software engineering community (software agents) (Wooldridge and Jennings, 1999). Both communities have influenced the mobile agent research, so a mobile agent is also a paradigmatic software abstraction and includes autonomous behaviour (intelligence). Mobile agents are proposed for different tasks such as network search (more recently e-commerce (Hulaas et al., 1999)), network management (Baldi et al., 1997) and network intrusion detection (Jansen et al., 1999).

On the network level, the emerging mobile code technology is called *active networking* (Tennenhouse et al., 1997, Calvert et al., 1998). The mobile code is often referred to as *capsule* and is directly integrated into the network traffic packets. Thus, the code flows directly on the communication path that is subject of the code's computation and it can be executed on a per-packet granularity. Here, the abstraction and intelligence aspect is secondary. The focus is on the interaction with the network infrastructure. Active network packets access the networking functionalities of the routers they pass (e.g. forwarding and routing) and change these functionalities for packets or classes of packets. Furthermore, performance is a crucial issue, since the code should be able to manipulate data at the line speed (in todays backbone network this can be up to several gigabits per second).

Still, there is no solid line between mobile agents and active networking. For example the active networking testbed ANTS (Wetherall et al., 1998) can also be seen as a mobile agent testbed, since capsules are Java objects, and the code is not included in network data packets but is dynamically loaded upon need. The approach that we describe in this paper cannot be classified clearly as mobile agents or active networking. Service delivery control agents examine network services down to the structure of forwarded network packets. Furthermore, the SDC agents don't necessarily need to be intelligent. They can just collect whatever data the customer is looking for and send it back to a customer application at home. Also, the performance of the SDC agents is an issue since one of their goals can be to monitor the network at wire speed. For these reasons, SDC agents seem to be an application of active networking. However, the control agents do not necessarily need to travel in capsules but can be transferred out-band of the data traffic, like mobile agents do.

2.2. WHY MUST THE SERVICE DELIVERY CONTROL BE MOBILE?

Mobile Code has the questionable reputation of being a solution in search of a problem (J. Ousterhout). In this section we will argue why mobility is necessary for service delivery control agents.

Generic interface. No provider will open its network administration system to the customers. By providing an agent platform at relevant sites (see section 3) the provider can give access in a controlled fashion. From the customer's point-of-view the control agents can perform whatever tests that the customer thinks is necessary. It is easy to introduce new tests for new network services.

Cross checking. A misconducting provider can easily fool a customer that relies on the measurements published by the provider. End-to-end measurements can in some cases indicate to the customer that the service is not delivered as guaranteed. But in the case of a multi-provider service such measurements cannot identify the misconducting provider. SDC agents can be sent out to perform active measurements by producing and measuring traffic at different sites. Mobility allows the agents to virtually 'track-down' the problem source.

In order to bring more substance to these theoretical advantages, we need to specify the architecture of the supporting infrastructure. Then we can demonstrate these theoretical advantages on concrete examples.

3. A SUPPORTING INFRASTRUCTURE FOR SERVICE DELIVERY CONTROL AGENTS

Like any other network monitoring system, the SDC agents need a supporting infrastructure. In this section we discuss the required components.

3.1. LOCATION OF THE CONTROL POINTS

The Internet is a heterogeneous network, it consists of thousands of administrative domains. The interior network of these domains is administered in different ways and consists of different kinds of networking technologies such as Frame Relay, ATM, MPLS or Sonet. This may render the access to the traffic inside of the domain very difficult (e.g. for optically switched technology). The least common denominator is the Internet Protocol (IP). The IP traffic is exchanged between the domains at so-called peering points, according to peering- or service level agreements. While the network engineering and management of the interior network of the domains is usually hidden, the peering points are by their nature open (at least to the peer). For service delivery control the peering points are thus of high interest. Note, that is suffices to track down a problem to a provider. Once the problem is found to relate to a given administrative domain, it is up to its administration to further locate the problem in the inside of their network, using the network management system of their choice. Therefore, the SDC agent nodes should be located at the peering points. This guarantees, that the control has access to the IP traffic and that the control can relate identified problems to a specific provider. Of course, a provider can also offer additional control points in the inside of its network as an additional service to its customers or for its own service control purposes.

Figure 1 shows SDC agents which were sent out by a customer application running on a machine owned by the customer. The customer application also coordinates the agents, processes their feedback and forwards the results to the user. The agents migrate to the peering points to perform particular local checks on the service.

3.2. NODE ARCHITECTURE

The SDC agents should be able to perform any kind of passive measurements, however they should not be able to eavesdrop or analyse traffic of other customers. To perform active measurements, the agents should be able to send traffic, but again, this should be traffic related to the specific customer. Spoofing of foreign IP addresses or denial-of-service attacks should not be facilitated. Given these requirements we foresee the following node architecture as depicted in figure 2: At the peering router, there is a *T-component* that serves as a high-performance and configurable packet copying mechanism. The T-component can be configured to copy network packets according to filtering rules based on IP packet information such as source and destination address (see section 3.3). It adds a high-accuracy time-stamp to the packet. For optimisa-

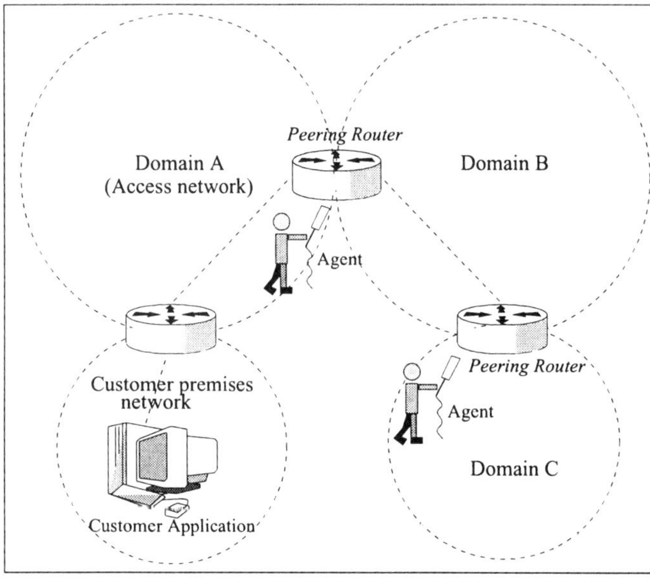

Figure 1 Measuring at peering points.

tion it can be told to only copy the packet headers. The T-component forwards the requested packet copies to the *Node environment.* Note, that for security reasons the agents do *not* have direct access to the T-component. If the peering router runs under UNIX, the `tcpdump` (Jacobson et al., 1989) command can be used as T-component.

The *node environment* is hosting and executing the SDC agents. It does not have to run on the peering router. In fact, no provider will want to run foreign code on such a crucial machine, so the node will probably run on a separate machine in a 'demilitarized zone', but has to be connected to the T-component. Customers push their agents into the node. The agent does not necessarily have be encrypted, but a strong authentication protocol is needed. This ensures that the node can properly authorise the agent. When the agent arrives at a node, it has to undergo a welcome procedure. After the authentication, the agent asks the node for resources (CPU time, memory and specific traffic). The agent also specifies a packet filter for the bypassing packets it is interested in (see section 3.3). Based on policies, the node authorises the agent for these resources. It provides an execution environment (user-level thread in a 'sand-box'). The agent execution environment also contains an inbound and an outbound packet queue, which is the only way the agent can receive and sent packets.

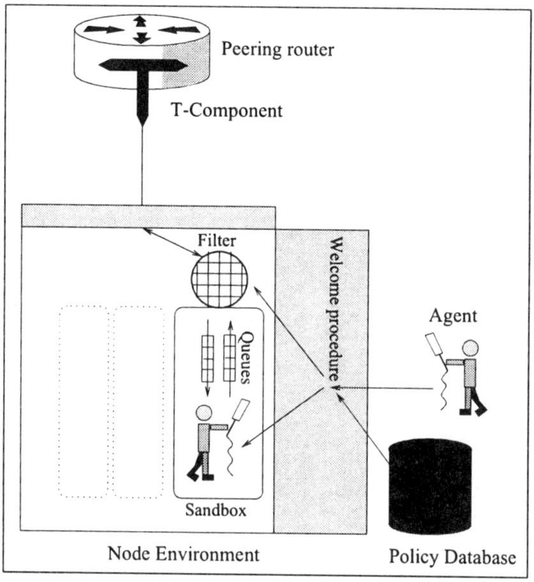

Figure 2 The node environment.

3.3. AUTHORISATION AND FILTERING

The agents are effectively separated from the network by filters on the in- and outbound queue. The agents carry a definition of the desired filters. The node environment has policies describing what filters are appropriate for what agent. We propose a mechanism using the mathematical cut between two sets. Both the agent and the node describe the filtering rules as a set of integers (addresses, protocol numbers etc.). We foresee three kinds of wildcards: an integer range, an integer list and 'any' (matching everything). While the agent holds a filter describing what specific traffic it is looking for, the node holds the most general filter it allows for that agent. The node uses the mathematical cut of both filters. The agent can query if the resulting filter is empty (matches no packets at all) or not equal to what is has requested, and react upon this (e.g. terminate gracefully). The node holds generic filters in its policies so that it does not need to keep a filter for each potential customer.

For example the university of Berne (Switzerland) owns the network 130.92.0.0/16. Suppose that the system administration of the university is interested in traffic that a specific institute with the subnet 130.92.-66.0/24 sends across the trans-atlantic link to the US. Thus, the administrator would send an agent to the peering point between the Swiss

University Internet Provider (SWITCH) and its US peer MCI Worldcom. The agent would be signed by the system administration of the university and contain a filter specifying the source address with a range wildcard which ranges from -2107883008 to -2107882753 (2-complement of the smallest and the biggest 32 bit address in 130.92.66). Suppose the node would contain a policy saying that the system administrator of the university of Berne is allowed to see any traffic originating form their network, thus its source filter entry is the range wildcard -2107899904 to -2107834369 (130.92.0.0/16). The mathematical cut of the generic node filter and the agent's filter will deliver the desired packets to the agent, since the range specified by the agent is a subset of the generic range assigned by the node. Note that the cut between lists and ranges is always empty or lists or ranges which facilitates the object-oriented implementation of the filters.

Our prototypical and object-oriented filter implementation features filtering for source and destination address, protocol number, the maximum number of packets to receive and the length of payload to be delivered to the agent. It supports the briefly described wildcard mechanism.

3.4. SECURITY ISSUES

The security of the proposed infrastructure bases on three concepts. First and foremost, the agents must authenticate themselves with strong cryptography. We do not foresee to implement this mechanism ourselves but rather rely on existing and stable technology such as PGP (Zimmermann, 2000), or built-in mechanisms of available agent platforms. Authentication allows the node to relate each agent to a customer, which is responsible for the behaviour of the agent. Second, the agents do not run on the controlled network devices but rather on a dedicated general-purpose computer. Third, the agents run in a sand-box. They have no direct access to neither node nor network resources. Their only communication mechanism uses the in- and outbound queues which are controlled by node filters. The cutting of agent filters with a default filter provided by the node assures in a convenient way that the agents cannot eavesdrop or spoof other peoples traffic.

This architecture is the basis for service delivery control with mobile code. Since the architecture is non-intrusive and only relies on basic agent mechanisms such as authentication and an execution sand-box, we believe that state-of-the art agent technology (Lange and Oshima, 1998, Vitek and Bryce, 1999, Fünfrocken, 1998, Gray, 1998) can provide most parts of such a platform, and that such a platform can be deployed in the Internet. The only missing key component is thus the T-component (and the agents themselves). The next two sections describe two specific applications of the platform.

4. CONTROLLING A VIRTUAL PRIVATE NETWORK SERVICE

Virtual private networks (VPN) for the Internet (Ferguson and Huston, 1998a, Ferguson and Huston, 1998b) provide a transparent and secure mechanism to interconnect remote sites with IP (see figure 3). IP packets are encapsulated in new IP packets when entering the Internet (tunnelling). The payload of the new packet (the original packet) is encrypted. Virtual private networks over the Internet are a cheap and secure alternative to leased line based private corporate networks. They take advantage of the ubiquitousness of the Internet and the trend towards Intranets[1]. The Internet Engineering Task Force (IETF) proposed the VPN standard IPSec (Kent and Atkinson, 1998), which is supported by many vendors nowadays. However, VPNs and especially their cryptographic mechanisms are difficult to understand and manage (Günter et al., 1999). Therefore, service providers begin to offer VPN services where they setup and manage the tunnel end-points for their customers. However, the security is transparent to the customer. The customer believes, that all the IP traffic that is leaving the network is encrypted. But encryption is computational intensive. How can the customer be sure, that the provider is really performing the IPSec protocol properly and not just e.g. compressing the payload?

A VPN control agent. Given an agent infrastructure as described in section 3 we have many possibilities to check whether the VPN provider is performing as promised.

- The customer can occasionally send out agents to the egress peering points of the access network (see figure 1). These agents check for traffic originating from internal network addresses of the customer's network. The agents should not find any such packets if the service is working properly, since the packets should be encapsulated.
- The customer can send out agents that monitor the IPSec protocol activities. Agents can for example monitor the presence[2] of the key

[1] Intranets: Corporate networks based on IP technology.
[2] The IKE protocol of IPSec is encrypted, therefore not much more than the presence can be monitored.

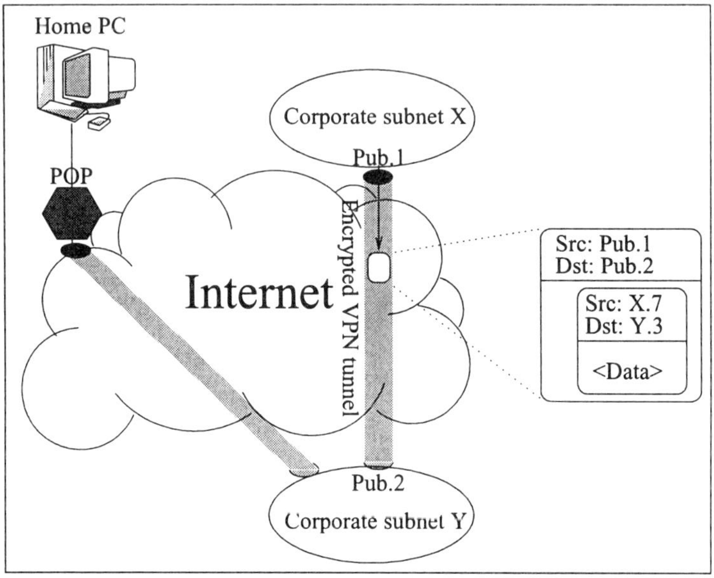

Figure 3 Application scenario for virtual private networks.

exchange protocol IKE (UDP to port 500). The agents can also analyse the packet structure to see if the proper tunnelling modes are used (examine the IP `protocol` field: 50 for ESP, 51 for AH (Kent and Atkinson, 1998)).

- The customer can validate the encryption using statistical tests. This is the most difficult task, thus we explain it in more detail.

Statistical encryption checks. The VPN delivery control agent has also (albeit limited) possibilities to validate the quality of the encryption. Such an agent requests from the node (some) packets that are IPSec encrypted (protocol=50, Encryption Security Payload). The nodes of VPN providers can accept that request, because the payload should be encrypted, so the privacy of the sender is not compromised. Paranoid providers may also limit such access to agents of their customers. The VPN delivery control agent can now apply statistical tests to the payload or to parts of it. The motivation behind this is that a good encryption scheme scrambles the bits so that they look random. Statistical tests can detect regularities in the payload which are a sure sign that no valid encryption scheme has been used.

So far, we implemented two simple statistical tests, namely (1) the byte frequency test which counts the occurrence of each byte value [-128..127] and tests for uniform distribution (Knuth, 1981). (2) The run-test divides the byte stream in sequences of increasing (decreasing) bytes. It counts the number of occurrences of sequences with lengths 1,2,3,4,5 and 6 or more (Knuth, 1981). For example the byte sequence $(-33, 104, | -45, 3, 34, |7, | -19, | -93, 1)$ contains two (increasing) runs of length one, two of length 2 and one of length 3. Both tests are evaluated by comparing the counts (bytes or run-lengths) with an expected distribution. The evaluation uses the well-known χ^2 statistics (Knuth, 1981).

The VPN delivery control agent requests encrypted payload and classifies the data into categories (byte values or run-lengths). After all packet data is classified and a significant[3] amount of data has been collected, the agent evaluates the test. Note, that all statistical tests only deliver probability values and not absolute values. The tests indicate the probability that the tested data was produced by a uniform distributed and independent random function. If the data is significantly off the expected distribution the agent can sent back an alert to the customer application. In case of doubt, the agent can also send back the last packet to be examined by a human expert.

The two proposed statistical tests are by far not the only possible tests. They suffice for our purpose since both tests can detect if compression instead of encryption is used (Braun et al., 2000). The genericity of the agent infrastructure allows the agent programmer to easily deploy other tests such as the Anderson-Darling test (Paxson et al., 1998).

Implementation and Evaluation. We have implemented a prototypical agent written in the programming language Java to perform the described tests. However, performance tests indicated some limitations, since statistical tests require significant computation. Running on a Sparc ULTRA 5 with a 269 MHz CPU, the agent could test 1.5 Mbps encrypted data with the run-test and 1 Mbps data with the byte-frequency test. In case the agent wants to monitor a line with higher throughput, it can choose not to analyse every byte in every packet, since sample testing will also suffice to detect misbehaviour.

It is important to note, that traditional customer premises and stationary control programs are under no circumstances able to perform the VPN checks we described in this section. The application would need to have insight in what is going on inside of the Internet, that goes far

[3]This is dependent on the test, e.g. for the byte-frequency test it is about 3000 bytes.

beyond from SNMP (Case et al., 1990) or web-based network management entries. This necessity of code mobility is not VPN specific as the example of the next section shows.

5. CONTROLLING DIFFERENTIATED SERVICES

DiffServ is a light-weight and scalable QoS mechanism proposed by the IETF (Blake et al., 1998). A single byte (DiffServ Code Point (DSCP), formerly called TOS) in the IP header is used to code different per-hop behaviours (PHB) that an IP packet can experience. Inside of a network, all IP traffic using the same code point is called a DiffServ behaviour aggregate and is treated the same way. Since there are only a handful of PHBs, the DiffServ architecture scales also to large core networks. To provide DiffServ across multiple administrative domains the DiffServ architecture proposes automated bandwidth brokers (Terzis et al., 1999) to negotiate service level agreements (SLA) between different autonomous systems. These agreements describe the volume of DiffServ traffic that can be exchanged between two domains and the price for that traffic. If all the domains between two end users have engineered their networks properly and have established SLAs for the DiffServ volume expected, the DiffServ architecture can guarantee end-to-end QoS. However, today's network engineering mechanisms work with overprovisioning or introduce large signaling overhead (Günter and Braun, 1999). Therefore, a provider may deliberately try to over-book its SLAs in order to save money. In case the misconducting provider looses in-profile premium packets due to internal network congestion and insufficient SLA provisioning, it can always argue that it has never received the packets or that another provider down the stream has lost the packets. If the customer has only end-to-end measurements available (s)he cannot detect which of several providers causes the problem.

Given the infrastructure described in section 3, the suspicious customer can send out traffic measurement agents. These agents report back current statistics, and allow the home application to track down a possible problem. Note that for this purpose, mobile agents are not really necessary. Publicly readable measurement information databases (e.g. SNMP) would also do. However, it is much more difficult for the misconducting provider to manipulate local measurement agents than to manipulate local SNMP tables. For the later, some simple `put` commands will cover the traces. For the former the provider needs to analyse the agent code to understand how to trick it into sending false data. The analysis must be done online, since the customer can upload the agent at any time. Even if the provider manages to trick the agent, the customer has tracked down the problem to either the misconducting provider or the neighbour provider, on which the former tries to put the blame on. With mobile agents we can now put both providers under the test using active measurements to single out the bad guy.

Active measurements. Instead of just measuring what is being sent end-to-end (passive measurements), the agents provide the customer with the unique ability to actively generate test traffic from within the network. To test a suspect provider the agents can surround the network of the provider migrating to the closest node that is not under the control of the provider. From there, they can inject small amounts of test traffic. This traffic may use source addresses of the domain of the customer (and the DSCP to be tested) to seamlessly merge with the regular traffic of the customer. The measurement results will now reveal if the provider causes the problem or not. Note, that a provider might manipulate the active measurement being conducted on its neighbour. In general, however there are more than one neighbours to the tested provider, thus such manipulation can be detected once the different measurement results are compared. The measurements have thus to be coordinated. This issue, however, is ongoing research. Nevertheless, it is obvious that the described active measurements combined with the mobility of the SDC agents is a much more powerful tool for performance tests than static methods such as SNMP requests.

Implementation and Evaluation. We implemented agents for local jitter and bandwidth measurements. To perform such measurements the agent needs only to requests the packet headers and the timing information provided by the T-component (see section 3.2). The local jitter can be calculated by evaluating timing information of consecutive packets. For through-put calculation the packet length field provides the necessary information. Since the agent evaluates packet headers only, Java agents perform sufficiently for backbone wire speed.

The implementation work is in an early stage. We foresee to implement agents to account packet loss, delay (coping with the clock skew problem) and packet corruption as well.

6. RELATED WORK

We already referred to work in the area of active networking and mobile agents. None of this work deals with service delivery control. In this section we compare our approach to traditional ways of service delivery control, which is based on measurements.

Network measurement is by its nature a distributed task. Even the old but nevertheless useful `ping` tool needs a source and a destination (to reflect the ICMP message). Today's measurements can be divided in two groups: *end-to-end measurements*, usually carried out by customers and *network traffic measurements* carried out by network providers.

Most end-to-end measurements on the Internet are performed at a small scale to find out local connectivity problems. Large scale measurements which can reveal insights in the Internet topology impose great logistical difficulties (Bolot, 1993). A state of the art approach is to off-line distribute measurement daemons, that are run by the local system administration (Paxson, 1997). The possibilities of mobile code is not exploited here.

Measurements inside of the Internet are carried out by almost any provider, in order to engineer their networks. Tools are often SNMP based, or specialised to the involved network equipment (e.g. Netflow (Cisco, 2000)). Often, providers hesitate to make any results of these measurements publicly available, because they fear to offer attacking points to their competitors. In academically influenced networks, network measurement data is sometimes available for via http, e.g. at the web site of the national laboratory for applied network research (National Laboratory For Applied Network Research, 2000). However, such data is too aggregated for service delivery control. Architectures for fine-grained traffic data repositories have been proposed (Kato et al., 1999). However, since they do not collect the data on a per-service and per-customer basis. All collected data enters the same database. Therefore, they need to scramble the origin of the data for privacy reasons. This will also render the data useless for some SDC applications.

To our knowledge, mobile service delivery control agents are a novel approach to an emerging problem of new Internet services, namely service delivery control. The most closely related work is probably the Q-bone measurement framework of the Internet2 (Teitelbaum et al., 1999). However, this work uses provider-oriented stationary measurements which are not intended for the use by the customer. Furthermore, the measurements focus on the supervision of one specific network service, namely Differentiated Services (see section 5).

7. CONCLUSION

This paper presents the concept, the architecture and the motivation behind using mobile code for the on-line control of the delivery of Internet network services. The approach is classified as a hybrid approach between mobile agent technology and active networking. The paper identifies the advantage of mobile service delivery control agents: 1) the generic and secure interface to measurements, which can be carried out inside remote networks, 2) the ability to actively perform cross checks inside remote networks. The proposed agent infrastructure is kept simple including a threefold security concept, thus keeping the overhead of mobile code low. The paper illustrates the advantages by presenting two examples of new network services proposed by the Internet Engineering Task Force: IPSec based virtual private networks (VPN), and differentiated services (DiffServ). Specialised measurement agents can test both services to a degree that is not feasible with stationary measurement techniques.

Acknowledgements. This work has been performed in the framework of the project Advanced Network and Agent Infrastructure for the Support of Federations of Workflow Trading Systems - ANAISOFT (SPP ICS ElCom Project 5003-057753) which is funded by the Swiss National Science Foundation (SNF). The SNF also funded the project Charging and Accounting Technology for the Internet - CATI (CAPIV 5003-054559/1 and MEDeB 5003-054560/1) which supported this work in its early stages. The authors would like to acknowledge all contributions from their current and former project and working colleagues. The implementation environment for the SDC agents is funded by the SNF R'Equip project no. 2160-053299.98/1 and the foundation 'Förderung der wissenschaftlichen Forschung an der Universität Bern'.

References

[Baldi et al., 1997] Baldi, M., Gai, S., and Picco, G. P. (1997). Exploiting code mobility in decentralized and flexible network management. In *Proceedings of the 1st International Workshop on Mobile Agents*, Berlin, Germany.

[Blake et al., 1998] Blake, S., Black, D., Carlson, M., Davies, E., Wang, Z., and Weiss, W. (1998). An architecture for differentiated services. RFC 2475.

[Bolot, 1993] Bolot, J.-C. (1993). End-to-end packet delay and loss behavior in the Internet. In *Proc. SIGCOMM '93*, pages 289–298.

[Braun et al., 2000] Braun, T., Günter, M., Khalil, I., and Liu, L. (2000). Performance evaluation of virtual private networks. Technical report, Institute for Applied Mathematics and Computer Science.

[Calvert et al., 1998] Calvert, K., Bhattacharjee, S., Zegura, E., and Sterbenz, J. (1998). Directions in active networks. *IEEE Communications*, 36(10).

[Case et al., 1990] Case, J., Fedor, M., Schoffstall, M., and Davin, J. (1990). A simple network management protocol (SNMP). RFC 1157.

[Chess et al., 1997] Chess, D., Harrison, C., and Kershenbaum, A. (1997). Mobile agents: Are they a good idea? In Vitek, J. and Tschudin, C., editors, *Mobile Object Systems: Towards the Programmable Internet*, number 1222 in LNCS, pages 25–45.

[Cisco, 2000] Cisco (2000). Network management. http://www.cisco.com/univercd/cc/td/doc/product/rtrmgmt/.

[Ferguson and Huston, 1998a] Ferguson, P. and Huston, G. (1998a). What is a VPN - part I. *The Internet Protocol Journal*, 1(1).

[Ferguson and Huston, 1998b] Ferguson, P. and Huston, G. (1998b). What is a VPN - part II. *The Internet Protocol Journal*, 1(2).

[Fünfrocken, 1998] Fünfrocken, S. (1998). Transparent migration of Java-based mobile agents: Capturing and reestabilshing the state of Java programs. In Rothermel, K. and Fritz, H., editors, *Proc. Mobile Agents MA '98*.

[Gray, 1998] Gray, R. S. (1998). Agent Tcl: A flexible and secure mobile-agent system. Technical report, Darhmouth College. PCS-TR98-327.

[Günter and Braun, 1999] Günter, M. and Braun, T. (1999). Evaluation of bandwidth broker signaling. In *Proceedings of the International Conference on Network Protocols ICNP'99*, pages 145–152. IEEE Computer Society.

[Günter et al., 1999] Günter, M., Braun, T., and Khalil, I. (1999). An architecture for managing QoS-enabled VPNs over the Internet. In *Proceedings of the 24th Conference on Local Computer Networks LCN'99*, pages p.122–131. IEEE Computer Society.

[Hulaas et al., 1999] Hulaas, J., Gannoune, L., Francioli, J., Chachkov, S., Schtz, F., and Harms, J. (1999). Electronic commerce of Internet domain names using mobile agents. In *Proceedings of the Second International Conference on Telecommunications and Electronic Commerce (ICTEC'99)*.

[Jacobson et al., 1989] Jacobson, V., Leres, C., and McCanne, S. (1989). Tcpdump. available via ftp to: ftp.ee.lbl.gov.

[Jansen et al., 1999] Jansen, W., Mell, P., Karygiannis, T., and Marks, D. (1999). Applying mobile agents to intrusion detection and response. Technical report, National Institute of Standards and Technology.

[Kato et al., 1999] Kato, A., Murai, J., Katsuno, S., and Asami, T. (1999). An Internet traffic data repository: The architecture and the design policy. In *Proc. INET '99*. http://www.isoc.org/inet99/proceedings/4h/4h_1.htm.

[Kent and Atkinson, 1998] Kent, S. and Atkinson, R. (1998). Security architecture for the internet protocol. RFC 2401.

[Knuth, 1981] Knuth, D. E. (1981). *The art of computer programming*, volume 2 Seminumerical Algorithms. Addison-Wesley, 2 edition.

[Lange and Oshima, 1998] Lange, D. and Oshima, M. (1998). *Programming and Deploying Java Mobile Agents with Aglets*. Addison-Wesley.

[Murch and Johnson, 1999] Murch, R. and Johnson, T. (1999). *Intelligent Software Agents*. Prentice Hall.

[National Laboratory For Applied Network Research, 2000] National Laboratory For Applied Network Research (2000). Network analysis infrastructure. http://moat.nlanr.net/.

[Paxson, 1997] Paxson, V. (1997). End-to-end Internet packet dynamics. In *Proc. SIGCOMM '97*.

[Paxson et al., 1998] Paxson, V., Almes, G., Mahdavi, J., and Mathis, M. (1998). Framework for IP performance metrics. RFC 2330.

[Teitelbaum et al., 1999] Teitelbaum, B., Hares, S., Dunn, L., Neilson, R., Narayan, V., and Reichmeyer, F. (1999). Internet2 QBone: Building a testbed for differentiated services. *IEEE Network*, 13(5):8–16.

[Tennenhouse et al., 1997] Tennenhouse, D. L., Smith, J. M., Sincoskie, W. D., Wetherall, D. J., and Minden, G. J. (1997). A survey of active network research. *IEEE Communications Magazine*, 35(1):80–86.

[Terzis et al., 1999] Terzis, A., Wang, L., Ogawa, J., and Zhang, L. (1999). A two-tier resource management model for the Internet. In *IEEE Global Internet'99*.

[Vitek and Bryce, 1999] Vitek, J. and Bryce, C. (1999). The JavaSeal mobile agent kernel. In *Proc Symposium on Agent systems (ASA '99) and Applications and Symposium on Mobile Agents (MA '99)*.

[Wetherall et al., 1998] Wetherall, D., Guttag, J., and Tennenhouse, D. L. (1998). ANTS: A toolkit for building and dynamically deploying network protocols. In *IEEE OPENARCH '98*. San Francisco.

[White, 1994] White, J. (1994). Telescript technology: The foundation for the electronic marketplace. General Magic White Paper.

[Wooldridge and Jennings, 1999] Wooldridge, M. J. and Jennings, N. R. (1999). Software engineering with agents: Pitfalls and pratfalls. *IEEE Internet Computing*, pages 20–27.

[Zimmermann, 2000] Zimmermann, P. (2000). Pretty good privacy. http://www.pgpi.org/.

Traffic considerations for security services in VoIP systems

Wilhelm Wimmreuter
Siemens AG
Information and Communication Networks
Hifmannstr. 51, D-81543, Munich, Germany
wilhelm.wimmreuter@icn.siemens.de

Key words: VoIP, Voice over IP, Security, Traffic, Authentication

Abstract: Voice over IP systems require a new infrastructure of security services. The demand is currently growing and deploys basic-authentication in the first step. Scalable implementations however further evolution of this technologies. This new authentication services using signed certificates and crypto-promise to solve current scalability and management issues. Additionally, if carefully implemented, they can serve for other applications besides VoIP as well. Certificate-authorities, directories and others backend services build the trust environment that delivers the required credentials to network elements, servers and user equipment. All this new backend-services generate a new category of IP traffic.

This paper roughly estimates the expected volume of authentication-traffic and emphasises the importance to select the right services, to handle the possible traffic surge.

1. INTRODUCTION:

Traditional public services telephony networks typically provide security based on "Trust by Wire" and therefore have no requirement to utilise other means of security besides the trust imposed by wiring. This is true between network elements and the connection to the customer premises. Of course, tampering with wires is of concern for telephony systems too, but besides that and with the exception of special cases for Legal Intercept and Malicious Call Tracing, there are no further provisions necessary.

This situation changes with the transition to the open Internet and *Figure 1* gives a brief overview on that.

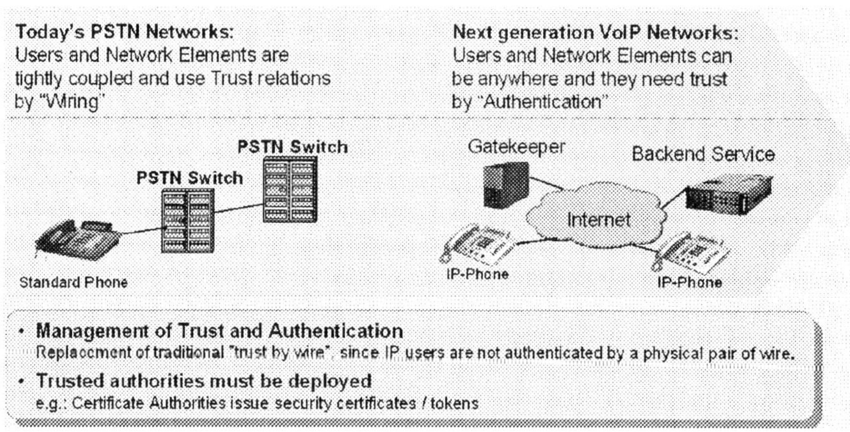

Figure 1. From Trust-by-Wire to Trust-by-Authentication

Public telephony services for VoIP networks on the other hand can not relay on "Trust by Wire", since the very nature of the Internet does not have the wired trust of traditional phone terminals and network elements. IP networks are build upon "Trust by Authentication". As a major implication, such networks must authenticate their users and their network elements. They must proof each others identity by some means of secure information exchange.

Of course, IP networks for telephony can be operated in secure environments and thus be operated as closed networks. But this will only provide transport technology an is against the goal to provide multi-service networks that can share their resources.

This paper analyses the traffic surge we might expect by introducing new security services for authentication for VoIP users in the internet. Therefore we concentrate to analyse authentication services that need information exchange for every call.

The paper does not consider security related traffic for network element security and other supplementary security services. This simply because of the fact, that these services will not generate bulk-traffic and therefore do not need that kind of attention for traffic engineering.

1.1 Elements of VoIP Systems

To get some common ground we summarise the key elements of VoIP systems.

Figure 2 shows a typical configuration for Carrier VoIP Systems with the following base components:

- **Gateways** translate media and signalling information between PCM based telephone systems and the internet.
- **IP-Phones and Terminal adapters** are provide the new customer terminal equipment for VoIP systems.
  "Endpoints" is a commonly used term that refers to both, **gateways** and **terminals**
- **Gatekeepers** handle call-control, authentication, authorisation, and billing.
- **Backend-Services** typically handle the storage of information such as billing, routing, configuration and authentication data for subscribers.
- **Clearing-House-Services** deal with resources and costs between operators. They are the merchants between multiple operators and fortunately will replace settlement contracts in future.

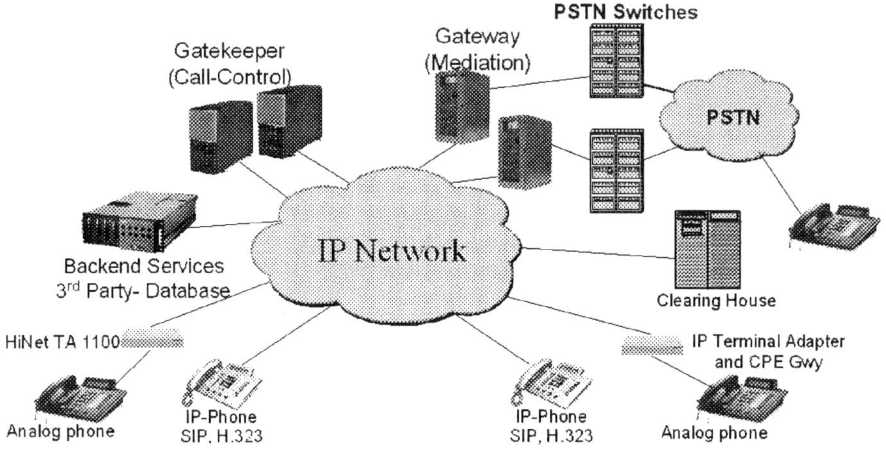

Figure 2. Basic elements of VoIP systems

IP-phones, Soft-PABX and VoIP-enabled-workstations are the new customer premises terminals for VoIP networks.

IP-terminal-adapters and customer-premises-gateways allow the use of traditional phones, faxes, modems to name a view.

Obviously, these components only provide limited security, and while various standards give guidance on how to implement security on this devices, there is almost nothing that provides a sound environment to support security throughout the network. The missing elements mainly consist of key-management and other backend-services that provide security support for and between above elements.

1.2 Intelligent Backend - Services for Security

Backend services for authentication support the basic requirements of security for VoIP systems and must be deployed on the internet. These systems provide all key- and certificate- credentials to authenticate and authorise calls within one and across multiple administrative domains.

Additionally, other security services for media- and signalling- stream encryption, collaborative-work, and transaction-support for E-commerce pave the way for new multi-service-networks. This services go beyond the authentication and authorisation of phone calls. Certainly, invocation of this services is not be done for every call and therefore we must not expect a huge overhead of network traffic generated by this services.

1.2.1 Scenarios for VoIP Telephony

As shown in *Figure 2*, there are basically five scenarios for telephony on the internet.

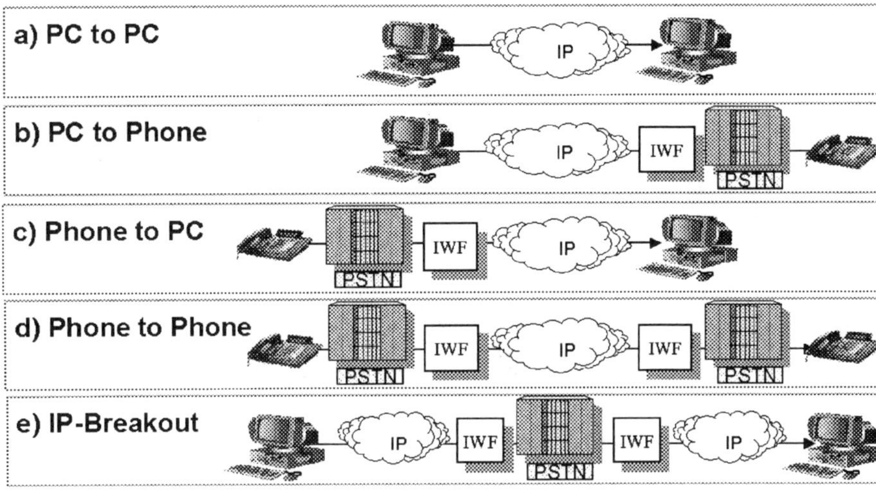

Figure 3. Scenarios for IP-Telephony

a) **IP-Endpoint to IP-Endpoint**
   The Typical PC to PC situation or the situation when VoIP replaces traditional TDM telephony.
b) **IP-Endpoint to Traditional Telephony Endpoint**
   PC or other IP-endpoints calling a standard telephone.
c) **Traditional Telephony Endpoint to IP-Endpoint**
   A telephone subscriber calls a IP-Based endpoint such as PC's or IP-terminals or IP-Phones
d) **Traditional Telephony to IP-Traditional Telephony Endpoint**
   The classical Long distance bypass to participate cheaper transmission offered on the internet.
e) **IP-Endpoint to IP-Endpoint via PSTN systems (PSTN Breakout)**
   IP endpoints configured to use the PSTN network in case the IP network is overloaded or no peering between IP networks is available.

At present, scenario b) and d) are likely to be the only ones commercially used for public VoIP telephony systems. This is because VoIP still has low penetration and the majority of telephone-subscribers (endpoints) reside in the traditional telephony environment. Of course, there are exceptions where campus-networks replaced their PABX and introduced VoIP in their company. However, this Soft - PABX's typically reside on isolated networks where trust and billing is not that much of a concern compared to public telephony.

1.2.2 Today's authentication for public VoIP telephony

Out of the major areas of security in VoIP systems, we focus on services for user-authentication. This because identification and authorisation of users is likely to generate an additional traffic surge if not carefully designed.

Other areas of security such as privacy and operational-security deal with encryption and attack prevention. Thus, they deal more with host related issues and may use CPU power, but will not create additional traffic load.

Service providers at present use subscription based authentication and in rare cases credit card based authentication. These services are offered based on prepaid accounts, or authentication and billing through the phone account of the callers local phone provider. Depending on the status of VoIP providers, they either act as licensed interconnect operator in the deregulated telephone market, or based on subscriber like interface in the "not jet" deregulated markets. Fully licensed operators are entitled to use carrier-selection-dialling and have the option to bill their users on the phone-bill of the (incumbent) local operator.

The following authentication methods are typically used in today's environments.

Authentication methods for PSTN originated calls:

- **Password authentication for a single call**
  Prepaid card services based in access code dialling with user-ID and password authentication for every call.
- **Password authentication for multiple calls**
  Prepaid card services based in access code dialling with user-ID and password authentication allowing multiple calls once logged in to the Telephony service provider.
- **Customer-line-identification based authentication**
  This method is based on the fact, that gateways and the users access (local telephone switch) maintain a trusted relation. Authentication is based on the service level agreement between the subscriber and the local telephony operator.
  Advantage: No extra dialling besides dialling an access-code
  Disadvantage: No user mobility possible

This authentication methods generate message flows between the PSTN-gateway, the call-control-instance, and backend-services of the VoIP operator (proxy, gatekeeper, backend-services). Today, this traffic has no effect on the open internet, since gateways and servers are typically interconnected in closed environments or isolated subnets.

As a consequence, no additional authentication messages are required for authentication based on customer-line-identification since the PSTN delivered CLI info is trusted by the VoIP operator.

Authentication methods for Internet originated calls:

- **Password authentication for a single call**
  User-ID and password authentication through service selection portals.
- **Password authentication for multiple calls**
  User-ID and password authentication through service selection portals.

For this basic password authentication methods, the overhead of message exchange adds up to at least 4 additional roundtrips for MD5 challenge / message-digest exchange. This messages add up to at least 1 k Byte of data transferred for authentication of every single call.

1.2.3 Authentication services for future demand security

Scalable systems will need to deploy other authentication services. This because management of trust and authentication must scale for complex interdomain scenarios and trust relations between an ever growing number of operators can not be managed by mutual agreements based on written contracts.

The following technologies promise solutions for the problem:

Security information:

- Public-key authentication.
- Tokens to the endpoint.
- Certificates (X.509)

Security Services:

- Certificate authorities / Trust centres
- Key management systems
- Key revocation systems
- Directories for certificate-authorities

Such systems can be based on the procedure, that a client authenticates himself with a request signed with the client's private key. The server can then discover the origin of the request if it has access to the public key. The public key shall preferably be signed by a trusted third party. Further, the server my issues a token with limited lifetime to the user. This could avoid re-registration and therefore additional authentication messages for every call.

Figure 4 shows backend security services that support endpoints and servers.

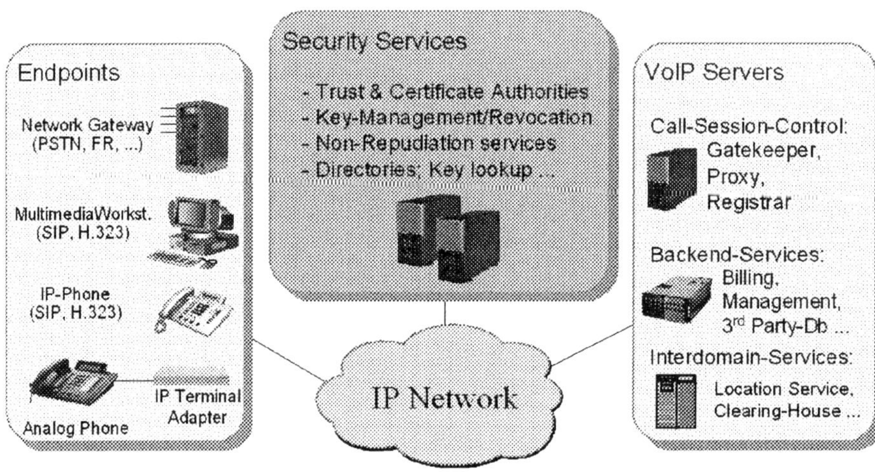

Figure 4. Authentication Services for IP Telephony

This new backend-services consume bandwidth for their required message exchange. But in case they are carefully designed, a portion of this traffic can be avoided by distributing re-usable information elements with limited lifetime or invocation counters.

Additionally, well thought implementation will open these systems for other applications besides VoIP as well.

2. ESTIMATED TRAFFIC GENERATED FOR SECURITY:

Based on the framework given in *section 1*, the following estimates are based on the average traffic produced in a city with about 250.000 traditional telephony subscribers. We compare basic signalling and the overhead of traffic expected by introducing IP-based authentication as mentioned above. It is assumed further, that authentication source of security traffic in VoIP networks. Signalling- and security- traffic adds to the top of basic traffic generated by media stream transmission.

2.1 Media stream traffic volume

The given call distribution as in *Figure 5* shows, that around 1/3'rd of the calls is to be served in the three hours around the busy-hour of the day. Thus, the major bandwidth concerns may arise at that time. This are the most important values to evaluate for network dimensioning. The Busy-Hour-Call-Attempts (BHCA) and the Mean-Hold-Time (MHT) are the key values in dimensioning this networks. All estimates is based on 200.000 BHCA as shown in *Figure 5* and all following bandwidth considerations are based on a given Mean-Hold-Time of 120 Sec.

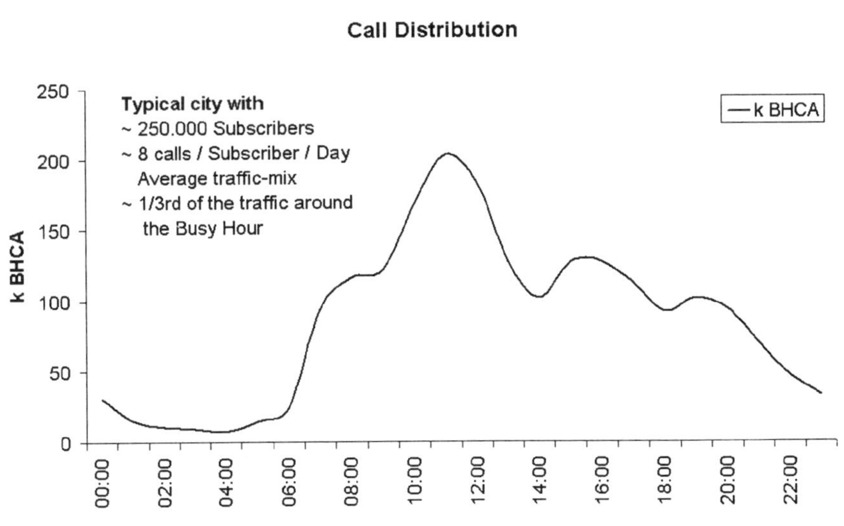

Figure 5. Traffic volume for 250.000 subscribers

Media-streaming traffic volume as shown in *Figure 6* is derived based on traditional telephony methods. Thus poisson distributed queuing for telephony according to "Erlang" is assumed to fit for IP-based voice streams as well.

The bandwidth consumption is based on the measurement in *Figure 5* with a set of bandwidths derived from the measured PSTN 64-Kbit/Sec stream.

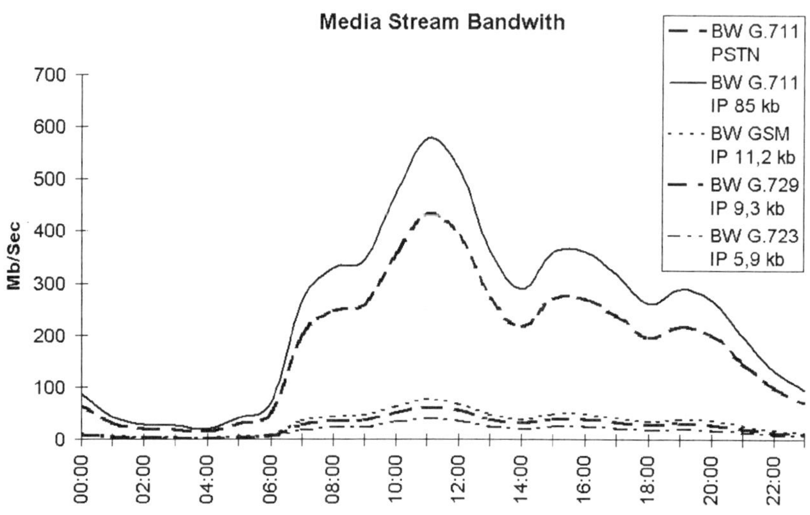

Figure 6. Media stream bandwidth requirement

Figure 6 shows the peak bandwidth requirement of the given PSTN network at about 450 Mbits/Sec.[1] In the busy-hour. VoIP telephony without compression adds to almost 600 Mbits/Sec. and compressed IP transmission goes down to about one tenth of the PSTN bandwidth trading in some quality to reach this goal.

At present, there is a tendency against compression, since bandwidth is available at large and impacts of compression to FAX and other modem traffic turns out to be a problem in telephony environments. Additionally, IPR and license fees are quite unpredictable for suppliers as well as for operators.

[1] Peak bandwidth in the given PSTN network does not imply the necessity to provide all the bandwidth. In fact, these systems are typically designed with about 30% of loss on a given peering. This can be done, since e.g. residential-areas might not have the same busy-hour as business-areas. Overflow through temporarily unloaded peerings can save in this case

2.2 Signalling traffic volume

Signalling traffic adds up to the VoIP media traffic budget. *Figure 7* shows the comparison of best- and worst- case IP based signalling scenarios with SS7 signalling[2].

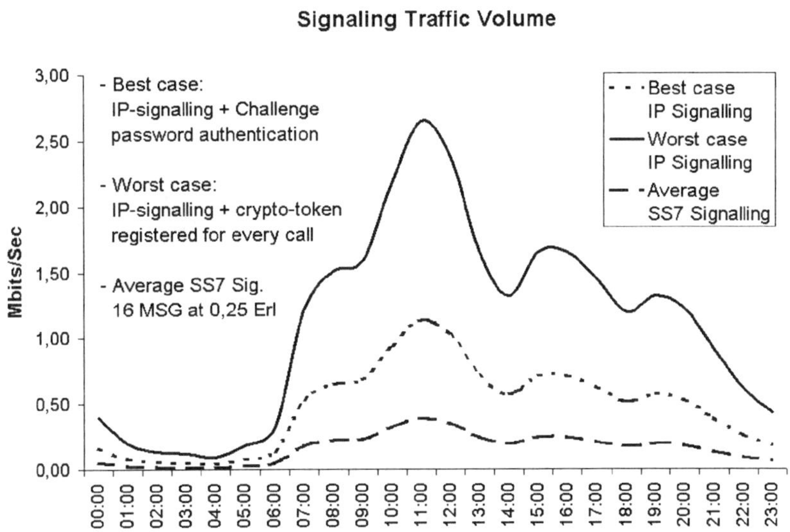

Figure 7. Basic-call signalling bandwidth requirement

Assumptions to simulate signalling for VoIP

- The difference between H.323 and SIP for basic-call signalling is small and therefore ignored (1020 vs. 1040 bytes).
- Only two cases of authentication are considered
  Basic authentication with challenged passwords for every call, and
  Crypto-token authentication with re-registration for every call
- 6 messages are assumed for IP-Basic-Call-setup and release
- 4 or 11 messages for IP-Authentication depending on services used.

The graph in *Figure 7* shows, that SS7 is the most economic signalling method. However, compared to the media streaming requirements, signalling traffic adds up to at most 0,5% of media streaming.

[2] Note: Since SS7 is considered pure interoffice trunk signalling and VoIP typically serves for both, trunk and endpoint traffic, a rather complex call was taken in to consideration for the SS7 part of signalling (Basic-call: 8 Msg with 27 Bytes; Used: 16 Msg with 90 Bytes). This shall cover quite complex subscriber signalling cases.

2.3 Authentication overhead

Looking at the worst case of authentication traffic, about 2 Mbits/Sec of additional traffic is needed for authentication with crypto-tokens that register for every call. Further it is assumed, that each token must be signed by the certificate-authority acting as an independent 3'rd party trust-centre. Of course, this is worst case and according to above traffic estimates, this might be unwise to implement. On the other hand, as long as volume of VoIP traffic is low, a limited lifetime of crypto-tokens is likely to run out anyhow.

Basic clear-password-authentication, and locally resolved challenge-authentication are not considered in this estimate. This because of low security for the clear-password method, and severe scaling problems for the locally resolved challenge method. Therefore 11 message-roundtrips are assumed for to invoke certificate-authorities. This seems to be the only way to design re-usable security services for other applications besides VoIP telephony.

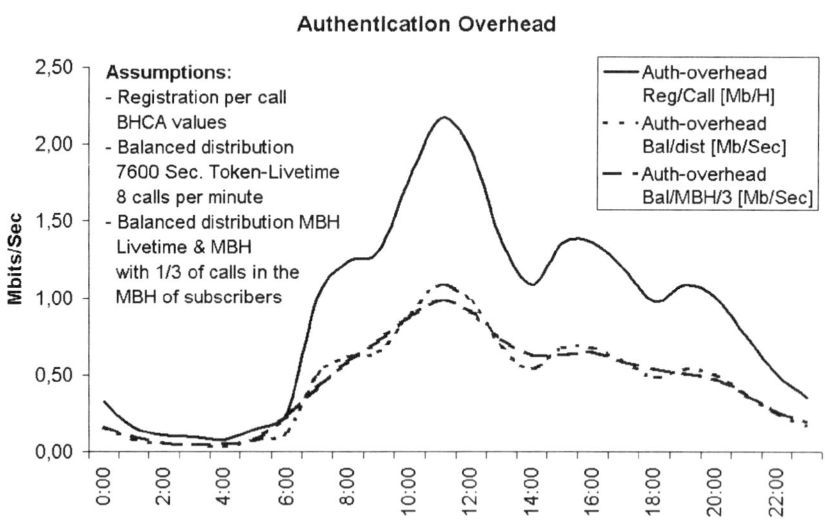

Figure 8. Optimising the authentication overhead

The scenarios according to *Figure 8* are based on 2 hour lifetime for the crypto-token, an average of 8 phone calls per subscriber per day, and the fact, that 1/3rd of the calls are done in the 3 hours around the busy hour of every subscriber.

Compared to the total volume of signalling-Traffic in *Figure 7*, the graph in *Figure 8* shows that the majority of signalling traffic for the basic-call is produced by authentication.

Graphs in *Figure 8* provide the following results:

- **Registration for each call with access to the certificate-authority:** This provides for the worst case with more than 2Mbits/second in the busy-hour.
  Of course this is likely the most secure method as well.

- **Balanced distribution of re-registration at the end of lifetime:** Generates less than 50% of the traffic compared to the previous one.

- **Balanced distribution of re-registration considering busy-hour:** Here another optimisation considered the fact, that $1/3^{rd}$ of the calls are done around the mean-busy-hour of the subscriber. Therefore a running-average forces (re-) registration at the begin of the mean-busy hour and therefore saves one registration around the busy-hour.
  This test did not result in real savings, since it only flattens the peaks a little bit. Further tuning to better guess the start of busy hour and the modification of the lifetime of tokens did not further improve results. In consequence, considering all the costs for implementation and management achieving this small the small gain seems to be not worth the effort.

3. CONCLUSION:

As shown, the transition to public VoIP networks has major impact to the network traffic. As a consequence, network design, service deployment and business cases must be adopted to this new environment.

Security services make up an essential piece of this puzzle. In fact, this services and authentication in particular produce about 10 times as much signalling traffic compared to signalling traffic in traditional telephony systems.

Trading in media stream bandwidth for signalling seems to be a winning solution. But we shall be careful since this might end in arbitrage business that lasts for a short while only. More advanced solutions will take the benefits of new business cases possible through advanced authentication as required for VoIP.

Since authentication services must be implemented for VoIP, scalability and re-use of authentication services for other applications such as E-commerce, banking and others is feasible.

Certificate-authorities and registration-authorities possibly managed by independent 3'rd party operators, public directories, and encryption are some of the basic services that open the way for multi-purpose authentication. Only this value add my justifies the extra cost of advanced authentication services.

Management of trust and authentication must become scalable and secure. This includes directory-lookup, key-generation, key-management and services providing key-revocation to resolve quite problems around compromised security information.

Finally, if we implement all this security-systems for VoIP without considering the re-use for additional applications, we will have to ask ourselves whether security is affordable for a pure phone call.

4. REFERENCES

IETF Working Groups
http://www.ietf.org/html.charters/
ETSI TIPHON documents (open access)
http://www.etsi.org/tiphon/
GMD Focus Collection of References & Papers
http://www.fokus.gmd.de/research/cc/glone/projects/ipt/
H.323 Protocol suite: Current documents of SG16
ftp://standard.pictel.com/avc-site/
ITU-T Recommendation H.235 Version 1, Security and Encryption for H Series (H.323 and other H.245 based) multimedia terminals, 1998.
SIP Web Page at Columbia University
http://www.cs.columbia.edu/sip/
Henning Schulzrinne et. al.
Telephone Traffic Theory, Tables & Charts
Siemens AG
ISBN 3-8009-1343-7

Part Two

Platforms for Advanced Services

AN ARCHITECTURE FOR PROVIDING ADVANCED TELECOMMUNICATION SERVICES

Yann Duponchel, Marcel Graf, Hong Linh Truong
IBM Research
Zurich Research Laboratory
8803 Rüschlikon
Switzerland

Abstract We describe a novel service architecture that allows service providers to deploy and provision telecommunications services in an easy and efficient way. In contrast to today's Intelligent Network (IN) specification, the new architecture includes mechanisms for the automatic deployment, modification, and provisioning of services. It exploits the convergence towards IP as the universal network infrastructure to provide means for combining both Web and telephony services into more sophisticated and advanced ones.

Keywords: IN, Intelligent Network, IP Telephony, Service creation, Service provisioning, Communications services

1 INTRODUCTION

To accommodate the explosive growth of Internet traffic, network operators are building a high-capacity IP infrastructure, and there is no doubt that telephony will be just another application running on the same infrastructure. The cost advantage of managing and maintaining a single network infrastructure for all types of traffic is particularly important for newcomers, but less so for incumbent operators. Furthermore, using a single infrastructure will make it easier to combine telephony with Web-based services, thus creating new kinds of advanced and sophisticated services that have not been possible in the past or are difficult to implement on separate infrastructures. Such advanced services are in general thought to be the new income sources for which the service providers are desperately looking.

Before discussing how advanced services can best be created and deployed in an IP environment, let us review the basic architecture of an IP telephony service. Such an architecture is illustrated in Figure 1. Although it is based on ITU-T recommendation H.323 [1],[2], it can easily be shown that it also applies — with minor modifications — to other architectures that employ, for example, the IETF protocols SIP [3] or Megaco [4]. The most important difference between IP and legacy telephony architectures (e.g. PSTN, ISDN) is that the former does not require dedicated voice switches, because both signalling control information and voice signals use IP as communication means. Voice signals are sent directly between terminals, reusing the same IP router infrastructure as for the exchange of signalling information.

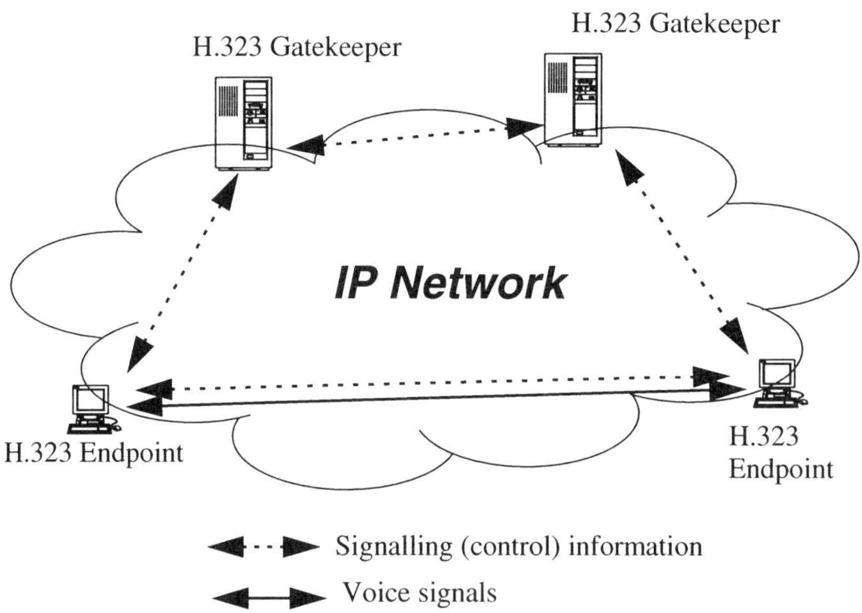

Figure 1: H.323-based IP telephony.

The gatekeepers shown in Figure 1 perform only control functions such as registration, access and admission control, address translation, etc. [2]. They have no voice switching capabilities and can therefore be implemented in software as applications servers.

Recommendation H.323 defines two methods to exchange signalling information, i.e. the information for controlling telephone calls:

1. *Direct method:* The signalling information needed for the control of a telephone call is exchanged directly between the endpoints involved. The endpoints may, however, need the assistance of a gatekeeper (or some other servers) for functions such as address resolution, access control, bandwidth management, etc.
2. *Gatekeeper-routed method*: In this case the terminals "talk" only to their gatekeeper, which now has control over the telephone call. The signalling information is exchanged between the terminals and their gatekeepers. The gatekeeper performs almost the same control functions as the legacy telephony switches, except that they do not switch voice signals.

In the context of service creation it is reasonable to foresee that such supplementary services as call forwarding or call transfer will be implemented directly in the endpoints rather than in the gatekeepers by exploiting the intelligence of the endpoints and the end-to-end capability of the direct method mentioned above. However the gatekeeper-routed method remains of interest to the service providers, not only because it gives them a control point for the telephone calls, but also because it enables them to offer more sophisticated services, e.g. by combining telephony with other Web-based information services. An example of such a combination is the stock alert service: the service subscriber is alerted, i.e. receives a phone call on any of his telephone sets (home, office, cellular, etc.), if the price of certain stocks crosses certain limits.

An incumbent telecommunications service provider, which already has a service creation infrastructure based on the Intelligent Network (IN) architecture [5], would certainly look for possibilities to reuse its existing IN infrastructure. Such a possibility is shown in Figure 2. The gatekeepers are considered Service Switching Points (SSPs), which means they intercept telephone calls and recognize when it is necessary to contact a Service Control Point (SCP) for further instructions on how to proceed with a call. The only difference between a gatekeeper and an authentic SSP is that the gatekeeper performs only call control, whereas a conventional SSP performs both call control and voice switching.

Although the architecture shown in Figure 2 allows incumbent telecommunications service providers to reuse their expensive SCP-based service creation infrastructure to extend the offering of traditional supplementary services to IP telephony users, it limits the range of offerable supplementary services to the capabilities of current SCPs and SSPs.

Another disadvantage of reusing the IN architecture is its lack of operation and management standardization: The IN architecture standardizes only the communication protocols between the SCPs and SSPs; it does not specify the mechanisms needed to "put" the service logics into the SCPs, to manage

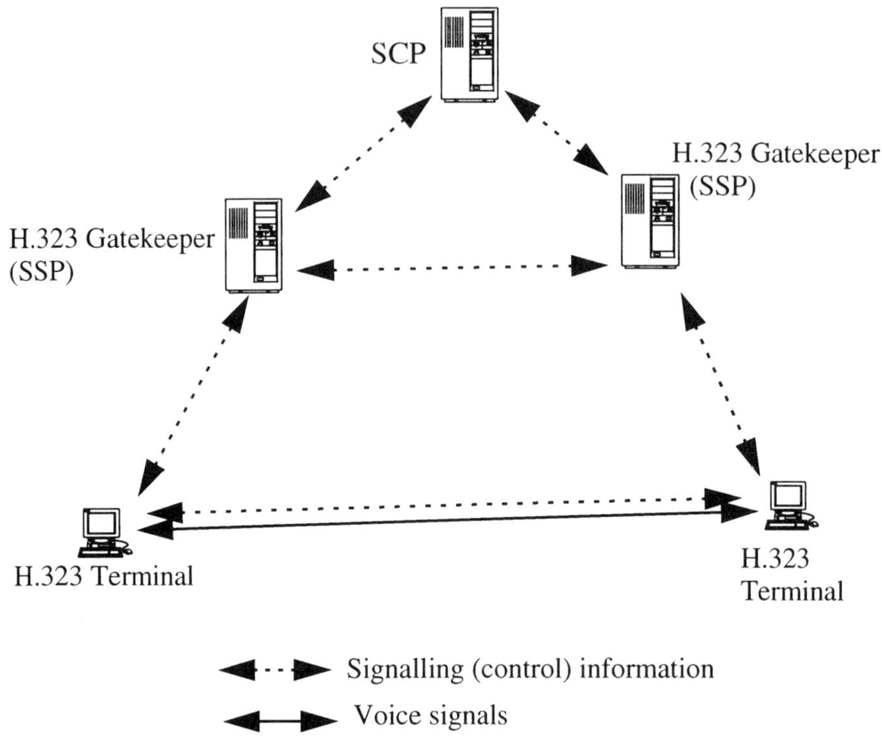

Figure 2: Applying IN architecture to IP telephony.

and customize them, and to set the required service triggers in the SSPs (e.g. recognizing the access prefix of a service). This lack of specification leads to incompatible and vendor-specific SCP/SSP platforms. Typically, a rapid deployment of services is only possible within a single vendor environment. Furthermore, it is not possible to port a service logic written for a specific platform to another, thus making it almost impossible to deploy and manage a service across multiple vendors' platforms.

In this paper we present a novel architecture that allows service providers to deploy and provision advanced telecommunications services in an easy and efficient way. In contrast to the IN architecture described above, the new architecture includes mechanisms for automatic deployment, modifications and provisioning of services. It provides to the service providers an easy and efficient means for injecting new service logics into the network and modifying them. It also allows service subscribers to directly customize their service profiles. Besides being very resistant to failure of single components, the architecture is also scalable in terms of number of users, calls and services.

We have organized the paper as follows. In the next section we discuss some service concepts that are important for understanding our architecture. The new service architecture is then presented in Section 3, with a detailed description of the components and their interactions. Section 4 shows how a concrete IP telephony service based on the ITU-T Recommendation H.323 could be integrated into the architecture.

2 CORE AND ADVANCED SERVICES

In our architecture we differentiate between what we call *core* and *advanced* services. Core services are services used as common building blocks to create advanced services. Examples of core services are

- telephony service,
- e-mail,
- short message service (SMS)[1],
- instant messaging (IM), and
- presence.

Presence and instant messaging [6] is a new mode of communication that has recently become very popular in the Internet. Presence is a service that allows user A to declare its interest in the presence states of another user B, e.g. whether user B is currently connected to the network and, if so, whether user B is actively using his terminal, etc. This service is typically implemented by an application running on the terminal of user B, which publishes presence information about user B; the service delivers notifications to all users subscribed to observing each time the presence state of user B changes. Knowing the state of user B, user A may then start an instant messaging session with user B in which both users exchange short instant messages. An instant messaging session is very similar to a telephone call in the sense that it is synchronous (instant), i.e. messages sent by a user are delivered almost immediately to the peer user.

Although both presence and instant messaging are currently offered in the Internet as a combined service (e.g. AOL IM, Yahoo! Messenger) we regard them in our architecture as two separate core services that can be used independently to create new advanced services.

---

1. In the GSM mobile network, this service allows a cellular user to send a short text message to another cellular user.

An interesting observation is that most of today's telephony systems already collect presence information about their users, but do not offer it as an explicit service to their users. The PSTN, for example, monitors the status of a telephone line of a subscriber and determines whether it is busy or free. The fact of a telephone line going from a busy state to a not-busy state is a piece of presence information. Some systems make use of that information to offer so-called "call completion" supplementary services [7], in which a calling user for example can request the network to monitor a busy called user and connect him to that user as soon as he becomes not busy. Another example is the mobile telephone network, which foresees a procedure for mobile phones to register with the network before they can make and receive calls. The network keeps track of the registration status. The fact that a mobile telephone is registered is again a piece of presence information. It is interesting because it indicates that there is a high likelihood that the user is able to accept a call. A similar registration procedure also exits in IP telephony.

In our architecture we propose that all these different pieces of presence information be integrated into a single service available to all other ones. Based on such a converged presence service, the presence information is no longer restricted to a specific application: a user is no longer *available with instant messaging* or *available with telephony* but rather *available at a certain terminal*. In this way we can exploit the multiservice capability of modern communications terminals, e.g. a mobile phone that is capable of receiving telephone calls, SMS messages, and in the near future also instant messages.

Regarding the telephony core service mentioned above, it should be noted that in our architecture we do not differentiate between the different types of network technology that can offer telephony services: i.e., there is only one telephony core service, which includes the PSTN, ISDN, cellular networks, IP telephony networks, etc.

As mentioned, taking the core services as common building blocks, one can create advanced services by

- either modifying the way a core service is invoked, e.g. call forwarding, call completion. In the legacy telephony world such advanced services are called "*supplementary services*";
- or combining several core services to provide more sophisticated ones. We call such advanced services "*hybrid services*".

Of course an advanced service may be both supplementary and hybrid.

Our notion of hybrid services is different from the one defined in [8], in which it is defined as "services which span many network technologies, such as the public switched telephony network (PSTN), cellular networks, and networks based on IP". According to this definition, the telephony core service we mentioned above would already be a hybrid one.

3 THE NEW SERVICE ARCHITECTURE

The main aim of our architecture is to define an environment that will allow a rapid and efficient creation and deployment of advanced services.

In designing the service architecture emphasis was placed on the following key aspects. First, it should provide the service provider with an easy and efficient means to deploy new service logics into the network and to modify them. Second, it should allow service subscribers to customize their profiles themselves (e.g. via the Web and not with the help of a call center agent) and have the changes propagated rapidly into the network. Third, it should be applicable to the creation of hybrid advanced services as defined in Section 2, and not restricted to supplementary services. And last but not least, it should be scalable to large numbers of subscribers and services, and resistant to components failures.

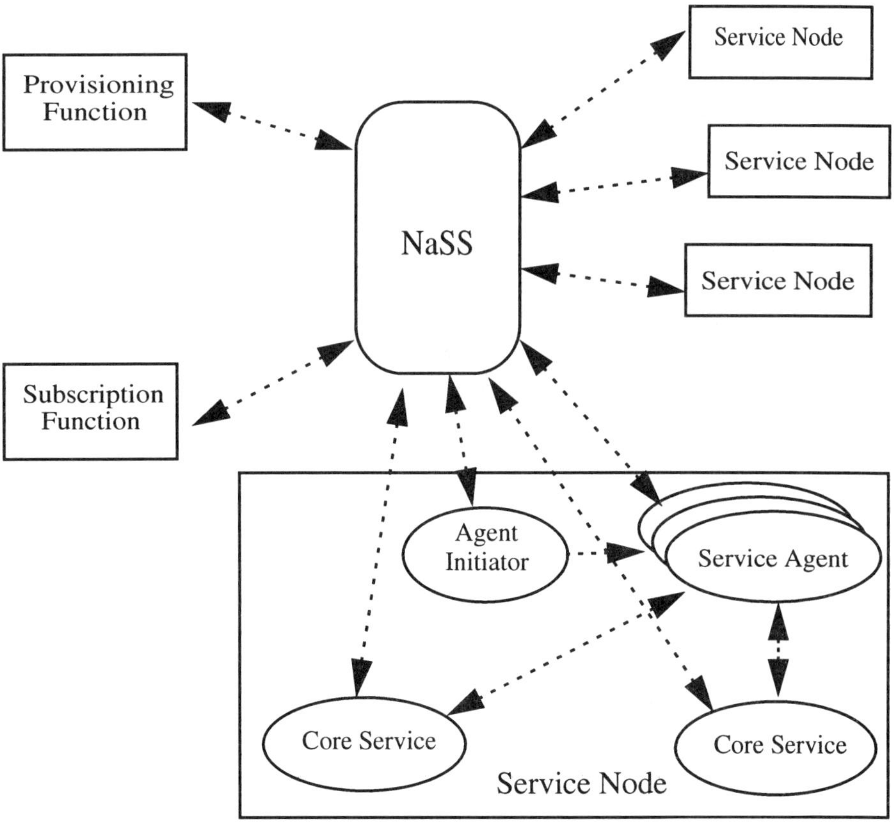

Figure 3: New service execution architecture.

The proposed service architecture is shown in Figure 3. It consists of the following components:

- Notification and Synchronization Service (NaSS),
- Provisioning Function,
- Subscription Function,
- Core Services,
- Service Agents,
- Agent Initiators, and
- Service Nodes.

We will first give a short overview on the functions of these components and their interactions. The follow-on subsections will then contain a more detailed description.

3.1 OVERVIEW

Typically there are four phases during the life cycle of a service [9]: creation, provision, subscription, and utilization.

During the creation phase, first the service is specified and designed, usually derived from a textual description, then the required service logics are developed, coded, verified and tested. Various formal languages such as SDL (Specification and Description Language) or LOTOS [10] can be of help during the specification and design steps. To simplify the coding tasks visual tools combined with reusable components such as the IN's concept of service-independent blocks (SIBs) or the object-oriented Java Beans can be used. The creation phase is, however, beyond the scope of our architecture, and for the remaining part of this paper, we assumed the existence of the service logics. Such service logics are named in our architecture *Service Agents*.

Having the service agents, the service provider now needs to distribute them to the *Service Nodes*, in which they will be loaded and executed during the service utilization phase. To be scalable our architecture supports the existence of multiple service nodes, which may or may not have the same functionality.

The *Provisioning Function* shown in Figure 3 helps the service provider to inject the service agents into the appropriate service nodes using the communication mechanisms of the so-called *Notification and Synchronization Service (NaSS)*. As will be described in more details in Section 3.2, the NaSS provides to all other components of the architecture an interprocess communication mechanism that is based on an anonymous publication and subscription paradigm. That means, a publisher does not need to know who will receive his notification, he just publishes it under a certain name; the NaSS will then deliver the notification to the components that have subscribed to that name.

In our example, with the help of the provisioning function the service provider publishes the service agents into the NaSS, without the need for knowing anything about the characteristics of the available service nodes. The NaSS will then deliver them to the service nodes that have subscribed for those service agents. This procedure will be described in more detail in Section 3.6.

Similar to the provisioning function, the *Subscription Function* in Figure 3 allows a service user to subscribe and customize his service profile himself (e.g. using the Web) and publishes the resulting data into the NaSS. The NaSS will then deliver those data to the service agents that have subscribed for those data.

As mentioned, service agents execute the logic that is required to implement a certain advanced service. For this purpose they typically wait for an event coming from a *Core Service*, at which they then interact with that core service and/or with additional ones. For example, a service agent implementing the supplementary service "Call Forwarding on No Reply" (CFNR) [11] only becomes active if there is an incoming call to the subscriber. At this event (which is generated by the telephony core service) the service agent starts the no-reply timer and watches the state of the call. It does nothing if the call is answered, otherwise at the expiration of the no-reply timer, it disconnects the ringing connection and redirects the call towards the diverted-to number.

To make the service node scalable in large numbers of active service agents, we define in our architecture the concept of an *Agent Initiator*, which watches the events on behalf of the service agents. At the occurrence of an event the agent initiator will load into memory those agents that are interested in the event and start them. It is also the responsibility of the agent initiator to remove an agent from the node's memory after that agent has done its job. The concept of the agent initiator is explained in more detail in Section 3.4.

3.2 NOTIFICATION AND SYNCHRONIZATION SERVICE (NASS)

The heart of our architecture is the *Notification and Synchronization Service* (NaSS), which provides an interprocess communication mechanism to all the other components. As illustrated in Figure 4 there are two kinds of NaSS users: (1) NaSS *Publishers*, which send notifications to the NaSS, and (2) NaSS *Subscribers*, which receive notifications from the NaSS. The main function of the NaSS is to deliver the notifications and attached objects sent by publishers to the corresponding subscribers. The association between publishers and subscribers is performed by means of a *Name*.

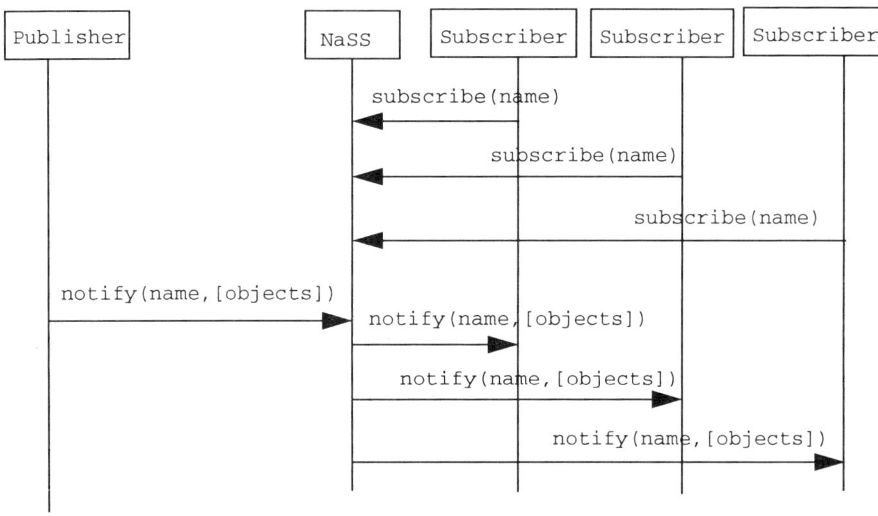

Figure 4: Notification and Synchronization Service (NaSS).

NaSS names have a hierarchical structure. This enables a subscriber to use wildcards to refer to an entire subtree of the hierarchy and thus to subscribe to several names with a single subscription.

Communication via the NaSS is *anonymous*, because to receive a notification sent by a certain publisher, a subscriber needs only know the name under which the publisher sends the notification (and not the publisher itself). The publisher also does not know who will receive its notification. Both publisher and subscriber do not even have to know whether the peer they want to communicate exists yet.

If there are no subscribers at the time of publication, the notification will be remembered by the NaSS for later delivery to new subscribers. Several publishers may publish notifications under the same name. All of them are delivered to the subscribers, if any. However, the NaSS remembers only the last notification, i.e. a new subscriber will receive only the last stored notification.

The above property and the anonymous communication are particularly important for the subscription phase of a service, in which a service subscriber modifies its profile and publishes the new data into the system using the Subscription Function (see Figure 3). This data is needed by a service agent activated, for example, when the subscriber is involved in a telephone call. The anonymous communication property allows the subscription function to inject the data into the system without the need to know which agents will get it and on which nodes they will run. The memory property (remembering the last notification) allows the service agents to always get the most up-to-date data.

The NaSS service is both synchronous and asynchronous. It is synchronous because it guarantees the latency with which notifications are delivered to subscribers. At at the same time it is an asynchronous service because it remembers the last notification and delivers it to a new subscriber.

In general, all components communicate with each other using the NaSS publication and subscription communication mechanism, without the need to know the identity and location of the peer component. The NaSS is also used for local communication. This permits the network operator to place any component anywhere in the network without the need for complicated configuration work.

The NaSS concept of anonymous and asynchronous communication described above is actually not a novel concept. It has its root already back in the 1980's in the field of parallel computing [12] and has recently become again an interesting research area in the field of Java-based distributed computing and mobile agents [13],[14]. New is however our proposal for applying this concept to the field of telecommunication service deployment and execution and its extension from an asynchronous messaging system to a synchronous one to fulfill the real-time requirement of telecommunications services.

3.3 CORE SERVICES

As mentioned in Section 2, *Core Services* are the building blocks for creating new and sophisticated advanced services. In our architecture there are two ways the other components can access the functionality of a core service:

1. Through the NaSS
   In this case the core service can be accessed by all components, independent of their location.
2. Directly
   In this case the core service is primarily accessible by components residing on the same service node (the concept of a service node is defined in Section 3.5). It may not be available to components residing on other service nodes.

The direct access type has the advantage of better performance in those cases where interaction with the core service requires an intensive exchange of notifications. If the component accessing the core service resided on a different service node, the resulting high network communication overhead would impose a significant performance penalty. An example of a core service that would need direct access is a call-control service using the Java Telephony API [15], which has a very detailed call model. In JTAPI, a telephone call is represented as a set of finite state machines that undergo state transitions. Each state transition is delivered using the "Event Object" pattern [16] to the service logic. The number of events generated by a call is very high when compared, for example, to the IN call model. Thus, a component that wants to control a call offered by a JTAPI interface needs be colocated with that JTAPI.

Providing access to a core service does not mean that a platform conforming to our architecture also has to implement that core service. If the core service is already provided by another system, then of course the platform will not implement that core service again but only provides the means to access and control it. For example, in the case of the telephony core service the platform will not implement the telephony service again, but merely provides the means for controlling telephone calls to its components. The telephony service itself is implemented elsewhere, e.g. by the PSTN, by a cellular system, or by IP telephony system.

3.4 AGENT INITIATORS AND SERVICE AGENTS

The main function of a *Service Agent* is to execute the logic that is specific to a certain advanced service. Service agents are designed and created by the service creator. In general several service agents are needed to implement one advanced service. The architecture does not assume any specific relation between a service agent and the phases of the service life cycle, i.e. during a certain service phase, multiple service agents may be executed in parallel, and an service agent may also cover multiple phases. The architecture also neglects the interactions that may arise between various service agents. It is the responsibility of the service creation to resolve the feature interaction problems [17] during the service specification and design phases.

Typically a service agent is in passive mode most of the time. It becomes active only at the occurrence of a certain event (e.g. an incoming call), executes the required actions, and returns to the passive mode. Because there may be a large number of service agents that need to listen for their triggering event, they all would have to be loaded into memory all the time, independent of whether they are passive or active.

To be scalable in a large number of service agents, we introduce in our architecture the concept of an *Agent Initiator* that listens for the relevant events on behalf of the service agents. At the occurrence of an event, the agent initiator loads the corresponding service agents into memory and initiates them. Once initiated, a service agent runs on its own, i.e. it interacts directly with the other components, without any help from the Agent Initiator. In addition, the agent initiator is responsible for unloading the service agents from memory when they have done their job and return to the passive mode.

The agent initiator needs the following information to be able to load and initiate a service agent:

1. The *events* at which the agent initiator has to load and initiate the agent;
2. the *code* to be loaded (this is the code that implements the logic to be executed by the agent once it becomes active);
3. the *data objects* needed by the agent to do its job, and
4. the *core services* the service agent needs to access to do its job, e.g. telephony core service. This information is important for core services that are accessible only directly.

Thus when creating the service agents the service creator not only needs to create the code of the agents, but also provide a so-called *Service Agent Descriptor* that describes these four pieces of information to the agent initiator. As in our architecture all information exchange is performed via the NaSS, a service agent descriptor contains the NaSS names under which the agent initiator has to subscribe to obtain the information required:

- Event names: the NaSS names of the events that will trigger the initiation of the agent;
- Code names: the NaSS names under which the agent code can be retrieved;
- Data names: the NaSS names under which the data objects needed by the agent can be retrieved, and
- Core services names: the NaSS names of the core services needed by the agent.

During the provisioning phase the service agents' codes and their descriptors are made available to the agent initiators by publishing them in the NaSS. As a consequence, at service node start-up time, the agent initiator subscribes to the NaSS for all service agent descriptors in order to get the service agents made available by the service provider. Section 3.6 provides more details on the use of NaSS during the provisioning phase of a service.

Upon receiving a service agent descriptor, the agent initiator checks the core services descriptor to find out whether it is able to support that service agent (it is assumed that the agent initiator knows the capabilities of the service node it resides on). If not, the published service agent will be ignored. Otherwise, the agent initiator subscribes to the NaSS or a core service interface (e.g. in the case of JTAPI) for the events specified in the service agent's event descriptor.

Upon occurrence of one of the events described in the service agent's event descriptor, the agent initiator

- subscribes to the NaSS to get the agent code (if this has not already been done);
- subscribes to the NaSS to get the data objects specified by the service agent's data descriptor, and
- starts the service agent's code with the data objects it gets from the NaSS.

As the NaSS functionality is available to all components of the architecture, the service agents can also access the NaSS directly to get the information they need to do their jobs rather than getting it from the agent initiator. Our architecture provides both mechanisms, and it is up to the service creator to select the one that is most appropriate to his services.

Note that the agent initiator is agnostic about

- the number of service agents a certain advanced service may need. It is the responsibility of the service creator to define the number of service agents needed by a service. For example, if the service requires actions at multiple events, then the service creator may define multiple agents, one for each of the events. He may also create a single one which listens to all events (except the first one, which is delegated to the agent initiator);
- the number of subscribers involved by a service agent (i.e. the logic executed by a service agent may be specific for a single subscriber, but also may be valid for a group of subscribers), and
- the relationship between a service agent and the service life cycle.

3.5 SERVICE NODE

A service node is a functional grouping that contains at least an agent initiator and is therefore able to initiate service agents. Furthermore it provides to the components it hosts access to the NaSS for communicating and exchanging information with other components.

In general, there are multiple service nodes within a system implementing our architecture. To be scalable in terms of the number of users it can accommodate, the processing load generated in the entire system needs to be spread over several service nodes. There are two possible approaches regarding the division of the load: by function (task) or by load-generating request. In the case of division by function a node specializes in performing only certain subtasks. To handle a request several service nodes need to communicate to coordinate the subtasks. In the case of division by load-generating request, a service node performs all subtasks for a request. As a request enters the system it is assigned to one service node.

The latter approach in general has some advantages over the former because (1) it has a lower communication overhead, (2) is more robust because the load of a failed service node can be taken over by the other service nodes, and (3) easier to manage when the system needs to be scaled up as another service node can be merely be added.

Our architecture allows the implementation of both approaches, even a combination of them. The concept of agent initiators and service agents, together with the NaSS communication mechanisms, permits the addition and removal of advanced services without the need for re-configuring and restarting the service nodes. New service agents are distributed automatically to the subscribed service nodes and started there without the need for complicated configuration work. It also allows the addition and removal of service nodes to distribute load over multiple machines, thus making the architecture scalable in terms of the numbers of users, calls and services that can be accommodated. It furthermore adds reliability to the system because a failing node can easily be replaced by another one.

3.6 PROVISIONING AND SUBSCRIPTION FUNCTIONS

Assuming the existence of the service agents coded as described in Section 3.4, the *provisioning function* shown in Figure 3 helps the operator inject the service agents into the network using the publishing mechanism of the NaSS (provision phase). For example, the operator can publish the service agent implementing the CFNR supplementary service [16] using the NaSS name `/serviceagent/descriptor/cfnr`. It is the responsibility of the service creator to choose the name such that it is unique in the system. With this naming scheme, the Agent Initiator would then have to subscribe to the name `/serviceagent/descriptor/*`. The wildcard `*` is used to refer to all notifications published using names that start with `/serviceagent/descriptor/`. In this way all service agent descriptors published via the provisioning function will be forwarded by the NaSS to the Agent Initiators.

Certainly, by giving a more complex structure to the NaSS names, specialized service nodes and agent initiators could be introduced into the network. In this case, these nodes would subscribe only to (and therefore get only) the agents of the services in which they were interested. This example shows the strength of our architecture: service logics can easily be transferred to the appropriate network nodes, without the need to know the location and configuration of the various nodes explicitly.

This also applies to the subscription phase, in which a user can subscribe to a service and customize his service profile according to his needs. The *subscription function* in Figure 3, besides interfacing with the user, publishes the most up-to-date subscriber data into the system. In this way any components in the system, e.g. service agents, currently subscribed to those data always get the most recent one from the NaSS.

Although the subscription function is shown in Figure 3 as a component different from a service node, it can certainly be designed in a very similar way. This means that it will also contain an agent initiator for initiating service agents, which, in this case, will run during the subscription phase and not, as in the case of the service nodes, during the utilization phase. These "subscription" agents are also created by the creator and injected into the network using the provision function described above.

4 INTEGRATION OF H.323-BASED TELEPHONY SERVICE INTO THE ARCHITECTURE

In the preceding sections the inclusion of the telephony core service was handled in a very generic way, i.e. without specific assumptions on how that service is actually implemented. Whether it is provided by a legacy PSTN or an IP-based network was not important.

Figure 5 now shows how the architecture looks if the core telephony service is based on H.323. The gatekeeper-routed call model is selected because we want to provide the service agents running in the service nodes with the capability of controlling the H.323 calls. The H.323 gatekeeper functions, as specified in [1], are split into two parts:

1. The *H.323 signalling proxy*, which mainly handles the signalling functions of the Q.931 and H.245 protocols and provides the telephony call control services to the other components of the service node, and
2. the *H.323 RAS Server*, which covers the functions defined by the H.225.0 RAS protocol such as address translation, admission control, etc.

The separation of the gatekeeper functions into a signalling and a RAS part allows the implementation of a very efficient load distribution mechanism. All service nodes use the NaSS communication mechanism to publish

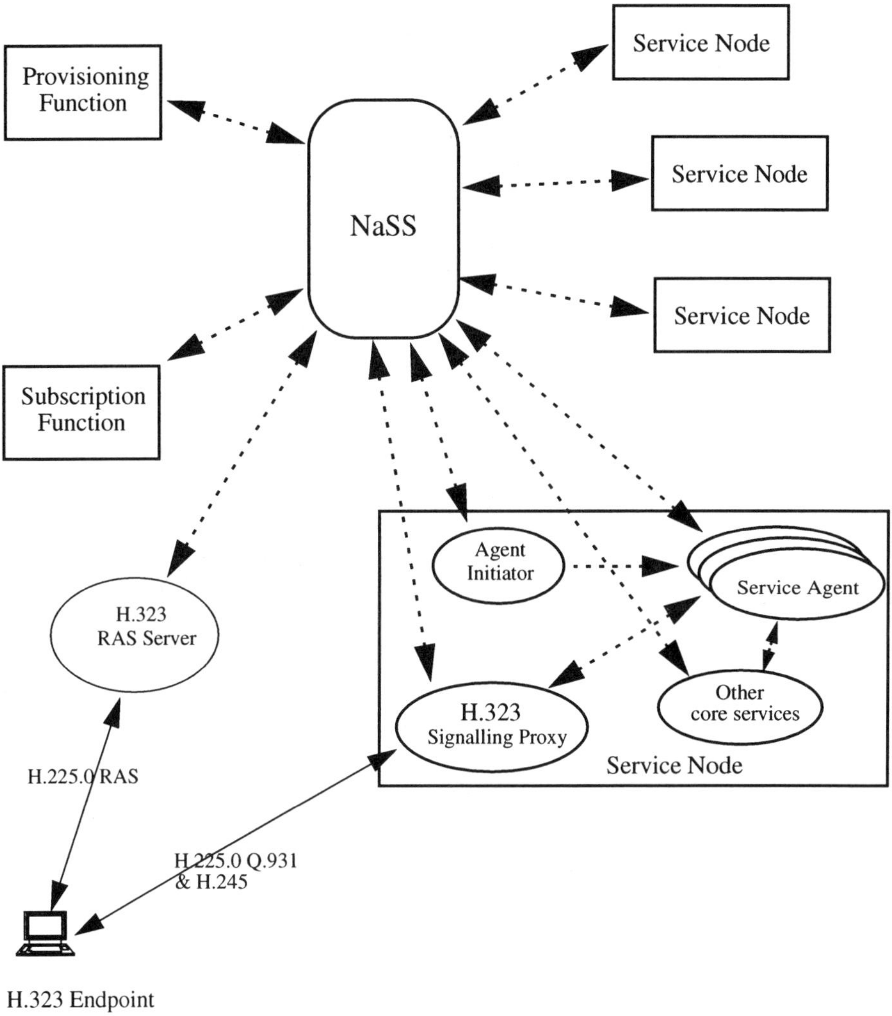

Figure 5: Integration of H.323-based telephony service into the architecture.

information about their availability and load situation. Because the communication via the NaSS is anonymous, they do not know the existence of the RAS server, and even do not know who will manage their load.

To get the load information of the service nodes, the RAS server responsible subscribes to NaSS with the name used by the service nodes to publish their load information. Based on the load information collected, the RAS server is able to construct an overview of the actual load situation of the service nodes and can therefore distribute new calls to the most appropriate nodes. The entire scenario is illustrated in Figure 6, which starts with the RAS server subscribing to the load notifications sent by the service nodes. According to Recommendation H.323, an endpoint needs to send an *admission request* (*ARQ*) message to the gatekeeper (in our case to the RAS server) before it can set up a call. The RAS server will then request the endpoint to send its Q.931 SETUP to the most appropriate service node. The IP address of the selected service node is given back to the endpoint via the *admission confirm (ACF)* message.

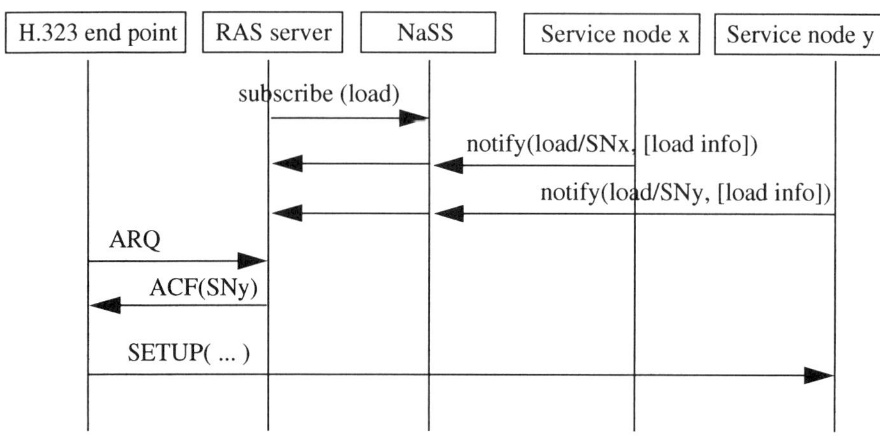

Figure 6: Load distribution.

The NaSS capability can furthermore be exploited to implement the H.323 address registration and translation. In H.323 an endpoint may be assigned an alias that is independent of its transport address. This is important for example for endpoints that access the Internet via a dial-up connection with dynamic IP address assignment. To be nevertheless always reachable under the same alias, the endpoint has to inform the gatekeeper about its current transport address using the so-called registration procedure.

Figure 7 illustrates how the H.323 address registration and translation can be implemented using the NaSS capability. The endpoint informs the gatekeeper (in our case the RAS server) about its current transport address by means of a *registration request (RRQ)* message. Upon receiving an RRQ from an endpoint, the RAS server publishes the received H.323 alias address and the corresponding transport address into the NaSS, thus making the translation immediately available to all components of the system. Any compo-

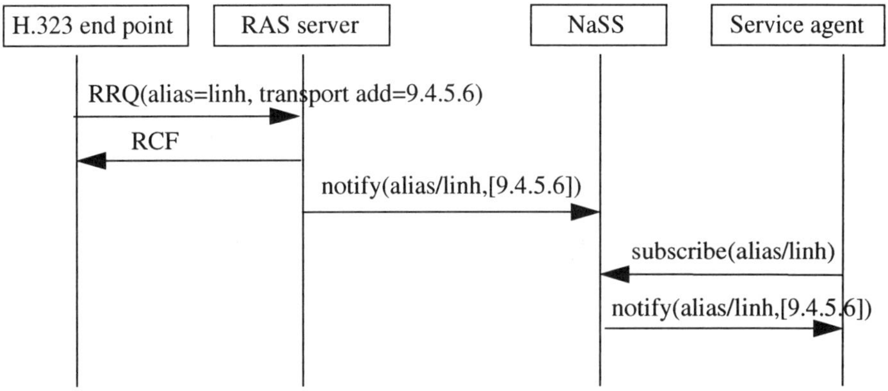

Figure 7: H.323 address registration and translation.

nent, e.g. a service agent that needs to set up a call to a certain endpoint, can retrieve the actual transport address of that endpoint simply by subscribing to the NaSS for the alias associated with that endpoint. To remove an association (e.g. in the case of an unregistration), the RAS server publishes an empty association, i.e. a notification that contains an H.323 alias but without a corresponding transport address. Note that all H.323 RAS specifics, e.g. time to live, etc., are implemented in the RAS server, not in the NaSS. The NaSS stores only the association between an alias and a transport address.

Although the architecture illustrated in Figure 5 was developed based on the ITU-T H.323 recommendation, it is obvious that it can also apply to the IETF SIP protocol by replacing the H.323 RAS server and signalling proxy by a SIP redirect and proxy servers, respectively.

5 CONCLUSION

We have described a novel architecture that allows the rapid and efficient creation and deployment of sophisticated advanced services. The core component of the architecture, the NaSS, provides the service creator with an easy and efficient means for injecting new service logics into the network and modifying them. It also allows service subscribers to customize their service profiles directly. The concept of service agents and agent initiators permits the addition or removal of services without the need to restart the service nodes. It also permits the addition and removal of service nodes to distribute load over multiple machines, thus making the architecture scalable in terms of the number of users, calls and services it can accommodate. The dynamic task assignment adds reliability to the system because a failing service node can easily be replaced by another one.

Our future work consists of building a prototype to assess the correctness of the concepts we have presented in this paper. We are evaluating various implementations of the NaSS, in particular regarding the fulfillment of the requirements described in Section 3.2. A performance evaluation of the architecture using a typical hardware and software platform also remains to be done. The results of these activities will be communicated in future publications.

ACKNOWLEDGMENT

The authors are thankful to Lucas Heusler for many insightful discussions and an in-depth review of the paper. They are also grateful to Asser Tantawi, Linda Steinmuller, Pnina Vortmann, Samuel Kallner and Igal Golan for useful comments.

References

[1] ITU-T Recommendation H.323, "Packet-based Multimedia Communications System", Version 2, Oct. 1999

[2] J. Toga and J. Ott, "ITU-T Standardization Activities for Interactive Multimedia Communications on Packet-based Networks: H.323 and Related Recommendations", Computer Networks 31, Feb. 1999

[3] H. Schulzrinne and J. Rosenberg, "Internet Telephony: Architecture and Protocols - an IETF Perspective", Computer Networks 31, Feb. 1999

[4] F. Cuervo et al., "Megaco Protocol", draft-ietf-megaco-protocol-07.txt, Feb. 2000

[5] I. Faynberg, L.R. Gabuzda, M.P. Kaplan, and N.J. Shah, "The Intelligent Network Standards", McGraw Hill, 1997

[6] M. Day, J. Rosenberg, and H. Sugano, "A Model for Presence and Instant Messaging", RFC 2778, Feb. 2000

[7] ECMA Standard 185, "Call Completion Supplementary Services", Dec. 1992

[8] C. Gbaguidi, J.-P. Hubaux, G. Pacifici, and A.N. Tantawi, "Integration of Internet and Telecommunications: An Architecture for Hybrid Services", IEEE JSAC, Vol. 17, No. 9, Sept. 1999

[9] ETSI Technical Report ETR323, "Intelligent Network (IN): Service Life Cycle Reference Model for Services Supported by an IN", Dec. 1996

[10] K.J. Turner (Editor), "Using Formal Description Techniques - An Introduction to Estelle, LOTOS and SDL", John Wiley & Sons, 1992

[11] ECMA Standard 173, "Diversion Supplementary Services", June 1992

[12] S. Ahuja, N. Carriero, and D. Gelernter, "Linda and Friends", Computer, pp. 26-34, August 1986

[13] E. Freeman, S. Hupfer, K. Arnold, "JavaSpaces(TM) Principles, Patterns and Practice (The Jini(TM) Technogy Series)", Addison-Wesley, Reading, 1999

[14] T.J. Lehman, S.W. McLaughry, and P. Wyckoff, "TSpaces: The Next Wave", Hawaii Int'l Conf. on System Sciences (HICSS-32), Jan. 1999

[15] "JTAPI Introduction", http://java.sun.com/products/jtapi/

[16] E. Gamma, R. Helm, R. Johnson, and J. Vlissides, "Design Patterns: Elements of Reusable Object-Oriented Software", Addison-Wesley, Reading, 1995

[17] F.S. Dworak et al., "Feature Interaction Problem in Telecomunication Systems", Proc. of the 7th Int'l Conf. on Software Engineering for Telecommunication Switching Systems, pp. 59-62, July 1989, Bournemouth, UK.

A Service Platform for Internet-Telecom Services using SIP

S. Bessler[1], A.V. Nisanyan[2], K. Peterbauer[1], R. Pailer[3], J. Stadler[4]

[1] *Kapsch AG, Pottendorferstrasse 19, A – 1121 Vienna, Austria*
{Bessler,Peterbau}@kapsch.net

[3] *SIEMENS AG Österreich,Gudrunstrasse 11, A – 1100 Vienna Austria*
agop-vart.nisanyan@siemens.at

[3] *Ericsson Austria AG, Pottendorferstrasse 25 – 27, A – 1121 Vienna, Austria*
Rudolf.Pailer@sea.ericsson.se

[4] *Vienna University of Technology, Institute of Communication Networks, Favoritenstrasse 9 / 388, A – 1040 Vienna, Austria Johannes.Stadler@tuwien.ac.at*

Keywords: Service Platforms, SIP, Multimedia Services, Convergence in Heterogeneous Networks, PARLAY

Abstract: This paper proposes the introduction of a Service Platform for the creation, execution and management of multimedia services in heterogeneous networks. Examining the business-roles, the actors in the Service Platform are identified. Furthermore several common building blocks for developing services for the Internet are described and a brief overview of some modern technologies for an object-oriented, component-based, distributed platform for multimedia services is given. The Session Initiation Protocol (SIP) has been identified as very useful to implement all functions according to the multimedia part of the Platform. The usage of the PARLAY API as an open interface between the Platform Services and SIP, the PSTN or the mobile network provides a lot of additional advantages. The interworking between SIP and PARLAY is shown in a call-routing example. Furthermore the call-setup for a multimedia (e.g. video) conference is explained. This should demonstrate the usefulness and the ability of this protocol for introducing a session concept. Finally an outlook of open research topics regarding this concept is given as well as a short overview of related work.

1. INTRODUCTION

The FTW research project 'Service Platform and Interoperability' proposes the introduction of a Service Platform (SP) for the creation, operation and management of real-time multimedia services in heterogeneous networks.

The SP is a middleware, which provides a distributed framework for services. The managed services shall be accessible from different types of terminals, carried over different types of networks and using different protocols.

During analysis of the requirements for the SP, the need for a multimedia session concept has been identified. Sessions are responsible for keeping track of the various data streams associated with one multimedia communication (e.g. data, voice, video streams). This paper proposes the usage of the Session Initiation Protocol (SIP [1]).

The paper is organized in six chapters. After this introduction, the second chapter introduces the concept of a Service Platform and describes the requirements it must fulfill. As an important technology for the architecture of SIP as an object-oriented, component-based system CORBA is identified.

The third chapter shows our architecture of the Service Platform. All necessary components are discussed and explained.

In the fourth chapter we answer the question why we use an API between the Platform and the underlying call control. We discuss the advantages of such an approach, and give an account of why using PARLAY [2]. An example of a call-routing process should prove this theory. Also the setup of a multimedia conference using SIP over Parlay is shown.

In the fifth chapter we provide an overview of open research topics regarding the SP concept and we give an outlook on the possibilities of implementing the SIP session control as part of the SP.

The sixth chapter presents a brief summary of related work.

2. THE CONCEPT OF THE SERVICE PLATFORM

2.1 The Business Case

The business roles model is defined in accordance with the suggestions and discussions of TINA-C, TSAS (OMG) and Open Marketplace (OMG). We identify the business roles end user, subscriber, service provider, 3rd party service provider and network provider.

The service provider runs and maintains the service platform. He maintains the data store with all the users and service logic as well as the contracts and configuration of the subscribers. Furthermore he creates new services, which are deployed on the Platform and offered to the subscribers. Also he provides connectivity and service agreement to network providers.

The SP has to provide mechanisms for registration of subscriber resources such as content and web-servers, as well as enterprise resources needed for the implementation of a certain service.

A subscriber is a user or a company that pays for a service offered by the service provider. Subscribers do neither host services nor have to maintain the service platform.

End users are registered on the behalf of the subscriber. The representatives of the subscriber (operators) have a subscription side management application to configure and manage the service of their users.

Depending on the service, a registered user can configure an application by himself, such as call forwarding, or the task can be done via subscriber side management, e.g. call center scenarios.

A 3rd party service provider provides content, which is offered by the Service Platform. A service provider could be 3rd party service provider too.

2.2 Basic Services

Several common building blocks for developing services for the Internet can be identified, including

- User and subscription management.
- Security management, enforcement of access control, management and distribution of certificates.
- Generic session management.
- Usage tracking, accounting, billing.

These common capabilities can be modeled as *basic* or *horizontal services,* which are always present at any host of the SP (billing and certificate management are outside the scope of this project).

Beside these functional requirements, several other basic services are valuable:

- *Registry and Lookup Service* used to register and locate services running at different hosts of the platform.
- *Transaction Service*, supporting two-phase-commit transactions.
- *Scheduling Service*, available to all services on the platform rather than just processes on a single host. Tasks can be scheduled and executed on the local or a remote host.
- *Event Service* as a publish/subscribe system decoupling the service, which is the source of an event from other services acting as "listeners".
- *Log Service* as a centralized facility, supporting internationalized log messages e.g. with timestamps or complete stack traces.

Logically these services can be seen as single instances in the whole platform. Transparent replication might be implemented solely for the purpose of increased availability.

2.3 Service Accessibility

Access can be provided to users for a variety of terminal types, e.g.

- PC with a standard Web browser for rendering HTML/XML/XSL documents, displaying multimedia data with plug-ins, and executing downloadable code such as Java applets.
- Multimedia PC running dedicated applications e.g. a fully featured SIP client.
- WAP/WTA-enabled terminals (Wireless Application Protocol/Wireless Telephony Application), such as mobile phones and personal digital assistants. These terminals might also be aware of their current location, e.g. by using the GPS satellite system.
- Black telephones for simple voice communication or interaction using DTMF signaling or voice recognition.
- Fax machines.

Some services will be dedicated to certain terminal types, but many services will be accessed from more than one terminal type with a varying degree of functionality. These services can share basic infrastructure provided by the SP, shielding the services from the details of the underlying network protocols.

2.4 Backend Systems

The SP encapsulates the services and provides controlled access to various backend systems, including relational and object databases, directory services, application software packages and other legacy systems (billing, customer care systems etc.).

We will focus on the access to those legacy systems, which are governed by open standards, e.g. relational databases.

2.5 Service Life Cycle

The life cycle of a service involves numerous steps, including design, implementation, testing, deployment, operation, management, maintenance, and removal.

The service creation process is beyond the scope of our project. Our major goal is to define an appropriate programming model for services, which is prepared for an integrated service creation environment.

2.6 MIDDLEWARE Technologies

There are many modern technologies for an object-oriented, component-based, distributed platform for multimedia services, for instance CORBA or Enterprise Java Beans. These technologies fulfill many of the requirements of the SP with respect to the identified basic services and may serve as 'glue' for the SP. For our architecture we identified CORBA as the best suiting possibility for our requirements.

CORBA is the most widely used and deployed distributed object technology. It is an open industry standard, providing an object-oriented middleware spanning over heterogeneous platforms, (packet-) network protocols and programming languages. A large number of today's modern application servers are based on CORBA as backbone of an enterprise's information service, used in the middle tier between legacy systems at the backend and typically conventional Web servers at the front-end.

The core of CORBA is the Internet Inter-ORB Protocol (IIOP), which enables different applications to communicate via application-specific interfaces.

CORBA specifies a set of horizontal and vertical services. Among the *horizontal services* are

- a *Naming* and *Trading Service* for service registry and discovery,

- an *Event Service* for pulling and pushing typed and untyped events,
- and a *Transaction Service* for coordinating distributed transactions across multiple databases and legacy systems.

Recently the OMG specified a component-based object model, defining interfaces between a server-side component and the server application in which it is contained, e.g. with respect to transactional behaviour and persistence.

Most notably among the *vertical services*, the *TSAS* (Telecommunication Services Access and Subscription) specification defines standardized interfaces enabling end-users to access and configure telecommunication services, as they prefer and to allow value-added service providers to offer their services.

The Parlay API from the Parlay consortium specifies a high level CORBA-API that allows service developers to access resources in PSTN, mobile and IP-based networks. Consideration of a Parlay interface to the service platform unifies the model and defines the responsibility of service provider and network provider. Since Parlay operates only at the enterprise level, the mapping to user granularity has to be made, possibly as proposed by TSAS.

3. SERVICE PLATFORM APPROACH

Fig. 1 illustrates our approach for the realization of a service platform, which integrates different (PSTN, PLMN, IP) networks and services.

Our Service Platform Solution is based on following principles:

- The customer (end user side) access works via HTTP and WAP. Thus both, the wirelined (LAN, WAN) and wireless mobile (GSM, GPRS and UMTS) terminals are supported.
- The requirements on the end-user-terminals are minimum. They need only a usual webbrowser (HTTP)/WAP and a SIP User Agent.
- The service provider who owns the service platform and who has contracts with 3rd party service provider is the SP operator. He holds all the subscriber data, which is the most valuable asset in the Internet world. This service provider, together with all other contracted 3rd party providers, build a virtual service domain.

- A central idea is one point authentication in this whole virtual service domain. Ways to ensure this must be still investigated. A possible (temporary) solution can be the use of certificates.
- The application / service logic is distributed among the 3rd party service providers. Every 3rd party service provider can provide services independent of other service providers if needed. He needs only a contract with the SP operator if he wants to use services of Service Platform as authentication (one-point-shopping), call control (connection provisioning), media streaming or location services.
- For the implementation of application / service logic servlets are intended.
- Every contacted 3rd Party Service Provider can access the service platform (usually via servlets) over CORBA.
- The Service Platform acts as controlling and switching center for cross domain (PSTN, PLMN, IP) network connections and intelligent communications services as call forwarding.
- The service platform uses PARLAY as the controlling interface to the underlying networks.
- The IP-signaling will be realized over SIP.
- In accordance to the decision to use PARLAY we will implement on the SIP-Proxy a PARLAY interface and check the effects.
- Only the control data passes the service platform. The user data is transferred transparently between end-points that means without any Service Platform interaction. An exception to this point is the IP-PSTN-Gateway, which will be part of the service platform because of the lack of such gateways in the used network.
- The SIP will be used not only for voice-communication, but for all types of multimedia communication as e.g. video streaming.
- It is intended to use JAIN as the PARLAY (Client) wrapper if available in time for the implementation.
- For the realization it is intended to use JAVA, JAVA Media Framework, CORBA, SIP, CPL [3] and the servlet API.

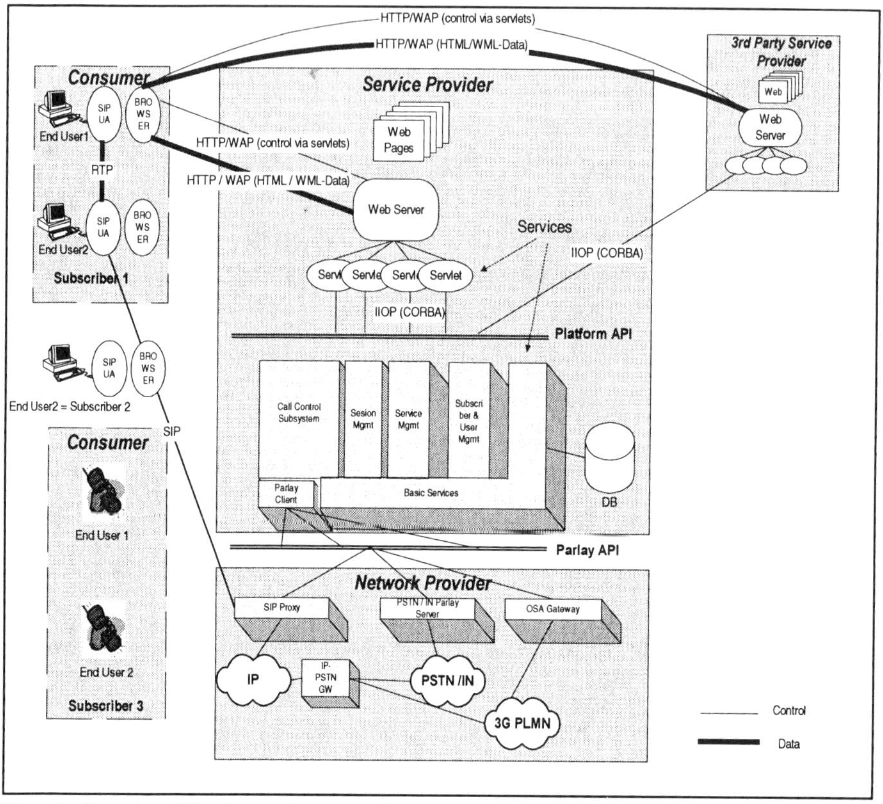

Fig. 1. Service Platform Approach

4. SIP AS BUILDING BLOCK OF MULTIMEDIA SERVICES

The IETF SIP protocol has been identified to be one of the most promising ways for a large-scale deployment of multimedia communications over today's data networks.

There are different entities defined in SIP. At first there are the SIP User Agents (UA), which could be separated in UA Servers and UA clients. UA clients initiate a call, while UA servers only react to calls. Both are statefull devices, because in the end terminal the state of a call has to be known. The second important entity in SIP is the SIP network server, which could act as proxy or as redirect server. Because of scalability the network server in the core net should be stateless, while the edge servers have to be statefull to provide all required functionality. Location server and register server complete this enumeration.

The session concept of SIP offers a generic way to establish, control and tear down multimedia sessions between communication end-points. Publications about SIP mostly concentrate on the usability of SIP to implement Voice over IP (VoIP) services. It is one of our research goals to show that SIP sessions can be used to establish multimedia streaming sessions as well as interactive multimedia communications.

In this chapter we show the way, how the interworking between SIP and Parlay could be performed, as well as some scenarios for multimedia functionality.

4.1 Third Party Call Control with SIP

The main task of the SP can be seen as a platform for the creation, management and execution of 'third party services'. In this paper a third party service is understood to be some kind of service logic that is located and runs outside of the trusted domain of the network resources that this service controls.

Third party control interfaces shall have the following important properties:

- The third party control interface offers an abstract view of the underlying network resource so that the features and functionality of the controlled resource can be used without detailed knowledge about the specific realization or implementation.
- The third party control interface offers security management, so that service logic running in an untrusted domain outside of the network resource can be authenticated and authorized to perform service specific tasks.

The PARLAY API 2.0 specification has been identified by our group as a good solution for above requirements. PARLAY is an industry initiative to create a CORBA based interface from an enterprise application (client) to different networks (PSTN, mobile, IP) and their resources e.g. voice- or mailboxes. Especially for the control of call processing in advanced mobile switches (e.g. UMTS), the PARLAY API seems to develop into an industry wide standard.

The Service Platform will control call processing in the SIP Domain Controller via a PARLAY interface in order to realize third party call control services. A call control service running on the Service Platform will therefore act as a PARALAY client, while the SIP Domain Controller has to implement the PARLAY server interfaces. To show the interworking between SIP and PARLAY the Use Case 'Routing of a SIP call' is described in detail.

4.2 Use Case 'Routing of a SIP call'

The router logic (e.g. a CPL script) can be configured by the users. The Service Platform only provides a simple interface that offers a method to notify the router logic of a new incoming call. A second method is used to inform the router logic of the result of a requested routing task. The platform service classes ServiceRouteNewCall_Manager and ServieRouteNewCall_Call have to implement the PARALY IpAppCallControlManager interface and the IpAppCall interface respectively. These central classes are responsible for the specialization of the general PARLAY functionality towards a well-defined feature, namely the routing of new incoming calls towards a new destination.

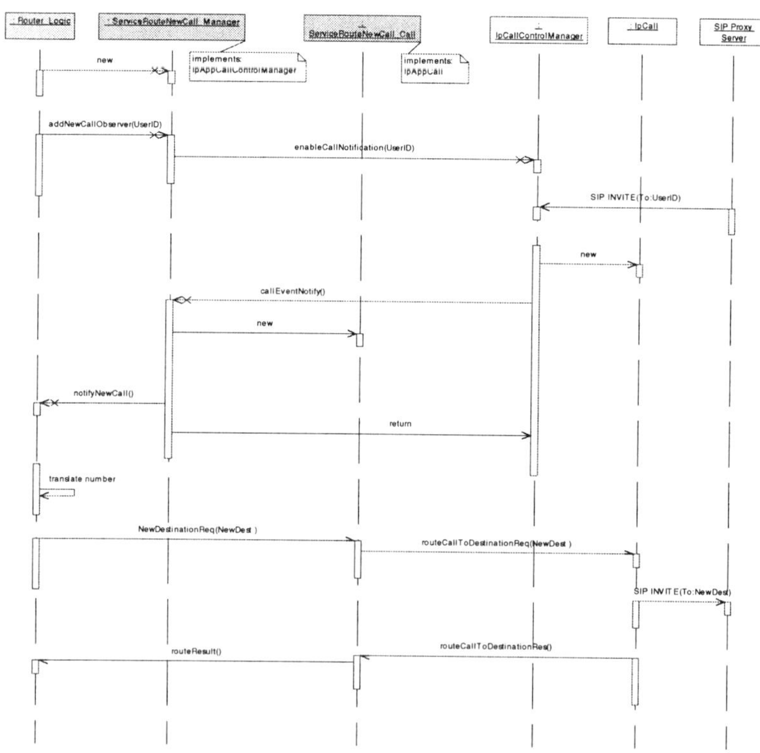

Fig. 2. Routing of a SIP call

The sequence of events is as follows:

1) The router logic registers to receive notification in case of a new incoming call.

2) The service manager adds the router logic to the listed observers and requests the notification from the PARLAY server. The PARLAY server interfaces are implemented by a SIP proxy server.
3) A new INVITE request is received and the PARLAY server notifies the service manager.
4) The service manager creates a service call object and notifies the router logic.
5) The router logic computes a new destination for the call.
6) The router logic has got together with the notification a reference to the service call interface, where it requests now the new destination for the call.
7) The service call object forwards this request to the SIP server, which proxies the INVITE to the new destination.
8) The PARLAY server responds with the result of the routing operation.
9) This result is sent to the routing logic.

4.3 Further Use Cases

The service platform covers following requirements:

- The user is able to receive multimedia streams, such as videoclips or audioclips, from a content server, located at the 3rd party provider premises and registered at the service platform. An additional goal is to easily adapt the existing content servers as well as the user terminals to support streaming.
- The user is able to build up a VoIP or videoconferencing connection to a 3rd party provider or another user. In case the 3rd party provider implements a SIP UAS too, a pure IP-based connection is set up. In case the 3rd party provider only offers a connection over the PSTN, a VoIP connection should be set up only with a media gateway. The connection to the PSTN is done using a media gateway controller, which transforms the SIP INVITE request to PSTN signaling. The data is transferred over the media gateway.

The cases above could and should be implemented as a 3rd party call to give the Platform the control over the call.

- The user is able to build up a VoIP connection to other users currently logged in.

- The user is able to allow the subscriber to send web pages directly to his desktop.
- A user that changes location or availability is able to define a redirect service to the new location, web page or messaging box.
- The subscriber is able to manage his user base himself.

4.4 Media Streaming

Representative for most of the use cases described above we want to show the sequence of a media streaming session (see Fig. 3). Maybe the user wants to stream down a video clip or wants to set up a VoIP connection to a 3rd party service provider. The iPApp in the following diagram represents the PARLAY client. The user clicks to a link on a website. This causes a HTTP-request. Via a servlet the request comes to the PARLAY Client, which generates a routeToOrigination-Request. The call setup works in this case like a 3rd party call.

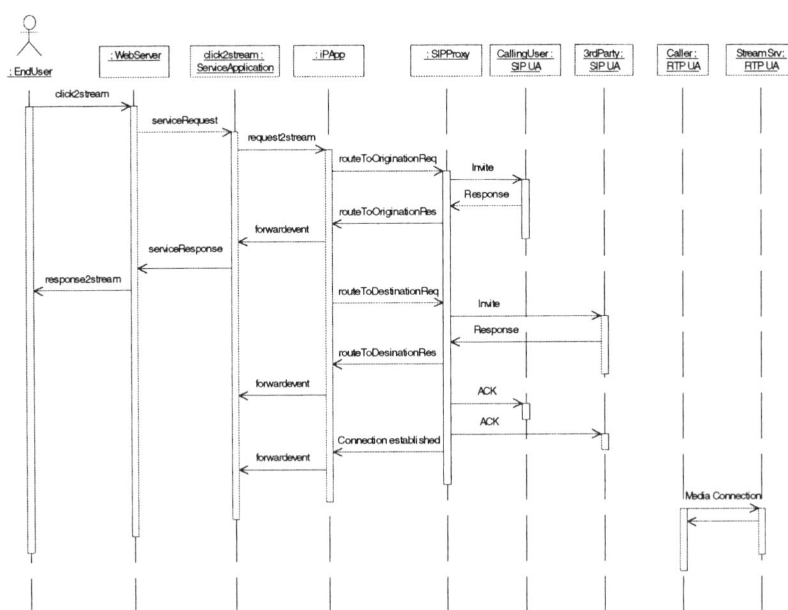

Fig. 3. Media Streaming

An INVITE request (without IP and Port of the second UA) is sent to the originating UA. After the response (with IP and Port of the originating UA as well as supported media types) the answer to the HTTP request is performed. If the originating UA would not be reachable, an error page could be sent back here. After this a routeToDestination-Request is sent to the SIP proxy, which generates an SIP INVITE with IP and Port of the originator. In the response the destination UA sends back its IP and Port, which reach the originator in the ACK message. After all ACK's, the SP is notified about the established connection and the media channel is set up. The transport of the media should be done using RTP [4]. We decided to use JMF, which provides good possibilities for synchronizing different media streams. JMF 2.0 includes already classes for the RTP transport.

5. OPEN TOPICS

In the following we list open research topics regarding the SP concept.

- Evaluation of the described technologies as middleware for the SP.
- Service creation and service interactions.
- QoS aspects.
- SIP enabled content server (e.g. video streaming).
- Definition and implementation of a generic session management API.
- Design and implementation of SIP enabled multimedia client software.

6. RELATED WORK

The idea of a service platform is not new at all. The most popular service architecture for the PSTN is the Intelligent Network (IN). The IN is provided with a Service Creation Environment (SCE) in which a service is created from a set of standardized service-independent building blocks (SIBs).

Gbaguidi [5] proposes a service platform for so called hybrid services, which is based on a Service Creation Environment (SCE) and a Service Infrastructure (SI). In the SCE a service logic is put together by the service creator using a set of service components as building blocks. The service logic is then sent to a service factory, which creates a new instance of service. The service instance is organized into service subsystems that are further uploaded into the system infrastructure, more specifically in the controllers of the system elements.

The Service Infrastructure is composed of a collection of controllers and the underlying system elements of the Intelligent Network, such as terminals, gateways, network nodes, etc. An event interface enables the data flow between the system elements and the service logic.

Manione and Renditore [6] describe a service platform based on a TINA architecture that has been simplified to cope with the requirements of internet-telecom service development: simplified session modeling, identity of the service provider business role with the retailer, enhanced subscriber management, logical separation between service logic (service environment) and generic platform services.

The WebCentric web call center application provides added value by offering VoIP call together with pushed WebPages to the caller, or other media combinations such as chat or email. The paper describes clearly the technology used for the different terminal types, without addressing generic interfaces to IN, or to go into the aspects of service creation and deployment [7].

REFERENCES

1 M. Handley, H. Schulzrinne, E. Schooler, J. Rosenberg: SIP: Session Initiation Protocol. Internet Engineering Task Force RFC 2543, March 1999

2 Parlay API Specification 1.2, User Guide. http://www.parlay.org, Sept. 1999

3 J. Lennox, H. Schulzrinne: CPL: A Language for User Control of Internet Telephony Services, Internet Draft, Internet Engineering Task Force, March 2000.

4 H. Schulzrinne, S. Casner, R. Frederick, V. Jacobson: RTP: a transport protocol for real-time applications. Internet Engineering Task Force RFC 1889, Jan. 1996

5 C. Gbaguidi, et al, Integration of Internet and telecommunications: An Architecture for Hybrid Services. IEEE JSAC vol.17, no.9, Sept. 1999

6 R. Manione, P. Renditore: A TINA Light Service Architecture for the Internet-Telecom Scenario, TINA 99 Conference, Hawaii, USA, April 1999

7 R. Manione, M. Festa, P. Renditore, D. Sereno, M. Spaziani:. Service Issues in Web Call Centers, Telecom 99 Forum, Geneva, 1999

TOWARDS DYNAMIC COMPOSITION OF HYBRID COMMUNICATION SERVICES

Jacqueline Floch and Rolv Bræk
Department of Telematics, NTNU
N-7491 Trondheim, Norway
e-mail: {Jacqueline.Floch, Rolv.Braek}@item.ntnu.no

Key words: Architecture, Telecommunication, Services, Service models, Service composition, Service execution framework, Roles, Role assignment, Role learning.

Abstract: Due to the deregulation of the telecom network and the Internet, users will have access to an increasing number of heterogeneous communication services and will need to adapt their services or learn new services in order to interact with other users and systems. We propose a dynamic composition method that enables services to be constructed dynamically or "on-the-fly" from existing functional elements (service roles). Roles and actors that play roles are key concepts in our approach. A service role is defined as the part an object takes in a service. Service execution requires that roles are assigned to actors in a coordinated way. Our approach enables the systematic and structured specification of services, and provides mechanisms for service composition and an execution environment.

1 INTRODUCTION

Since short time to market for new services is increasingly important, the telecom operators and manufacturers are in constant search for better frameworks and methods for the rapid construction and deployment of services. Frameworks such as the TINA Service Architecture [1] or the programmable architecture from EPFL [2] have been proposed. We believe that a traditional approach to service life-cycle as proposed by TINA with the associated scenarios for service deployment, user subscription and service withdrawal, is not flexible enough to future market expectations and service providers needs. Users expect to access a similar set of services independently of what network they happen to use, they expect to get access to new and useful services as they become available, and they expect to be able to communicate seamlessly with other users and applications regardless of what service provider and network operator they use. Conversely, service providers need to develop and deploy services across the network technologies at a pace and with a degree of harmonisation that keeps up with user expectation. As the open world of Internet enables several parties to develop and provide rapidly new services, as new actors are emerging in the deregulated telecom network, users will have access to an increasing number of heterogeneous services and will need to adapt their services or learn new services in order to interact with other users and systems. The mobility of users also creates a need for adaptation. For example, UMTS (Universal Mobile Telecommunication System) introduces the Virtual Home Environment (VHE) and service portability. This allows users to access their personalised services in any visited network in the same way as in their home network, even if the services are not actually offered in the visited network.

Hybrid communication services are sometimes defined as services that span different network technologies including the public switched network (PSTN) and the Internet [3]. We also consider as hybrid services, services that are provided by interactions between *heterogeneous* service components, developed by different parties or made available by different service providers. In our work, the latter kind is in focus. Our goal is to define a framework for service execution and methods for service design that enable services to be designed separately and then composed dynamically using Plug-and-Play techniques [4].

Service design is complex. Communication services normally require the coordinated effort of several distributed components, where some of the components may be involved in several services. Services involve several users where some of them may participate in several services. In addition to this intricate structure of services, the need for service inter-operability across several operator domains and for service availability over different network technologies also contribute to the complexity of service design. Our approach to service design is based on *service roles* [5]. We define a service role as the part an entity (an object in our framework) takes in a service, and we consider a service as a collaboration between roles. By using services roles, we are able to break down the complexity of service specification, and also to combine roles and provide new services in a flexible way.

Dynamic service composition enables hybrid services to be constructed dynamically from heterogeneous service roles. The method supports service adaptation and learning, and thus provides service users and providers with the ability to build new services "on-the-fly".

This paper presents our approach to service development and execution based on role modelling and dynamic service composition. The next section introduces the basic concepts: *roles* and *actors*. In Section 3 we present the service execution framework in which we are experimenting with the dynamic composition method. The purpose of the framework is to enable the systematic and structured specification of services by defining concepts for service modelling. The framework also provides basic mechanisms for service composition and an execution environment. In Section 4 we present some examples that illustrate how dynamic composition may be used during service execution. The examples also introduce the concepts of *role assignment, negotiation* and *learning*. Section 5 presents the main ideas in our approach and discuss potential solutions. We identify the requirements that the approach sets on the service development. In Section 6 we discuss composition correctness. Finally we compare our approach to some related work.

2 ACTORS AND ROLES

Actors and service roles are key concepts in our framework. A service role is the part an entity (an object in our framework) takes part in a service. Thus a service is seen as a collaboration between roles [5]. Service execution requires that roles are assigned dynamically to objects (that we denote as actors) in a coordinated way.

Roles may be introduced in a rather intuitive way. The concept of role is used extensively every day, either to describe relations between persons, for example family roles such as mother and daughter, or to describe functions and responsibilities, for example organisational roles such as professor, secretary, librarian, student. Also in telephony, it is customary to refer to the A-subscriber and B-subscriber, or the caller and the callee in a call.

The concept of role was already introduced in the end of the 70's in the context of data modelling [6] and has emerged again in the object-oriented literature. Roles are used both for data modelling [7] and functional modelling [8], [9], [10]. In our approach, service roles are functional roles that encapsulate the functional properties of components involved in a service.

As services involve several distributed components where some of them may be involved in several services, using roles enables to focus on single "slices" of behaviour. This is illustrated in Figure 1. This figure shows a service role collaboration diagram for basic call. It also shows an underlying object structure implementing the roles. In this figure two User Servers are involved that play the subscriber roles "caller basic call" and "callee basic call". The User Servers may play different roles in other services as shown in Figure 2. Here one of the User Server plays the role "call forward". This service role collaboration may occur if the user B initially called by A is busy or does not reply.

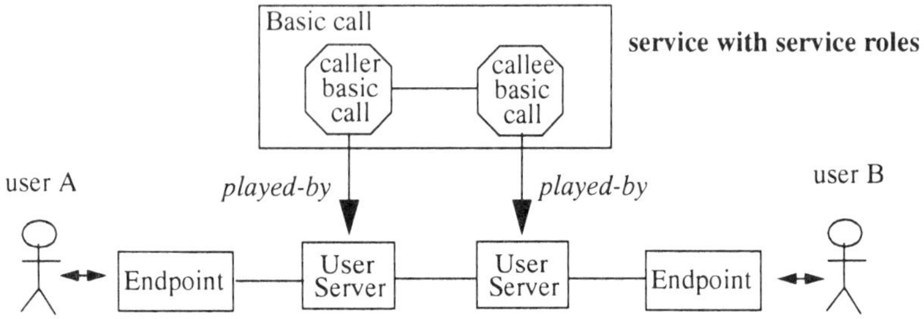

Figure 1. Service role collaboration for basic call.

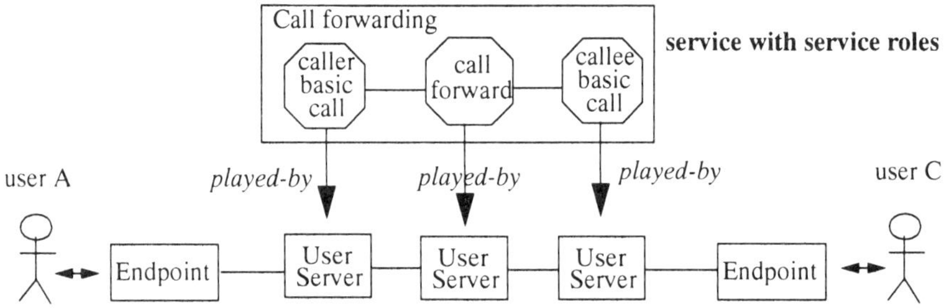

Figure 2. Service role collaboration for call forward.

These examples illustrate that roles may be combined in different manner. The role "caller basic call" may both collaborate with "callee basic call" or "call forward". The roles must be specified such that they collaborate in a consistent manner. In addition the collaboration of roles should not conflict or interfere in an undesirable manner. In the call forward example, we may check that the user A does not object to be forwarded to C.

While these two first examples show quite simple service role collaborations, some service demand more complex collaboration structures. This is shown in Figure 3. Here one of the User Server plays concurrently the two roles "callee basic call" and "call waiting" towards the other User Servers. Some coordination between these roles may be required.

Although these examples use telecommunication services, role modelling can also be applied to information or management services. Telecommunication services are especially interesting because they usually involve several interacting roles played by concurrent objects.

Role modelling is not new. The originality of our work is that we use roles for communication services and that we assign roles and compose them dynamically. In our approach, actors are objects that are created and assigned service roles dynamically on demand. A user may participate in several independent services or service sessions. In that case, several concurrent actors are created and allocated the independent services roles. Actors may change

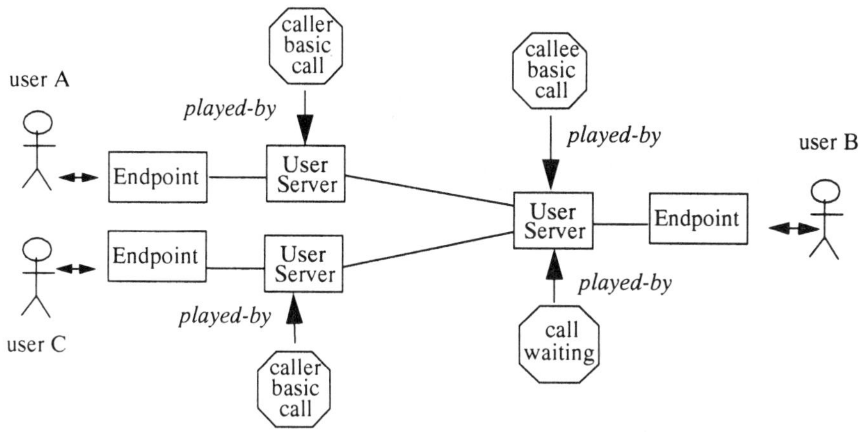

Figure 3. Service role collaboration for call waiting.

roles during their lifetime. Within each service session, an actor may also play several roles. These roles are usually functionally related or they may share some common resources (e.g. a terminal).

The assignment and composition of roles must be coordinated in order to ensure a correct behaviour. For this purpose, actors are created with some intrinsic behaviour required for role coordination. We distinguish between the intrinsic behavioural properties of an actor i.e. the properties of the actor itself, and the extrinsic properties that are obtained through the role assignment [9].

In resume, the main issues in the work presented here are the definition of rules for the specification of composable service roles, the assignment of roles to actors and their coordination within one actor.

3 EXECUTION FRAMEWORK

We have introduced a simple execution framework in which we are experimenting with dynamic composition. The purpose of this framework is twofold: 1. to enable the systematic and independent specification of services by defining concepts for service modelling; 2. to provide basic mechanisms for service composition and execution.

Our framework is inspired by the Telecommunication Information Network Architecture (TINA) [11]:

- We use a wide definition for service that includes telecommunication, management and information services.
- We separate between service logic and network resources and protocols.
- The components of the architecture are defined and implemented as interacting objects, following a distributed object oriented approach.

- We use the concept of session where a session is a means for representing a relationship between a group of objects collaborating to a service.
- We assume that the components execute in a distributed processing environment that support locating objects and interaction between objects.

In order to focus on the composition problem, we have introduced several simplifications to TINA. While TINA proposes a network-centric approach where the central service components belong to the provider domain, we do not make any assumption about the domain where our components are being deployed. We also simplify the TINA session model and let the user session components play the role of the session manager. This latter simplification restricts mobility and multi-party support, but it also considerably simplify the service usage scenarios and facilitate our experimentation.

Figure 4 shows a simplified UML collaboration diagram where the main computational service components (or objects) in our framework are introduced. The components are defined in a generic manner and need to be specialized for specific services. The generic functionalities of the components are defined as follows:

- The **User Endpoint** represents different interfaces that provide the user with service access and usage such as a telephone set, a personal computer or any other terminal equipment. The user may be a human user or an application.
- The **User Agent** is a service independent component that represents the user and the user profile in the network. It records information about the user subscription to services, the endpoints where the user can be reached, the user preferences and the user participation into services. The User Agent is involved during the instantiation of new service sessions and is in charge of role assignment. The User Agent becomes increasingly important when services become more personalised.
- The **Session Actor** is a service independent component that can play different service roles. The Session Actor coordinates the execution of concurrent roles. A service role is assigned to a Session Actor either when a user requests to participate in a service session (e.g. "caller" role in a basic call), or when a user is invited to participate in a service session (e.g. "callee" role in a basic call or "call waiting" when busy), or when some special events happen during a service session (i.e. "billing" after session control negotiation, or "announcement" before the initiation of call forward).
- The **Communication Endpoint** is a service independent component. It coordinates the establishment of stream flow connections. This component is not detailed in our framework. It represents the interfaces to the transport network resources.

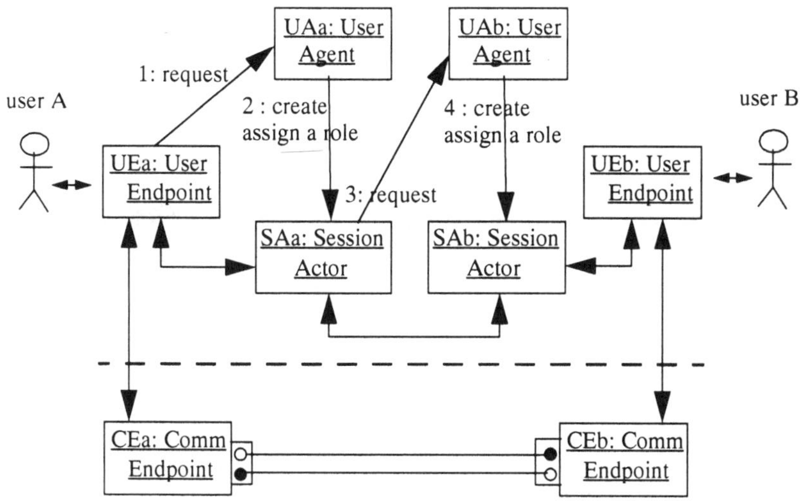

Figure 4. Service session model.

The computational objects execute in a distributed processing environment (DPE) such as CORBA or Java RMI that mainly support communication and binding with remote objects. Through the definition of the User Agent and Session Actor, and the allocation of responsibilities to these components, we confine service specification enabling the service designer to focus on service logic, role interaction and consistency. In this way our framework provides additional functionality to a DPE.

4 EXAMPLES

Our aim with the following examples is to:

- demonstrate the potential of dynamic composition,
- describe different forms of composition (e.g. sequential or parallel composition),
- identify the basic elements or steps in the composition process and,
- capture the basic requirements to the composition method.

We present three examples. The first example introduces the concept of role *assignment*, the second the concept of *adaptation* and the third the concept of *learning*. The examples are presented using UML sequence diagrams that describe the interactions between the system components [12]. Several alternative interactions are often possible; we present one of them. The aim is not to propose an optimal service design, but rather to identify particular needs in the composition approach. Again here as in Section 2 and for the same reasons, we use simple telecommunication services.

4.1 Role assignment in a Basic Call

This example is given as introduction to the concept of role assignment to Session Actors. It shows that we consider a service as a collaboration between roles and that these roles are assigned to actors at service execution. The example involves two users A and B where user A is making a basic call to B. See Figure 5.

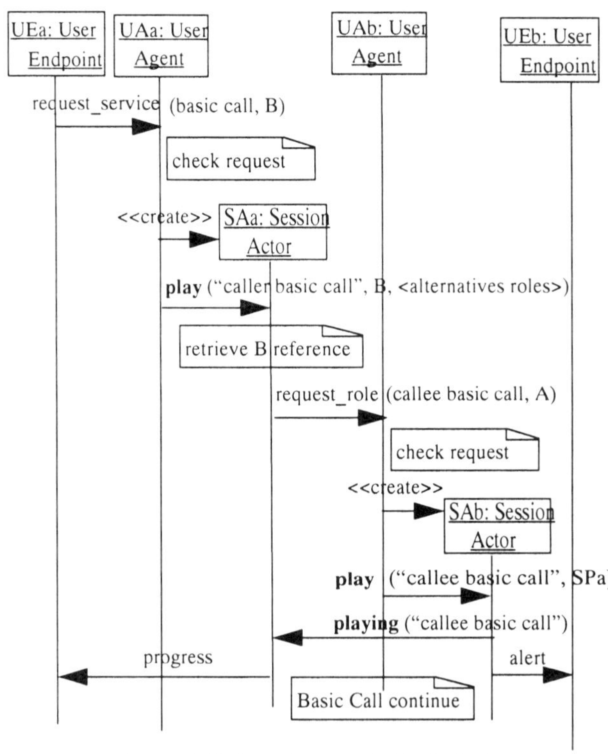

Figure 5. Role assignment in Basic Call.

In this example a Session Actor is created and assigned the "caller basic call" role by the User Agent on A side when the user A requests to participate in the service. The retrieval and loading of role behaviours are basic functionalities supported by the distributed processing environment. The new Session Actor locates the User Agent that represents B and invites the user B to participate in the service. A new Session Actor is created and assigned the requested "callee basic call" role by the User Agent on B side. Then the service can continue.

The assignment of roles must be coordinated in order to avoid inconsistencies or undesirable service interaction. The User Agent serves to coordinate the assignment. It also checks the request against the user's preferences and subscription profile.

When assigning the role to Session Actor on A side, the User Agent may specify a list of alternative roles. This information indicates to the Session Actor which other roles may be played if the current request cannot be honoured. Role compatibility information, conditions and priorities may be associated to alternative roles. For example an alternative role may be associated to the condition "B busy". Using the specification of alternative roles, the Session Actor may decide on behaviour changes without involving the User Agent. This will be shown in the next example.

The messages "play" and "playing" are service role independent. We will introduce other role assignment related messages and interactions in the next examples. We define these interactions such that role assignment can be clearly distinguished from playing the role itself.

4.2 Call Forward when Busy

This example illustrates the ability to adapt a service session to a specific situation by using an alternative behaviour. It involves three users A, B and C. The user A wishes to establish a basic call to a busy user B. Incoming calls to B are forwarded to C if B is busy. The example shows that several Session Actors may co-exist and execute in parallel on one user side (here on B side).

On request from A, the User Agent for B determines that the "call forward" role should be played (see Figure 6). This role is played concurrently from the role currently played at B and is assigned to a new Session Actor. The Session Actor on the A side is informed about B playing this new role; this gives the Session Actor the opportunity to accept or reject the role, but also to check for call barring cases, etc. before service execution continues.

We assume that the Session Actor on A side was given alternative roles during its initial role assignment (see Figure 5). Thus it knows that the current role "caller basic call" is compatible with the new role "call forward" and that the assignment of a new role on A side is not necessary. We will see in the next example that A may also reject the role proposed by B and negotiate other roles with B.

4.3 Learning alternative behaviour

This example illustrates the ability to enhance a service by learning a new behaviour. It involves two users A and B. The user A wishes to establish a basic call to a busy user B. A and B negotiate a behaviour for handling the

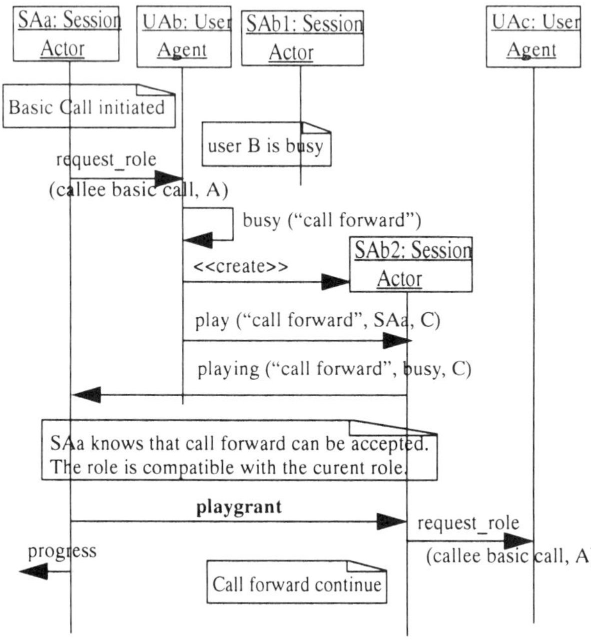

Figure 6. Service adaptation and parallel role assignment.

busy situation. We assume that different services are available on A and B sides. This would be the case when A and B are subscribers in different networks.

On request from A, the User Agent for B determines that the "call waiting" role should be played (see Figure 7). The "call waiting" role permits to notify the user B about the incoming (waiting) call and to let user B choose to accept or reject the waiting call. This role has to be tightly synchronized with the role currently played by B, so it is allocated to the existing Session Actor. The Session Actor on the A side does not agree on this role (the user A may wish to not interrupt B) and proposes to initiate the alternative "call back" role. This service role alternative is not available on B side where the Session Actor has to be learned the new service role. Learning includes the downloading of code to the Actor.

We assume that both Session Actors were informed about alternative roles and assignment conditions during their initial role assignment and are able to decide about behaviour changes without involving the User Agents. The Session Actors inform the User Agents about behavioural changes (sending the message "playstatus"). In some cases the Session Actors may need external support to make decision about the assignment of a new role. Decision support may be provided by the User Agent, some other specialized agent in the network or the end user.

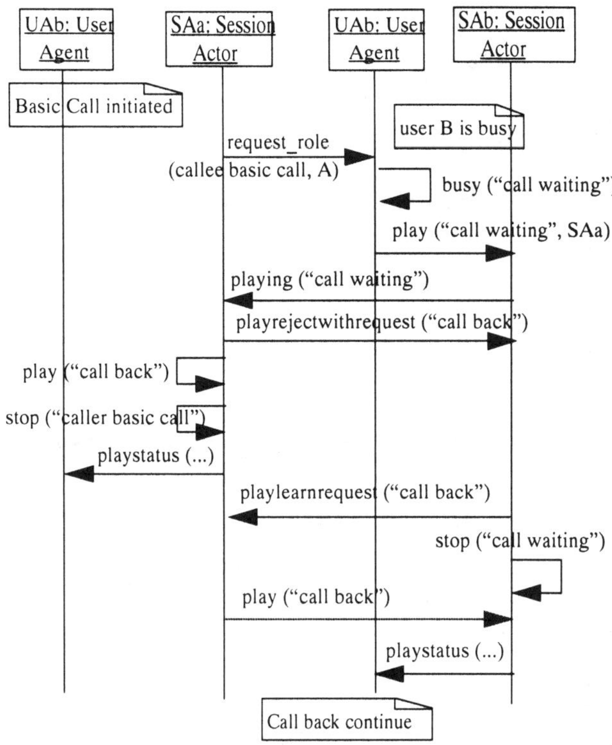

*Figure 7.*Learning a new role.

The "call back" service role has to be coordinated with the role currently played by the Session Actor on B side. The activation of "call back" is delayed until some synchronization point is reached in the role played by B, for example the suspension of the role or its completion. Some conditions may be attached to the synchronization points. A synchronization condition could specify that any voice based service should be delayed after completion.

4.4 Other examples

The composition approach may also be used in other situations and applied to information and management services:

- Filters or translators may be used in order to adapt a service to a user or an application. For example, a user may prefer not to be notified of the arrival of electronic messages in the morning, except from those sent from some selected sources.
- Services may also be adapted to handle some specific type of information. For example, the information retrieved from a name directory may have to

be re-formatted. Translation from voice to text or from text to voice may also be needed.

- A service may be adapted to some particular equipment. A subscriber of a PSTN that is using a traditional telephone set with no processing capability may need to interact with an electronic service support application that was designed for interaction with remote PC users. Some adaptation function based on an IVR may be introduced between the telephone user and this electronic service.
- Billing, logging and other management functions may be designed as service roles that are composed with telecommunication and information services.

5 ROLE BASED SERVICE DEVELOPMENT

5.1 Specifying roles as state machines

Service roles make it possible to break down a complex behaviours into simpler ones. A service role is the part that an object plays in a service. Each object plays (communicates) with other objects that play dual or consistent service roles. By dual roles we mean that roles complement each other and can play together in a correct way, e.g. "caller" and "callee" roles in a call. By consistent roles, we mean that roles can be aligned in order to provide a consistent behaviour (see Section 6).

We use the same mechanisms for designing role types as for classes. Interactions between roles can be modelled using message sequence charts and their detailed behaviour using state machines. For example the behaviour of roles may be specified using MSC [13] and SDL [14]. A main difference between role types and classes is that role types cannot be instantiated on their own but need to be assigned to objects. Role assignment sets requirements on the structure and behaviour of the object being assigned a role. This will be discussed later in this paper.

5.2 Forms of composition

Service roles are composed in order to provide advanced service functionality. We distinguish between different forms of composition:

- Sequential composition: service roles are executed in a sequential order. Information may be transferred between two consecutive roles. For example, the "basic call" and the "call back" service roles may be composed serially.

- Parallel composition: service roles are executed in parallel. The roles may be executed for the same period of time or they may start and end at different times. The "billing" and "call basic" service roles may be composed in parallel with billing being started when "call basic" reaches a connected state.
- Alternating composition: service roles are executed in an alternate manner i.e. only one role is active while the others are passive waiting for some event that enables them to be activated. Synchronization points are points of the role execution where the activation, suspension and resumption of the role execution can take place. Conditions may be associated with the synchronization points. In that case, the activation (or suspension and resumption) of the role execution can only take place if the conditions are fulfilled. Actions to be performed at role activation (or suspension and resumption) may also be attached; for example, communication resources may be released at suspension and reallocated at resumption. The "basic call" and "call waiting" service roles may be composed in an alternative way.

5.3 Composition behaviour

Our approach requires that roles are designed in order to enable composition. Roles must be prepared to report and catch events related to composition, and to coordinate with other roles they are composed with. We define separately the composition behaviour and the service behaviour. Roles are assigned, negotiated, learnt, suspended and resumed using interactions specific to the composition behaviour and independent from the service role behaviour.

Role composition is triggered and controlled by service dependent events, by user states (e.g. idle, connected, busy) and user preferences. In order to generalize composition triggering and to define patterns for role design, we need to identify and classify triggering events. We reuse the knowledge from other frameworks where important service events have been identified, e.g. Parlay [15].

Role assignment, negotiation and suspension may be attached service dependent information; this is done using service independent mechanisms. As shown in the previous examples, a list of alternative roles (and associated conditions and actions) may be specified for role negotiation. The contents of the list depend on the service role, but the mechanisms defined for specifying and interpreting the list are service independent. For example, in "call forward when busy", we may specify that a check for call barring cases (this is a service dependent action) has to be performed before accepting the "call forward".

5.4 Synchronization

Alternating composition requires roles to be tightly coordinated with other roles. Coordination occurs at synchronisation points. As dynamic composition may be applied to service roles that are designed by different parties and at different times, it may not be possible, when designing a role, to foresee the behaviour of the roles this role will be composed with. In order to generalize the coordination of alternate roles, we need to identify potential synchronisation events. Here again, we reuse the knowledge from other frameworks where important service events have been identified, e.g. Parlay [15].

5.5 Role manager

The role manager encapsulates the intrinsic behaviour of a Session Actor. It supports the assignment of service roles to actors, and coordinates the negotiation and the learning of new roles. The role manager checks and initiates the conditions and actions attached to service assignment and negotiation. In addition, it coordinates the synchronisation of alternate roles.

While the User Agent coordinates a set of actors acting for a user, the role manager coordinates a set of roles within an actor. Thus our framework introduces two levels of control, one at the actor level and one at the role level.

5.6 Negotiation decision

In the examples presented in Section 4, we have assumed that actors were able to decide whether a role should be granted or rejected, and to negotiate alternative roles. This decision was based on the interpretation of a list of alternative roles provided by the User Agent. The list is generated from a role map that describes relations between roles and correct execution orders (see Section 6). The role map can be adapted to user's preferences. For example, a user may prefer not to use "call waiting" when busy although this is a correct combination of service role and state.

If the information contained in the list of alternative roles does not permit the Session Actor to make a decision during negotiation, the Session Actor may require external support. Decision support may be provided by the User Agent or some other specialized agent in the network. The users may also be involved in the decision process. As some terminals with limited capabilities make it difficult to involve the users, the capabilities of the user terminals have to be taken into account. The preferences of the users involved in the service must also be significant. In our third example, the user A rejects "call waiting" because A does not wish to interrupt B; when A proposes the "call back", it would be incoherent to interrupt B in order to let him decide if the role can be agreed.

5.7 Several actors and actor types

The examples presented in this paper have focused on the collaboration of roles assigned to two different Session Actors. We plan to develop cases where more than two Session Actors are involved.

In our work, we have chosen to first concentrate on the assignment of roles to Session Actors. We will further develop the method in order to support the assignment of roles to the other components of the framework e.g. the User Endpoint.

6 COMPOSITION CORRECTNESS

Although our work concentrates at the moment on the service design aspects, we also have in mind the need for ensuring service correctness. Service correctness is not a particular requirement set by the composition method. We meet similar problems in traditional service development and deployment approaches where the completeness and consistency of service specifications, and the conformance of implementations to specifications have to be ensured. However the assignment of roles to actors and their composition in our method require special attention.

Development based on formal methods enable the verification of the specifications, the automatic generation of code and test cases. It seems that few attempts to use formal methods on large scale have been done and that the methods are usually applied in general, not to the special case of hybrid communication services [16]. By breaking down complex service behaviours into small behavioural roles where each role can be specified using some formal method (e.g. SDL [14]), and by constraining the ways these small behaviours can be composed together, we intend to produce specifications and executable services that can be more easily verified and validated.

The role of the User Agent is crucial with respect to behavioural consistency. The User Agent coordinates the assignment of roles in order to avoid inconsistencies or undesirable service interactions. It also checks that the assignment is consistent with respect to the user's preferences and subscription profile. We can distinguish between two problem areas:

- the role alignment i.e. ensuring that the roles assigned to interacting Session Actors or negotiated between Session Actors collaborate in a consistent manner leading to the execution of a correct service. For example the

"caller basic call" role should interact consistently with the "callee basic call" role.
- the role execution ordering i.e. checking that roles are assigned and played in a correct order. For example, the call forward role can only be assigned in relation with some other role assignment request such as call diversion.

6.1 Role alignment

Each Session Actor plays (or provides) a role that requires the interacting Session Actor to play a consistent role. An association between actors is valid if the provided roles "contain" the required roles [5]. Roles being assigned should be aligned in order to ensure containment. We distinguish between three levels of assignment:

- Validation. Checking that the required roles are contained in the provided roles. Validation is illustrated in our first example where the "caller basic call" role and "callee basic call" role must be validated.
- Adaptation. The roles are negotiated and adapted. Adaptation is illustrated in our second example where the "call forward" role is adapted to the "caller basic call" role.
- Learning. A role can be learned when the existing roles on one side are insufficient. Learning is illustrated in our third example where the "call back" role is learnt by the B side.

Ideally, role alignment should be performed dynamically. This would enable new services to be produced in a flexible manner during service adaptation and service learning. In practice, the alignment is done during system design. When a service is designed, new service roles are defined that complement each other (dual roles) or new roles are designed to play with existing roles.

6.2 Role execution ordering

Service roles cannot be composed in any order. When a new role is introduced, it is necessary to relate this new role to other existing roles and describe execution constraints. We call this description a role map. The map specifies the allowed role sequences. A notation was introduced in [9] for specifying allowed role sequences. This notation needs to be further developed in order to enable annotations that describe events, conditions and actions to be associated to the sequences in the map.

7 OTHER ISSUES

We are aware that several other issues would need consideration. We have chosen to concentrate on the functional service aspects. We list some other interesting issues:

- Role discovery and downloading. Existing frameworks such as Java and Jini support retrieval and downloading. The Plug-and-Play project proposes a basic infrastructure for dynamic plug-in and develops a demonstration platform based on Java that enables dynamic component plug-in [4]. Our work is based on the concepts introduced by Plug-and-Play.
- Service subscription. The possibility to learn new service roles at execution time changes the way of thinking about service subscription (i.e. the right to access a service). Operators and services providers may wish to limit the access to new services.
- Service billing. In traditional telecommunication networks, billing is related to service subscription and use. Dynamic composition makes traditional billing difficult. Rules for billing services that are not part of a subscriber repertoire but are learned during service execution may be designed.
- Failure management. The services roles downloaded from other parts may lead to errors. Failure management support is needed that permits the control of failures for downloaded code.
- Upgrade. There may exist several versions of a service role. Support for deploying new versions, for identifying specific versions is needed.
- Security. The operators and service providers will require the control of new service roles with respect to security.
- Performance. The dynamic composition should not introduce waiting delays that are longer than tolerated by the service users.

8 RELATED WORK

As explained in Section 3, our framework is inspired by TINA [11]. We extend this approach in order to provide a flexible composition of services. We have introduced simplifications to the TINA session model, that facilitate our experimentation.

The notion of role is currently used in several object oriented methodologies. [9] describes the properties of roles. It introduces the concept of "roleification" and compares it to other abstraction concepts as specialization and aggregation. It discusses the specification of behavioural roles and present how roles can be attached to objects. It also defines a notation for the representation of allowed role sequences. We propose to further develop this notation in order to specify role maps. OORASS is an object oriented methodology based on role analysis and synthesis [8]. Roles are used in the analysis that allows the designer to break the total problem into sub-problems. Synthesis combines the role in order to produce objects. While OORAS has been mainly applied to information systems, we focus on communication services. In addition, we propose a dynamic synthesis of roles.

A service composition method is proposed in the CANES project [17]. The method addresses transport services in Active Networks i.e. Internet based networks where packets may contain executable programs that are delivered to network elements. The services discussed in CANEs are related to transport, e.g. packet forwarding, packet filtering, transcoding. Our work focuses on high level services that have a more complex structure than transport services. The "hybrid" aspect is not really important in CANEs and is not discussed.

Several execution frameworks are available (e.g. CORBA, Jini, DCOM) that propose general solutions to component based systems. In these frameworks, the components are described by static interfaces. Thus the interfaces do not cover the dynamic behaviour and cannot be checked in order to guarantee the correctness of interactions between components that are dynamically bound together. [18] introduces to several architecture concepts and investigate the properties that interfaces should provide in order to guarantee the correctness of connections.

Feature interaction is a fundamental problem in service creation. The problem arises from the extension of telecommunications system functionality, feature by feature. In [19] the authors try to define a modular approach to service specifications. Each feature is implemented by one or two components types, and each external call is processed by a dynamically assembled configuration of components. The proposed configuration is limited to an assembly of pipes and filters, and the interactions modules are strictly limited by the architecture. Our architecture is less restrictive than the pipe-and-filter based architecture. We believe that the coordinated composition of service roles and the definition of role maps are means for avoiding undesirable interactions.

9 CONCLUSION

We have proposed an approach and a service execution framework that enable services to be constructed dynamically from heterogeneous functional components or service roles. Using roles, we are able to break down the complexity of service specification. The service execution framework supports service negotiation and learning. Compared to traditional service development and deployment methods, our approach offer several potential advantages:

- Roles encapsulate a limited set of functional properties. They are small and can be easily understood.
- The method enforces systematic design. Roles are defined using generic composition patterns.
- Service role implementations may be shared between users.
- Service behaviour may be negotiated and adapted to the special needs of users.
- New service roles can be easily be tested.
- Rapid and incremental deployment of new services is enabled.
- The allowed sequences of roles can be described. These sequences can be adapted to user's preferences. Undesirable service interactions may also be avoided.
- The approach is applicable to different types of services such as telecommunication, information and management services.

ACKNOWLEDGMENT

Our work is done in the frame of the Plug-and-Play project supported by the Norwegian Foundation for Research (NFR) [4].

We would especially like to thank you Humberto Castejon for inspiring discussions. Humberto, MSc. student at ETSIT, Universidad Politécnica de Valencia, has been working with service role modelling and implementation in our institute in the frame of his MSc. thesis.

REFERENCES

[1] TINA-C deliverable: Service Architecture Version 5.0. Available at http://www.tinac.com/ (accessed February 2000).

[2] C. Gbaguidi, J.P. Hubaux and M. Hamdi. "A programmable Architecture for the provision of Hybrid Services," IEEE Communication Magazine, July 1999.

[3] G. Vanececk et al. "Enabling Hybrid services in Emerging Data Networks," IEEE Communication Magazine, July 1999.

[4] F.A. Aagesen et al. "Towards a Plug and Play Architecture for Telecommunications," in Proc. of the Fifth International Conference on Intelligence in Networks, November 1999. Kluwer Academic Publishers. Information available at http://www.item.ntnu.no/~plugandplay/ (accessed February 2000).

[5] R. Bræk. "Using Roles with Types and Objects For Service Development," in Proc. of the Fifth International Conference on Intelligence in Networks, November 1999. Kluwer Academic Publishers.

[6] C.W. Bachman and M. Daya. "The role concept in data models," in Proc. of the Third International Conference on Very Large Databases, 1977.

[7] R. Wueringa and W. de Jonge. "The identification of objects and roles - Object identifiers revisited-," Technical Report IR-267, Faculty of Mathematics and Computer Science, Vrije Universuteit, Amsterdam., December 1991.

[8] T. Reenskaug and al. "OORASS: Seamless support for the creation and maintenance of object oriented systems." Journal of object-oriented programming, 1992.

[9] B.B. Kristensen and K. Østerbye. "Roles: Conceptual Abstraction Theory and Practical Language Issues," Theory and Practice of Object Systems, Vol. 2(3), 1996.

[10] M. Mezini and K. Lieberherr. "Adaptative Plug-and-Play Components for Evolutionary Software Development," in Proc. of the 13th Conference on Object-Oriented Programming Systems, Languages and Applications, October 1998.

[11] TINA. Information available at http://www.tinac.com/ (accessed February 2000).

[12] OMG Unified Modeling Language Specification. Version 1.3 June 1999. Available at http://www.omg.org/ (accessed December 1999).

[13] Recommendation Z.120, ITU Message Sequence Charts (MSC), October 1996. ITU-T.

[14] Recommendation Z.100, ITU Specification and Design Language (SDL), June 1994. ITU-T.

[15] Parlay API Specification 2.0: Generic Call Control Service Data Definitions. Available at http://www.parlay.org/ (accessed February 2000).

[16] X. Logean, F. Dietrich and J.P. Hubaux. "On Applying Formal Techniques to the Development of Hybrid Services: Challenges and Directions," IEEE Communication Magazine, July 1999.

[17] E. Zegura (lead author). "Composable Services for Active Networks. Georgia". Architecture document from the CANEs project, Georgia Institute of Technology. Information available at http://www.cc.gatech.edu/projects/canes/ (accessed Mars 2000).

[18] D. Luckham and al. "Three Concepts of Architecture". Stanford University Technical Report CSL-TR-95-674, July 1995.

[19] M. Jackson and P. Zave. "Distributed Feature Composition: A Virtual Architecture for Telecommunications Services," IEEE Transactions on Software Engineering, October 1998.

IT-Technology for High Available Solutions in the Telco Environment

RTP - Reliant Telco Platform

Konrad Wiesneth and Stefan Arntzen
Fujitsu Siemens Computers

Abstract: This paper gives an overview about a software approach to use standard computer technique in the telecommunication environment. The most important requirements in this sector are highest availability together with short response time. Servers which fulfill these requirements are often called "Carrier Grade Servers". A special implementation for such a Carrier Grade Server is the Reliant Telco Platform (RTP) based on UNIX - or NT-Cluster- Technology and an appropriate middleware layer.

Keywords Reliable service platform, Cluster-system, Availability, Scalability

1. BACKGROUND AND PARADIGM SHIFT

Back in the 80s and early 90s most countries had a state-owned or semi-state telecommunications company which, in conjunction with the national government, dictated prices and kept them constant for relatively long periods. This monopoly resulted in a high price policy for telecommunication services and large profits for the network providers. Thus in turn they could afford proprietary and expensive network infrastructures. Technically, this period was characterized by the dominance of fixed networks and proprietary equipment. Also telecommunication networks and the upcoming data networks of the IT-world were considered as entirely separate worlds.

But then happened two things which changed the whole telecommunication world: Firstly technically: People began to look at data and voice networks as one infrastructure; the market for mobile networks exploded; there was a dramatic need for so-called "intelligent services" like freephone, follow-me or prepaid; VoIP became more and more important. Secondly financially: Within a very short time deregulation broke down monopolies and new competitors arose.

At the same time the triumphant march of the Open Systems, e.g. UNIX- or NT-Systems, has begun in the IT-world. This resulted also in two big outcomes: Firstly technically: It becomes very easy for customers to opt for a different server manufacturer and to use heterogeneous components and software products in virtually any combination.. Secondly financially: This competition lead to considerable price drops and a development dynamic never dreamt of before.

In addition to these points Internet has become one of the most important communication networks. Therefore exchange of data get extremely cheap and the request to have the same features with the same price advantages grows. Now the time has come where telecommunications and information techniques will grow together even faster.

2. REQUIREMENTS FOR A TELECOMMUNICATION PLATFORM

The complex business environment of telecommunications imposes a lot of requirements on the systems used in the telecommunication area:

- High availability,
- Scalable performance,
- Short response time,
- Connectivity to different networks,
- Disaster recovery,
- Integration into different network management systems.

Another very important item is the relationship between price, availability and performance (Fig. 1). To solve these requirements in an economic way an open system architecture can be the solution. Since several years clusters are the response of the IT-world to the requirements availability and scalability and have in the meantime a high maturity.

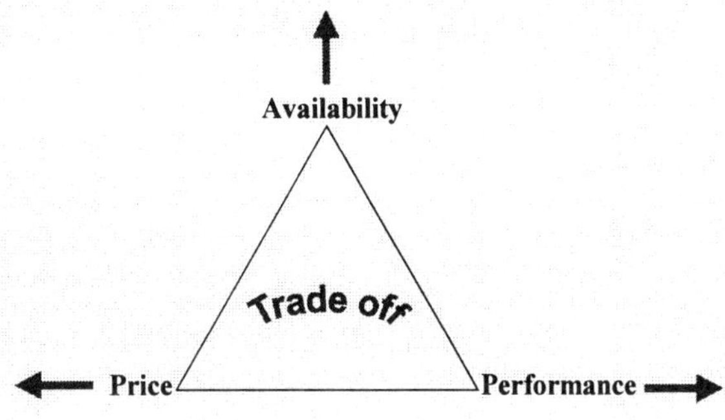

Figure 1. Fundamental Requirements

3. CLUSTER-TECHNIQUE

Although some hardware and software measures for increasing the availability of individual single systems has been implemented, only an availability of standard servers between 99.9 and 99.99% could be reached. This is not enough in the Telco-environment. In this area, an availability of greater than 99.999 % - that means an outage of less than five minutes per year - is required. There are two approaches to improve the availability above 99.999 %. Firstly, Fault Tolerant (FT) servers with a special non-standard HW-technology or secondly, clusters of standard servers combined with a special middleware layer.

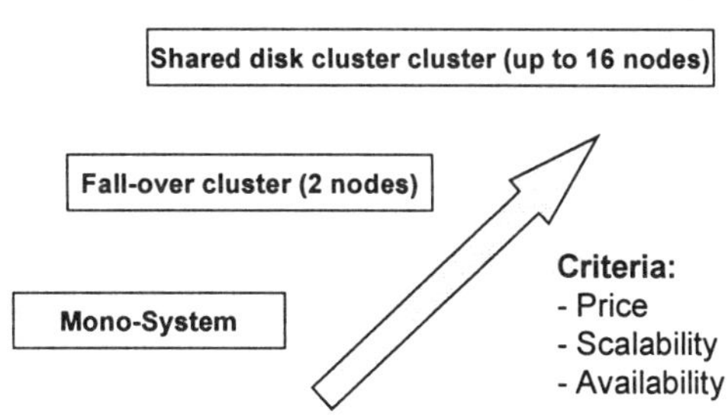

Figure 2. Price, scalability, and availability

FT-servers are normally proprietary high-price solutions and often restricted with regard to scalability and protection against context-specific software errors. Clusters are a bundle of individual servers (nodes) connected via

via a very fast cluster interconnecting and working with a common set of peripherals, especially disks.

A cluster offers two dimensions for improving performance and availability. Firstly, more processors and more availability functions (e.g. hot plug functions) in the individual servers based on a symmetrical multiprocessor technology (SMP-systems) and secondly more nodes within the cluster. Fig. 2 illustrates scalable system-configurations with increasing performance and availability with an optimal cost/performance ratio.

Because several instances of operating systems (one per node) and applications are involved, protection against context-specific software errors by changing the environment and continuing the application on a different node is possible and functions like "rolling upgrade to new versions of the operating system or the application" can be realized.

However programming clusters, especially parallel clusters, is normally a complicated and error-prone process. One has to think about load distribution, nonstop programming, inter-cluster communication, data locking, upgrade mechanism, changes of the configuration (e.g. adding or removing nodes) and so on. And last but not least one has to control the nodes states and in case of node-problems one has to switch the application to the surviving node.

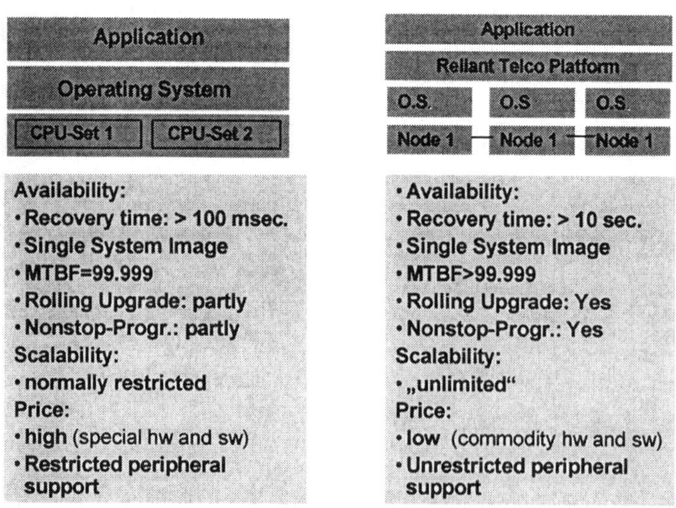

Figure 3. Fault Tolerant Systems versus Cluster Systems

Our approach to hide these problems from the application is to provide a new middleware-layer, called RTP (Reliant Telco Platform). This platform provides a single-system-image to the telco-specific application and hides the complex cluster environment completely. In Fig. 3, a comparison is given between a single fault-tolerant system and a cluster solution based on RTP.

4. SINGLE-SYSTEM-IMAGE

To protect the application developer from all these nasty effects of a cluster-environment a single-system-image API (Application Programming Interface) is necessary. The view of the application of a cluster should look like a single high-available computer, no matter of how many nodes the cluster consists of (Fig. 4). For example the application should not be aware of the location of its resources and devices. The application sees only a logical cluster-wide addressing scheme; the transition to the physical addresses of the resources is task of the RTP

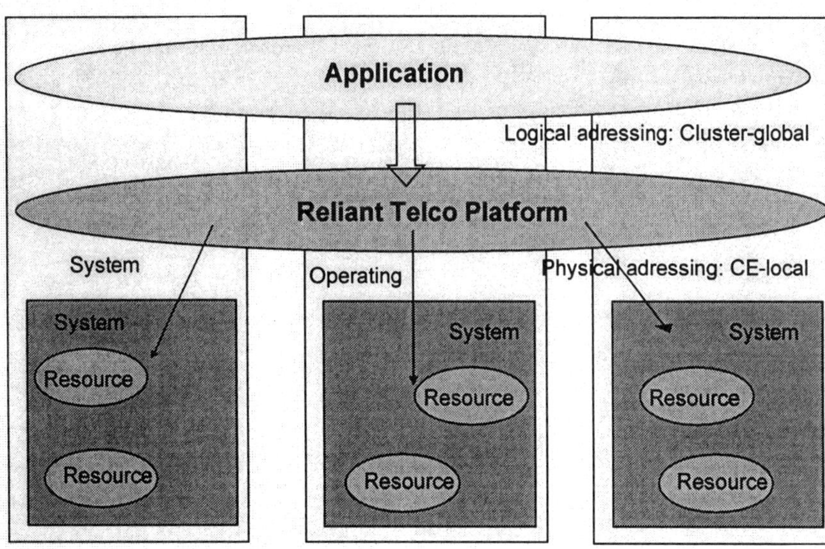

Figure 4. Single System Image

5. DESIGN PRINCIPLES OF THE RTP

For better understanding of the following some basic design principles for the RTP-cluster approach will be explained:

DOUBLE ERRORS:
One of the main basic principles is the provision of a fault-tolerant system which protects against single errors. Double errors are also tolerated in certain components (e.g. failure of two cluster nodes), but they may result in data and call losses. A concept that protects against multiple errors is extremely complicated and error-prone, as procedures for handling complex processes can themselves result in new error sources. Hardware expenditure would increase dramatically with negative impacts on performance. Such a concept has therefore not been developed.

SOFTWARE ERRORS:
Software cannot be written error-free but fault-tolerant. Many software problems are transient; that is, the problem is caused by an unusual environment state typically resulting from a transient hardware problem, a resource limit is exceeded, or a race condition occurs. In such cases, reinitializing the program state to an earlier point and resuming execution often works successfully because the environment is different. Nevertheless it is essential that applications are tested in a non-stop environment as far as possible.

ERROR DETECTION:
Error detection methods represent the main difference between special fault-tolerant systems and the cluster concept. The cluster concept cannot protect against processor errors. Although plausibility checks are used to try to detect and correct error states during program execution and periodic audit processes, there can be no 100% protection against processor errors. Special-purpose systems attempt to protect themselves against such errors by having several processors working in clock-controlled synchronism execute the same commands and detect errors by comparing the results. Experience has shown that this type of processor error is extremely rare and (once it occurs) can be very easily detected by the above-mentioned plausibility checks.

LIMITING THE SPREAD AND EFFECTS OF ERRORS:
Errors must be detected as early as possible and prevented from spreading out to other components, and in particular throughout the entire cluster. The design principle of combining local operating systems into a cluster by means of a narrow cluster package offers effective protection against the spread of errors. The cluster concept enables any component to be replaced online; the

appropriate node is shut down, the component is replaced, and then the node is reintegrated into the cluster. With this procedure, however, errors have a relatively strong impact on performance and availability. Therefore, methods such as hot pluggable devices and I/O boards are intended to minimize the effects of errors.

ROLLING UPGRADE:

In order to meet availability requirements, it must be possible to carry out all administrative activities online, i.e. without having to shut down the entire system. The following sections describe important techniques used for this purpose, e.g. online data backup. In particular, they include installing patches and updating to new versions of the products affected. Once again, the cluster concept offers the best solution here. It enables a cluster node to be shut down, the new or corrected product version to be loaded, the node to be reintegrated back into the cluster, etc.. However, it must be ensured that all affected products are capable of interacting with partner products of a different version on other nodes, at least for a brief transition period.

NON-STOP PROGRAMMING:

An application – particularly a client/server application – can be regarded as a continuous sequence of consistent states. The consistent states may be defined as contexts describing the corresponding process environment or global states. In the event of an error, it is possible to resume processing at the last consistent state, provided the corresponding context is still available.

REENTRANT PROGRAMMING:

Cluster nodes are normally SMP (Symmetrical Multi-Processor) systems. The performance and scalability of an SMP system essentially depends on whether or not there are sufficient executable processes available at a particular point of time. This is the only requirement an SMP system requests from the application; in all other points a SMP system looks like a simple monoprocessor system. On the other hand, for performance reasons, it is not advisable to generate a separate process for each request. The generation and deletion of processes results in considerable overheads. A process pool represents a compromise between these two requirements. Working with process-pools or thread-pools requests a reentrant programming technique. That means one process can handle several dialog steps (call or call-segments) in parallel.

COMBINING REENTRANT AND NON-STOP PROGRAMMING:

The RTP attempts to fulfil both requirements, reentrant and non-stop programming, with the same programming model. The consistency points of the application are the beginning and end of a dialog-step (or call-segment). The associated context at these points of time is stored in a so-called context data pool, which is mirrored for availability reasons on a backup system within the cluster. All the programmer needs to do is to retrieve the context data at the beginning of a dialog step and to save it again at the end; the RTP takes care of the rest. This technique makes the occasionally complex reentrant and non-stop programming transparent to the applications programmer. In case of a node error, the dialog can be continued on the node where the context data are mirrored by the previous dialog-step.

STORING DATA

The simplest way to store data in a cluster is to save them in a cluster wide available database. This is the simplest but not the most performant way. Depending on the purposes, the RTP uses therefore different techniques for storing data

- shared memory which allows the access to the data from several processes within a node,
- mirroring of the data to a second node which makes it possible to address the data from the original and the buddy-node, and
- a cluster-wide available and accessible database.

6. IMPLEMENTATION OF DETAILS OF THE RTP

The RTP consists of the following units:

- Connectivity-Functional-Unit: Responsible for the integration into the network;
- Snmp/Corba-Functional-Unit: Responsible for the integration into the network management;
- Data Base-Functional-Unit: Used for RTP- and User-Data;
- Scalability-Functional-Unit: Responsible for load distribution and dispatching;
- Availability-Functional-Unit: Responsible for Node- and Process-Management and to allow Reentrant & Nonstop-Programming
- Data- and Utilities- Functional-Unit: Responsible for Trace, Ticket-Handling, Object-Management, Statistics, etc.;
- Manageability-Functional-Unit: Platform-Management, especially "Rolling Upgrade" with zero call-interruption time.

Fig. 5 shows the architecture of a RTP-based cluster-system indicating the position of the diverse functional units, particularly for the requirements scalability, availability, data and utilities, and manageability.

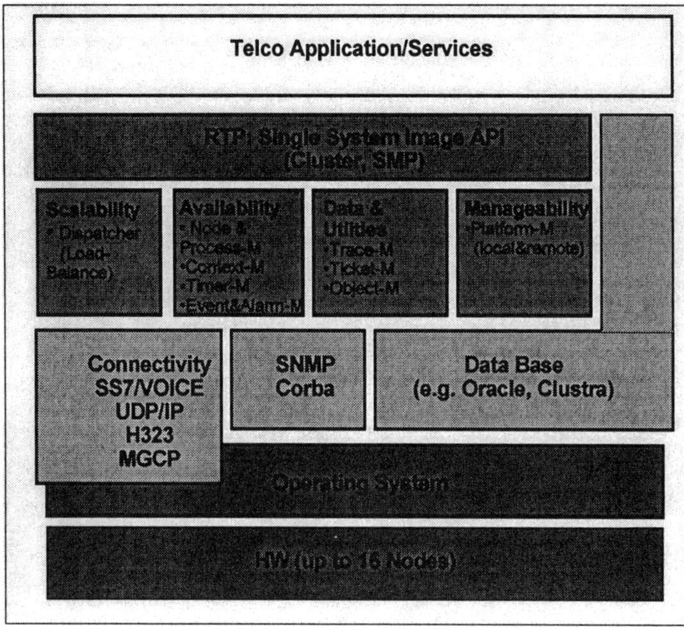

Figure 5. Architecture of a RTP-based Cluster System

Such an architecture can be highlighted by the following properties:

- High availability thru cluster-technology
- Single system image
- Nonstop programming
- Load distribution
- Connectivity to Telco-networks
- Mated pair configuration (Disaster recovery)
- Event and alarm management
- Integration into different network management systems
- Scalability; predictable performance, and availability

In the following some components are give in more detail:

6.1 Super Node Manager

The Super Node Manager is responsible for starting, supervising and re-starting the functional units mentioned above. It is also responsible for node management and control. For this purpose, a health check mechanism is implemented and the node state distributed within the cluster

6.2 Signalling System No.7 (SS7)

In the voice-oriented telco-industry, the Signalling System No.7 (SS7) plays an important role. A cluster as possible destination of a SS7-message (signaling message) must be seen from the outside world as a whole and requires therefore a special implementation of the SS7 protocol-stack. This implementation is related on the mirroring of the addressing information and protocol states within the cluster to guarantee a common addressing scheme (one point code) or a suitable load distribution and recovery procedures in case of node errors. By a time out mechanism and by repetition of messages call losses can be avoided.

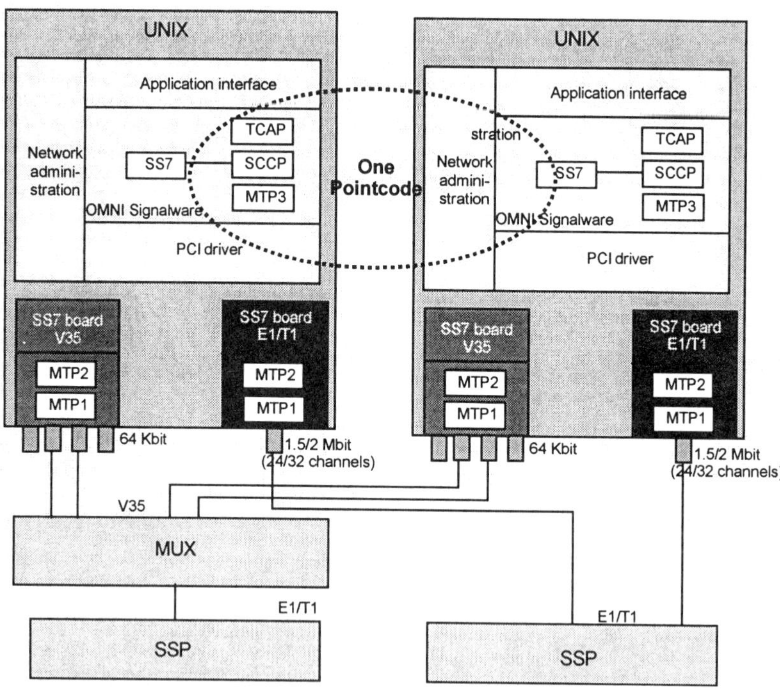

Figure 6. Signalling System No.7 (SS7)

Fig.6 illustrates the necessary protocol stack for interfacing with SS7 which uses at the lower three protocol layers the Message Transfer Part modules MTP1 to MTP3. The module SCCP (Signalling Connection Control Part) performs connection management and the module TCAP (Transaction Capability Application Part) handles packet communications. The two units SSP (Service Switching Point) shown in the lower part of the figure are the interface modules located at the switching nodes of the network.

6.3 Distribution of processes

One of the main objectives of the Reliant Telco Platform is to provide the applications programmer with a single system image. Because the underlying operating systems work locally, i.e. they are not aware of the existence of the cluster, whereby the Reliant Telco Platform is responsible for building the cluster accordingly. Some of the most important objects of the application toolkit are the processes which actually carry out the work. These processes are distributed throughout the cluster, and can thus reside on different nodes.

The Node Manager is responsible for the startup and shutdown, restart and monitoring of processes on one particular node. The monitoring is done by issuing health check messages in regular time frames.

The processes of the Reliant Telco Platform communicate with each other via the communication manager. Messages are the heart of the distributed environment of the RTP. Local systems like an operating system generally recognize two distinct worlds: the local environment and the network. The Reliant Telco Platform, however, sees only one world: the world of messages. In this environment, a process can send a message to another process on the same node just as easily as to a process on another node. Through corresponding addressing mechanisms, the actual location of the process is fully transparent to the caller.

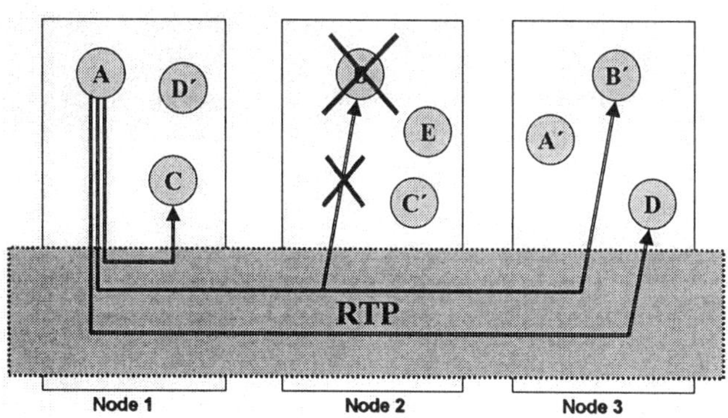

Figure 7. Application/Communication Manager with backup processes

To achieve this, processes can be combined into alias groups, which are used for IPC (Inter Process Communication) addressing. This can be performed on the basis of m:n relations; a process can be a member of several groups. Each group has a cluster-wide logical name; its members are described in a list.

Two IPC addressing modes are supported for the alias groups: (1) (local) equal load distribution; (2) backup.

The mode "equal load distribution" means that processes in the group are addressed one after the other in a round robin procedure; so this mode guarantees an equal use of CPUs and nodes.

Extremely effective for non-stop programming is the use of backup alias groups. Backup alias groups are represented by a list of processes. One process in the list is known as the primary process. This may or may not be followed by several backup processes, which take over the responsibilities of the primary process should it fail. Fig. 7 illustrates that a failed process B in node 2 can immediately be taken over by the backup process B' in node 3.

Based on these backup alias groups, a concept of process pairs can be easily implemented. These process pairs communicate with each other internally and can be addressed externally as a single process.

6.4 Dispatcher/Load Balance

In a large cluster-environment not every node is connected to the network (e.g. to the SS7-network; at least two connections for redundancy reasons are necessary); therefore a dispatcher is necessary, to achieve reasonable scalability. This can be reached by distributing the load equally throughout all cluster-nodes. Fig. 8 illustrates load balancing with can be performed with different strategies.

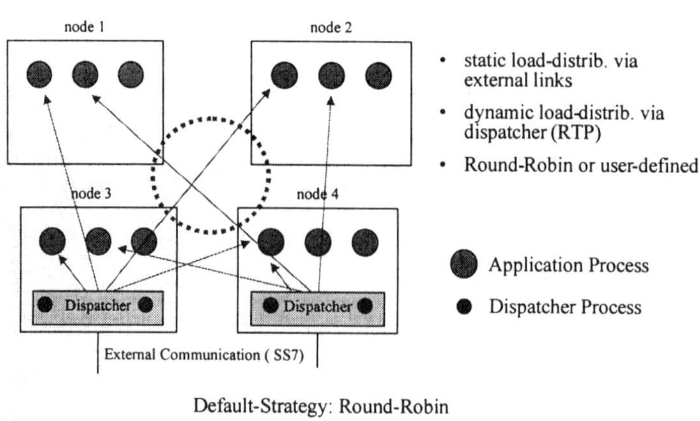

Figure 8. Load Balance

6.5 Context-Manager

The context manager plays a central role in implementing non-stop and reentrant programming models. The application contexts (checkpoints) must be stored safely to allow the continuation of the process in case of different errors, especially of process- and node-errors. In principle the space for storing the context can be seen as a cluster wide available shared memory. In principle such a shared memory could be realized via the data base; but this is would generate to much overhead. Therefore via the context-manager a technique is used which supports the context store as a mirrored shared memory. Fig. 9 shows an active and a backup process in each node with asynchronous mirroring of call contexts to the other node. In case of a node failure, a call in progress will automatically be continued by the backup process in the other node.

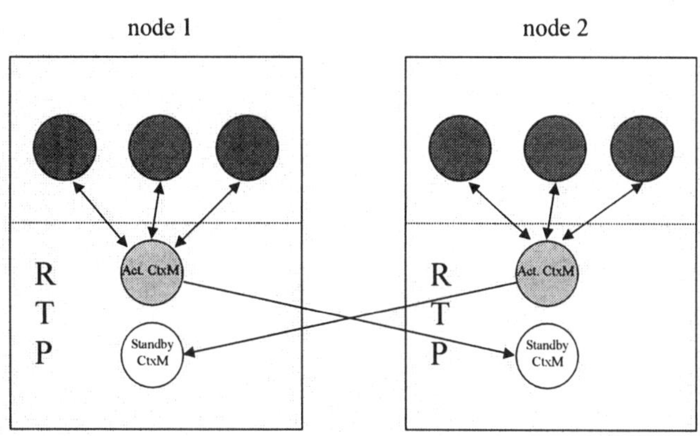

Figure 9. Context Manager

6.6 Rolling Upgrade

Rolling upgrade is only possible if the application, the middleware layers and the operating systems fulfill the compatibility requirement of versions n to n+m (normally m=1). This requirement is hard to realize in a distributed environment. Not only the APIs must be compatible but also the communication mechanism and the contents of the messages. Furthermore, the upgrade for the hardware and software components require a computer element (CE) restart has to be done stepwise for each CE, always performing the whole cycle. In Fig. 10 the procedure for a rolling upgrade is given.

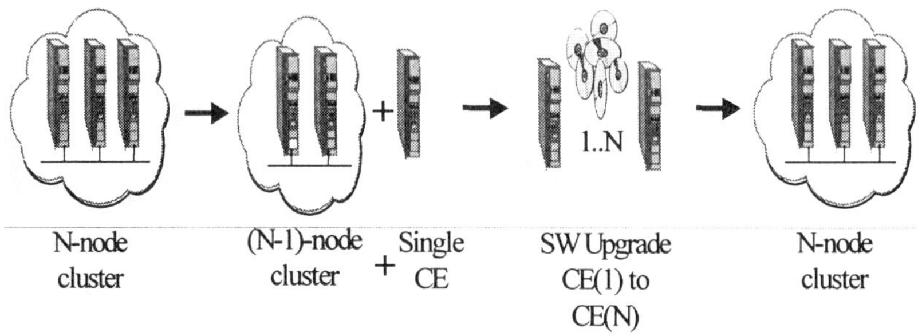

Figure 10. Rolling Upgrade

6.7 Disaster Recovery

The scalability of shared-disk and shared-nothing-database architectures highly depends on the latency of the messages sent to the different instances of the database-manager. The latency of the messages relates directly to the distance of the nodes of the cluster; the closer the node the better the latency. Therefore the distance of the node is restricted to several hundred meters. Disaster configurations which protect against global disasters like earth quakes require different technologies with asynchronous replication methods for the appropriate data.

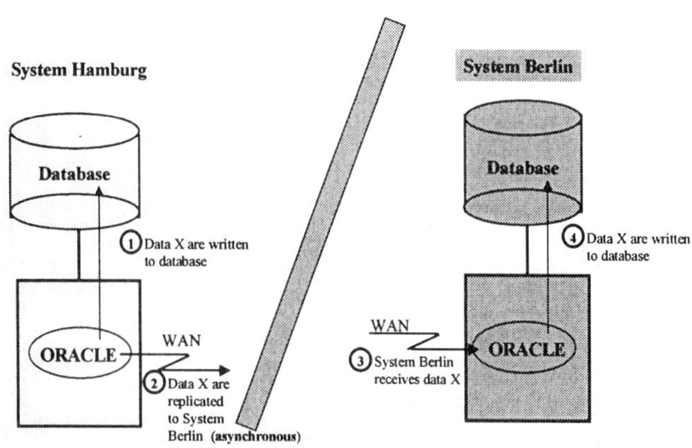

Figure 11. Mated Pair / Disaster Recovery

The replication can be done by the logical disk subsystem or by the databases. Such configurations are often called Mated Pair Concepts. Updating between to distributed nodes over wide area networks (WAN) is shown in Fig. 11.

7. SUMMARY

The Reliant Telco Platform RTP is not a telecommunication solution but a telecommunication "middleware" or "toolset" withthe highlights: (1) Leading edge system based on a cluster technology; (2) Application programming interface (API) with a single system image; (3) Highly available (continuous operation, no planned or unplanned interruption of service); (3) Predictable performance, (4) Disaster recovery. This allows the application development to concentrate on their task to develop Services and not to think about handling a cluster especially in case of failure.

A main goal for the future is, to provide the RTP for the use in the JAIN environment. This will lead to a standardised application.

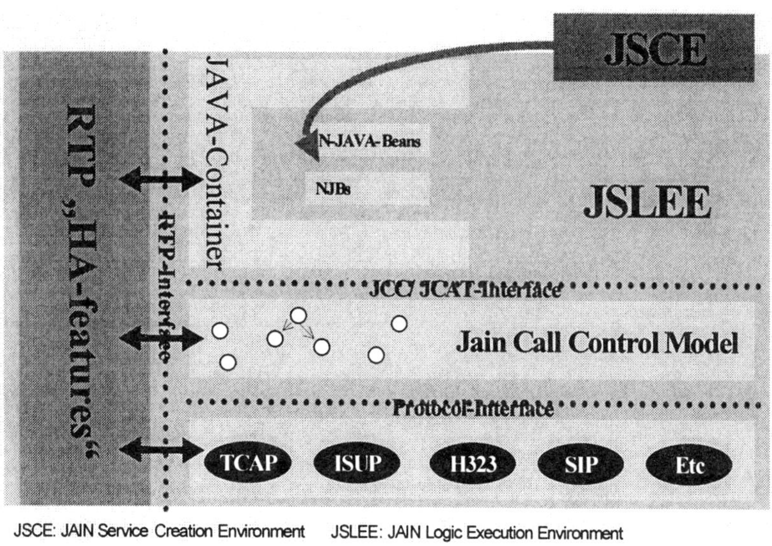

Figure 13. JAIN-Outlook

ACKNOWLEDGEMENT

Special thanks to Prof. Dr. Dieter Carl (Head of Siemens IN-develop-ment) and Dr. Ralf Rieken (Siemens Network Architecture), whose platform concept was the starting idea of our work.

Part Three

Active and Programmable Networks

LOCATING SERVICES IN PROGRAMMABLE NETWORKS WITH AMNET-SIGNALLING

Anke Speer, Ralph Wittmann, Martina Zitterbart
Technical University of Braunschweig
[speer|wittmann|zit]@ibr.cs.tu-bs.de

Abstract AMnet provides a framework for flexible and rapid service creation. It is based on Active and Programmable Networking technologies and uses active nodes (AMnodes) within the network. These AMnodes are executing service modules for the provision of individual communication services. Using on-demand loadable service modules helps to enhance the functionality of intermediate systems without long global standardization processes.

Placing application-dedicated functionality within the network requires a flexible signalling protocol to locate and announce as well as to establish and maintain the corresponding services. The AMnet signalling protocol discussed within this paper uses active packets – evaluation packets – for service location. The AMnet service control architecture further comprises periodic service announcements to make available services known to receivers.

Keywords: Active Multicasting, Service Location, Active Signalling, Active Networking, Programmable Networking

1. INTRODUCTION

In the past decades a tremendous growth in the use of computer networks in general and the Internet in particular can be noticed. Besides the provided interpersonal communication, like e-mail, the networks advance in becoming the media for distributed computation, telecollaboration, distance learning, e-commerce and the like. Many of these applications are inherently based on group communication. Therefore, efficient and scalable group services are needed for proper communication support. The particular group service to be provided highly depends on the application itself and on the media transported, e.g., text, audio or video. Issues that vary within these group services are reliability and quality of service support. It is important to note that no single group service will be able to serve the high variety of existing and emerging group applications.

With respect to group members, situations may appear in which different group members would like to experience a different group service leading to a *heterogeneous group service.* This can be motivated due to a low-bandwidth access link (e.g., a wireless link) or due to a lower monetary budget of a group member. Such group members should not be able to degrade the overall service quality of a collaborating group. However, in most current approaches the service provided to individual group members is penetrated by the group member with the lowest service capabilities. Such an approach is not acceptable for multimedia and collaborative applications in heterogeneous networking environments.

The AMnet approach makes user-tailored data streams available to individual receivers and, thus, enables heterogeneous group services [15]. It therefore utilizes dedicated support within the network and applies the ideas of Active and Programmable Networking to avoid global standardization processes. *Service modules* are dynamically placed on active network nodes (AMnodes) dependent on user requirements. These service modules then individually tailor data streams according to the group members service requirements. Typical examples of such service modules are QoS-filter and modules for error control.

Figure 1 shows a tele-collaboration scenario – a typical example for possible AMnet support – where the participants differ in their Quality of Service (QoS) requirements. According to the AMnet approach the AMnodes are responsible for adapting the original data stream in accordance with the desired QoS-levels.

It should be highlighted that AMnet provides a general framework for heterogeneous group communication that is not limited to service modules like QoS-filter or error control. In contrast, AMnet provides an open framework where new services can easily be incorporated. An active signalling protocol is used for the on-demand placement of service modules. It is discussed in detail within this paper.

Current Heterogeneity Support. Currently, different approaches to signal and provide heterogeneous communication services are developed. The provision of heterogeneity can be distinguished in end-to-end support and network support.

To provide heterogeneity, a possible action of the sender is hierarchical coding of the output stream. Receivers can select different media qualities by joining appropriate multicast groups related to the coding levels (e.g., [13]). In [1] a communication model is described where hi-

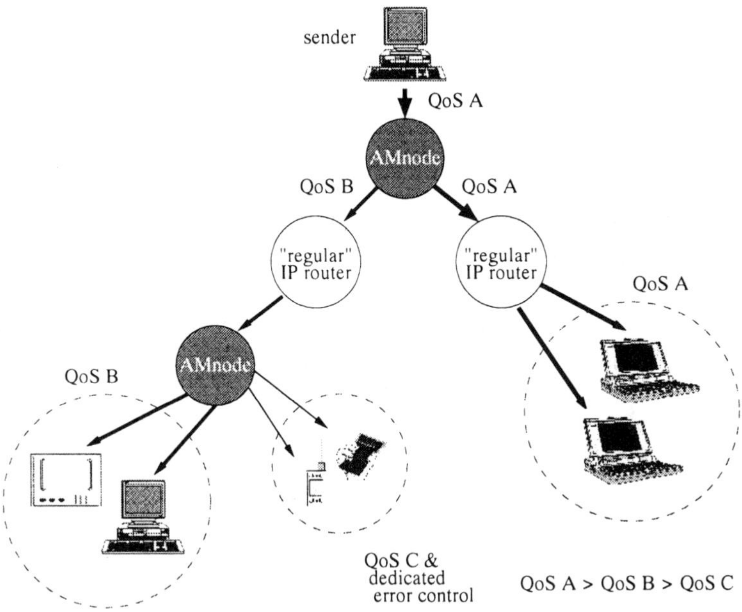

Figure 1 Basic AMnet scenario: Tele-collaboration

erarchical coding into one output stream allows for lossy decompression at the receiver. Dependent on the loss ratio the original signal can be restored on a corresponding quality level.

The model of integrated services within the Internet [5] is a step towards heterogeneous services. Using the RSVP protocol heterogeneous resource reservations are possible. However, current flow specifications limit heterogeneity support to network performance parameters.

A well-known way to place enhanced functionality within the network is the establishment of transport level [4] or application level gateways [3, 7]. Common to those solutions is the withdrawal of transparent end-to-end network operation. The gateway system which hosts the additional functionality is the peer entity of both sending and receiving system.

Active and Programmable Networking. Based on the idea of placing additional functionality inside the network the concepts of Active and Programmable Networking were developed. These concepts allow for application level processing within network nodes, such as IP routers. This is done with capsules [18] in the Active Networking approach and via programmable switches in the Programmable Networking approach

[22]. Capsules carry themselves the code to be executed on intermediate systems [8]. Whereas, programmable switches execute pre-loaded application code on packets flowing through them [22]. Both approaches help to enhance the functionality of the network without long and intractable standardization processes.

According to different research work the provision of group services with Active and Programmable Networking technology appears to be a promising idea. Not only capsules [20] but also programmable switches [2] and in particular, sophisticated acknowledgment [6] and congestion control [9] are proposed to provide basic multicast control mechanisms. However, these approaches do not consider the support of heterogeneous group communication.

To put it in a nutshell, it can be said that the potential of group communication with heterogeneous services has not been investigated in detail in the context of Active and Programmable Networking.

In contrast, our work focuses on service provision in heterogeneous group communication environments. The establishment and maintenance as well as the location and announcement of corresponding services requires a flexible signalling protocol which is presented throughout this paper. The paper is organized as follows. Section 2 introduces the AMnet approach for the provision of heterogeneous group communication. In section 3 the service concept of AMnet is described. Section 4 presents the general concepts of the active signalling protocol with basic implementation approaches. Section 5 comprises some results of `ns`-simulations performed to analyze the protocol behavior. Section 6 closes with a summary and an outlook on ongoing research.

2. AMNET

In the following, a brief introduction into the AMnet service model is given. A more detailed discussion of AMnet can be found in [15]. Basically, AMnet is based on IP-multicast and the group concept in general.

AMnet aims at the provision of scalable heterogeneous group communication with efficient and rapid service creation. It is based on the placement of additional functionality inside the network by *service modules*. Service modules are responsible for the adaptation of data streams to specific service demands. These modules are dynamically loaded by Active Multicasting Nodes (AMnodes) which form the core building blocks of AMnet [14] and operate on the communication path between sender and receivers. In this sense AMnet represents a Programmable Networking approach.

A goal of AMnet is to support service heterogeneity transparently to the origin of a data stream as well as to the receivers. Thereby, AMnet follows a receiver oriented approach. Loadable service modules with out-of-band signalling are used for service creation instead of capsules. The reasons are two-fold. Firstly, capsules have to be provided by the sender, i.e., the desired sender transparency and receiver orientation could not be achieved with capsules. Secondly, hardware support with dedicated service modules for performance improvement becomes possible for AMnet with out-of-band signalling whereas a combined HW/SW solution is envisioned. As the structure of the hardware dependent modules differs significantly from software-based modules, a capsule-based approach is not easily applicable.

Our approach for rapid service creation uses programmable switches, the active signalling protocol described in this paper, however, uses capsules to select appropriate service providers. These capsules are called *evaluation packets* and contain an evaluation program to be executed on the AMnodes in order to determine whether the AMnodes satisfy the qualifications for providing a service individually for a receiver.

Placing application-dedicated functionality within the network raises some questions: where should those services be located, how should they be established and maintained, how should a receiver be associated to a dedicated service and how should different services be managed within a *session*? In this context, a session refers to a communication scenario where a designated sender issues a data stream which can be received from several communication participants without or after adaption, thus, different service levels are provided. Altogether the foregoing questions have to be solved by a flexible signalling protocol which provides a receiver-controllable localization of the service modules.

In a first prototype implementation RSVP was used as signalling protocol [21]. With this prototype QoS filters were requested by receivers via RSVP. To this end appropriate parameters are added to RSVP reservation messages. But RSVP does not offer the flexibility to support optimal allocation of service modules, because RSVP tends to locate the service as close to the sender as possible. While this is desirable for media translation, protocol adaptation can not be realized this way. In the following sections the model of a new active signalling protocol to be used within AMnet is explained.

3. AMNET SERVICE CONCEPT

This section presents first how AMnet is able to provide heterogeneous group communication. Then the mechanisms for service location and service announcement are described.

3.1. SERVICE LEVELS

Service heterogeneity within a session needs to be bound to a manageable degree of diversity. Therefore, one concept of AMnet service control is to logically group receivers with similar service demands into distinct multicast receiver groups – the *service level groups* (see Figure 2), distinguishable by their multicast-addresses. The receivers join the corresponding group on demand through IGMP [10].

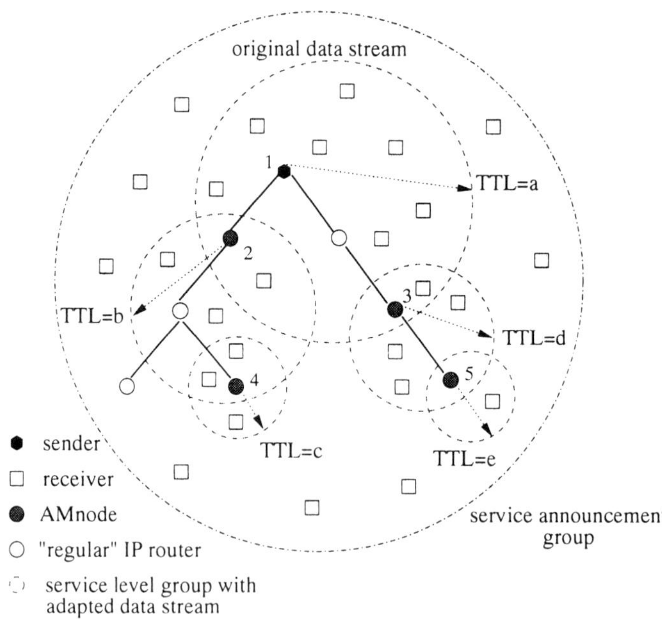

Figure 2 Example multicast tree

Each service level group within a communication session represents all receivers whose service demands can be resolved with a single group service. Therefore, each group represents a different view onto the same original data corresponding to the adaptation within the AMnodes. The communication service offered by AMnet can be described by a hierarchically ordered tree of service level groups (e.g., Figure 2). The service of a group is supported by an AMnode through the use of appropriate service modules. The actual service may be derived from the processing of the original data stream (see AMnode 1 in Figure 2) or from the service of another service level group (see AMnode 2 in Figure 2). The scope of an AMnet service level group is limited by the actual TTL value assigned to packets issued by the corresponding AMnode. The TTL value of every group must not exceed the scope of the service announcement group (see below).

Furthermore, the establishment of service level groups permits the provision of different service qualities within one region without the services interfering with each other. Two data streams with different media formats or individual error control must sometimes coexist on a communication link and have to be distinguishable by the appropriate receivers. Therefore, all packets are explicitly assigned to corresponding service level groups by their multicast destination address.

The hierarchical order of service level groups allows for an efficient establishment of different quality levels within a session. One distinct service quality might be easily derived from another already available quality level if, for instance, only a different (weaker) error control policy for network overload conditions has to be inserted into the already adapted data stream.

The service quality experienced at the receiver is a function of the service level of the group and the current network conditions between group source and each individual receiver. The service within a group can only differ in performance-oriented, packet-based service parameters such as delay or loss probability. Other parameters which define the content-based nature of the service are homogeneous (e.g., media format, acknowledgment strategy, error control) within a service level group. However, the distinction of services can be triggered by both: performance oriented parameters and content. Consider two receivers with very different loss probabilities. In that case different acknowledgment strategies and error control mechanisms might be necessary which require different service level groups.

The hierarchical ordering of services does not automatically imply hierarchical degradation of all service parameters. Some parameters can be provided unchanged, other parameters even improve. As an example, the insertion of a new service can improve media playback quality due to less jitter at the cost of higher, but uncritical delay. This could be useful for video distribution, for example.

3.2. SERVICE LOCATION AND ANNOUNCEMENT

In contrast to the capsule approach, AMnet's service creation is not based on executable code to be transmitted in data packets. Instead, the AMnet signalling provides procedures to locate and announce service modules and to establish and maintain appropriate services on the AMnodes. These signalling procedures utilize Active Networking technologies. Control of AMnet services is maintained completely out-of-band.

The following components are part of the service control. Figure 3 shows their correlation:

- Session announcement,
- service announcement,
- service module repository, and
- service access.

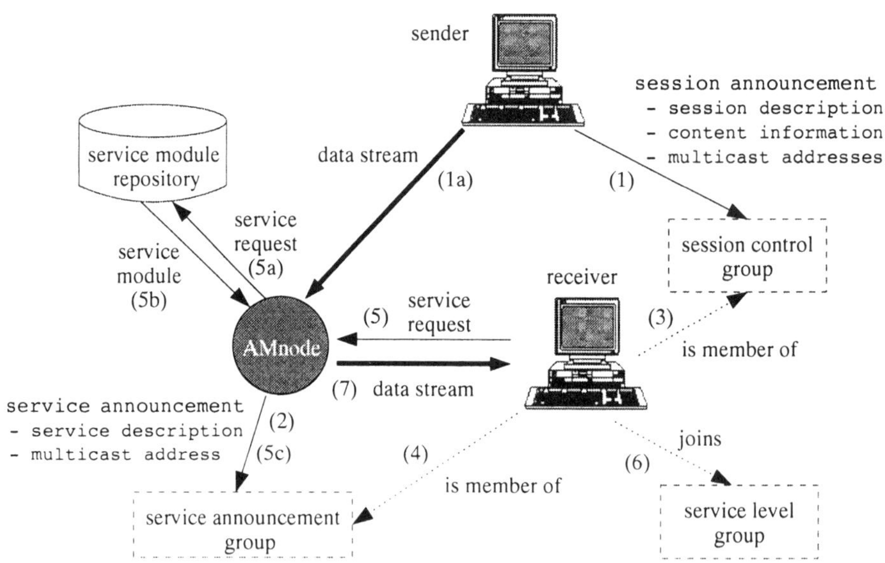

Figure 3 Service Control Architecture

Session Announcement. A session announcement is advertised by a sender on a separate multicast group – the *session control group*. In this group every AMnet session is announced similar to the well-known Session Announcement Protocol (SAP) [12] of the MBone. SAP itself was not utilized to avoid the use of its standardized port and because the overhead of introducing the needed new data-fields and functionalities was considered as too much.

The session announcement contains a description of the session, including information such as bandwidth and delay requirements, and content specific information, e.g., data format and compression scheme. Moreover, the path the session announcement took from the session sender to a specific receiver is included as well in order to reconstruct the path (see below). Additionally, the description contains the multicast address of the original data stream and the multicast address of the service announcement group. Based on the session description receivers decide whether to join a session.

Service Announcement. Whereas the session control group provides information of all available AMnet sessions, the *service announcement group* forms a data base of the available AMnet services for a given session. This group is used for the signalling of heterogeneous service capabilities and demands, i.e., all available service level groups are announced within this group. All session participants are member of the service announcement group to be able to learn about available services. Because if a new service is established by an AMnode, the node advertises the appropriate service description within this group.

Service Module Repository. The service module repository is a distributed data base which contains service modules and their description. The purpose of the repository is to make service modules available to an AMnode in case that a service module is not already cached at that node. Service modules can be stored in the service repository by AMnet users and network management procedures.

Service Access. The mechanism for service access follows the same principles as the current practice of the MBone. If a participant wants to use an AMnet service, e.g., media translation or enhanced error control, the participant checks by joining the session's service announcement group whether the desired service is already available. If the service exists, the participant simply joins the associated group of the best matching AMnode that provides that service. If the service does not exist, the participant may ask an appropriate AMnode to provide and announce that service.

4. SIGNALLING AMNET SERVICES

This chapter describes first the concepts of the active signalling protocol, afterwards the prototype implementation is presented.

4.1. GENERAL CONCEPT

An important aspect of the AMnet service control is the selection of an appropriate AMnode for service provision. This selection depends on the type of the requested service and on the metrics propagated by the receiver of the session towards the AMnode. For example, one receiver may choose to use a media translation service which has to be placed as close to the session sender as possible, whereas another receiver may choose a protocol adaptation service which has to be placed near by the receiver itself. For the purpose of selecting an appropriate AMnode, each session receiver comprises a *performance meter*. According to the metrics and according to the description of the desired service the performance meter rates the service announcements the receiver gets from the joined service announcement group in order to determine whether an appropriate service is already in operation.

For the selection procedure, two cases need to be distinguished:

- the desired service is already provided at the time a receiver joins the session
- no appropriate service is available

In the first case the selection takes place within the set of AMnodes that already provide the desired service, in the second case it is selected within the set of AMnodes that fulfill the propagated metrics and, therefore, potentially can provide the requested service.

In both cases active packets – so-called *evaluation packets* – are used for the selection. These packets contain the following data:

- the path the session announcement took from the session sender to the receiver; this information is included in the session announcement and is needed because every AMnode has to know its parent AMnode to correctly forward the evaluation packet (see below). Caching this information in the AMnode is considered as too much overhead.
- the address and the result of the best matching AMnode so far resp. the address and the result of the actually evaluated AMnode (see below)
- the description of the desired service

- the evaluation program to be executed on each AMnode which receives such an evaluation packet to determine whether it satisfies the requirements of being a service provider

The way the packets are forwarded and the AMnodes are evaluated slightly differ if there are already providing AMnodes or if an AMnode has to be found which fits into the metrics and can provide the desired service:

If one or more AMnodes currently provide the desired service, the performance meter of the session receiver issues the evaluation packet to each of these AMnodes in order to evaluate them according to the propagated metrics. This is done per unicast since it is assumed that there is only a very limited number of corresponding AMnodes available. The AMnodes attach the evaluation results to the evaluation packet and return it to the receiver. After the performance meter has collected all responses, it selects the best matching AMnode among the queried ones and makes the receiver join the appropriate service level group.

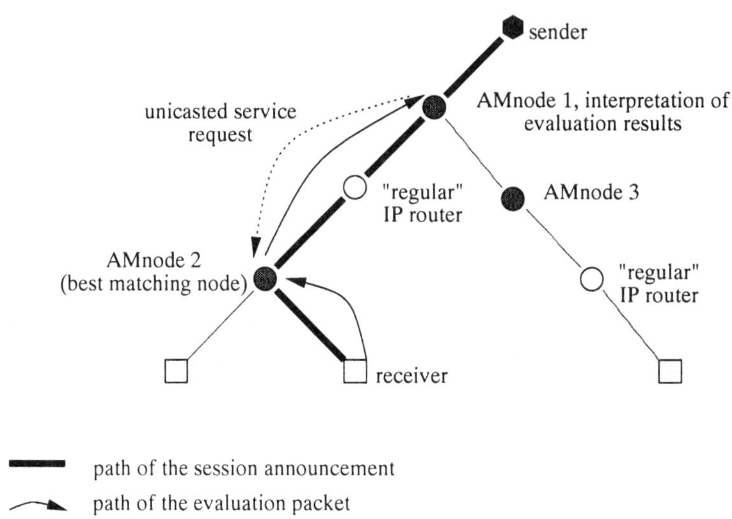

Figure 4 Scheme of the basic tree search

In case no AMnode provides the requested service yet or no providing AMnode fits to the propagated metrics a slightly different selection procedure is used. That means in the worst case, both selection procedures take place one after another. This selection procedure uses a different mechanism – the so-called *basic tree search* – to determine a suited AMnode. Again, the performance meter of the session receiver issues an evaluation packet. This is forwarded to the closest AMnode

(AMnode 2, Figure 4). This node can be identified by the path-entry included in the evaluation packet and known from the session announcement. According to the description of the desired service contained in the received evaluation packet and according to the available node resources, the AMnode management decides whether the node is capable of providing the service. In the positive case, it executes the evaluation program and incorporates the result into the evaluation packet. This, however, is only performed if the actual AMnode fits better than the AMnode previously noted in the packet – if there is any. Therefore, the evaluation packet contains at most one address: the address of the best matching AMnode so far. If the service can not be provided, because of limited resources or similar reasons, or if a better matching AMnode is already cached in the evaluation packet, the packet remains unchanged.

With the basic tree search the actual evaluated AMnode does not directly respond back to the session receiver. In contrast, it forwards the evaluation packet to its parent AMnode on its way to the session sender (AMnode 1, Figure 4). At this node the admission test and evaluation takes place again. This is repeated until the evaluation packet arrives at the last AMnode in front of the session sender (AMnode 1, Figure 4). This AMnode is responsible for the interpretation of the evaluation result carried by the evaluation packet. This last AMnode, then, unicasts a service request for the desired service – known from the description in the evaluation packet – to the best matching AMnode (e.g., AMnode 2, Figure 4).

After the selected AMnode received the request to provide the service, it downloads the appropriate service from the service module repository and announces the establishment in the service announcement group. Since the session receiver which started the whole procedure is a member of the service announcement group it learns about the availability of the desired service and subsequently can join the corresponding service level group.

The session announcements are sent periodically. Therefore, it is possible to notice newly available AMnodes – they will be included in the path-description of the session announcement – and it is as well possible to include them dynamically into the evaluation.

Since the basic tree search mechanism is limited to AMnodes being located on the multicast tree, an *extended tree search mechanism* is under development. AMnodes beyond the path might as well be able to provide the service and fulfill the metrics even in a better way than the AMnodes on the path. The extended tree search can possibly use the concepts of IP-multicast and forward the evaluation packets to the neighbor AMnodes of the AMnodes on the multicast tree. Therefore, the

AMnodes have to be reachable over a known multicast address. Then, for example, a ring search evaluation can be processed on the AMnodes beyond the multicast tree limited through a TTL. The value of the TTL has to be decreased with increasing distance from the receiver.

It is obvious that the evaluation packets are capsules as introduced in the context of Active Networking. Therefore, overall it can be noticed that AMnet is based on Programmable Networking technologies with service creation through loadable service modules, but as well on Active Networking technologies (i.e., capsules) for the location of probable service providers.

4.2. PROTOTYPE IMPLEMENTATION

Currently, an implementation of the signalling protocol is in progress. This subsection gives a short overview about the basic design considerations.

Java is the chosen programming language with the Java Development Kit 1.2 of SUN because of its platform-independent facilities. The session announcement and the service announcement group are realized through IP-multicast groups. The signalling components on the session sender, on the receivers and on the AMnodes are implemented as Java classes.

The analysis of the requirements for the service module repository showed that they can be fulfilled widely by an LDAP server. Therefore, we selected LDAP for the design and implementation of the repository. Other reasons for utilizing LDAP were its easy use and its extensibility as well as the availability of Java implementations (e.g., from Netscape). In [16] a suggestion for recording Java objects into an LDAP directory information tree is given. Table 1 gives the attributes which describe the service modules. With this description the modules can be stored and recorded directly in the LDAP server.

5. SIMULATION EXPERIMENTS

In order to evaluate the signalling protocol outlined above, simulation experiments were carried out in [17] using the simulator `ns-2` [19] from U.C.Berkeley/LBNL. Issues of interest throughout the simulations include the frequency of AMnode setups, the delay between service request and provision and the like.

The network topology used for the simulation experiments consisted of 250 nodes. It was automatically generated by GT-ITM [11]. The backbone is constructed out of 10 nodes with up to 3 belonging subnets with 8 nodes at an average (see Figure 5).

Attribute	*Meaning*
moduleName	name of the service
serviceType	e.g., QoS filter or error control
description	description of the service
os	operating system needed by the service
osVersion	version number restriction for the operating system
format	data format in which the service is stored within the directory information tree
moduleType	mode of the service (e.g., Java, shared library)
moduleData	service data
objectClass	if the service is a Java module, objectClass has to give the name of the executing class
className	if the service is not just Java code, className has to give the name of the class which starts that service

Table 1 Description attributes for service modules

The simulations described in [17] were all performed on the same network topology (see Figure 5) with the same receiver distribution and requirements but with different input parameters. As routing protocol, a DVMRP-like protocol was used. The metrics propagated by all receivers was the minimization of the distance between the receiver and the AMnode which provides the desired service.

The variable configuration parameters were:

- the lifetime of a service module on an AMnode,
- the time for loading and installation of service modules on AMnodes,
- the maximum allowed distance between the AMnode with the appropriate service and the receiver,

In the following, selected results of the simulations are presented. In every simulation 42 AMnodes are available. They are located on edge routers interconnecting the backbone with the local subnetworks (e.g., AMnode 162 in Figure 5). Moreover, every node in the backbone can operate as an AMnode.

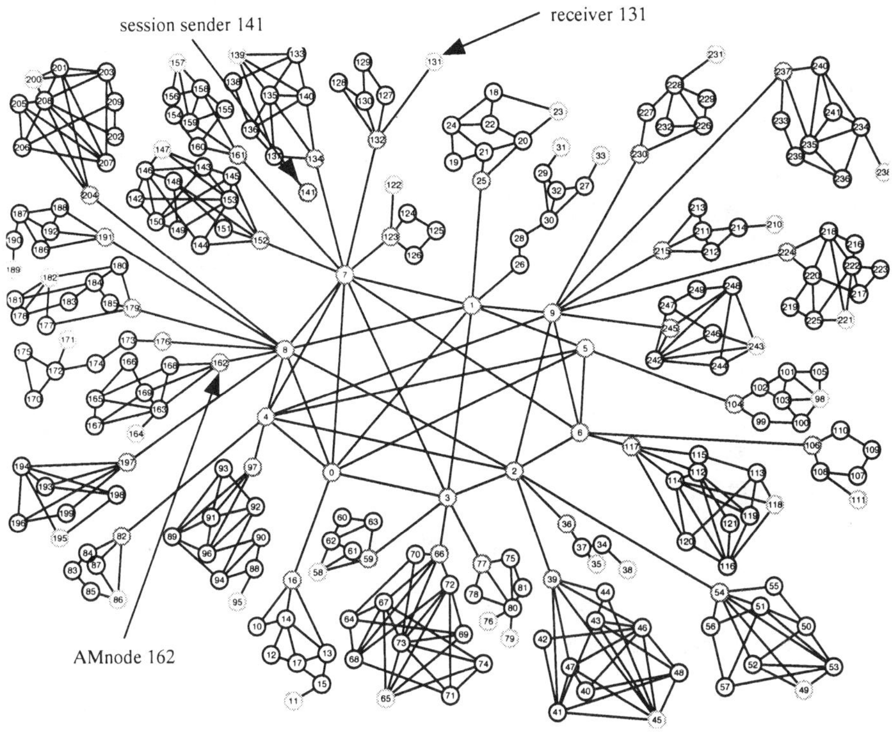

Figure 5 Simulation topology

In some arbitrary chosen subnets additional AMnodes are placed. 33 receivers with 10 different requests are distributed across the 30 subnets. They join the session successively every 50 ms. However, every second receiver just announces its desired service and leaves the session right afterwards. This is done to simulate dynamic scenarios and to simulate situations in which service modules start to operate without having an actual user. The session announcement is repeated every 10 ms as well as the service announcement of every available service.

Before the first receiver joins the session no service module on an AMnode is available except for the session sender where the original data stream is produced. Therefore, the first ten receivers with different requests have to start the basic tree search to find an appropriate AMnode capable to provide the desired service. Thus, every simulation has a starting phase of about 5 seconds until the first ten different receivers have joined. Whether the next joining receiver has to start the basic tree search, too, depends on the lifetime of the services – if the appropriate services are still in operation, the receiver can join the corresponding service level group – and on the distance between the AMnode which already provides the desired service and the receiver – the distance has to be smaller than the maximum allowed distance.

For every simulation the following items were observed:

- the delay at the receiver in receiving the desired data
- the number and distribution of AMnodes with active service modules
- the cumulative number of signalling packets sent/received from/at the nodes
- the number of signalling packets sent from two distinct nodes
- the number of signalling packets received at two distinct nodes

As an example, the observed results for receiver 131 are presented (see Figures 6 to 8). The observed period of time is 17 seconds, the first receiver joins the session at 0.5 seconds. The number of packets it received during different simulation experiments – dependent on the variation of the maximum allowed distance between the AMnode with the appropriate service and the receiver – is depicted in Figure 6. The influence of the lifetime of a service module on an AMnode is shown in Figure 7. The loading and installation overhead of service modules on AMnodes is shown in Figure 8.

Figure 6 shows that the reduction of the maximum allowed distance between the receiver and its corresponding AMnode increases the number of packets received at node 131. This is due to the fact that similar service requests of AMnodes which have a distance of less than twice the allowed distance from each other can be combined. Therefore, less AMnodes have to provide the desired service and less service announcements are sent. If the maximum allowed distance is very small, just a few similar requests can be combined. That means, another AMnode has to provide the service and additional service announcements have to be sent, so the number of received packets at the receiver increases. The results of the simulations with a distance of 10 respectively 15 AMnodes are very similar because of the diameter of the chosen network. The diameter is between 10 and 18 nodes. Thus, 10 is a good choice for the distance because then the best combination of similar service requests is achieved. The delay time between the service request and the receiving of the data is equal in the simulation experiments with 10 and 15 nodes as maximum distance. However, the delay times decrease with decreasing distance because the service providers are closer to the receiver.

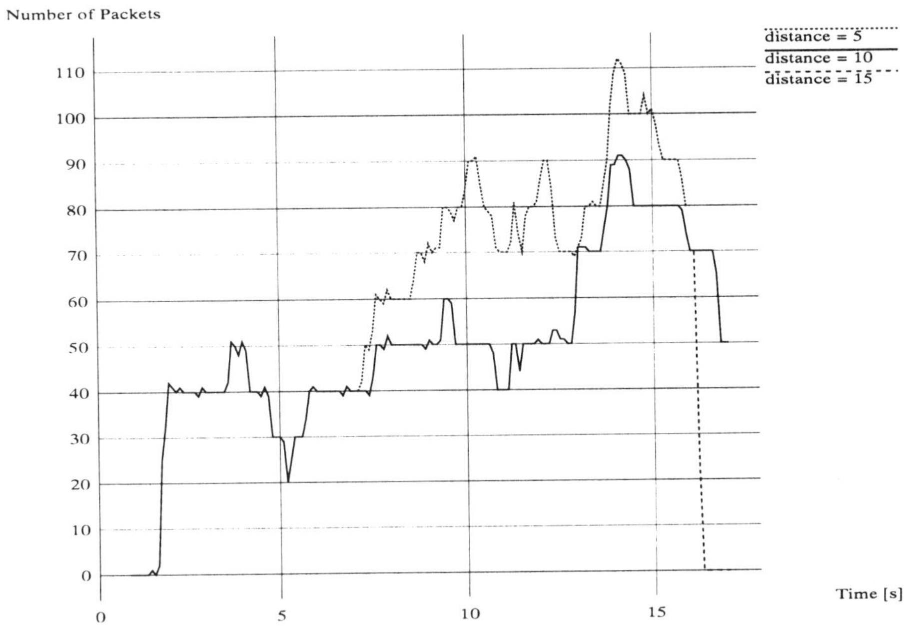

Figure 6 Influence of the maximum distance

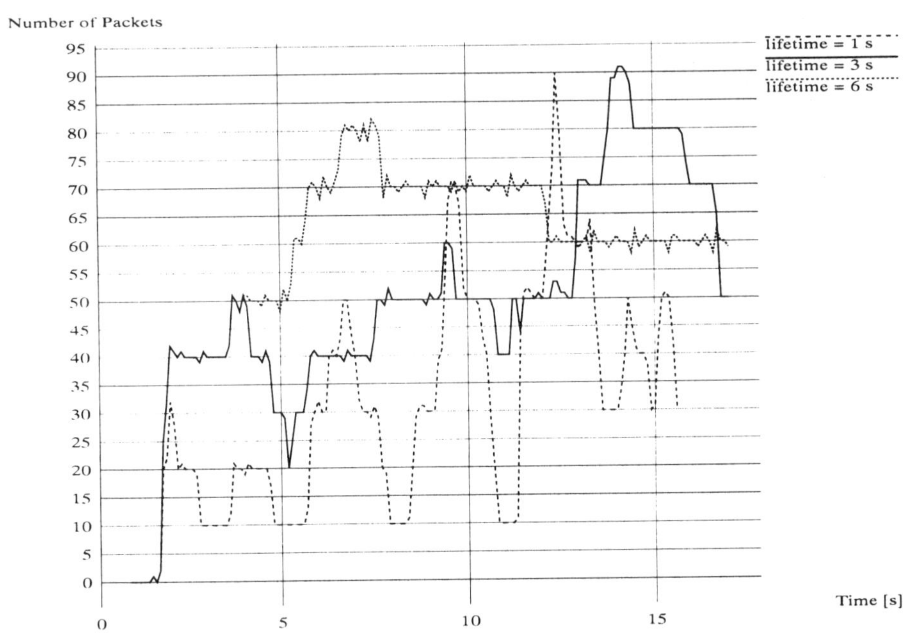

Figure 7 Influence of the lifetime of the service modules

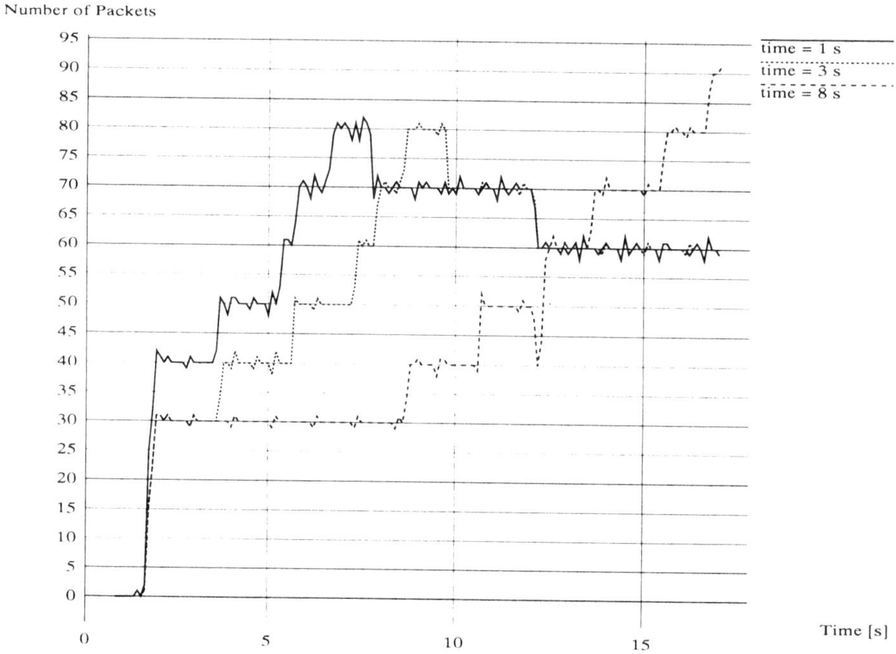

Figure 8 Influence of loading and installation overhead

Additionally, the tree search needs not to look further than 5 nodes up in the multicast tree.

Figure 7 shows the effect of a longer lifetime of the service modules. If a service module has no users, it operates just for as long as its lifetime is. If a receiver starts to use this service during its active time, the service is refreshed and can again operate for its lifetime. Because even modules which do not have any users operate for the whole lifetime of 6 seconds, receiver 131 gets a lot more packets than if the lifetime is just 3 respectively 1 seconds. The disadvantage with short lifetimes is that the receiver may have to require its desired service frequently. The delay times for the receiving of the required data are very similar because the lifetime of the service modules do not have any influence on that.

Figure 8 underlines that the loading and installation overhead of service modules has strong effects on the number of received packets. The graph belonging to the simulation with 1 second loading time firstly is very steep. The number of received packets increases very fast due to the fact that service requests are fulfilled rapidly and, therefore, many AMnodes start sending right after the receivers' request. The graph belonging to the simulation with a loading time of 8 seconds is considerably less steep. For the first 8 seconds just the session announcement packets were measured. Only after 8 seconds when the first service starts to operate the number of packets increases. Clearly, the delay times for receiving the required data increase similar to the growth of the loading and installation overhead.

The simulations resulted in the following conclusions:

- The longer the lifetime of a service module the rarer do the receivers have to look for a new service provision. Thereby, the time while a receiver has to wait for its data and the signalling effort which is necessary for the search of a new service provider decreases. However, long lifetimes of service modules produce high overload on the AMnodes: If the appropriate group stays without any participants, the resources of the AMnode are unnecessarily wasted.

- The greater the allowed distance between sending AMnode and receiver the fewer sending AMnodes are necessary because distributed but similar requirements can be combined and fulfilled by the same AMnode.

- The longer the time for loading and installation of service modules on AMnodes the more identically service provisions are requested. There is still no mechanism which indicates to a new joining receiver that its desired service was already requested and that the search is still in progress.

6. SUMMARY AND OUTLOOK

In this paper an outline of the AMnet approach for flexible and rapid service creation in IP based networks was presented. The AMnet signalling protocol for locating and announcing services provided by active intermediate nodes (AMnodes) was discussed in some detail. Programmable Networking technologies are used with service creation through loadable service modules. This mechanism helps to enhance the functionality of the network without long global standardization processes. Active Networking technologies are utilized for signalling purposes by active packets – the so-called evaluation packets. Simulation experiments showed the influence of the propagated metrics, the lifetime and performance of service modules and the influence of loading and installation overhead.

Future work concerning the signalling protocol will focus on further experiments, followed by the development of merging and reconfiguration mechanisms for service modules and their localization inside the network as well as improvements to the presented mechanisms. For example, service requests in progress should be cached to avoid any unnecessary requests and, thus, to reduce signalling overhead. Beyond it, the implementation of the signalling protocol will be followed up and a larger testbed initially spawning four German universities will be set up.

References

[1] A. Albanese, J. Bloemer, J. Edmonds, M. Luby, and M. Sudan. Priority Encoding Transmission. *IEEE Transactions on Information Theory*, 42(6):1737–1744, November 1996.

[2] D. S. Alexander, W. A. Arbaugh, M. W. Hicks, P. Kakkar, A. D. Keromytis, J. T. Moore, C. A. Gunter, S. M. Nettles, and J. M. Smith. The SwitchWare Active Network Architecture. *IEEE Network*, 12(3):29–36, 1998.

[3] E. Amir, S. McCanne, and H. Zhang. An Application Level Video Gateway. In *Proc. of ACM Multimedia '95*, pages 255–265, San Francisco, California, November 1995.

[4] H. Balakrishnan, S. Seshan, E. Amir, and R. Katz. Improving TCP/IP Performance over Wireless Networks. In *Proc. of 1st ACM Conference on Mobile Computing and Networking*, pages 2–11, Berkeley, California, November 1995.

[5] R. Braden, D. Clark, and S. Shenker. Integrated Services in the Internet Architecture: An Overview. RFC 1633, ISI, June 1994.

[6] M. Calderon, M. Sedano, A. Azcorra, and C. Alonso. Active Network Support for Multicast Applications. *IEEE Network Magazine*, 12(3):46–52, May 1998.

[7] Y. Chawathe, S. Fink, S. McCanne, and E. Brewer. A Proxy Architecture for Reliable Multicast in Heterogeneous Environments. In *Proc. of ACM Multimedia '98*, pages 151–159, Bristol, UK, September 1998.

[8] J. V. Guttag D. J. Wetherall and D. L. Tennenhouse. ANTS: A Toolkit for Building and Dynamically Deploying Network Protocols. In *Proceedings of the IEEE OPENARCH*, pages 117–129, April 1998.

[9] Theodore Faber. ACC: Using Active Networking to Enhance Feedback Congestion Control Mechanisms. *IEEE Network*, 12(3):61–65, 1998.

[10] B. Fenner. Internet Group Management Protocol, Version 2. RFC 2236, IETF, November 1997.

[11] GT-ITM. Georgia Tech Internetwork Topology Models. http://www.cc.gatech.edu/fac/Ellen.Zegura/graphs.html.

[12] Mark Handley. SAP: Session Announcement Protocol. Internet Draft, Internet Engineering Task Force, November 1996. Work in progress.

[13] S. McCanne, V. Jacobson, and M. Vetterli. Receiver-driven Layered Multicast. In *ACM SIGCOMM '96*, pages 1–14, San Francisco, USA, August 1996.

[14] Bernard Metzler. *A Flexible and Efficient System Support for Open Communication Services Platforms.* PhD thesis, Technische Universität Braunschweig, December 1999. in German.

[15] Bernard Metzler, Till Harbaum, Ralph Wittmann, and Martina Zitterbart. AMnet: Heterogeneous Multicast Services based on Active Networking. In *Proc. of the 2nd Workshop on Open Architectures and Network Programming (OPENARCH99)*, pages 98–107, New York, NY, USA, March 1999. IEEE.

[16] V. Ryan, S. Seligman, and R. Lee. Schema for Representing Java Objects in an LDAP Directory. RFC 2713, IETF, October 1999.

[17] A. Speer. A flexible Signalling Protocol for AMnet. Diploma thesis, TU Braunschweig, July 1999. in German.

[18] D. L. Tennenhouse, J. S. Smith, W. D. Sincoskie, D. J. Wetherall, and G. J. Minden. A Survey of Active Network Research. *IEEE Communications Magazine*, 35(1):80–86, January 1997.

[19] UCB/LBNL/VINT. Network Simulator ns-2. http://www-mash.cs.berkeley.edu/ns/.

[20] D. Wetherall, U. Legedza, and J. Guttag. Introducing New Internet Services: Why and How. *IEEE Network Magazine Special Issue on Active and Programmable Networks*, July 1998.

[21] R. Wittmann and M. Zitterbart. AMnet: Active Multicasting Network. In *Proc. of Intern. Conf. on Communications (ICC'98)*, pages 896–900, Atlanta, GA, USA, June 1998. IEEE.

[22] Y. Yemini and S. daSilva. Towards Programmable Networks. In *Proc. of IFIP/IEEE International Workshop on Distributed Systems: Operations and Management*, L'Aquila, Italy, October 1996. http://www.cs.columbia.edu/dcc/netscript/Publications/publications.html.

Open Programmable Layer-3 Networking

Hardware Approach for Full Active Network

Koji HINO, Takashi EGAWA, and Yoshiaki KIRIHA
NEC Corporation

Key words: Active Network, Programmable Layer-3, High speed Active Layer-3 Processor

Abstract: This paper discusses high speed (multi-gigabits/second) active networking techniques used to introduce active functionality into layer-3 processing in order to provide wide-ranging flexibility to internetworking. We propose here StreamCode, a compact object code for layer-3 programming, a StreamCode Processor that achieves a high-speed execution of the object code, and an architecture of StreamCode-based high-speed active networking node. The processor has a unique instruction fetch mechanism to prevail over promiscuous instructions and data flows inside the processor, and has resource management function to execute StreamCode programs safely. Our FPGA based prototype system with sample applications confirmed the feasibility of proposed StreamCode based programmable Layer-3 networking.

1. INTRODUCTION

An active network technology is aimed to bring remarkable flexibility and extendibility to future networks, and many discussions from the several technical viewpoints such as language, node OS, security, and performance questions, have been made [1-5]. Some studies suggest controlling layer-3 functions (i.e. routing table searching, packet forwarding, and queueing management, etc.) in an active networking framework. The authors also believe that layer-3 activeness must be seriously considered. This is because most of current and proposed network services (ex. QoS control[6], VPN, IPv6[7], DiffServe[8]) require some modification to layer-3 functionality on many network nodes.

Yet, however, research on active layer-3 control, especially for a high speed (multi-gigabits/sec) networking environments has been limited. To make future layer-3 networks active, and to utilize the activated layer-3 networks into practical as current IP networks, we have to introduce new layer-3 networking systems whose performance (packet/sec, bandwidth for actual application contents) is comparable to non-active networking systems, with enough programmability, security, and stability.

We propose here a hardware-based layer-3 active networking system, aiming to achieve multi-gigabits per second link throughput. It consists of two sections: a packet-by-packet active code execution environment (an active layer-3 section), and an ordinal active network execution environment (an active node section). We are focusing on the active layer-3 section, and interaction mechanisms for these two sections in this paper.

For the active layer-3 section, we have designed an object code, 'StreamCode,' and a processor for it. StreamCode is a machine language level binary object code with which users can program layer-3 actions according to packet sender's idea, and the object code is executed by dedicated StreamCode Processor. StreamCode fragments are executed shortly, only while that packet (which contains the StreamCode fragment) is stayed at an input buffer of StreamCode Processor. So functionality and complexity of each StreamCode program is essentially limited. To compensate for this limitation, we have introduced interaction mechanisms between StreamCode execution and active node section (on network nodes, servers, and terminals).

Section 2 of this paper, we discuss related research, in Section 3 we introduce the basic concept of an active layer-3 networking architecture, and in Section 4 we describe our StreamCode based networking architecture in detail. In Section 5, we introduce the StreamCode Processor, which is dedicated to activating layer-3 functions, in Section 6, we briefly describe our hardware and software based prototype system, and in Section 7, we summarize our work.

2. RELATED WORK

There are basically two approaches to active networks: the programmable switch approach, and the capsule approach [3]. In the programmable switch approach, layer-3 functionality is classified as a target of control for active programs or open programmable APIs, and actual switches are ATM or IP switches with externally controllable interfaces. Current research efforts are focused on controlling existing layer-3 parameters. With respect to the capsule approach, researches are currently focused on program execution itself, however, the next step will have to deal with autonomous layer-3 functionality, and capsule execution environment will require layer-3 activeness.

This paper introduces StreamCode, which is very influenced by capsule approaches. StreamCode has roughly the same packet format as does the capsule approach, with programs located in each packet, and nodes have execution units for these program pieces. There is a difference, however, in their respective execution environments. Capsule runs on top of general purpose unified operating environment, while StreamCode runs inside an active layer-3 execution environment.

Four studies related to active layer-3s are particularly important with respect to our work. First, [9] proposes to download transport active modules into network nodes. Although this basic concept is the same as ours, the approach is different in the sense that active modules are defined as node-vendor-supplied proprietary code modules, which offer standard API to modify its transport behavior. In our approach, the sender of code implementing active L3 module is defined as end user's applications, including servers and clients. Second, [10] describes importance of active-execution's performance for practical active network. The results verify that using processor's native object code will boost three times faster than Java bytecode. This fact shows that indirect execution (i.e. interpreting portable object code by general purpose processor) of active programs is slower than hardware based native execution, and we believe the fact proves the appropriateness of our hardware based active networking approach. Third, [14] presents Field Programmable Gate Array (FPGA) based protocol processor. This idea will provide programmable hardware inside layer-3 function with reasonable performance: node vendor and/or network provider can change hardware function, and end user can request to select hardware functions. However, FPGA has a limitation in its reconfiguration delay, thus packet-by-packet level activeness is hardly achieved with this approach. In our approach, each packet has different active program to realize application-driven network progression (uncountable kind of active programs), in expectation of high-speed wired-logic processors. Nevertheless, we are very interested in the study, because our design will eventually need programmable co-processor framework. Fourth, [11] introduces high performance active node targeting multi gigabits/sec level link speed.

Especially, fourth one's target environment is very close to ours, except that they assume that 'flow' can be managed by each node and use these flow information to bypass active processor. Their architecture may be resulted from their precondition, active packet to non-active packet ratio is not so high, and / or each flow shares its active algorithm. However, we expect that active network aware application should use a different kind of active programs even for single flow, according to their need. Furthermore, we assume that there will be too many (ex. billions) flows aggregated on core routers, to be distinguished in a timely manner. Taking into account such assumptions, our approach chooses to abandon flow detection and all packets must contain program to declare its behavior. Each router's layer-3 function is 'passive' execution engine, and has no autonomous functions for flow detection, no search functions for packet handling program/data, no application-specific programmed behavior, etc.: all actions of the router are described in each incoming packets.

3. HOW TO ACHIEVE PROGRAMMABLE LAYER-3

To make layer-3 functions active network friendly, we have to develop interfaces from active modules to layer-3 functions, as shown in Figure 1. In the figure, stars are some part of layer-3 functions, which serve active network aware actions. Here, we call these as active L3 modules. Active L3 modules must be under control of a node OS, and their control interface will be provided by the node OS, as these functionalities are thought as important network resources. There are three possible ways to implement active L3 modules.

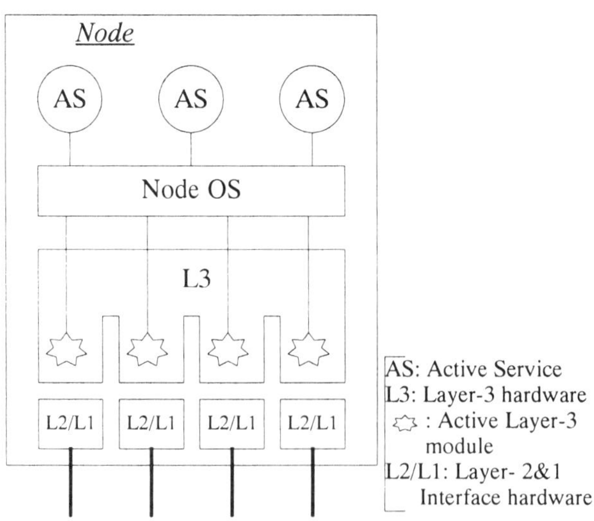

Figure 1: Active Layer-3 module

a) Stop improving layer-3: A functionality and interface of an active L3 module are specified and fixed forever. Any active services can change layer-3 behaviors through the interface of node OS. Active L3 module itself is developed and installed by the node vendor. This plan is the most practical and rational answer, and upgrades current network to active network, but we hope no future functional improvement of layer-3 except for its speed and cost. Ironically, as active network spread widely (we hope so), functional improvement of layer-3 will be discouraged, and would not be allowed. As a result, active service cannot use these improvements.
b) Drop-in layer-3 module: Although an active L3 module's functionality is not specified, standard interfaces for installing new active L3 modules, controlling these modules, and uninstalling these modules will be introduced. Functions of each active L3 module will, as same as network service build upon IP, consist of standard functions, private functions, and to-be-standard functions. However, since active L3 modules are depending on each node's architecture (i.e. proprietary), they are not reusable over the whole network. This is a good answer to balance flexibility and performance, because node vendor can tune these active L3 modules to their products. This plan will be implemented by network processors like TOPcore[13]. On the other side, if a node vendor would not want to make active L3 modules, or another node vendor would discontinue their business, no more active L3 module will be developed.
c) Make active layer-3 execution environment: A new active networking environment for active L3 modules will be introduced. It should be deployed not on the top of the node OS, but inside node's layer-3 functions. Active L3 modules are written in standard languages, and the execution environment must provide abstract and standard functions for active L3 modules to describe specialized layer-3 functions. With the environment, although we can achieve maximum flexibility, standardization effort on a language and API for active L3 modules should be necessary. Active L3 modules can be downloaded through node OS, or can be transferred as a program piece in each packet like capsule-type active network. An active L3 module, as a part of layer-3 functions, will be selected and executed upon arrival of each packet, so severe requirements for module selection delay and module execution throughput are exist. These requirements disallow us to integrate active L3 module execution environment to ordinal active node OS environment in a unified fashion. Because of this, two execution environments will have to coexist and interact with each other.

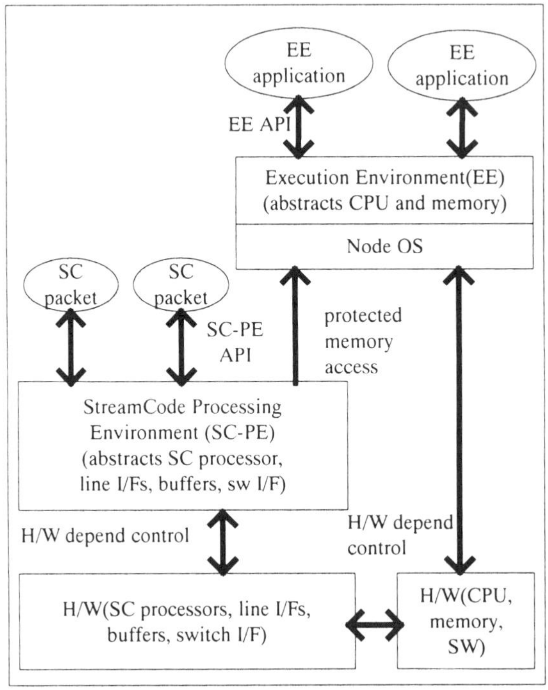

Figure 2: Functional Architecture of StreamCode Node

Taking into account the above each consideration, we have chosen answer c, because of its maximum flexibility and challenging goal. In order to shorten active L3 module selection delay to appropriate level such as sub-microsecond order, we have to decide how the implemented code of active layer-3 module will be delivered to each router. There are three options: 1) manual downloading by applications, 2) automatic fetching from code servers, 3) attaching code to every packet. Option 1 has a scalability problem because manual downloading implies that node has to manage code memory in a hard state manner. Option 2 introduces inadmissible delay before starting execution, if there were the unlimited number of code modules. Option 3 has a problem on code size overhead, but has no delay to start execution. We decided to take option 3 as a result that we regard the code execution speed as principal, although we should persevere with traffic inefficiency caused by each code fragment. We note that decreasing transmission cost and longer MTU of layer-3 packet will help to ease this disadvantage, and to make fruitful active network, millions kind of code fragments should be used over the network.

4. STREAMCODE BASED ACTIVE NODE ARCHITECTURE

Based on the previous discussion, we propose an active L3 system, named 'StreamCode' based active node. A StreamCode program implements an active L3 module, which is included in every packet and is executed by dedicated hardware (StreamCode Processor) in a non-blocking fashion. Due to this processing principle, service specific calculation only implemented by long-lived algorithms (ex. specialized routing table management) cannot be maintained by StreamCode. To support more flexible active L3 functionality, only one execution environment (i.e. StreamCode Processor) is not enough for an active node, and another execution environment for the above long-lived algorithm supporting active L3 functionality is required.

Taking into account the above requirements, our StreamCode based active node is organized as illustrated in Figure 2. In this figure, bottom two boxes represent hardware resources. The left box represents hardware of each line interface including layer 1 and layer 2 interface, StreamCode Processor, packet buffer, and interface to packet switch. Right box represents a central hardware including node's main CPU, main memory and packet switch. The packet switch is passive backplane on which packets are exchanged between line interfaces. Each part of hardware provides software execution environment for active network programs. StreamCode Processor controls layer 1 and 2 line interface, buffers, and switch interface to provide StreamCode Processing Environment (SC-PE). SC-PE is an abstract environment for StreamCode program. Right side of the figure uses general active networking terms[1], and the concept is the same as other active networking systems, except that the memory access from SC-PE must be managed by node OS, as shown at the center of the figure. Concerning the interaction among two execution environment, we provide a mechanism to the StreamCode Processor's MMU to view a specific part of node processor's main memory. Although a straightforward definition of these interface may be (extendable) function call through the node OS to the specific application running on the node OS, we did not adopt these interface because of calling delay and uncertainness of its completion of execution in sub-millisecond. These memory areas have to be managed by node OS as an interface window to layer-3 code processor, and the access right should be well managed by the node OS and each MMU. These shared memory may be eventually allowed to write from the StreamCode Processor, but that situation will require complex distributed shared memory management system, so we choose to provide read only access, at this time. Therefore, there are no direct interface from the StreamCode Processor to node OS, but the StreamCode can send information to the node OS by making and tossing a small packet to the node OS as usual packet transfer. A detailed mechanism for such protected access is discussed in the next section.

Table 1: StreamCode Instructions

Instruction	*Assembly code example*	*Meanings*
MOV1, MOV2, MOV4,..	MOV4 r(1) => r(2) MOV2 r(100) => m(r(2),r(4))	load and store primitives
ADD, SUB,..	ADD r(1), r(2) => r(3)	arithmetic/logical calculation
SKIP	SKIP EQ, r(1), r(2) -> label	conditional/unconditional jump
FIN	FIN	finish execution

Table 2: Co-processor Related Instructions

Instruction	*Assembly code example*	*Meanings*
FUNC	FUNC OUT_SINGLE, r(1), r(2), r(4),... => r(9)	invoke co-processor's function
WAIT	WAIT r(9)	check co-processor's status
FIN	FIN r(9)	wait co-processor, then finish

The rest of this section, we focus on the design of StreamCode. Our system is mainly committed for hardware execution, so code should be executed by hardware processor directly. This leads that a code set of the processor must be standardized, and should be carefully designed. Generality of a code set, compactness of written codes, execution speed, and the complexity of processor must be balanced. Our first design follows RISC type instruction set which should provide proven performance. One typical feature of StreamCode is that we have introduced extensible instructions to invoke co-processor's function asynchronously. Routing table search function, and data path (inside the processor) control function are major functions implemented as co-processor. There is another way to define primitives by presuming target applications, then divide sequence to reusable functional components, like traditional intelligent network approach. We did not deploy this approach, because this approach limits future development of unpredictable, forthcoming applications. Primitive instructions are shown in table 1, and co-processor related instructions are listed in table 2. Actual bitmap assignments for each instruction code are still under optimizing phase, and not shown here.

Another feature of StreamCode specification is that we did not provide automatic code caching, nor external code fetching (i.e. from outside of node). If code caching is required by a StreamCode application programmer, then the programmer may explicitly write StreamCode program to load pre-stored StreamCode program fragment from on-node memory, on his/her own responsibility for loading delay and for existence of the program fragment on every StreamCode-capable nodes. This design eliminates code finding/fetching mechanisms and its delay from StreamCode Processor's performance-critical part.

In current networking environment, application programmer can make applications behave contents sensitive, but network does not provide enough information and control to these applications. By introducing proposed StreamCode based active nodes, everything will change. All packet forwarding is under control of StreamCode inside each packet, and application programmer can change these in a StreamCode packet-by-packet basis. Furthermore, by using ordinary EE and interaction mechanisms between EE and SC-PE, more efficient and flexible networking environment can be achieved.

5. STREAMCODE PROCESSOR ARCHITECTURE

According to the StreamCode specification, we have designed a StreamCode Processor. Basically, primitive instructions are simple enough to be executed at one clock due to pipeline techniques. However there are some time consuming algorithms in a layer-3 functionality, like searching routing table, transferring packet contents to output interface, and calculating checksum, hash value, or CRC. To deal with this, co-processor is utilized and co-processor related instructions have been introduced. Based on these instructions, StreamCode Processor can invoke co-processor by specifying known co-processor identification number, with formulated number of arguments. Argument list is specified for each co-processor function, and invoking instruction has generalized argument passing mechanism for generality and future extendibility. Co-processor works in parallel with main processor, and returns a result value to the specified (as argument) register asynchronously. Main processor also has instructions like checking co-processor is still running or finished, blocking further execution and wait for co-processor to finish. Some of co-processor functions have to be standardized, and others are free to use. Standard ones include Content Addressable Memory (CAM) type table search, IPv4 and IPv6 types of routing table lookups, packet contents transfer engine. Non-standard co-processor may be implemented as node specific ASIC, or as configurable device like FPGA, and their existence can be located on network wide directory system as specialized resource functions. These co-processors may include media-specific transcoding processor, media-specific compression/decompression, and cryptographic engine.

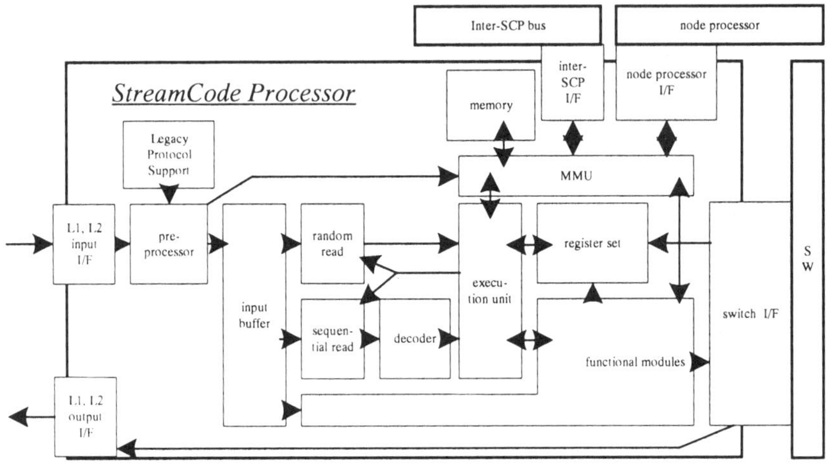

Figure 3: Line Interface Module with StreamCode Processor

Figure 3 shows functional modules of a line interface module, including StreamCode Processor, layer 1 and layer 2 input and output interfaces (L1, L2 input I/F and L1, L2 output I/F), switch interface, node processor interface and interconnection interface between StreamCode Processors on the same StreamCode node (inter-SCP I/F). L1, L2 input I/F analyzes input signal and assembles layer-3 packet according to physical/link layer protocols. The figure also includes packet switch (SW), node processor, and inter-SCP bus, which may be outside to the line interface module.

StreamCode Processor is constructed from functional modules described below:

- Input Buffer: The packet contents including StreamCode program are stored. This buffer has a valid pointer inside, and maintains how much contents are already stored from a L1/L2 input interface. If a sequential read function, a random read function, or one of functional modules try to access beyond that valid pointer, the access will be delayed until target region of the buffer will be filled by actual packet contents or StreamCode program. This blocking function will be reset if an arriving packet is shorter than expected, and at that time, whole execution is of course abandoned. As this figure shows, input buffer has one write interface and multiple read interfaces. These multiple access should be simultaneously allowed to avoid a pipeline stall situation. Size of input buffer follows expecting largest MTU size used in the network, and may be as 16K bytes to 32K bytes. Regarding to the current microprocessor's implementation, these size of cache can catch up processor's core clock without any delay, thus, this input buffer can be thought as the first level cache.

- Sequential Read Function: This reads StreamCode program from input buffer sequentially, and feeds instructions to a decoder. This can be thought as the internal (i.e. inside StreamCode Processor) instruction fetch unit. If branch has occurred, branch target should be set by execution unit, and this function will read the next instruction from that point. Many general processors have special functions with the instruction fetch unit to implement a speculative execution, expecting not making cache miss which make the total execution be slow. StreamCode Processor does not have so many disadvantages even if branch has occurred unexpectedly. Because all instructions are already on input buffer, or, sequential read function has no ways for waiting these pieces of StreamCode program if requested instructions are still on the wire and cannot be fetched.
- Decoder: This reads the output of the sequential read function, and decodes it as StreamCode instructions'. Finally feeds this analyzed information to the execution unit.
- Random Read Function: This is driven by the execution unit to read random place of input buffer. An access method is the same as the sequential read function except that reading target memory can be randomly specified.
- Register Set: Group of registers. Before execution unit starts processing an incoming new StreamCode program, all registers are reset to zero. In StreamCode Processor, the memory access is relatively slow because of its protection mechanism, and the amount of memory is limited for temporal use of each StreamCode program execution. To help StreamCode program run faster, we provide 256 registers, at an architecture level. Each register has currently 32 bit width, and serves free read and write operations to execution unit. Additionally, each register has an extra synchronizing support bit outside 32 bits, which reflect whether co-processor has finished the job or not. This bit is set as "not finished state", when the register is designated as storing a space of co-processor execution. Accessing to a register which has a synchronizing bit set as not finished, will cause the read access to block, and after the completion of co-processor's execution, the read operation continues.
- MMU: Every access to memory requires two parameters; one for specifying the memory space identification number, and another for offset inside that memory space. Memory spaces can be obtained through MMU requested by the node OS and that request may be invoked by EE application. When the node OS allocates memory area, it generates an authentication key value and sets it to MMU, and returns that information to the requesting EE application. Later, the EE application sends the key information to an end user's application, and the application sets the key information to StreamCode programs to use that memory space. The

actual memory space identification number is a StreamCode Processor dependent value, and not usable on other StreamCode Processors. It is an unrealistic idea to carry every memory identification number in a StreamCode packet; we make one level indirect mapping, or translation table, of the memory space identification number. At first, network wide application generates the pseudo number which may be identical to that one time, then use EE applications to allocate memory spaces with the generated number as a tag number. Each node OS on network nodes will allocate requested memory, and store a key-value pair to MMU, the tag number as a key, and the physical memory area identification number as a value. After the allocation, StreamCode program can access memory area by showing the tag number, and then get the memory identification number, and StreamCode program finally asks MMU to open access to that memory area. The authentication key value can be the same value all over the network, if EE application is written to do so. Another implementation may store the authentication key of the next node, inside the locked memory area. In that case, StreamCode program will be written as follows: open the lock, read key for the next node, replace packet contents to record that key, then go to the next node.

- Memory: This is a passive storage, managed by MMU. It can be implemented as uniform memory spool, but we have designed memory to have hint information, when it is allocated. We suppose hint can take value like 'temporal small buffers, disposable,' 'packet buffers to remember whole packet,' and so on. StreamCode Processor can be implemented to utilize these hint values in order to allocate disposable region on memory with 'clear all' functions, and to allocate packet buffers on memory with sequential read/write interface.
- Execution Unit: This controls entire execution of the processor according to a decoded StreamCode program sequence. This unit contains Arithmetic and Logical Unit (ALU) which performs calculation between registers, and short size data move (currently, 1 to 4 octets at once) between registers, from memory to register, or from register to memory. If an instruction requires access to memory, this unit checks the operand's memory space index number, next, if it is zero (it means that input packet should be read), this unit will trigger random read function with the specified offset. If memory space index is not zero, then the execution unit will access to MMU. This execution unit also controls invocation of co-processor, shown as functional modules. If some exception (ex. memory access violation, illegal instruction, etc.) is noticed by this unit, then total execution for the packet is simply abandoned. Implementing exception-handling mechanism is our future work.
- Functional Modules: These are functional logics of co-processors. These provide time consuming or complex functionalities to StreamCode

programs. A 'FUNC' instruction can be used to invoke each co-processor, and this instruction supports variable length argument list. All co-processors' argument information are registered at the execution unit, and when FUNC instruction is executed, their arguments are checked at run time.

- Pre-Processor: This checks layer 1 and layer 2 information of each incoming packet, and expects pre-configured routine jobs. Their jobs include invoking legacy protocol support function, reset processor after the end of packet detection, and fill input buffer by incoming packet contents.
- Legacy Protocol Support Function: This is invoked by a pre-processor to support legacy, i.e. not StreamCode based, layer-3 protocols. Actually, this module is a dictionary of StreamCode programs, indexed by legacy layer-3 protocols, and the stored StreamCode programs can be used to emulate each legacy layer-3 protocol. Pre-processor inserts this emulation program to each legacy packet, and the packet is executed as the StreamCode based packet.

One of the most important features of StreamCode Processor is a resource protection mechanism that is mainly achieved by MMU. We will discuss the resource protection with its related functionality in the rest of this section.

It is not allowed for any active program to eat up processor's resource, run out of network resource, nor access to other service's information. StreamCode may come from unknown end user's application; strict protection mechanism should be implemented in the processor. Code verification techniques are often proposed in active network research area, but these methods are suitable for the limited number of code sender, implying exchanging and storing trust information which introduce delay and storage scalability problems with our on-the-fly StreamCode execution system. We adopt rather hardware-oriented approach, again. StreamCode executions are forcibly terminated just after the final bit of the packet, which contains the executed StreamCode, is ready to read at the input buffer of the processor. StreamCode has to deal with this limit, or packet will be lost. Registers are always reset before new packet arrival, and memory can be accessed only through memory management unit (MMU), which requires some authorization key to open access rights. Each co-processor invocation is also managed by hardware, and has a default value for a number of invocations. To break these limits, authorization key is requested also. Furthermore, arguments to co-processor function, like routing table identification with table search function, output interface identification with packet output function, are also restricted by default. Some of them have free access, and others not. For example, default IPv4 routing table and default IPv6 routing table may allow free access to all StreamCode programs, but other IPv4 routing tables specific to the VPN service, will request the authorization key to search that table. As a result, none of vicious StreamCode can break active L3 environment at a packet-by-packet execution level. At a network level, vicious StreamCode still have a method for denial of service attack, by looping around the network only using freely available resources. To deny that attack, disable free access for packet output co-processor function, or believe other network nodes that they honestly decreasing hop-count like resource usage statistics recorded on each packet.

6. PROTOTYPE SYSTEM

In this section, our proof-of-concept prototype system is discussed. The prototype includes software-implemented StreamCode interpreter, hardware-implemented StreamCode Processor built upon Field Programmable Gate Array (FPGA), and StreamCode over UDP/IP network, and sample applications running on this environment.

6.1 Software based StreamCode Processor

Before developing a hardware version of StreamCode Processor, we had built a software version of the processor to verify our concept of networking architecture. The software version additionally has a debugging functionality for StreamCode application programs. The software processor handles StreamCode over UDP/IP, by using specified UDP/IP connections as unreliable layer 2 links.

6.2 Hardware based StreamCode Processor

As our project aims to achieve hardware based active layer-3 functionality, hardware implementation feasibility must be demonstrated. We have developed FPGA based StreamCode Processor to show that. Brief implementation notes are described bellow.

Current implementation is at its very first step, and we intend to examine its usability, stability, and execution bottleneck of implemented logic. Due to its FPGA based implementation, we can easily change its hardware logic, to tune up our implementation, to test several ideas, and to improve our architecture.

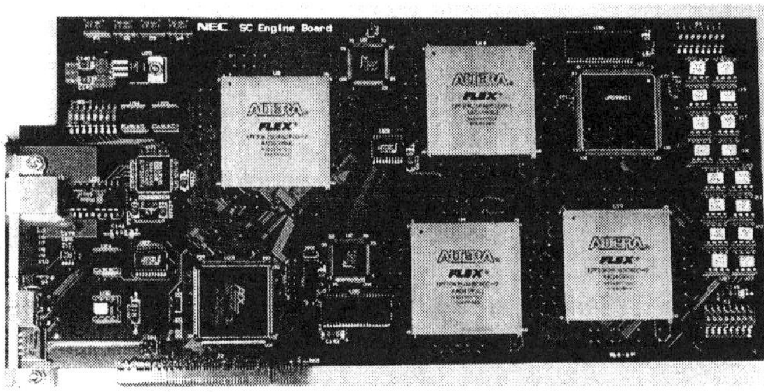

Figure 4: Prototype StreamCode Processor Card

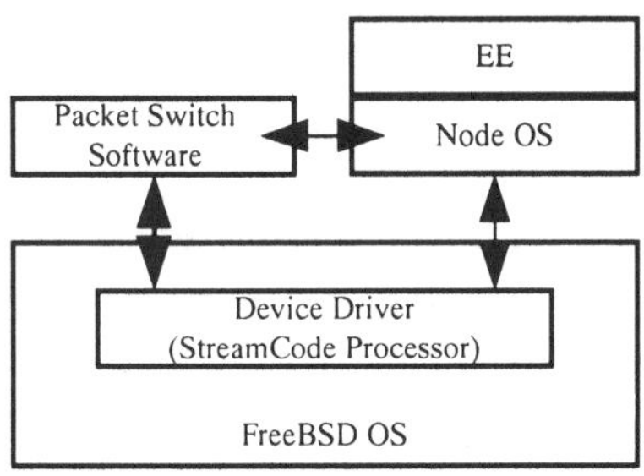

Figure 5: Supporting Software modules

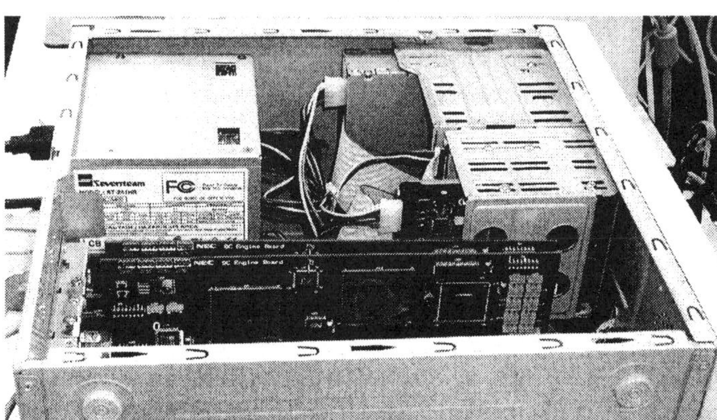

Figure 6: Mounted Processor Card

- Processor Card: We have implemented our processor as PCI card, and one PCI card works as one StreamCode Processor equipped with line interface. Figure 4 shows the appearance of the card, and Figure 5 describes software components on FreeBSD supporting these cards.

Multiple cards can be mounted to a usual personal computer as shown in Figure 6. An application software running on EE can control each card and packet switch software through the node OS. An incoming StreamCode based packet is processed by hardware StreamCode Processor, then, as a result of executing StreamCode program, the packet (may have some modifications by its StreamCode execution) is transferred to software packet switch with desired the destination interface identify number. The software packet switch will send each packet to the specified interface.

- Processor Core: Basic execution units of processor core are implemented on FPGA, including input buffer control unit (fill it, read it sequentially or randomly), decode unit, execute unit, co-processor interface, and registers. Input buffer is implemented by dual-port SRAM chip, and memory unit that can be accessed from execution unit, is implemented by SRAM chip. First implementation had dropped legacy protocol support unit, MMU, and pipelined execution.
- Co-Processors: We have implemented one co-processor, multipurpose data copy function. This function unit is implemented on FPGA, and can be used as FUNC OUT_SINGLE instruction. This unit can access input buffer, memory, register, and main memory of personal computer. Main memory is accessed to transfer packet content to/from the software packet switch. We prepared a space to mount our routing table search chip [12], which provides the longest prefix match search function in less than 100 nanoseconds (10 million search/sec).
- Outer Interface: There are two outer interfaces in this card, 100Mbit/sec Ethernet interface and PCI interface. Layer 1 handling of Ethernet packet is done by a dedicated chip, and layer 2 function is implemented on FPGA, which is now configured to handle StreamCode over UDP/IP. Hardware based processor and software based processor use the same packet format, so they are interoperable and can treat the same StreamCode packet. PCI interface is also implemented as a dedicated PCI bridge chip and some control functions on FPGA logic. DMA can be used between FPGA logic and the main memory of a personal computer, through the PCI interface chip.

6.3 Sample Applications

We have developed two sample StreamCode applications to check processors and the concept of StreamCode based active networking architecture.

First application is simple IPv4 like packet forwarding application. The code was written as to search destination address by table search function, then decrement hop count and forward entire packet to the output interface according to the search result. StreamCode object for this application occupies 106 octets of each packet.

Second application is multi-QoS multicast application. For this application, a multicast user table is prepared on memory area of each StreamCode Processor. A server machine sends multiple MPEG streams simultaneously, each stream contains the same contents but with different quality (one packet contains no more than one quality). All streams are packetized with StreamCode object and traveling through network belonging to the prepared multicast table. The table also contains maximum quality levels for each destination interface, which is calculated by end user's and network administrator's request separately. On each node, if a stream has higher quality than requested, or if the destination interface is congested to satisfy the quality, this stream will be dropped. On a node just before the receiving terminal, only one stream with the highest quality, among available streams that have a right to the final destination, is selected and forwarded to the terminal. The server machine attaches a different StreamCode object to each packet, according to a target picture's quality, importance, and contents meaning. The difference of each StreamCode object will play different behavior of each packet, like how congestion level is seriously evaluated for the packet/picture, which (user's or network administrator's) request is precedent to others, etc. All this functionality is written in less than 400 octets StreamCode object.

7. SUMMARY

We proposed architecture and its first implementation of hardware based active layer-3 networking system, which consists of object code specification of StreamCode, StreamCode Processor, and active network nodes. Our implementation is still in progress, however, examined applications validate our system's ability to execute complex application using StreamCode Processors. While our FPGA based StreamCode Processor currently runs at only 16MHz clock with its inefficient implementation, it can read packets with short benchmarking StreamCode object at 90Mbits/sec speed, and handles complex multi quality multicast application at 10Mbits/sec speed.

We are currently redesigning StreamCode instruction code to shrink the size of object code and remove bottleneck of execution, and improving FPGA implementation for completing its full functions. As an interim result of our efforts, now the processor core's performance for code fetching and execution is estimated at 22.5MIPS / 650Mbps (average score) with 27MHz processor-clock (FPGA-simulator's estimate). This result shows that for 100Mbps link, even in hardest situation like packets fully-filled with StreamCode program are arriving at full link speed, the processor can execute StreamCode programs in each packets 6 times (as iteration) within the packet's lifetime. Next, we have to develop programmable packet-by-packet buffer management mechanisms, i.e. QoS control system suitable for our platform, where execution effectiveness, functional flexibility and network-wide stability must be balanced. At the same time, we should refine protocol stack / middleware on terminals, to provide fully activated layer-3 network functions to end user's applications. Application programmer's commitment to active network is a key to promise a future networking environment.

REFERENCE:

[1] Calvert, K., "Architectural Framework for Active Networks Version 1.0," http://www.dcs.uky.edu/~calvert/arch-1-0.ps , July 1999

[2] Yemini, Y., et al., "Towards Programmable Networks," in Proc. NOMS 96, October 1996.

[3] Tennenhouse, D., et al., "A Survey of Active Network Research", IEEE Comm. Mag., January 1997.

[4] Wetherall, D., et al., "ANTS: A Toolkit for Building and Dynamically Deploying Network Protocols," in Proc. IEEE OPENARCH'98, April 1998.

[5] Alexander, S. et al., "The SwitchWare Active Network Architecture," IEEE Network Mag., pp. 29-36, May/June 1998.

[6] Braden, B., Ed., et. al., "Resource Reservation Protocol (RSVP) - Version 1 Functional Specification", RFC 2205, September 1997.

[7] Deering, S., Ed., et al., "Internet Protocol, Version 6 (IPv6) Specification", RFC 2460, December 1998.

[8] Blake, S., Ed., et al., "An Architecture for Differentiated Services", RFC2475, December 1998.

[9] Campbell, A. T., et al., "The Genesis Kernel: A Virtual Network Operating System for Spawning Network Architectures," in Proc. IEEE OPENARCH'99, March 1999.

[10] Nygren, E L., et al., "PAN: A High-Performance Active Network Node Supporting Multiple Mobile Code Systems," in Proc. IEEE OPENARCH'99, March 1999.

[11] Decasper, D., et al.,"A Scalable, High Performance Active Network Node," In IEEE Network, January/February 1999

[12] Kobayashi, M., et al, "A Longest Prefix Match Search Engine for Multi-gigabit IP Processing," IEEE ICC 2000, June 2000

[13] EZchip Technologies, "Network Processor Designs for Next-Generation networking Equipment," http://www.ezchip.com/, Dec 1999.

[14] Hadzic, I., et al., "On-the-fly Programmable Hardware for Networks," in Proc. IEEE Globecom 98, November 1998

Active Network Management via Agent Technology

Yasin Kaplankiran, Alexander Keiblinger, Hermann Többen
Institute for Systems Analysis and IDP, TU-Berlin/DAI-Lab
Franklinstraße 28/29, 10587 Berlin, Germany
`{kaplan,keiblin,toebben}@cs.tu-berlin.de`

Key words: Intelligent Agents, VPNs, active networks, IP-tunneling, marketplaces

Abstract: Active networks represent a significant step in the evolution of packet-switched networks, from traditional packet-forwarding engines to more general functionality supporting dynamic control and modification of network behavior. This paper describes a basic framework for an active network management system based upon agent technology, which can provide different views on a network, each related to a specific problem domain. Exemplarily the domain of VPN provisioning will be presented by a VPN management service covering all kind of aspects needed to provide VPNs for a wide quantity and variety of customers.

1. INTRODUCTION

Telecommunication providers change from voice carriers to suppliers of differentiated services. They need to leverage their existing network infrastructure to manage it more efficiently and enhance it for service provisioning of today's and future services. Active Networking is a promising new technology, delivering a distributed infrastructure for service provisioning. Active Networks promise to fulfil both requirements, the efficient and flexible management of network elements and networks as well as the fast introduction and provision of new services.

This paper describes an agent based Active Networking architecture. In addition we use the JIAC (Java based Intelligent Agent Componentware)

agent platform, which has been developed at the DAI-Labor. JIAC is an open and scalable agent architecture. Intelligent agents are pieces of software which act autonomously on behalf of a user or another agent. Agents have the ability to find and filter information, negotiate for services, easily automate complex tasks or collaborate with other software agents to solve complex problems. Mobile agents can even migrate between network nodes to accomplish a task nearby the necessary resources.

We use the setup and administration of VPN's respective their virtual links as a scenario for agent based active networking in this paper due to the overlap of VPN management needs and agent technology promises.

While the demand for VPNs grows rapidly, the supply lacks. VPNs are still very hard to set up and to maintain. To be able to satisfy the future demand, it will be necessary to greatly reduce the effort for VPN provision demand and customizable network services. The need for detailed control down to the single network element has led to the concept of Active Networks. This trend is further supported by ever falling costs for computing power, enabling the provision of mightful computing resources on network elements [3].

Active Networks remove the distinction between the network and connected computing devices by enabling the dynamic execution of program code on the network elements themselves. With the network becoming programmable, users can dynamically customize the network resources to their needs. Hence the dynamic programing of network nodes enables network as well as service providers to integrate new services into their networks easily and quickly and to manage their network efficiently [1].

2. ACTIVE NETWORKING BASED ON AOT

Agent technology is a new approach of designing software. Agents are active software components, which can either reside at specific locations in the network (stationary agents) or travel between network nodes (mobile agents). Agents have the ability to find and filter information, negotiate for services, automate complex tasks and collaborate with other software agents to solve even more complex problems.

The agent oriented technology (AOT) is well suited for network management [2]. It can be effectively used to decentralize network management activities [5]. In particular the characteristic mobility, co-operation and autonomy of the agents emphasize the advantages of AOT in distributed management applications. AOT is an approach to fulfill the prerequisites of active nodes and mobile program code in active networking.

3. JIAC - JAVA BASED AGENT ARCHITECURE

An agent based solution confirming to the structure of JIAC consists of **Agents** and **marketplaces** as the two elementary components. According to their ability to migrate agents can be further divided into two basic flavors: **stationary** and **mobile agents**. Marketplaces as the basic agent platform provide the environment in which stationary agents reside, offering their services to other stationary or mobile agents visiting the marketplace. JIAC marketplaces need a JAVA execution environment, namely a Java Virtual Machine (JVM). Every marketplace contains a **manager agent**, being responsible for administering the marketplace and providing basic administrative services to the other agents on the marketplace.

The **stationary agents** reside on a certain marketplace, which means they are constantly assigned to that marketplace. The following types of agents may be considered stationary agents: the marketplace manager agents, service provider agents and content provider agents. **Mobile agents** have the capability to travel from one marketplace to other marketplaces in an effort to find a suitable service provider agent. Mobility is one of the features supported by the use of marketplaces. It enables an agent to travel across the network to a marketplace where the resources it intends to use are locally accessible, thus reducing network load and communication latency (see also [4]). Given the ability of the active network components as described above, we envision the following management architecture (c.f. 1).

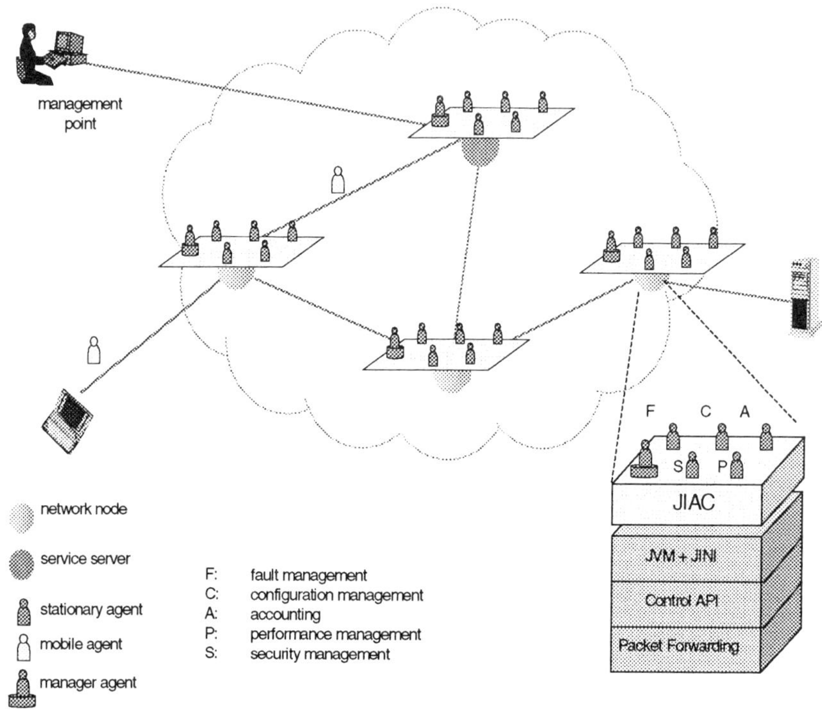

Fig. 1 Agent-based network management

On every active network node there will be established an agent platform. The agent platform consists of a marketplace (the yellow box labeled JIAC) with a manager agent and some stationary agents, providing standard functionality and mobile agents as well. Naturally the agents can communicate with each other via the network itself. Migration as a process is embedded with the communication. The stationary agents provide the basic network management functions known as FCAPS in the OSI-Management area. The brokering between agents' services and legacy systems can be provided via a JINI environment [4].

3.1 Capabilities of JIAC-Agents

The JIAC agent architecture provides the designer of agent systems with a broad range of distinct capabilities. On one hand agents are able to communicate and to cooperate with each other on the base of speech-acts and interaction protocols. Within their activities JIAC agents make up their decisions upon knowledge, which is represented in an adequate language called JIAC Agent Description Language (JADL). Moreover the agents are equipped with sophisticated planing capabilities. The latter gives benefit towards the configuration problems concerning the setup of connections within a heterogeneous network.

4. VPN MANAGEMENT SERVICE

In this chapter we will present Virtual Private Networking (VPNs) as a possible „killer application“ for agent based Active Networking. To achieve this goal, we will present an agent based design for the dynamic creation as well as the management of VPN's.

We understand a VPN as a private network, connecting remote locations using public networking infrastructure like the Internet. Using packet tunneling abstracts from the underlying network infrastructure and guarantees data security with encryption mechanisms.

4.1 Motivation

We regarded VPN provisioning from a global point of view to identify the main aspects. Regarding not only the operational phase of VPNs, but the complete process from planning to deployment, including maintenance and expansion of the VPN.

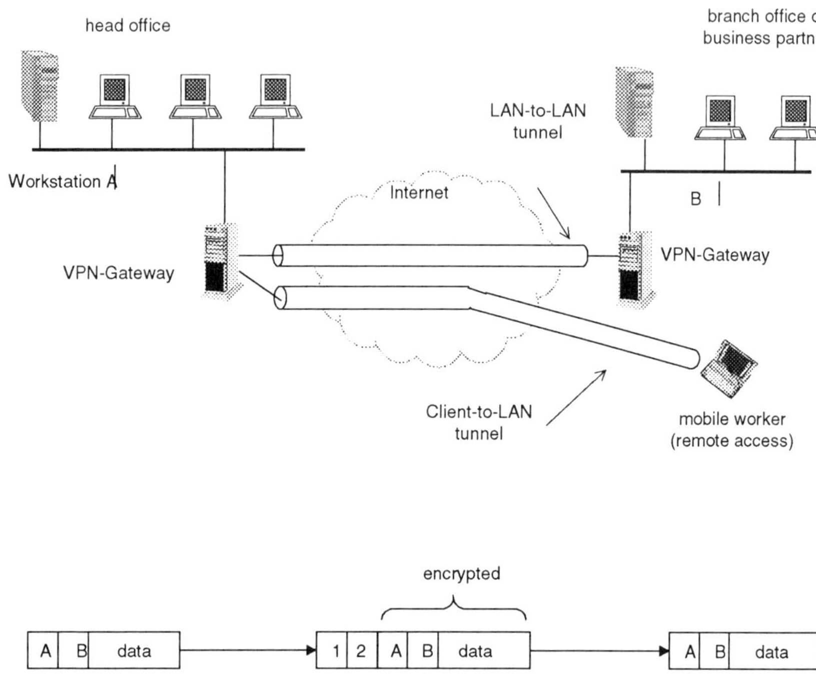

Fig. 2: Example of a VPN scenario

Figure 2 shows a usual VPN scenario. There is a virtual corporate network connecting a headquarter with branches and business partners using secure VPN lines. In addition, there is a remote workforce connected to the Intranet business data of the company using remote access software also using VPN technology.

4.2 The business model

Life-cycle of a corporate VPN installation. Business model characterizes the main four phases of this process: the specification (Agents S) of the customers demand reflected by the Service Level Agreements (SLAs), the configuration of the involved network-components (Agents C), the accounting and billing (Agents A), the testing and finally the operational use (Agent F, c.f. 3).

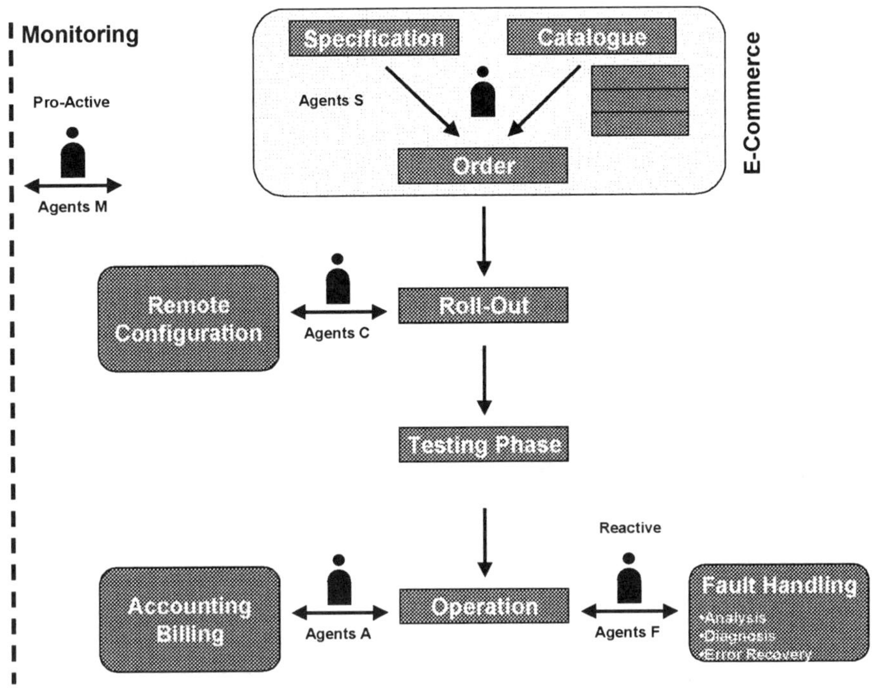

Fig. 3: VC-Market

Due to the above defined phases we have identified five areas where agents could be applied within this process:

– Specification Agents (area S):

Agents are used to help in the specification of a new VPN. This is done by taking customer requirements and existing networking hardware into account and mapping those onto predefined VPN templates of the provider. This process is iterative and relies expert knowledge from a database. Human experts are consulted if necessary. As an example you could imagine an agent based e-commerce environment, using a web front end as an automated customer point of contact and a call-center staffed with experts for customer advice in the background.

– Configuration Agents (area C):

Agents will be used in the roll-out phase, when newly installed VPN hardware must be configured to the specific needs of the customer. Remote configuration agents will take care of configuration issues, taking the burden off the shoulders of the installing technicians or customer staff. When the VPN has passed the test phase and is in production use, the next two areas in which we would employ agent technology are:

- Fault-Management Agents (area F):
  Agents of this kind would be triggered when a problem occurs. They analyze the problem, diagnose the cause and attempt to solve it. These agents would be equipped with components to cope with many of the standard problems. They contact specialized agents or as a last option human network managers for unknown problems and unusual error conditions, which they are unable to resolve.

- Accounting and billing agents (area A):
  Agents are responsible for accounting and billing. They collect accounting information in their decentral location and map the accounting data on billing items. In addition to the standard accounting and billing components, providers can add their own agents to facilitate special accounting schemes and billing rules.

- Monitoring agents (area M):
  In parallel to all these process steps, monitor agents will check for conformance to the specification, errors or performance problems. The monitor agents provide data which will be used by other agents or they spawn the fault management agents described above on detection of errors.

5. CONCLUSIONS

Agent oriented Technology seems to be well suited for network management. It can be effectively used to decentralize network management activities. In this paper the described an agent based framework for active networking. Within this approach every active network node will be equiped with an agent marketplace encompassing stationary agents which provide basic OSI-management functionalities and a manager agent for administrating the marketplace itself. As a possible „killer application“ for agent based Active Networking we described the dynamic configuration of an VPN. Telcos will benefit directly from the results of the research and the possible deployment of the developed technology in their networks. Agent based network management can help to reduce their network management costs, increase flexibility and offer new business opportunities by enabling new types of services.

6. REFERENCES

[1] Sahin Albayrak: "Intelligent Agents for Telecommunications Applications", IOS Press, June 1998

[2] Sahin Albayrak, Francisco J. Garijo (Eds.), „Intelligent Agents for Telecommunications Applications", Second International Workshop, IATA'98, Springer, July 1998

[3] David L. Tennenhouse, et.al.: "A Survey of Active Network Research", IEEE Communications Magazine, pp.80-85, Vol. 35, No. 1, January 1997

[4] Sun® Microsystems: "JiniTM Technology", http://www.sun.com/jini/

[5] R. Weihmayer, R. Brandau: „A Distributed AI Architecture for Customer Network Control" in „IEEE", pp. 656-662, 1990

Part Four

Dynamic Switching and Network Control

Smoothing Algorithms for DTM Networks

Csaba Antal (1), Sándor Molnár (2)
(1) Traffic Analysis and Network Performance Laboratory Ericsson Telecommunications Ltd.
e-mail:csaba.antal@eth.ericsson.se
(2) High Speed Networks Laboratory, Dept. of Telecommunications and Telematics, Budapest University of Technology and Economics
e-mail:molnar@ttt-atm.ttt.bme.hu

Abstract: Previous studies of Dynamic Synchronous Transfer Mode (DTM), which is based on multi-access dual fiber bus, showed that it could be a viable technology in the future. This paper points out that the characteristics of active and passive nodes are not balanced and the performance of the network can be increased by decreasing the need for slot allocation during call set-up. We propose a new type of smoothing algorithms, which is a new channel allocation algorithm that improves the fairness of DTM network with proper priority settings and the performance of channel allocation in ordinary operating conditions. Results are based on simulation, which was carried out by a self-developed simulator.

Key words: distributed channel allocation, fast circuit switching

1. INTRODUCTION

Dynamic Synchronous Transfer Mode (DTM) [5,10] is an attempt to build the next generation of local and metropolitan area networking technologies on fast circuit switching basis. It is a new broadband network architecture developed at the Royal Institute of Technology in Stockholm (KTH). The technology is in the focus of several Swedish companies, because it is the competitor to technologies like Asynchronous Transfer Mode (ATM) in special application areas.

ATM transfer capabilities define service classes, which are appropriate for several application types. However, ATM is not the most natural solution for applications with low delay and delay variance requirements due to its inherent cell (packet) switching characteristics. Applications with high bandwidth and low delay variance needs (e.g.: video on demand, video telephony) are more suited to the philosophy of circuit switched networks.

On the other hand ATM is a connection-oriented protocol. Therefore, interoperation with IP networks needs the emulation of the operation of connectionless networks. Interoperation with IP also limits the usage of the quality of service features of ATM.

DTM inherently supports real-time guarantees, which are crucial for services like voice or streaming video, and it was designed with IP technology in mind. Therefore it enables the efficient integration of voice and data traffic in the same network. DTM is therefore a candidate technology serving as a fast backbone for IP networks. DTM may also utilise wavelength division multiplexing (WDM). Therefore, it is one of the candidate technologies that may allow IP over WDM to evolve [10].

As DTM is a fast circuit switching technology, DTM connections are used only for the duration of data burst. If there is an idle period, resources are released. When data transmission starts again a new DTM connection is established. That is, user connections are split to several successive DTM connections.

The most important performance characteristics of the network *are connection set-up time* and *blocking probability*. If different bursts have considerably different set-up times and bursts are blocked within the call then there is less QoS guarantee for the whole connection. That is, the main benefits of circuit switching (like low and deterministic delay during the connection) are lost. Consequently, optimizing the mentioned characteristics is advantageous for burst-switching as well. This work is aimed at improving the performance and fairness of DTM networks via improving set-up time and blocking probability.

DTM is not a well-known technology, therefore before we present our results we shortly describe its main characteristics and the most important results of its performance evaluation.

DTM is designed for unidirectional medium with multiple access. The total medium capacity is shared by all connected nodes using time division multiplexing. Previous proposals and implementations [5,6] are based on dual-bus topology.

The total capacity of the bus is divided into cycles of 125 microseconds, which are further divided into 64 bit long slots. The sequence of slots at the same position in successive cycles is called DTM channel or simply slot.

There are two types of slots (and so DTM channels): data and static ones. Data slots are used for data transfer. For each DTM channel there is a token, which is assigned to one of the nodes. Both free and used data channels are assigned to nodes. Each channel has exactly one owner at a time. If a node has the right to use a channel (i.e. it has the token), then it has full control on it: it can setup a connection on it, send data within the connection, release a connection using the channel, or give the channel ownership to another node.

At system start-up, data channels (tokens) are allocated by the nodes but they can be transferred dynamically during operation. *Nodes can ask others for free data channels if they do not have enough to serve a new request.* To facilitate channel reallocation, nodes maintain status tables that contain the number of free channels at other nodes. This procedure is called *set-up time channel allocation or set-up time slot allocation.*

The other type of slot, called static slot, is used for broadcast control channels between the nodes. Nodes send control information in their static slots and listen to all the other static channels to receive control information.

In the basic set-up time channel allocation algorithm of DTM (KTH algorithm) [5,6], the request or of nodes during channel allocation is based on the closest first algorithm. We proposed two other request orders in [1,3,4] and carried out a performance evaluation study focusing on the fairness of the three variants, which are:

- Closest First : KTH-CF algorithm [6]
  It operates according to the closest first rule. That is, the node that received a connection request when having too few free slots sends the first slot request to the closest node out of nodes with free slots according to the status table. If the some of free slots at these two nodes is not enough for the pending request, the second closest node with free slots is

also asked for slots. If further slots are needed the closest non-requested nodes are contacted.

- Logical Ring : KTH-LR algorithm
  Nodes are ordered into a logical ring [2]. The order of channel requests is based on the location in the logical ring instead of the bus. That is, closest node is the first ring neighbour, second closest is the second ring neighbour, etc.
- Random : KTH-RA algorithm [3]
  Nodes choose the next node randomly to ask for channels out of nodes with free slots according to the local status table.

In all of the three above algorithms the number slot requests that a node can send out during a call set-up (retry limit) can be limited. The request order and the retry limit will be denoted in the name of the algorithms in the followings. For example, KTH-RA algorithm with retry limit of 10 is denoted by KTH-RA-10.

We have analysed the algorithms in two extreme conditions for short bus-length in [3]:

- Model 1 – status tables are *up-to-date* and nodes only try to get free channels from nodes with free channels
- Model 2 – status tables are *useless* and nodes try to get free channels from others without any apriori knowledge

Algorithms with up-to-date status tables (i.e. operating according to model 1) are denoted as KTH-S while the ones with useless status tables (i.e. operating according to model 2) are simply denoted by KTH (e.g. model 1: KTH-S-CF, model 2: KTH-CF).

Based on the study of the performance of set-up time slot allocation algorithms we showed in [2] that the addition of a new smoothing algorithm, which is called Background Channel Allocation algorithm (BCA), improved the performance of algorithms without status tables. In this paper we add another point of view for the motivation of smoothing algorithms, which is the lack of the balance between the performance of very active and relatively passive nodes. We also propose a new algorithm, which is called Set-up time Smoothing Algorithm (SSA), to improve the BCA algorithm. We show by simulation that the SSA algorithm is able to improve the fairness of the DTM network and it is also able to improve the performance of the DTM protocol. The analysis of the new SSA algorithm was carried out by self-developed simulation software.

The structure of the paper is as follows.

First, we present the motivation of smoothing algorithms. Then in Section 2, we describe the BCA algorithm and we also propose the new SSA algorithm. Simulation results can be found in Section 4. Finally in Section 5, we conclude the paper.

2. MOTIVATION

2.1 Asymmetry

In set-up time slot allocation algorithms, nodes initiating calls more often than the others have better performance characteristics. This effect is due to the cache-like behaviour of the KTH channel allocation protocol of DTM. That is, active nodes can reuse the free channels that are produced by local terminated connections. The effect can be observed at both algorithm types (i.e. with and without status table). In network where all - so-called client - nodes communicate to a dedicated - so-called server - node in the middle of the bus, the server node has lower blocking probability and shorter set-up time than clients have. KTH protocol is not efficient due to the asymmetry of the directions in a bi-directional client-server connection. The same effect can be interpreted as unfairness.

Table 1 illustrates the above property of KTH algorithm. Instead of displaying blocking probability or set-up time, the average number of slot request retries - needed for a successful connection establishment - are shown in the table. *Table 1* shows the average number of retries for KTH-RA algorithm (without status table) separately for the client-to-server, server-to-client directions and the average for all calls. The averages are calculated for different network loads. *Table 2* shows the same values for KTH-S-RA algorithm (with status table).

Table 1. Asymmetry of server-to-client and client-to-server directions, KTH-RA-10 without status table

Avr. # of slot req. retries vs. offered load	50%	70%	90%	110%	130%
Client-to-server	0.416	1.145	2.474	3.541	4.072
Server-to-client	0.009	0.033	0.121	0.26	0.389
All unidirectional	0.213	0.588	1.255	1.696	1.842

Table 2. Asymmetry of server-to client and client-to-server directions, KTH-RA-10 with status table

# of slot req. retries vs. offered load	50%	70%	90%	110%	130%
Client-to-server	0.193	0.438	0.73	0.905	0.969
Server-to-client	0.004	0.013	0.06	0.187	0.29
All unidirectional	0.099	0.226	0.395	0.547	0.629

It can be seen that *due to status table* the average number of retries is less than 1 also at 130% offered load. *Without status table* an average client node needs to request slots from 4 other nodes to collect the slots for a unidirectional connection to the server in average.

The asymmetry is also reflected in these numbers. E.g. at 50% offered load, 46 times more slot requests are needed to establish a client-to-server connection than for a server-to-client connection at both algorithms. Though this ratio decreases at higher offered load, it can not be neglected.

As this phenomenon may be the cause of unfairness, an algorithm is needed that makes balance between the characteristics of very active and less active nodes.

2.2 Too many slot requests

In addition to asymmetry a possible *improvement opportunity of algorithms without status table* can be seen in *Table 1* and *Table 2*. The number of slot requests needed to set-up a client-to-server connection is very high even at moderate loads. It is 1.145 at 70% load, which means that an average node asks slots for a successful connection more than once, which is a high value. Note that connections can be set up without slot request if the source node has free slots. The probability mass function of the same random variable (number of retries needed for a successful connection) can be seen for client-to-server connections in *Figure 1* and *Figure 2*.

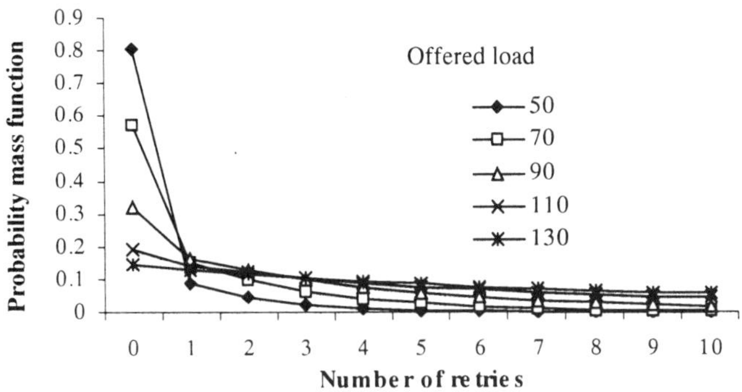

Figure 1. Probability mass function of number of slot request retries during a successful connection set-up, KTH-RA (without status table)

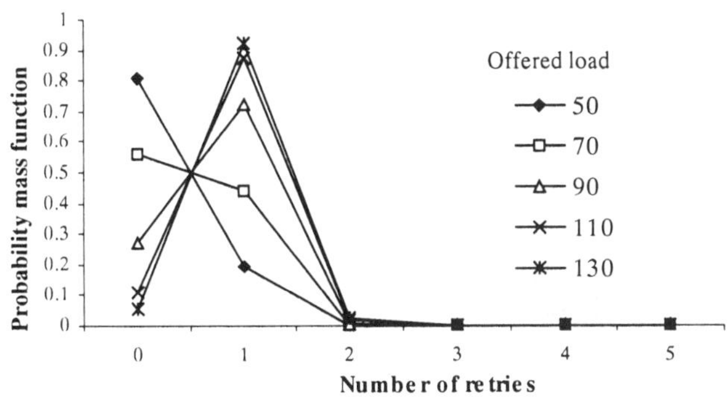

Figure 2. Probability mass function of number of slot request retries during a successful connection set-up, KTH-S-RA (with status table)

It can be seen that KTH-S-RA algorithm manages to collect the slot without slot request or with one slot request, as the probability of 2 or higher retries is very low. At the algorithm without status table, if the offered load is as low as 70% the probability that slots are needed from other node is 100-55=45%.

3. DESCRIPTION OF SMOOTHING ALGORITHMS

The goal of smoothing algorithms is to decrease (or eliminate) the need for slot allocation during call set-up and to balance the characteristics of active and passive nodes. To reach the goal smoothing algorithms are trying to distribute free slots amongst nodes according to their loads. We proposed BCA algorithm in [2] to decrease the need for slot allocation, i.e. to improve performance. Here, we propose a new and improved version of this algorithm: the Set-up-time Smoothing Algorithm. This section describes both algorithms

3.1 Background Channel Allocation Algorithm

Background Channel Allocation algorithm (**BCA algorithm**) [2], transfers slots between nodes in the background, independently of set-up requests coming from hosts. It can work parallel with any set-up-time algorithms. It is based on logical ring architecture.

In the algorithm nodes regularly exchange free channels with their direct neighbours along the logical ring.

The goal of the exchange *in the case of homogeneous network load* is to distribute free channels *evenly* amongst nodes. In order to achieve this goal, nodes check regularly if there is any difference between the number of local and neighbouring nodes' free channels. This process provides that neighbouring nodes have nearly the same number of free channels at any time instant, thus free channels are always distributed almost evenly amongst nodes.

This idea can be extended to a real algorithm, which considers the case of normal operation when the load is different at each node. A priority value for buses in both directions reflect the difference between nodes. That is, each node has a priority number for each bus, which depends on the traffic load sent to the given bus. Priorities can be constant or can change dynamically when adapting to the actual load of the network.

Exchange of free channels depends on the number of free channels and the value of priorities. Node i initiates slot allocation with its ring neighbour - node $i+1$ - if expression

$$\begin{aligned}&|(\text{free channels of node } i)(\text{priority of node } i+1)-\\&(\text{free channels of node } i+1)(\text{priority of node } i)| \qquad (1)\end{aligned}$$

can be decreased by slot allocation. The amount of slots to be transferred is determined so as to minimise (1) and considering that only free channels can be transferred. Node i asks slots from node $i+1$ if the first term of expression (1) is below the value of the second term and it transfers slots if the first term is the higher one. That is, in the case of equal priorities, node i transfers one channel to node $i+1$ if its number of free channels is higher by 2 than the ones of node $i+1$.

Node i calculates expression (1) whenever a local connection was set up or released (number of local free channels changed).

If the priority of a node is equal to zero then it is left out from the ring. The next successive node is the exchange partner instead. For example if the priority of node $i+1$ is 0 for the corresponding bus then node $i+2$ is the partner of node i for the allocation of free slots on that bus.

BCA algorithm is based on the comparison of the amount of local and neighbouring free channels. This is why it requires a very small status table where nodes keep a record of free slots of direct neighbouring nodes on the ring. Nodes send administration messages to the first upstream neighbouring node along the ring after each change in the number of local free channels in order to provide information for maintaining up-to-date tables.

Priority defined above does not effect directly the amount of bandwidth available for a node. It is rather related to the possibility of setting up a channel without slot reallocation, independently of the bandwidth used. This definition of priority can be used for optimising the network utilisation and channel set-up times.

Priorities can be dynamic and static as well. In the case of dynamic priorities, a traffic estimation procedure modifies the priority of the node. Estimators use parameters of previous connections (e.g.: amount of required bandwidth and interarrival times) to calculate the current priority. If the characteristics of the traffic are known effective estimators can be constructed. However, estimators can be built without preliminary information about the traffic. Dynamic priorities are not used in this examination, as offered load during a simulation run was static.

The other solution is to assign static priorities to nodes where priorities are changed at management level. In this case the basis of priority assignment can be the role of the node in the network or the price paid by the customer of the node.

3.2 Set-up-time Smoothing Algorithm

Set-up-time Smoothing Algorithm (SSA) is our new and improved version of BCA. In BCA, slot exchange is performed always between the same nodes: only direct neighbours along a logical ring transfer slots between each other in the background. The advantage of this operation is that nodes only have to store status information about their ring neighbours. However, it has drawbacks too in real implementations. In certain cases, distribution of free channels may differ significantly from the priority distribution. If – for example – there are a few nodes with very low activity between nodes that have many free slots and nodes that have high priority, BCA does not transport free slots to high priority nodes.

In SSA, free slots can be exchanged between the parties of connections during connection set-up and release procedure. The rules of slot exchange are the same as it is in BCA: nodes transfer free slots if expression (1) can be decreased.

Exchange partners of SSA are always different, and we assume that SSA needs to operate without status tables. Therefore, nodes have to add additional information into DTM Control Protocol (DCP) Announce and Attach messages.

The SSA procedure during connection set-up is described below:

1. The sender node sends the number of its free slots in the DCP Announce message.
2. The receiver compares the number of its free slots and that of the sender node.
3. a, If the receiver node should send slots according to (1), it includes the number of slots to be transferred into the DCP Attach message, and initiates a slot transfer procedure.
   b, If the sender node should send slots, the receiver puts the number of slots it asks from the receiver into the DCP Attach message. When the sender received the DCP Attach message, it initiates a slot transfer procedure with the receiver.

Note that definition of sender and receiver node is based on the data transmission roles (not on the transmission of control information). As connections are uni-directional at DCP level, one of the parties of the connection is always sender, and the others are receivers.

Operation of SSA algorithm can be described in the same way for connection release procedure.

The main differences between BCA and SSA are summarized in *Table 3*.

Table 3. Differences between BCA and SSA algorithms

	BCA	SSA
Who are slot request partners?	neighboring nodes along logical ring	parties of connections
When do slot requests occur?	any time	during connection set-up and release
How to send information about the number of free slots?	in separate messages	in modified connection set-up and release procedure

4. SIMULATION RESULTS

We proposed BCA and SSA algorithms to balance performance characteristics of active and passive nodes and to improve performance of the network.

The primary goal of this subsection is to examine whether the above design goals of BCA and SSA algorithms are fulfilled or not. The effectiveness dependency on burstiness of traffic and bus-length is also investigated.

We used two types of traffic generators during the simulations. The first one is based on WWW traffic sources [7,8]. The inter-arrival time of connections is based on Weibull distribution. The holding time of a request is modelled by Pareto distribution. The second traffic generator model is based on an aggregated voice model. Traffic generators are modelled with Poisson arrival processes and exponential holding times. The bandwidth of connections is deterministic in both models: 1 slot in each cycle.

We use the so-called "**client-server**" network configuration [1,2] in the paper, where each client node initiates bi-directional connections to the special "server" node, which is located in the middle of the dual-bus. We assume that all traffic generators have the same traffic parameters, and the number of generators connected to client nodes is the same. In the simulations to be presented, there is no client-to-client traffic.

In this paper, we examine the operation at two extreme bus-lengths. The so-called short dual-bus is 1 km long with 100 nodes. That is, the inter-node distance is 10m. At the long dual-bus the inter-node distance is 10 km, so the full bus-length is 1000 km.

The basis of our simulation work is the *steady-state mean of call blocking probability and connection set-up time.*

Connection set-up time consists of two parts:

- round-trip delay of the set-up message
- delay coming from slot requests

The first part depends on the location of the other party of the connection. This factor is proportional to the distance of client nodes from the dedicated server node. As differences caused by the physical distance of the other party of the connection are usually respected by customers, the delay of set-up message and its acknowledgement is subtracted from the connection set-up time when the bus is long. The resulted new characteristic is referred to as *average slot request time*.

4.1 Short bus, WWW traffic

First, the short bus and WWW traffic case is examined. Both smoothing algorithms perform best when using together with set-up-time slot allocation algorithms. As BCA uses logical ring during background allocation, it is applied together with KTH-LR. SSA is used with KTH-RA algorithm.

4.1.1 Performance

Performance of smoothing algorithms is simulated with the following configuration:

- Retry limit = 10
- client-server load profile
- priority settings : nodes with any activity have 1 as priority, idle nodes have 0 priority (separately for both directions)

Simulation results are displayed in *Figure 3* and *Figure 4* that can be used to define the application area of smoothing algorithms, to select the most appropriate offered load range and to study the effect of status tables.

Figure 3 and *Figure 4* show that smoothing does not improve the performance of algorithms with status table. Though a very small improvement can be noticed in set-up time at 50%, 70% and 90% load, at higher loads the performance of algorithms with status table degraded due to smoothing algorithms. The main performance gain of smoothing algorithms is that they could decrease the number of slot allocation retries needed for a successful connection. As *Figure 2* shows, the number of slot allocation retries is low when status tables are present, so there is no room for this kind of optimisation.

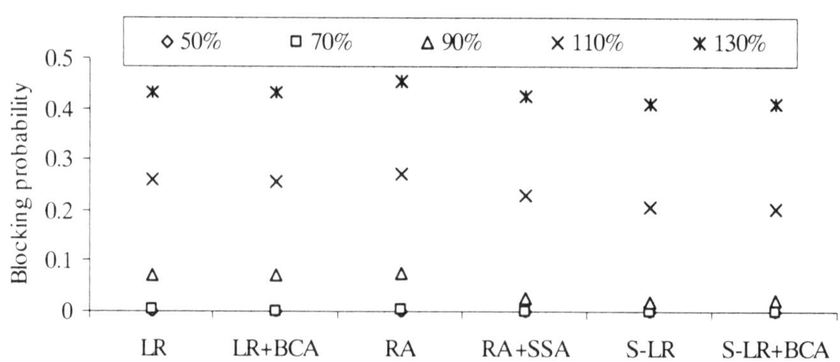

Figure 3. Blocking probability of smoothing algorithms

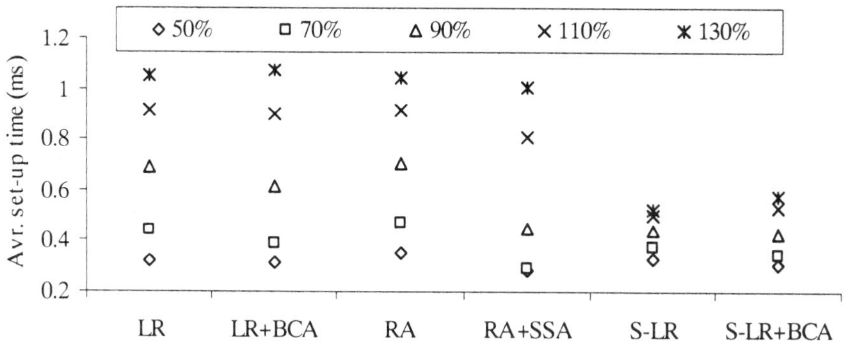

Figure 4. Average set-up time of smoothing algorithms

Improvement of set-up time of algorithms without status table is, however, significant if the network is not overloaded. Among KTH-LR, KTH-LR+BCA, KTH-RA and KTH-RA + SSA algorithms the last one is the best. SSA algorithm is most effective in 50%-100% offered load range. The highest improvement on set-up time is 0.25 ms at 90% load and KTH-RA + SSA algorithms. That is, SSA decreased average set-up time by 35 %. At 130% offered load, the gain of SSA algorithm is only 0.03 ms. Less effective ranges of smoothing algorithms can be explained with the followings:

Above 100% load, "there is nothing to smooth", i.e. there are very few free slots in the system.

Below 50% offered load, "there is nothing to optimise". i.e. there are many free slots in the system.

Blocking probability of the system was decreased due to SSA algorithm at each offered load, but its effect is small. The highest improvement is at 90% offered load, where blocking is lower with 0.05 (from 0.07 to 0.02).

The range between 50% and 100% offered load is the most important one for well-designed networks. If offered load is higher for longer time, the network is mis-dimensioned, and it is not able to provide efficient services. Lower offered load - for a long time - means that the network is over-dimensioned and the operator paid for unused bandwidth.

It is also interesting that at 50%, 70% and 90% offered loads KTH-RA with SSA have nearly the same performance as KTH-RA with status table.

4.1.2 Priority settings

In the previous section performance was examined with fixed priority settings. Now, we show how to find an optimal priority settings at client-server configuration taking into account the viewpoint of *symmetry and performance*.

Priorities of client nodes in this section are the same as they were before. Priority of server node, however, is varied between 50 and 0.05. The effect of priority settings on client--to-server connections, server-to-client connections and bi-directional connections can be seen in *Figure 5* and *Figure 6*. The first item is KTH-RA algorithm without smoothing at each offered load. At KTH-RA + SSA algorithm, the ratio of server priority and client priority is displayed after the name of SSA algorithm.

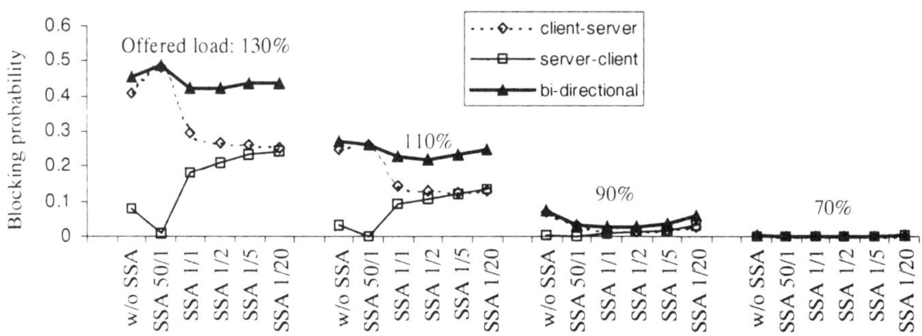

Figure 5. Effect of priority settings on blocking probability

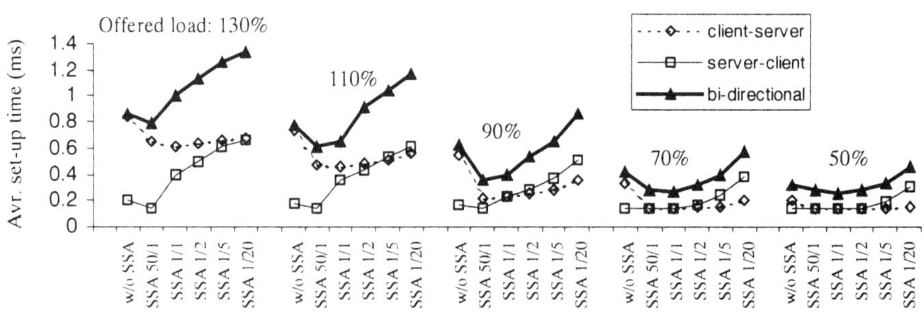

Figure 6. Effect of priority settings on average set-up time

First, let us take the *viewpoint of symmetry vs. priority ratios. Figure 5* shows that the lower the priority of server is the closer the blocking probabilities of client-server (uplink) and server-client (downlink) connections are. Blocking of uplink connections decreases and blocking of downlink connections increases when the priority of server node decreases.

Blocking probability of uplink connections, however, is always above that of downlink connections.

Average set-up time of uplink connections can exceed that of downlink connections at certain priority ratios. The cross-point of uplink and downlink curves depend on the offered load of the network. At 130% load downlink/uplink ratio is 1/20, at 90% load it is 1/1. Priority setting of server node mainly influences the set-up time of downlink connections. Set-up time of uplink connections does not decrease in case the server has lower priority.

Both figures show that priority is an effective tool to balance the characteristics of nodes with different load of connections.

The next question to answer is *how performance of bi-directional connections depends on priority ratio.* Based on *Figure 6*, optimal priority is the one, which strengthen cache effect, i.e. the higher the priority of server the lower average set-up time of bi-directional connections is. Priority of server does not effect so significantly the blocking probability curve, except that blocking increases if the server has too high priority.

4.2 Effect of long bus and less bursty Poisson sources

The previous subsection analysed SSA algorithm in detail. This section examines the effect of longer bus-length and less bursty traffic sources. Main configuration settings are the same as in the previous section:

- Retry limit = 10
- client-server load profile
- priority settings : nodes with any activity have 1 as priority, idle nodes have 0 priority (separately for both directions)

4.2.1 Long bus

Table 4 shows the performance of smoothing algorithms. The same conclusions can be drawn from this table as we obtained for short bus. Namely, smoothing does not improve significantly (rather degrade) the performance of slot allocation algorithms with status table. Set-up time of algorithms without status table, however, is decreased significantly in 50%-100% offered load range. SSA improved slightly blocking at almost every offered load. The largest improvement of set-up time is at 90% offered load: it is equal to 0.81 ms, which is 40% of the set-up time of KTH-RA algorithm. The largest improvement of blocking is 0.05 at 70% and 90% load.

Table 4. Performance of smoothing algorithms, long bus, bursty traffic

Load(%)	50	70	90	110	130	50	70	90	110	130
Algorithm	Blocking probability					Avr. slot request time				
KTH-LR	0	0.004	0.07	0.262	0.438	0.832	0.977	1.28	1.572	1.753
KTH-LR+BCA	0	0.007	0.075	0.252	0.441	0.822	0.885	1.194	1.54	1.764
KTH-RA	0	0.005	0.076	0.273	0.466	0.994	1.358	2.04	2.596	2.933
KTH-RA+SSA	0	0	0.025	0.227	0.427	0.786	0.808	1.23	2.325	2.89
KTH-S-LR	0	0	0.018	0.205	0.42	0.835	0.891	1.004	1.163	1.251
KTH-S-LR+BCA	0	0	0.019	0.203	0.423	0.81	0.842	0.99	1.196	1.315

4.2.2 Poisson traffic

Our last topic is the effect of burstiness on the effectiveness of smoothing algorithms. Poisson traffic is usually less demanding for the network, so the performance of set-up time algorithms improved compared to WWW traffic.

Table 5. Performance of smoothing algorithms, short bus, Poisson traffic

Load (%)	50	70	90	110	130	50	70	90	110	130
Alg.	Blocking probability					Avr. set-up time				
KTH-LR	0	0	0.005	0.202	0.419	0.266	0.274	0.388	0.813	0.999
KTH-LR+BCA	0	0	0.007	0.204	0.418	0.263	0.268	0.352	0.832	1.038
KTH-RA	0	0	0.006	0.219	0.451	0.263	0.272	0.398	0.745	0.878
KTH-RA+SSA	0	0	0.001	0.204	0.43	0.263	0.263	0.314	0.854	1.011
KTH-S-LR	0	0	0	0.186	0.413	0.265	0.271	0.319	0.438	0.463
KTH-S-LR+BCA	0	0	0	0.185	0.412	0.263	0.267	0.307	0.481	0.518

Simulation results in *Table 5* show, however, that SSA algorithm is more effective at bursty traffic than here. SSA decreased blocking of KTH-RA without status table at any load. It also decreased set-up time below 100% offered load. When, however, offered load is above 100% SSA and BCA increased set-up time, so it is worth to switch off smoothing when the system is overloaded. The performance of algorithms with status table is not improved with smoothing algorithms.

5. CONCLUSION

We have shown in this work that there is an asymmetry in set-up-time slot allocation algorithms of DTM. We have also shown that it can be corrected with smoothing algorithms using proper priority settings. As set-up-time algorithms provide better service for active nodes than for less active ones, their priority should be set to a smaller value to reach balance.

We have also shown the exact dependence of asymmetry on priority values. Simulation results show that - in the case of client-server network configuration – blocking probabilities of the down-link (server-client) and up-link (client-server) connection are equal if the priority of the server is 1 and that of the clients is 20. The priority ratio, where average set-up times of up-link and down-link connections are symmetrical, depends on the offered load in the network. E.g. at 130% offered load the symmetrical server/client priority ratio is 1/20; at 90% offered load it is 1/1.

Simulations also proved that adding smoothing algorithms to set-up time algorithms without status table improves the performance of the DTM dual-bus if offered load is between 50% and 100%.

It can be concluded that the priority that provides the smallest average set-up time for the bus is the one that strengthens cache effect. That is, the higher the priority of the server is the lower the average set-up times of bi-directional connections are. Priority of server does not effect the blocking probability curve significantly, except the case when it is too high.

It is also shown that smoothing algorithms improve the performance of the system more efficiently in the case of bursty sources than with Poisson traffic.

6. REFERENCES

1. Cs. Antal, J. Molnár, " Fairness of a DTM Dual-bus with Large Inter-node Distances", 8th International Conference on Telecommunication Systems, Modeling and Analysis; March 9-12, 2000, Nashville, Tennessee, USA, pp. 414-420
2. Cs. Antal, J. Molnár, S. Molnár, G. Szabó, "Performance Study of Distributed Channel Reallocation Techniques for a Fast Circuit Switched Network", Computer Communications Vol. 21 Num. 17, November 1998, Special Issue on the Stochastic Analysis and Optimal Management of Communication Systems, pp. 1597-1609
3. Cs. Antal, J. Molnár, S. Molnár, "Fairness Issues in a New Dual-Bus Networking Technology"; Proc. 7th IFIP/ICCC Conference on Information Networks, Aveiro, Portugal, June 1998, pp. 83-98
4. Cs. Antal, "Performance Analysis of the DTM Fast Circuit Switched Networking Technology", Ph.D. dissertation, Technical University of Budapest, Hungary, 1999

5. C. Bohm, M. Hidell, P. Landgren, L. Ramfelt, P. Sjödin, "Fast Circuit Switching for the Next Generation of High Performance Networks", IEEE J. on Selected Areas of Communications, Vol. 14., No. 2, February 1996, pp.298-305
6. C. Bohm, P. Lindgren, L. Ramfelt P. Sjödin, "Resource Reservation in DTM", 1st IEEE Symposium on Global Data Networking, Cairo, Dec. 1993, pp. 191-197
7. M. E. Crovella, A. Bestavros, "Self-Similarity in the World Wide Web Traffic: Evidence and Possible Causes", IEEE/ACM Transactions on Networking, 5(6): pp.835--846, December 1997
8. S. Deng, "Empirical model of WWW document arrivals at access link", IEEE International Conference on Communications, vol.3, ICC'96, 23-27 June 1996, Dallas, TX, USA pp. 1797-802
9. A. M. Law, W. D. Kelton, *"Simulation Modeling & Analysis"*, McGraw-Hill International Editions, 1991
10. P. Lindgren, "A Multi-Channel Network Architecture Based on Fast Circuit Switching", Ph.D. thesis, ISRN KTH/IT/R-96/08-SE, Stockholm, May 1996

EFFICIENCY OF THE PREDICTION OF HIGH PRIORITY TRAFFIC IN ENHANCING THE RATE BASED CONTROL OF LOW PRIORITY TRAFFIC

Kimmo Pulakka and Jarmo Harju

Tampere University of Technology. Telecommunications Laboratory. P.O. Box 553 FIN-33101 Tampere. FINLAND tel. +358-3-365 2111. fax +358-3-365 3808

e-mail: {pulakka, harju}@cs.tut.fi

Key words: Flow control, traffic prediction, packet switched networks, fuzzy logic

Abstract: In this paper, we study how effectively the rate based control of a low priority data transmission service in a packet switched backbone network can be implemented, when the control decisions are based on the predictions of the amount of high priority traffic. ANFIS (adaptive neuro-fuzzy inference systems) predictors are used for traffic prediction. In our service model, all control functions of the low priority data transmission service are distributed to the edge switches of the backbone network. The routes and data rates of the low priority data flows are iteratively controlled by the control system, according to the predicted data rate variations of the high priority data flows. The efficiency of the service model has been tested by simulations. The emphasis is on the effects of different traffic distributions of the high priority data flows and on the impact and importance of the amount of overhead caused by the traffic prediction and the transmission of control information.

1. INTRODUCTION

In the Internet, a packet switched backbone network of one operator transfers data between access networks with end-users' computers and backbone networks of other operators. Data transmission services of future backbone networks can be roughly classified to the prioritised, controllable and best effort transmission services. The edge switches of the backbone networks classify the incoming data flows into different service categories, and the core switches of the networks just transfer the data flows through the networks according to the service category classifications [1]. The classification is mainly based on the negotiated service level agreements of the data flows.

The three above-mentioned service categories have their own individual characteristics. The data flows of the prioritised services are served by giving privileges over the other traffic in the network. The prioritised services can also include guaranteed services. For such services, the network guarantees a certain quality of the transmission services for these traffic flows. The controllable and best effort traffic can use the free capacity of the network, left over by the prioritised data flows. The data flows of the controllable service categories are always somehow controlled by the network so that the data flow spesific or general control targets are reached. Traditionally, the best effort traffic is not controlled in any way in the network level.

In the transmission environment where the different data flows should be served differently and the RTT (Round Trip Time) of the control actions may be very long, some kind of traffic predictions have proved to be useful in order to reach the near-optimal utilisation of the capacity of the network [2]. From this point of view, the main target of the prediction is to try to reduce the bad effects of large RTT of the control actions. These predictions can be used for example in CAC (Call Admission Control), UPC (Usage Parameter Control), traffic shaping, routing and data rate control functions. A common prediction strategy is to predict the behaviour of certain type of traffic flows so that some other type of traffic could be controlled optimally.

Traffic prediction can be made by using several prediction methods. However, a common characteristic for almost all prediction methods is to use the time series for the prediction. The next value or values of the certain characteristic of the traffic flows are predicted using the measured past characteristics of the traffic flows. During the past few years, several studies have been published where the behaviour of the VBR (Variable Bit Rate) type of traffic has been predicted using LMS (Least Mean Square), RLS (Recursive Least Square) or fuzzy logic based methods (e.g. [3] [4] [5] [6]).

It is very difficult, if not impossible [7], to predict *real* data traffic in the network. As concluded in [7], the exact statistical characteristics of the predicted traffic should be known to produce good predictions. In real network environment, it is even impossible to collect the data fast enough to produce the exact statistical model of the predicted traffic. Hence, it is very important to think, how much computation power of the network components is used for different prediction tasks. In some situations, valuable predictions can be made using simple prediction algorithms.

However, in the huge backbone networks with numerous different transmission services highly automated management tools are needed to control the network at least nearly optimally [8]. To implement these automated management tools properly, prediction algorithms with effective computational intelligence will be needed.

In this paper, we describe our own data rate and multipath route selection protocol for the low priority controllable (LPC) traffic, based purely on the traffic load predictions of the prioritised traffic. The main idea is to develop a low priority data transmission service, which does not require complex control functions all around the network, but which still would ensure better transmission service than the standard best effort data transmission service. The free capacity of the network, left over by the data flows of the prioritised data transmission services, is predicted and shared between the LPC data flows. All prediction functions are located at the edge switches of the network, and the core switches only collect the data rate information for the prediction tasks of the edge switches. From the network operator's point of view, this kind of service would use the free capacity of the network more effectively than the best effort service. On the other hand, the network applications, which traditionally use the best effort transmission service, would get faster transmission service. However, the network would not guarantee anything and the LPC sources would not be forced to behave exactly according to the control of the backbone network.

The main question is, what kind of prediction system is effective enough to be implemented instead of traditional best effort service. We use several performance parameters to study how effectively the LPC data transmission service could be implemented. Especially, we study the effect of the frequency of the predictions and the control operations with various traffic models of the prioritised traffic. The adaptive neuro-fuzzy inference system (ANFIS) [9] has been used as the prediction engine. The reason for this selection was that ANFIS has been found to be effective in modelling the behaviour of highly non-linear dynamic systems [10].

The developed data transmission service model is tested by the simulations. In these simulations, the prioritised data traffic was generated by three different models: 1) *the first-order autoregressive process [11]*, 2) *the measured sample data rates of the H.261 video transmissions*, and 3) *the fractional gaussian noise self-similar process with strictly defined value of the Hurst parameter*. The simulated backbone network model does not follow any existing standardised network architecture. However, the modelled backbone network contains the typical elements of connection oriented high speed packet switched networks.

In Section 2 we explain the basic principles of our data transmission service, concentrating on the prediction operations of the high priority traffic. In Section 3 simulation results are given and analysed. Finally, some conclusions are drawn.

2. DESCRIPTION OF THE CONTROL SYSTEM

2.1. Basic principles

The studied backbone network model consists of edge and core switches. The access networks are connected to the backbone network via edge switches. The data packets coming from the source access networks flow across the source edge switches and travel through the core switches to the destination edge switches. The destination edge switches send the data packets to the destination access network (see Fig. 1). In this study, we expect that access networks are also some kind of packet based data networks, and it is possible to transfer special control information, related to our data transmission service, between the LPC source stations and edge switches of the backbone network. In the backbone network, data is transferred using constant length packets (cells). In the switches, every input and output link, connected to the switches, is handled by individual input and output ports. Output ports have a scheduler for the control traffic, prioritised traffic and LPC traffic. The control traffic has the highest priority and the LPC traffic has the lowest priority (see Fig. 2). Traditional best effort traffic is not included here, but if it were, its priority would be the lowest. In this study, we do not consider how the prioritised data flows are controlled. We just assume that their priority is always higher than the priority of the LPC data flows and their data rates are variable.

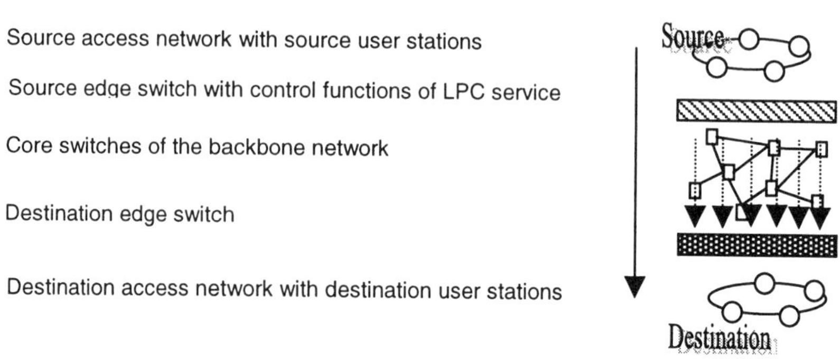

Figure 1. The data flow through the backbone network

At the beginning of every LPC data transmission, the LPC source stations use certain nominal data rates, and the data packets of these data flows are transferred through the network using certain nominal routes between the source and destination edge switches. After that, the control system of the LPC transmission service iteratively adjusts the routes and the data rates of the active LPC data flows based on the predicted loads of the prioritised data flows. After every control moment, the edge switches inform the LPC source stations about the predicted optimal data rates and the corresponding optimal routes by using a special control protocol. Then it is the responsibility of the data transmission applications of the source stations to adjust their data rates according to received control information. Between two control moments, the LPC source stations send data using the optimal data rates of previously selected optimal routes, and the edge switches take care that the incoming data packets of the LPC data flows are sent through the backbone network using the previously defined optimal routes.

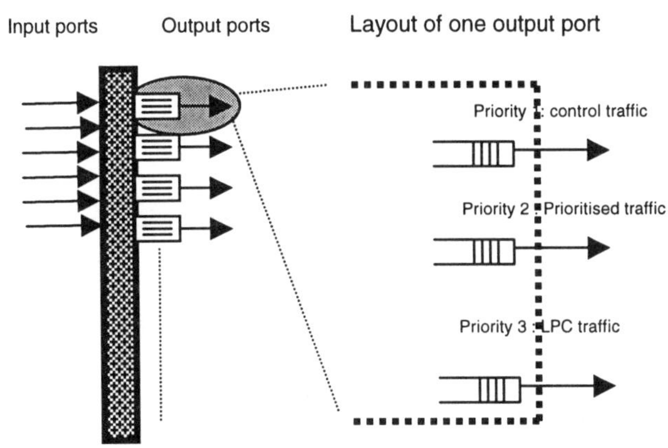

Figure 2. Layout of the core switches

The definitions of the optimal routes and optimal data rates of the LPC data flows are based on the predictions of the ANFIS predictors. For these predictions, the core switches of the backbone network measure the total data rate of the data flows of the prioritised transmission services in the output ports of the switches. To collect this information and to distribute it to all edge switches, the edge switches send periodically ERM (Edge Resource Management) packets to the network using a constant time interval t_s. These packets are flooded through the core switches to the other edge switches of the network, and the total data rate values of the prioritised traffic and the number of the active LPC flows in the most heavily loaded output ports of the switches in each route are copied to the ERM packet. After receiving any ERM packet, the edge switch copies the control information of the ERM packet to its special data base.

At a constant time interval t_c, the ANFIS predictor systems of the edge switches copy the latest control information from the special data base for the prediction operations. The ANFIS predictions are made in the constant time interval t_p ($t_p = n^*t_c$, $n \in Z_+$). In this way, the ANFIS predictors collect the set of the past data rate values of the prioritised data flows, calculate the values of the input variables of the ANFIS predictors using these values, and predict the *maximum* data rate values of the prioritised traffic flows in the near future. The prediction time is equal to t_p and the predictions are done individually for every route between the edge switches. At the end of the predictions, the optimal routes and data rates for the LPC sources are calculated, and the control operations described above are done. The control loop of the LPC data transmission service is described in Fig. 3.

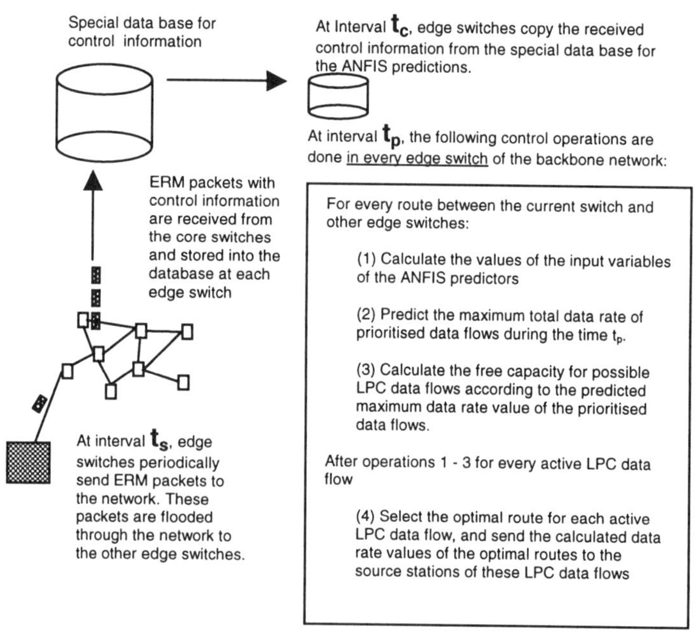

Figure 3. Control operations of the LPC data transmission service

It is important to notice that the values of the interval parameters t_s, t_c and t_p define how strictly the LPC data flows are controlled. If short intervals are used, the data rates of the prioritised data flows can be predicted more accurately, and the data rates of the LPC sources are controlled nearly optimally. The price to pay is the huge amount of the control operations of the network components. On the other hand, using long intervals reduces the computaional as well as control traffic overhead, but at the same time the accuracy of the control operations can be lost. We assume that the performance of the ANFIS predictors decrease when the prediction time increase. Finally, the operators of the backbone networks should decide how much resources of the network components is profitable to use to control the LPC data flows, and set the values of the interval parameters according to this decision. In Section 3, we will study how the values of the interval parameters affect the performance of the LPC data transmission service and how much network capacity is used in each case.

Before the ANFIS predictors can be used, their parameters must be set for the current prediction case. The parameters are always set in the off-line mode. At first, the statistically representative sample values of the data rates of the prioritised data flows are collected in the routes between the edge switches. After that, the input-output data pairs are constructed and the parameters of the ANFIS predictors are set by the special hybrid learning algorithm [9] of the ANFIS network. As a result, the ANFIS predictors can predict the data rates of the prioritised data flows in the near future. It is important to notice that the exactness of the ANFIS predictions depends strongly on the 'goodness' of the sample data rates used for learning. The performance of the ANFIS predictions decrease, if they should do the predictions for the cases, which are statistically far from the prediction cases used for the parameters setting processes. In Section 3, we will also study how exact predictions the ANFIS predictors calculate in the cases where statistical characteristics of the data rates differ from the statistical characteristics of the data rates used for the parameter setting process of the ANFIS predictor.

2.2. Description of the ANFIS predictors

The ANFIS network is an adaptive network-based fuzzy inference system, which can perform the functions of the Sugeno or Tsukamoto fuzzy models. The important characteristic of the network is the capability to learn the behaviour of the dynamic systems by training data pairs. The network includes a set of linear and non-linear parameters. The values of the parameters are adjusted by a special hybrid learning rule, which is a combination of the gradient method and the least squares estimate (LSE). The ANFIS network can include several input variables but only one output variable [9].

In our predictions case, three input variables have been chosen to describe the dynamic characteristics of the prioritised data flows. These three variables are 1) *the maximum of the data rate values of the prioritised data flows during the measurement time* t_p, 2) *the mean of the data rate values of the prioritised data flows during the measurement time* t_p, and 3) *the difference between the last two data rate measurements of the prioritised data flows during the measurement time* t_p. The output value of the ANFIS predictor is the maximum predicted total data rate of the prioritised data flows in the currently predicted route.

The configuration of the ANFIS network affects to the performance of the ANFIS predictions. The values of several factors used during the parameter setting process and the number of the membership functions of the input variables can be changed. For example, the increase of the number of the membership functions usually allows more exact prediction results. However, this also requires more sample data pairs and computation time to set the parameters of the ANFIS predictors properly. Finally, the configuration of the ANFIS predictors should be selected according to the characteristics of the predicted processes. The selections can be made by the experience of the system manager or by the special algorithms like CART (classification and regression tree) [13]. In section 3, we will compare the prediction results of two ANFIS predictors with different number of the membership functions of the input variables.

2.3. Selection of optimal routes and definition of optimal data rates

At the end of each control moment of the LPC data transmission service, the control system selects the optimal routes for the active LPC data flows and defines the allowed data rates for these flows. The route selections depend on 1) *the current and predicted loads of the prioritised and LPC data flows, and the current load of the control traffic in the routes*, 2) *the number of active LPC data flows in the routes*, and 3) *the randomised orders of the route selections of different LPC data flows.*

At first, the order in which the optimal routes between the LPC data flows are selected is randomised. After this, the optimal routes for LPC data flows are selected in the randomised order.

During every route selection, the route with the largest equally shared free capacity is defined as an optimal route and the current value of the equally shared free capacity is defined as an optimal data rate of the LPC data flow. After the selection, the equally shared free capacity of the selected route must be decreased to ensure that the next route selections of other LPC data flows are based on the right information.

$$\text{EQS capacity} = \frac{\text{Link capacity - (CT + Pred.PRIO + Curr.LPC + Pred.LPCC)}}{\text{\# of the active GC traffic flows}} \quad (1)$$

Where :

EQS capacity = Equally shared free capacity

CT = Current control traffic

Pred.PRIO = Predicted load of prioritised traffic

Curr.LPC = Load of current LPC traffic

Pred.LPCC = Predicted load changes of LPC traffic

3. SIMULATIONS

3.1. Description of the simulated network environment

The simulated network environment consisted of one backbone network and three access networks. The capacity of the transmission links between the switches of the backbone network was 35 Mbit/s, and the transmission capacity of the access networks was 100 Mbit/s. The high transmission capacities of the access networks ensure that no congestion can occur in the access networks in any case.

The backbone network was loaded by *five* continuously flowing prioritised traffic flows and *four* LPC data flows with randomly alternating idle and active periods. Poisson and exponential distributions were used to define the lengths of the active and idle periods of the LPC data flows. The feasible mean lengths of the active periods were 5, 10 and 20 seconds, and the mean lengths of the idle periods were 4, 8 and 12 seconds. Transmission rates of five prioritised data flows were generated by 1) *the first-order autoregressive process*,, 2) *measured sample data rates of the H.261 video transmissions*, or 3) *the fractional gaussian noise self-similar process*. The value of the Hurst parameter in the self-similar process was 0.9.

Data packets of all data flows were of equal length. The constant data packet size was 500 bytes. The optimal lengths of the LPC buffers and the distances between the switches of the network were also constant. The buffer size was 7000 packets and the distance between the switches was 500 km. The control system tried to share the free capacity between the LPC traffic flows so that the network would have been loaded optimally. Another control target was to prevent congestion in the network. The structure of the simulated network environment is described in Fig. 4.

The aim of the test series was to 1) measure the 'goodness' of the ANFIS predictor system compared to two other predictor systems, and 2) to study the effects of different values of the interval parameters t_s, t_c and t_p to the performance of the LPC data transmission service. The test series includes seven test cases. The test cases differ from each other by the used prediction methods and the data rate distributions of the prioritised traffic flows. Each test case includes four different test phases. The test phases differ from each others by the value sets of the interval parameters t_s, t_c and t_p. The simulation time of each test phase was 250 seconds.

The same value sets of the interval parameters and traffic scenarios of the LPC data flows were used for the test phases of the all test cases. Hence, it is possible to compare the performance results of different test cases. The characteristics of the simulation tests are summarised in Table 1.

In the tests, the distributions used to set the parameters of the ANFIS predictors were different from the distributions used to measure the performance of the ANFIS predictors. In this way, we tried to study what would be the realistic prediction capability of the ANFIS predictors. The data rate distributions of the prioritised data flows, measured in the output ports from the edge switch one to the core switches two, three and seven, are described in Figure 5. The data rate generation processes of the prioritised data flows are described in Table 1.

Table 1. Description of the tests

Issue	Characteristics
data-rate generation process 1	**Aim**: Generate data rates of prioritised data flows to measure the performance of the ANFIS predictors. **Data rate generator**: fractional gaussian noise self-similar process. The value of the Hurst parameter was 0.9. **Mean and std values, measured in the output ports between the edge switch 1 and core switches 2,3, and 7:** mean= 5.76, std = 2.67
data-rate generation process 2	**Aim**: Generate data rates of prioritised data flows to measure the performance of the ANFIS predictors. **Data rate generator**: First-order autoregressive process together with measured data rates of H.261 video transmissions. **Mean and std values, measured in the output ports between the edge switch 1 and core switches 2,3, and 7:** mean= 3.43, std = 1.55.
data-rate generation process 3	**Aim**: Generate data rates of prioritised data flows to set the parameters of the ANFIS predictors. **Data rate generator**: fractional gaussian noise self-similar process. The value of the Hurst parameter was 0.9. **Mean and std values, measured in the output**

Issue	Characteristics
	ports between the edge switch 1 and core switches 2,3, and 7: mean= 4.07, std = 3.20
Active periods of the LPC data transmissions	Variable in different test phases. Randomised using the Poisson distribution. Mean values 5,10 and 20 seconds.
Idle periods of the LPC data transmissions	Variable in different test phases. Randomised using exponential distribution. Mean values 4,8, and 12 seconds.
Packet size of the backbone network	500 bytes
Output buffer sizes of the LPC traffic	7000 packets
Distance between the switches of the backbone network	500 km
Test phases	Test phases are the same for every test case. 1: $(\mathbf{t_s}, \mathbf{t_c}, \mathbf{t_p})$ = (0.01, 0.01, 0.05) seconds 2: $(\mathbf{t_s}, \mathbf{t_c}, \mathbf{t_p})$ = (0.1, 0.1, 1) seconds 3: $(\mathbf{t_s}, \mathbf{t_c}, \mathbf{t_p})$ = (0.3, 0.3, 2) seconds 4: $(\mathbf{t_s}, \mathbf{t_c}, \mathbf{t_p})$ = (0.5, 0.5, 2) seconds Corresponding test phases in different test cases have the same traffic scenarios of the LPC traffic flows.
Test case 1	**Predictor engine:** ANFIS, # of membership functions = 3. **Data-rates of the prioritised data flows:** by data rate generation process 1. **Data rates used to set the parameters of ANFIS predictors:** by data rate generation process 3.
Test case 2	**Predictor engine:** ANFIS, # of membership functions = 6. **Data-rates of the prioritised data flows:** by data rate generation process 1. **Data rates used to set the parameters of ANFIS predictors:** by data rate generation process 3.
Test case 3	**Predictor engine:** Linear predictor **Data-rates of the prioritised data flows:** by data rate generation process 1.
Test case 4	**Predictor engine:** Current values **Data-rates of the prioritised data flows:** by data rate generation process 1.
Test case 5	**Predictor engine:** ANFIS, # of membership functions = 6. **Data-rates of the prioritised data flows:** by data rate generation process 2. **Data rates used to set the parameters of ANFIS predictors:** by data rate generation process 3.
Test case 6	**Predictor engine:** Linear predictor **Data-rates of the prioritised data flows:** by data rate generation process 2.
Test case 7	**Predictor engine:** Current values **Data-rates of the prioritised data flows:** by data rate generation process 2.

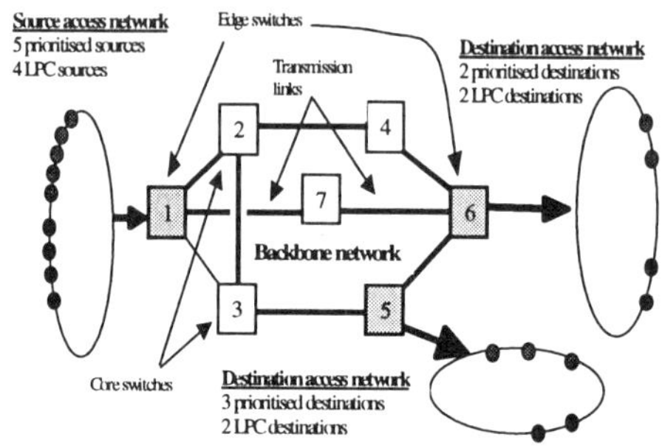

Figure 4. Simulated network environment

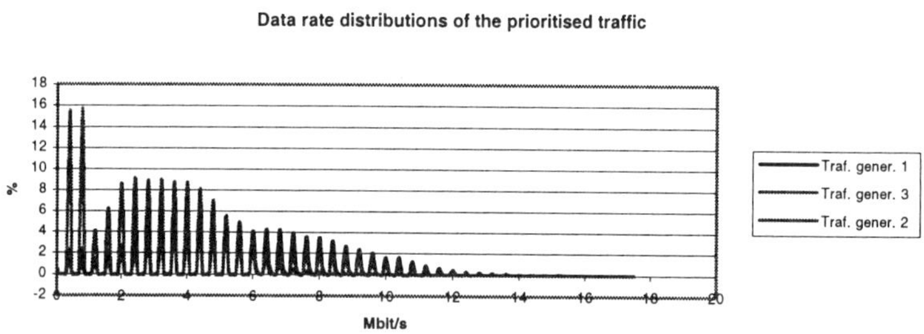

Figure 5. Alternative data rate distributions of the prioritised traffic flows

3.2. Performance variables

The performance of the LPC transmission service is tested by five performance variables. These variables are: 1*) the average occupancy of the LPC buffers in the three most heavily loaded output ports*, 2) *the standard deviation of the occupancy of these LPC buffers*, 3) *the packet loss ratio*, 4) *mean throughput*, and 5) *the average and the standard deviation of the prediction error of the data rate values of the prioritised data flows*. The values of the mean throughput were obtained by calculating, for each LPC data flow, the ratio of the achieved throughput to the theoretical throughput value given by the max-min fairness principle [12] under the assumption that 100% of the available capacity of the network was used. The prediction errors were measured between the predicted and measured data rates of the prioritised data flows.

The combination of the high throughput values and zero packet loss indicate the high performance of the system. The values of the average occupancy levels and standard deviations of the LPC buffers indicate how the buffers of the output ports are used. In our simulated network environment, we have measured the performance values for the output buffers from the switch one to the switches two, three and seven. These output ports should always be used heavily, because all incoming LPC traffic injected to the backbone network flows through the switch one. The single output port is used optimally, if the average value of the occupancy level of the LPC buffer is between zero and the optimal size of the LPC buffer, and the value of the standard deviation is low. In our control system, the low values of the standard deviation cannot be reached. The control system adaptively changes the routes of the LPC traffic flows, and this causes quite high variation of the occupancy levels in the output buffers of the switch one.

On the other hand, the small values of the interval parameters t_s, t_c and t_p require high overhead for the control of LPC data flows. In these tests, the prediction error values should be lower and throughput values should be higher than in the tests with large values of the interval parameters.

3.3 Description and analysis of the test results

At first, we will concentrate on the test case 1 (see Table 1). We will study how the different value sets of the interval parameters t_s, t_c and t_p, used in four separate test phases, affect to the performance results. Secondly, we will measure the performance differences of seven test cases. According to these test results, we will study how valuable the ANFIS predictors are to implement LPC data transmission service.

The test results of the test case 1 are summarised in Table 2. According to the test result, the mean throughput decreases when the values of the interval parameters increase. However, the difference between the throughput results of the best and worst tests is only 19 %, although the values of the interval parameters vary heavily. At same time, the difference between the prediction errors and the packet loss ratios of different tests is small. Obviously, the average and standard deviation values of the LPC buffers increase, when the values of the interval parameters increase. In these cases, the LPC sources are controlled slower and the buffer occupancy levels can vary heavily, before the control operations of the LPC data transmission service affect to the occupancies of the LPC buffers.

The performance results of seven test cases are summarised in Table 3. The values of the performance variables in Table 3 are the average performance results of four different test phases (see Table 1) of the test cases.

Table 2. Effects of the values of the interval parameters

Test phases (1-4) (t_s, t_c, t_p)	Mean throughput	Packet loss ratio	Prediction error (mean, std) [Mbit/s]	Occupancy of LPC buffers (mean, std) [Packets]
0.01, 0.01, 0.05	0.92	0.0009	1.61, 1.68	162, 795
0.1, 0.1, 1	0.78	0.0022	1.58, 1.63	482, 1258
0.3, 0.3, 2	0.77	0.0068	1.73, 1.74	511, 1418
0.5, 0.5, 2	0.75	0.0074	2.07, 2.00	569, 1470

According to the test results of test cases 1-4, we could conclude that the LPC data transmission service implemented by the well constructed ANFIS predictor produces better transmission service than the other alternatives. The performance results of the test case 1 are generally better than the performance results of any other test case. However, the throughput value in test case 2 is same as in test case 1, although the ANFIS predictor used in test case 2 includes more membership functions of the input variables than the ANFIS predictor used in test case 1 does. At the same time, the prediction error values of test case 1 are clearly better than in the error values of test case 2. This result indicates that more sample data pairs are needed to set the parameters of the ANFIS when the number of parameters in ANFIS increases. In these two cases, exactly same data pairs were used to set the parameters of the ANFIS predictors of both test cases. The ANFIS predictor in test case 2 would need more data pairs and calculation iterations of the hybrid leaning method to set the parameters of the ANFIS predictor to reach the optimal prediction results.

Although the performance results of test cases 3 and 4 are worse than the result of test case 1, the performance differences are small. It is important to notice that the performance results of test cases 3 and 4 are reached by clearly less computations than needed for ANFIS predictor, even if the ANFIS predictor would not include huge number of parameters.

In test cases 5-7, the data rates of the prioritised data flows were generated by the non self-similar process (Table 1, data rate generation process 2). In these test cases, the data rates should be easier to predict than the data rates used for test cases 1-4. The average and standard deviation values of the prediction errors in these cases are smaller than the error values in test cases 1-4. However, the performance differences between ANFIS prediction based data transmission service (test case 5) and the other services, based on simpler prediction methods, are smaller than those measured between test cases 1-4. The prediction error values of ANFIS predictor are even larger than the error values of simpler prediction methods. In this point of view, the value of the ANFIS predictors decreases compared to the simpler prediction methods when the data rates of the prioritised data flows are easier to predict.

In the case of the prediction of the maximum data rate values of the prioritised data flows, the optimal prediction results of the data rates do not ensure the optimal throughput results. In some load situations, the good throughput and packet loss results can be reached even if the prediction error values are large.

Table 3. The performance results of different test cases

Test cases (1-7)	Mean throughput	Packet loss ratio	Prediction error (mean, std) [Mbit/s]	Occupancy of LPC buffers (mean, std) [Packets]
(1) anf3, self-similar prior. traffic	0.81	0.0043	1.75, 1.76	431, 1235
(2) anf6, self-similar prio. traffic	0.81	0.0049	2.38, 2.24	441, 1227
(3) Linear predictor, self-similar prio. traffic	0.75	0.0058	2.14, 1.96	329, 961
(4) Current values, self-similar prio. traffic	0.74	0.0029	2.02, 1.93	262,831
(5) anf6, autoreg. and H.261 based prio. traffic	0.78	0.0064	1.77, 1.82	415, 1257
(6) Linear predictor, autoreg. and H.261 based prio. traffic	0.74	0.0061	1.38, 1.23	123, 631
(7) Current values, autoreg. and H.261 based prio. traffic	0.72	0.0038	1.25, 1.17	310, 943

4. CONCLUSIONS

In this paper, the low priority data transmission service is described. The transmission service is based on the iterative route selections and data rate adjustments of the sources. The control operations of the transmission service are based purely on the data rate predictions of the prioritised data flows. The predictions are done by the ANFIS predictors.

The performance of the transmission service is tested by the simulations. In these simulations, the performance of the transmission service is measured as a function of several parameters. The simulation results indicate that the transmission service can adapt to different situations without heavy variations of the performance of the service. It is also found that ANFIS predictors perform better than the tested simpler predictors do. The ANFIS predictors are better especially in those prediction cases where the data rates of the prioritised data flows are difficult to predict. However, the test results show that the maximum throughput difference between the ANFIS predictor and tested simpler predictor systems is only 8.6 %. At same time, the implementation of the ANFIS predictors is a much more complex task than the implementation of the simpler prediction methods. The ANFIS predictors must also be configured carefully, before they produce good performance results. However, ANFIS predictors require much less computations than the prediction methods based on some other computationally intelligence algorithms (e.g. Multi-Layer Perception networks) [13].

REFERENCES

[1] Rosenberg, S.; Aissaoui, M.; Galway, K.; Giroux, N. , Functionality at the edge: designing scalable multiservice ATM networks, IEEE Communications Magazine, pages: 88 - 90, 95-9, May 1998, Vol. 36

[2] Ahmet Sekercioglu, Y.; Pitsillides A., Intelligent control techniques for efficient requlation of ABR queue length in ATM switches, Proceedings. Second IEEE Symposium on Computers and Communication (Cat. No. 97TB100137), p 80-84

[3] Randhawa, T.S.; Hardy, R.H.S. Estimation and prediction of VBR traffic in high-speed networks using LMS filters. International Conference on Communications, 1998. ICC 98. Pages: 253 - 258 vol.1

[4] Randhawa, T.S.; Hardy, R.H.S. Application of AR based model in proactive management of VBR traffic. 1998 1st IEEE International Conference on ATM. ICATM-98. Pages: 234 - 241

[5] Bin Qiu; Liren Zhang; Wu, H.R. Fuzzy multi-step ahead prediction of VBR video sources. 1997 International Conference on Information, Communications and Signal Processing, 1997. ICICS. Pages: 1623 - 1626 vol.3

[6] Qixiang Pang; Shiduan Cheng; Peng Zhang. Adaptive fuzzy traffic predictor and its applications in ATM networks. 1998 IEEE International Conference on Communications, 1998. ICC 98. Pages: 1759 - 1763 vol.3

[7] Hall, J.; Mars, P. The limitations of artificial neural networks for traffic prediction. Third IEEE Symposium on Computers and Communications, 1998. ISCC '98. Pages: 8 – 12.

[8] Callon, R. Predictions for the core of the network. IEEE Internet Computing. Jan.-Feb. 2000. Vol. 4. Pages: 60 – 61.

[9] J.-S. R. Jang. ANFIS: Adaptive-Network-Based Fuzzy Inference System. IEEE Transactions on Systems, Man, and Cybernetics. Vol. 23 no. 3. May/June 1993. Issue 3, Pages: 665 – 685.

[10] J.-S.R. Jang, C.-T. Sun, Predicting chaotic time series with fuzzy if-then rules. Second IEEE International Conference on Fuzzy Systems 1993. Pages 1079-1084. vol 2.

[11] Mischa Schwartz. Broadband Integrated Networks. Prentice-Hall, A Simon & Schuster Company, Upper Saddle River, New Jersey. Section 3.5

[12] D. Bertsekas, R. Gallager. Data Networks. Second edition. Prentice-Hall, Englewood Cliffs, New Jersey, 1992. Section 6.5.2.

[13] J.-S. R. Jang, C.-T. Sun, E. Mizutani. Neuro-Fuzzy and Soft Computing – A Computational Approach to Learning and Machine Intelligence, Prentice Hall, Upper Saddle River, NJ 07458.

ADVERTISED WINDOW-BASED TCP FLOW CONTROL IN ROUTERS

O. Pop , I. Moldován, Cs. Simon, J. Bíró
High Speed Networks Laboratory
Department of Telecommunications and Telematics
Budapest University of Technology and Economics
H-1117, Pázmány P. s. 1/D. Budapest, Hungary
Phone: +36 1 463 3110 Fax: +36 1 463 3107
{pop, moldovan, simon, biro}@ttt-atm.ttt.bme.hu

A. Koike, H. Ishii
Information Sharing Platform Laboratories
Nippon Telegraph and Telephone Co., Japan
3-9-11, Midori-cho, Musashino-shi, Tokyo 180-8585, Japan
Phone: +81 422 59 4445 Fax: +81 422 60 7372
{koike.arata, ishii.hiroshi}@lab.ntt.co.jp

Abstract Nowadays, TCP is basically the only protocol that is used to provide a reliable data communication over the Internet. In recent research of TCP great effort has been put on enhancing the flow and congestion control mechanisms of this protocol. The most relevant of these mechanisms are the RED-like intelligent queue management techniques that try to avoid the congestion keeping the average queue size low while allowing occasional bursts of packets in the queue. However, these techniques notify TCP connections about congestion by dropping packets, thus reducing their goodput. Another possible technique is ECN (Explicit Congestion Notification) which may help avoiding the packet losses but it needs changes in the TCP implementations of the end systems.

In this paper we propose a TCP flow control technique for routers, based on modifying the window size advertised by the receiver. According to the congestion level at the routers we overwrite the corresponding bytes in the acknowledgement (Ack) packet headers. In this way we can match the amount of data sent by the TCP sources to the available buffer capacity, keeping the average queue length below the congestion level without buffer overflows, thus improving the goodput of the TCP connections. Moreover, this technique does not require any modification of the existing TCP protocol implementations, unlike the

ECN method. Per active-flow accounting of TCP connections is used in order to avoid 'window shrinking'. The performance evaluation of this technique is presented using measurements and simulation. The effects of the maintenance of per-flow information in the routers are also analyzed.

Keywords: TCP, flow control, advertised window

1. INTRODUCTION

TCP is the most widely used protocol that ensures error-free data communications between end systems in the Internet. TCP assumes no reliability on underlying network and includes flow and congestion control mechanisms to transport data in an efficient manner and efficiently utilise the network resources. All control is done by end-to-end basis. This nature is based on the history of TCP that interconnects various kinds of computers in different network environments. As a result, many people implement TCP/IP in almost all kind of operating systems (OSs) and therefore TCP runs on various kinds of networks. This enables application developers to use TCP/IP as a core protocol for their applications.

High-speed network age, however, reveals limitations of end-to-end TCP control. In what follows in section 2 we present the drawbacks of the TCP protocol and review the literature related to the methods used to improve its performance. The third section describes in detail our window based router algorithm. Our algorithm was thoroughly tested, using measurements and simulations. We have built a prototype IP network for measurements, using the most popular TCP implementations in the end-systems. The prototype implementation and the results of the measurements are presented in section four. To have a complete view about the behaviour of the algorithm, extensive simulations have been used, which are also described in section four. Finally we conclude our work. We attached two appendices as well, presenting the measurements regarding the processing overhead generated by the per-flow information handling in the routers, and the formal description of our algorithm.

2. BACKGROUND

In order to adapt the sending rate of the source to the available bandwidth TCP uses a sliding window based flow control [5, 18]. Remember that, with the sliding window protocol the window size is equal to the maximum number of unacknowledged data that a source may send. TCP defines a variable called congestion window computed by the source; the window size is the minimum of the congestion window and the window advertised by the destination. The evolution of the congestion window in a TCP source consists of two phases: slow-start phase and the congestion avoidance phase. The slow-start phase occurs during startup, as well as when a packet loss is detected by a timeout at the source. During the slow-start phase, the congestion window essentially doubles during each round-trip time (RTT), until a threshold window size known as slow-start threshold (ssthresh) is reached. At this point the host enters the congestion avoidance phase where TCP is probing for additional bandwidth by increasing the window more slowly, at the rate of one segment per RTT. The packet losses can be detected either by expiration of a timer at the source or when three duplicate acks are received. In the latter case we speak about the Fast Retransmit mechanism that is further followed by the Fast Recovery mechanism. The Fast Recovery mechanism avoids slow-start in such cases by setting the congestion window to approximately half of its current value and keeping the connection in the congestion avoidance phase.

The main drawback of this protocol is that, to detect network congestion, in most of the TCP implementations (except TCP Vegas), TCP sources rely mainly on packet losses, which triggers then a decrease of the sender's congestion window size. The TCP performance depends on a lot of factors such as the packet dropping strategy used by the router, the flow and congestion control mechanism of the TCP protocol or the nature of the network (high or low speed, large RTTs or not). Using Drop-Tail routers (packets are dropped when the buffer is full) losses occur in bursts causing timeouts, since the Fast Retransmit and Fast Recovery mechanism can be used only once in a given window. It is desirable especially in high-speed networks (backbones) to avoid timeouts, since they introduce burstiness and goodput degradation of the TCP connections.

More recently, random packet dropping strategies have been introduced (Random Early Detection, RED, and its variants) [2, 3, 4] which try to prevent congestion, rather than just reacting to it, by dropping packets before the buffers of the router are totally exhausted. These techniques try to spread the packet losses (avoiding bursty losses), therefore the early-dropped packets usually are Fast Retransmitted and no timeout occurs. Moreover, RED tries to keep the average queue length as low as possible while maximising the throughput of TCP flows. It is undesirable to have large queues because this significantly increases the average delay in the network. However, the number of packet losses is even higher than in the case of Drop-Tail routers reducing the TCP goodput especially when the roundtrip time of the connection is large.

Other, more explicit methods are considered, as the Explicit Congestion Notification (ECN) [6] different Ack delaying approaches [10, 12, 15, 16], or different explicit window based methods [13, 19, 20]. The ECN uses the TOS field in the IP header to notify senders about congestion without packet dropping. The disadvantage of this method is that it needs modification to the existing TCP/IP protocol suite.

The Ack delaying algorithms try to control the flow of Ack packets according to the level of congestion at the routers. In [10, 12, 16] this technique is used especially for regulating TCP traffic over ATM networks with the goal of minimizing the network-edge buffer requirements while maximizing TCP goodput. In [12] for every TCP connection there is a different ACK buffer that is drained according to the ABR explicit rate and a different indicator that memorise the state of the TCP sender, i.e. congestion avoidance or slow start. A more general method that can be used in common IP router and do not need to maintain per-flow information is also presented in [15]. The performance of these Ack-delay algorithms is high in a small network, but its scalability is highly affected by the number of active flows and their window size. They can introduce high delays by increasing the RTT of the TCP connections, and also the reaction to the congestion is slower than in case of the proposed algorithm.

Explicit window based flow control methods modify the TCP sender window by receiving explicitly the congestion status from network. In [19] the presented technique, is used for adjusting the TCP rate over ATM, by changing the congestion window according to the ABR explicit rate. In [20] in order to avoid the layer violation and re-computation of the CRC the congestion status is transmitted in the IP header of the Ack packets (in case of IPv6) or in special ICMP messages in case of IPv4. As in the case of ECN both approaches need to modify the TCP protocol in the end system. Explicit Window Adaptation (EWA) method [13] is very similar with our algorithm. Both EWA and our proposed algorithm try to prevent the congestion by modifying the advertised window size of the receivers encoded in the acknowledgement packet. But EWA did not take into account the window-shrinking phenomena [5]. It can be avoided only by maintaining per-TCP flow information and suppose some constrains on the way of changing the advertised window size. As in the ECN case this router algorithm explicitly notifies the sender about the congestion, but it does not need any changes in the TCP end systems.

3. ALGORITHM DESCRIPTION

3.1. NETWORK MODEL

Before describing our proposed algorithm, first let us consider the following network scenario. We have a router that connects two Ethernet segments of 100Mbps and 10Mbps as presented in Figure 1. The bottleneck is the router, therefore packets coming from the high-speed side will be discarded only at it. Although this is a simplified scenario, it can be regarded as a first step towards investigations of ingress routers at the border of a larger administrative domain in the Internet. Based on the above considerations we have also chosen this configuration as a model for our prototype system in order to make the performance evaluation of the algorithm. We have implemented our algorithm in the router, and deployed several TCP versions in the end systems.

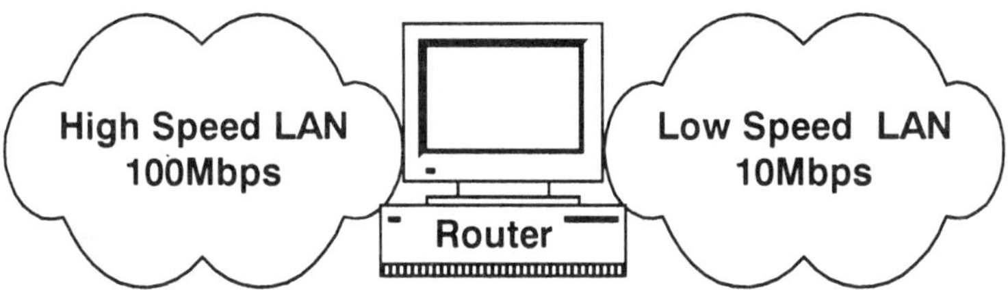

Figure 1 Network configuration

3.2. PROPOSED WINDOW ALGORITHM

Window Algorithm concept. The algorithm tries to match the sending rate of the TCP sources to the available bandwidth by adaptively controlling the advertised window field of the acknowledgement packets. In other words it tries to adapt the advertised window size of the connections by changing it as a function of the 'forward queue' length, i.e. the queue where the router puts the packets coming from the high speed network to the low speed network. We assume that we have only unidirectional TCP connections with the data packets travelling from the high speed network to the low speed network. In order to control the connections in a fair manner the algorithm changes the advertised window for each connection to the same target window. Thus, we avoid situations when the advertised window size is decreased drastically only for some connections while for other ones the advertised window size remains unchanged. In fact, the advertised window of the Ack packets is modified according to this target window, thus, indirectly as a function of the forward queue.

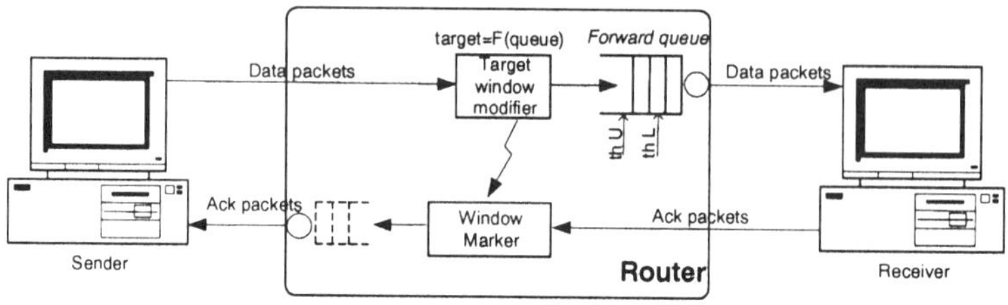

Figure 2 Algorithm concept

The point in the router where our algorithm is invoked is illustrated in Figure 2. The algorithm has been divided into two logical components. The Target Window Modifier changes the target window as a function of the forward queue length. We have also defined two thresholds on the forward queue, an upper one and a lower one. These two queue thresholds depend on the size of the queue and they are parameters of the algorithm. If the length of the forward queue is above the upper threshold the target window is decreased. Similarly, if the queue length is below the lower threshold, it is increased, while if the queue length is between the two queue thresholds, the target window remains unchanged.

The Target Window Modifier uses an additive increase and a multiplicative decrease of target window. At each packet arrival if the queue length is above the upper threshold all the bytes of packets are summed up in a counter, which contribute to this large buffer occupancy. If this counter reaches a certain given constant parameter (div_const) then the target window is divided by 2. The other important case is when the queue length is below the lower threshold and the target window is to be increased by a value which is proportional to the packet length (in bytes) just arrived. The proportionality coefficient is add_const. The add_const is always a power of 2, therefore the whole calculations can be achieved only with adding and shifting operations.

The other logical component, which is called Window Marker, modifies the advertised window of the incoming acknowledgement packets to the target window. Of course, the header of the incoming Ack packets is never overwritten with a greater value than advertised window size, otherwise we would overflow the receiver. A minimum window size is also enforced, since setting advertised window size smaller than the maximum segment size (MSS) negotiated during connection establishment can lead to starvation.

Before presenting the way of decreasing and increasing the advertised window size, we describe the window shrinking phenomena and its constraints imposed on our algorithm.

Window Shrinking. Window shrinking occurs when the right edge of the window moves to the left, i.e. the window size is reduced by a number higher than the currently acknowledged data. The Host Requirements RFC 1122 strongly discourages this, but also says TCP must be able to cope with a peer that does this. However, making measurements we have found that some TCP implementations are not able to do that. For a better understanding of window-shrinking phenomena and its influence we are going to present an example.

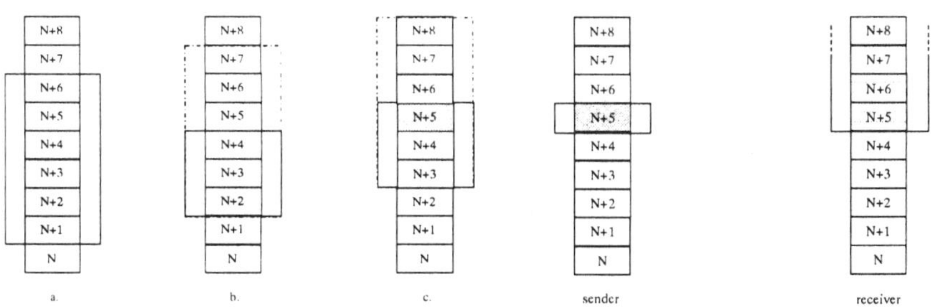

Figure 3 Window shrinking and its effect

Suppose that the congestion window is larger than or equal to the window advertised by the receiver. The sender has a window size of 6 packets (N+1, N+2,. , N+6) transmitted and is waiting for Acks (Figure 3.a). When it receives an Ack packet that acknowledges packet N+1, the window slides up. If we reduce the advertised window size in the router, for instance, to 3 packets, the N+5 and N+6 packets will go out of the window (Figure 3.b) and the window shrinking phenomena appears. Note that if the advertised window remains unchanged the sender transmits the packet N+7. This is marked with dotted line on the figure. In case of window shrinking, when the transmitter receives another Ack packet it moves up the window and packet N+5 will reappear in the window again. Now, there are two different situations. When the sender is able to cope with the window shrinking, it will not retransmit this packet. If not it will transmit it again (Figure 3.c).

If the gateway's algorithm does not avoid the window shrinking problem and the sender TCP is not able to cope with this, the situation in Figure 3 may appear.

Now let us assume that the sender has the window size further reduced to 1 packet. Packets N+5 and N+6 were transmitted but because we shrunk the window packet N+5 is retransmitted. When the receiver gets the packet N+5, it will transmit an Ack packet, which acknowledge the packet N+6. But because packet N+6 is outside the window, the sender will drop this Ack and will transmit again the packet N+5. This leads to an infinite loop.

We have experienced this symptom with the TCP implementation of the Linux 2.0. In this case the window shrinking leads to periodic retransmission of the last acknowledged packet.

Avoiding Window Shrinking. The window shrinking is avoided by decreasing the advertised window at once only by the size of the last acknowledged segment. It means that we have to keep the last acknowledged sequence number for every connection, and we reduce the advertised window size only by the difference between the actual Ack sequence number and the last Ack sequence number. For example in Figure 3.b we could decrease the window size only by one packet.

In this way we can get the desired window size without window shrinking by gradually decreasing the window as we receive the Ack's. Thus, when the Window Marker receive an Ack packet the target window is always compared to the advertised window of the previous transmitted Ack packet belonging to the same connection. In our case the per-flow maintenance of the advertised window of the previous transmitted Ack packet for every connection is also a subpart of the Window Marker module, but it may also be implemented as a different component.

Finally, we note that the advertised window size is only an upper bound to the congestion window size of the sender. Thus, when the congestion window on the sender side is smaller then the advertised window on the receiver side, e.g. in the slow start phase, the decreasing of the advertised window may be without effect, until the congestion window will be larger than the advertised window. We emphasize that the fastest way to reduce the sending rate of a TCP source would be to directly decrease its congestion window size. However this parameter is controlled by the TCP end system, its value is not advertised to the other TCP hosts, hence it can not be directly changed in the routers.

4. PERFORMANCE EVALUATION

In order to make the performance evaluations of this algorithm two possible approaches have been followed: simulation and benchmarking (measurements). Both of them have advantages and disadvantages. The simulation approach is straightforward since several simulation tools for protocol evaluation are available. However, it may eliminate important elements of TCP specifications and implementations. The performance of the TCP is affected not only by its protocol specification and internal algorithms, but also by its implementation, memory management mechanism in its underlying operating system and system architecture. On the other hand, the benchmarking method has also limitations because it is not practical to build and use a complex network scenario only for testing purposes.

4.1. MEASUREMENTS

In the test network configuration we have used different operating systems at the receiver and sender side. We have tested the algorithm with TCP versions implemented in the following operating systems: Linux 2.0, 2.2, NetBSD 1.3, SunOS Release 5.6, FreeBSD 3.0, Microsoft Windows 95/NT. All of these operating systems have different TCP implementations. At the router we have used Linux operating system because its source code which was to be modified is freely accessible.

Measurement with Drop-Tail Router. For comparison purposes, the measurements for performance evaluation were also performed for cases when there is no control algorithm in the router. All measurements presented in this paper are based on sender and receiver hosts with Linux 2.0. However, we observed the same characteristics for all other OSs.

The upper part of Figure 4 shows the relative sequence number plot in packets while the lower part shows the forward queue length when we have ten simultaneous connections, transferring 1Mbytes each. The forward queue size is set to 50 (dashed line).

When the forward queue length grows to 50, the arriving packets are dropped (Drop-Tail routers). You can follow the typical behaviour of the TCP Reno sender: the queue length grows until packet loss occurs, then queue length drops, and then quickly increases again causing other losses. There were 164 buffer overflows causing the sources to retransmit quite often. Although the TCP senders start to transmit simultaneously and transfer the same amount of data they do not finish at the same time. The time difference between the first and the last connection finished is about 2.5s, and the sending rate of each connection varies (the gradients of the connections are not constants - there are retransmissions and timeouts). These fluctuations lead to degradation of the fairness between the active TCP connections through the router.

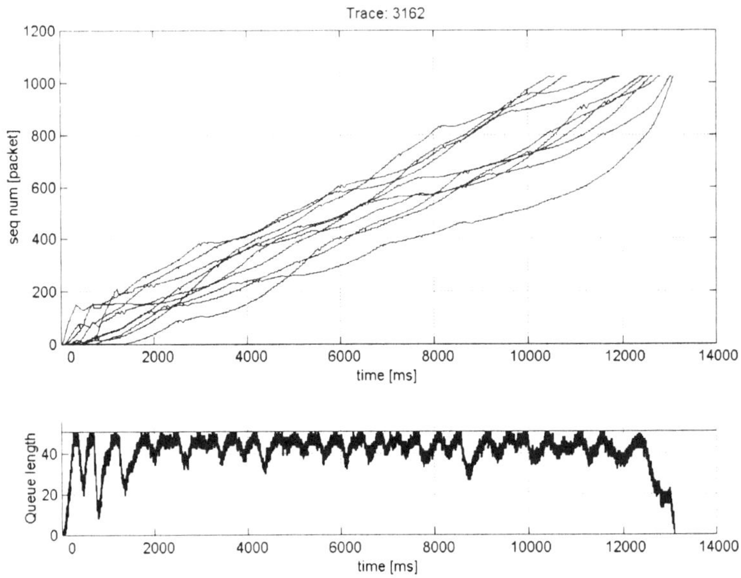

Figure 4 Ten connections with Drop-Tail

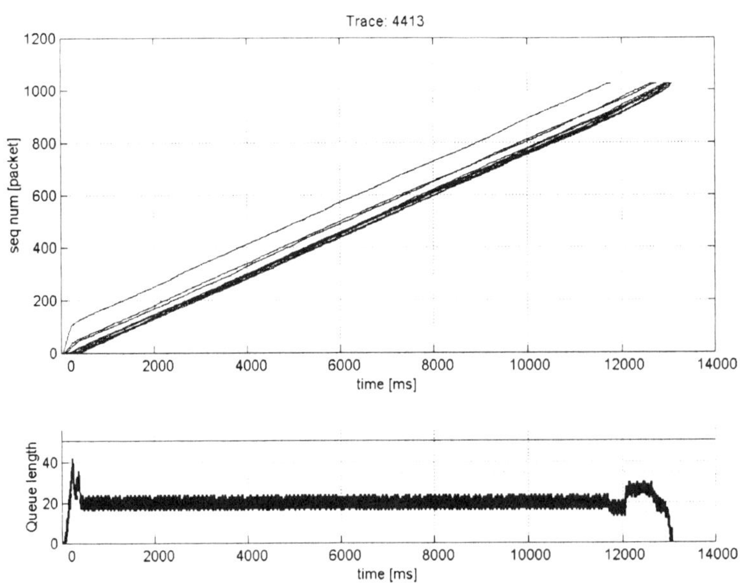

Figure 5 Ten connections with algorithm

Measurement with algorithm. We have tested the algorithm, with several operating systems and with different traffic generators and the results were very good in all cases. Goodput was slightly increased because packet losses were eliminated, and the fairness was also improved. For fairness calculation we have used Jain's measure [8].

We can see on Figure 5 the same measurement scenario as for the Figure 4, but with the proposed algorithm in the router. The forward queue size was also set to 50 packets (dashed line), the upper threshold was set to 35 packets, the lower threshold to 15 packets (dash-dotted lines), the "add_const" parameter to 15 packets and the "div_const" to 64 (the packet size is not considered).

After the slow start period of the connections, after about 200ms the forward queue length is kept below the upper threshold at an almost constant level, eliminating completely the packet losses. The connection, which started first, gets more share of initial throughput, but after the slow-start period all the curves of connections are parallel, which means that the bandwidth is shared equally as the sources transmit at the same rate.

By observing the results in Figure 4 and 5, we find that the throughput differences are small. This is because we measured in small RTT environment and even if we observe packet loss for Figure 4 (with no control algorithm), the TCP sender can perceive the packet loss immediately. Therefore although the fairness among connections for figure 4 is worse, the throughput degradation is small.

4.2. SIMULATION

One lack of our prototype network [Figure 1] is that the round trip times (RTT's) of the TCP connections were small having the same values.

TCP connections with large roundtrip time may much dramatically influence (decrease) the overall throughput when packets have been lost, that is to say, throughput of a TCP connection is inversely proportional with its round-trip time value. Moreover, in the measurements presented above we have found that the goodput of the TCP connections is not affected greatly by packet losses which reason is also the small RTT values.

In order to see how our algorithm increases the goodputs of connections with large roundtrip times we have made simulation investigations using the ns network simulator [17].

Network and traffic scenario. We consider simulation of the network from Figure 6. The link between the two routers is heavily loaded with background traffic only in one direction from Router1 to Router2. In order to model the background traffic we use FTP connections with large advertised window sizes and small roundtrip times In the simulator FTP sources always have a packet to send as soon as the congestion window allows them to do so. A sink at the receiver side immediately sends an Ack packet when it receives a data packet. The FTP sources rely on Reno version of the TCP protocol, i.e. the TCP sender side includes the slow start, congestion avoidance, Fast Recovery and Fast Retransmission mechanism, which are the same as the TCP mechanisms we used in the measurements.

Each FTP sender and FTP receiver is connected to Router1 and Router2, respectively through 10Mbps links. The TCP connections, which the FTP are using, have a maximum window of 64 packets while the delay between the FTP senders and Router1, and FTP receivers and Router2, respectively is chosen from the uniform distribution on [1ms, 3ms]. Moreover, they transfer 3 Mbytes data each starting uniform randomly between 0 and 120s.

Connection from node FTPSource generates the traffic under test (TUT). The window size for this connection is also 64 packets but the roundtrip time is around 322ms as the delay between the FTPReceiver and Router2 is 150ms. It starts when the link is already congested (at about 60s, to affect also its slow start mechanism) and transmits until the simulation ends (180ms). The packet length of both the background traffic and TUT is 1000 bytes. The queue size in routers was set to 100kBytes. Due to the random characteristic of the starting times of the connections and links delay, we have run each simulation 10 times.

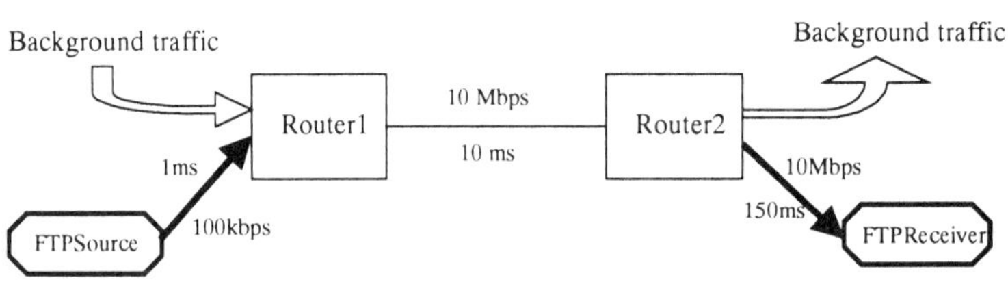

Figure 6 Network scenario for simulation

Simulation with Drop-Tail Router. Figure 7 shows the packet number growing of the tested connection and the queue length variation of Router1 using Drop-Tail. We can see that the queue is heavily congested between 50s and 180s and there are a lot of packet losses (3421). Therefore the packet number growing of the TUT is not linear, there are a lot of retransmissions, the roundtrip time is large and the goodput of the connection is decreased (71.65kbps).

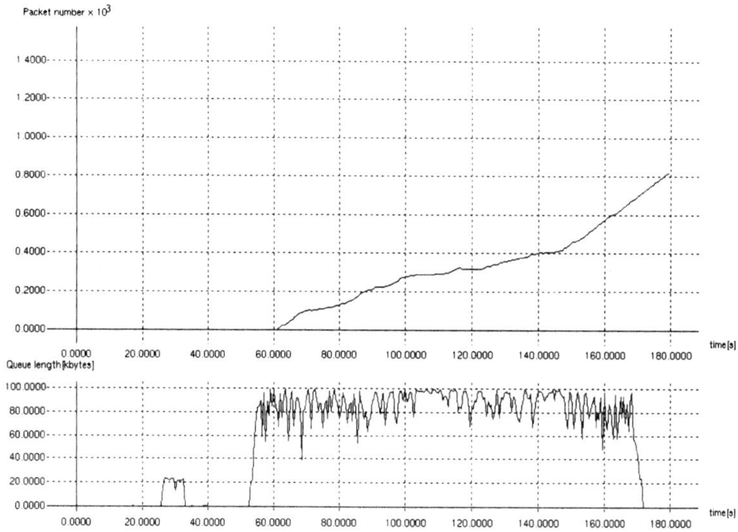

Figure 7 TUT with Drop-Tail

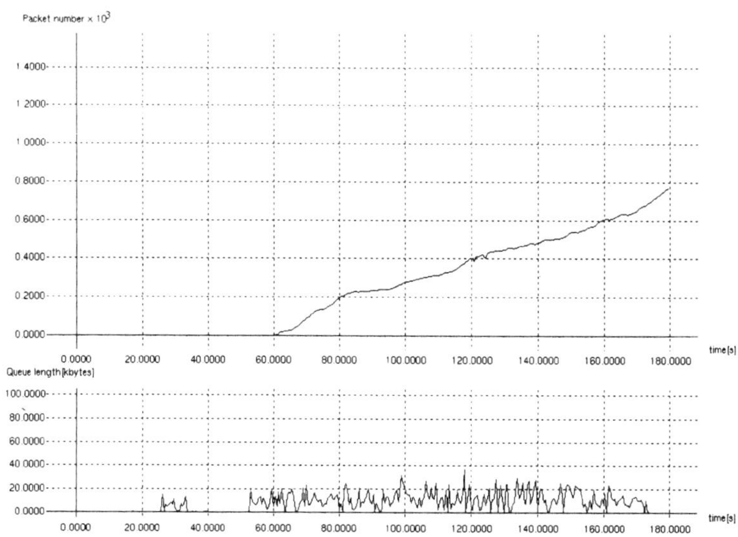

Figure 8 TUT with RED

Simulation with RED Router. The case of RED routers is shown in figure 8. The RED parameters are set as follow: wq=0.002, lower threshold 5 packets, upper threshold 15 packets, and maxp=1/50. In this case the queue length is kept low, the congestion is avoided but the number of losses is much higher than in the case of Drop-Tail (10588). These losses cause performance degradation only for the tested connection and not for the background traffic. This is evident since the background traffic has low latency connections and the RED algorithm spreads the losses, which induce Fast Retransmissions and Fast Recovery as we have mentioned in the introduction. Moreover, the TUT has a high-latency connection when the performance is affected much more by the number of losses and not only by their burstiness. The goodput of the tested connection was 70.217kbps.

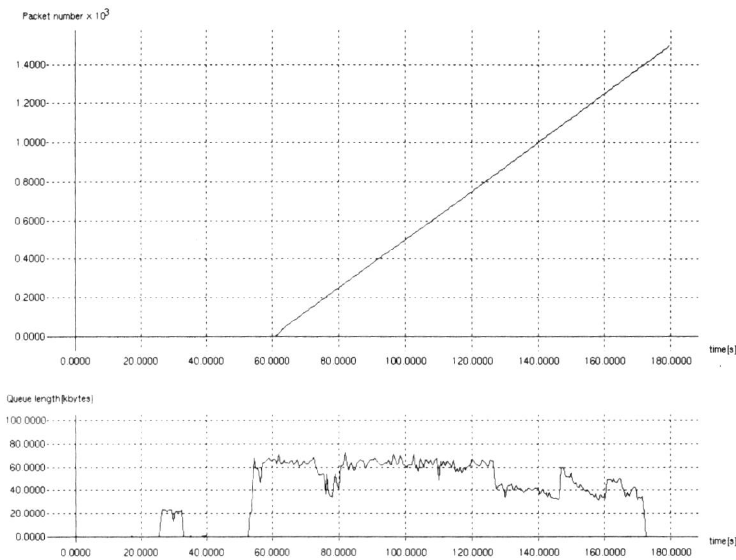

Figure 9 TUT with algorithm

Simulation with the algorithm. Figure 9 shows the case of our algorithm. We have used the same parameter setting for the Window algorithm as in the case of measurements, exception the upper queue threshold (70 packets) and the lower queue threshold (30 packets). We can see that there are no packet losses and the packet sequence number growing is almost linear. Thus, the goodput of the tested connection is much higher than in the case of Drop-Tail and RED, i.e. 99.65kbps. Since the link speed between the FTPSource and Router1 is 100kbps, we can find the TUT in Figure 9 fully utilize the link bandwidth without any bandwidth fluctuation.

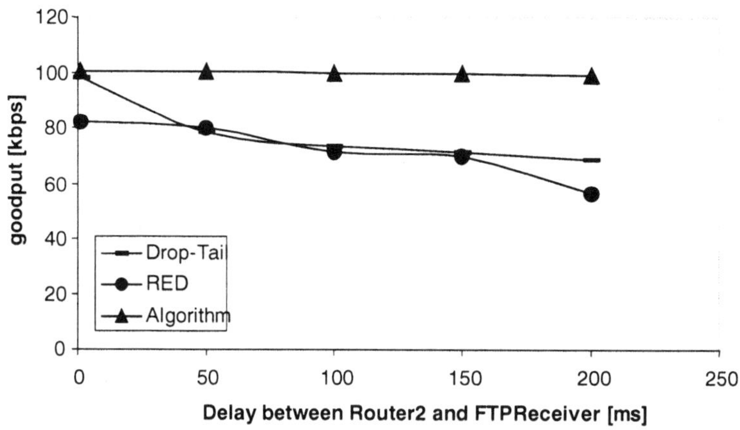

Figure 10 Goodput variation

In figure 10, we show the goodput in function of the delay between the Router2 and FTPReceiver for the tested connection. By looking at Figure 10, we can find that when we add delays for the tested TCP connection, our algorithm can maintain the goodput. However, without our algorithm TCP goodput decreases as the delay increases. This result clear shows the effect of avoiding packet loss in our algorithm.

5. CONCLUSION

We have developed and implemented a prototype system with an advertised window based algorithm in the router to eliminate buffer overflows. The prototype system with different operating systems can reveal the performance of our algorithm under different TCP implementations. We have also made simulations to test the proposed algorithm with more complicated network and traffic scenarios. In the simulation environment the undesirable effect of large RTT's on the TCP performance can also be tested. Our simulation investigations validate that the window-based algorithm can maintain the throughput of TCP connections with large RTT's as opposed to RED or Drop-Tail queue management. Both measurements and simulations of our algorithm have shown that it is possible to reduce the number of overflows significantly.

Basically, in the Internet the data packets and the corresponding Acks may travel the IP network through different routes. Therefore the Ack packet based flow control router algorithms can be used only in access routers. This is valid for our proposed algorithm, as well. These access routers connect for instance the private IP network of business users to the ATM backbone of an Internet service provider. Moreover, since this algorithm needs to maintain per-flow information its applicability is restricted to such routers, where the per-flow data processing is within acceptable bounds.

Acknowledgments

This work has been done in the framework of a joint project between Department of Telecommunications and Telematics, Technical University of Budapest and NTT, Japan. The continuous, encouraging and very useful support of our Japanese colleagues is highly appreciated.

References

[1] S. Floyd, K. Fall, "Router Mechanisms to Support End-to End Congestion Control", http://www-nrg.ee.lbl.gov/floyd/papers.html

[2] S. Floyd, V. Jacobson, "Random Early Routers for Congestion Avoidance", http://www- nrg.ee.lbl.gov/floyd/papers.html

[3] D. Lin, R. Morris, "Dynamics of Random early Detection", SIGCOMM'97

[4] W. Feng D.D. Kandlur, D. Saha, K.G. Shin, "Techniques for Eliminating Packet Loss in Congested TCP/IP Networks", Technical Report CSE-TR-349-97, University of Michigan, Dept. of Electrical Engineering and Computer Science, Nov. 1997

[5] R. Stevens, TCP/IP Illustrated, Vol. 1: The Protocols, Addison Wesley, 1994

[6] K.K. Ramakrishnan, S. Floyd, "A Proposal to add Explicit Congestion Notification (ECN) to IP", RFC 2481, January 1999.

[7] J. Nagle, "Congestion Control in IP/TCP Internetworks", RFC896, January 1994

[8] R. Jain, "The Art of Computer Systems Performance Analysis", John Wiley and Sons, Inc., New York, 1991

[9] The Linux Documentation Project (LDP), http://sunsite.unc.edu/pub/Linux/docs/LDP/

[10] A. Koike, "TCP flow control with ACR information", ATM Forum/97-0758R1

[11] Y. Murayama, S. Yamaguchi, "DBS: a powerful tool for TCP performance evaluations", Proc. of Performance and Control of Network Systems, SPIE Volume 3231, pp. 570-581, November 1997

[12] P. Narvez, K.-Y. Siu, "Acknowledgement bucket scheme for regulating TCP flow over ATM", Computer Networks and ISDN systems 30, pp. 1775-1791, 1998

[13] L. Kalampoukas, A. Varma, K. K. Ramakrishnan, Explicit Window Adaptation: A Method to Enhance TCP Performance", IEEE INFOCOM'98

[14] K. Fendick, M. A. Rodriguez, A. ACM SIGCOMM '92, Baltimore, MD, pp.136-148, August 1992. Weiss, "Analysis of a rate based control strategy with delayed feedback", Proc.

[15] I. Moldován, G. Bajkó, O. Pop, J. Bír, A. Koike, "TCP Flow Control in Routers", ICCC'99, Tokyo, Japan

[16] I. Moldován, G. Bajkó, O. Pop, J. Bír, A. Koike, "Rate controll algorithm for TCP over ABR", IFIP'99, Antwerp, Belgium

[17] http://www-mash.cs.berkeley.edu/ns/

[18] http://www.cis.ohio-state.edu/htbin/rfc/rfc1122.html

[19] Jae-il Jung, Inhwan Lee, "TCP window control scheme using ER value in TCP over ABR services", Electronics Letters, vol. 35 No.7, April 1999

[20] M. Gerla, R. Lo Cigno, S. Mascolo, W. Weng, "Generalized Window Advertising for TCP Congestion Control", Technical Report No. 990012, University of California Los Angeles, Computer Science Department, 1999

Appendix A

Processing overhead

Our algorithm uses per-flow statistics and we expect that the processing costs are proportionally higher as the number of active flows increase. Since the performance characteristics are crucial in this context, we considered the cost estimation an important issue. We have identified two components of the costs, the RAM space occupied by the flow-table and the CPU time used to look up this table. We considered that in the current implementation the storage capacity is not a bottleneck. Even if we store all the information needed for one connection, i.e. TCP/IP port and address field (12 bytes), advertised window size (2 bytes) and Ack sequence number (4 bytes), with the current RAM prices we can handle more than one million flows. The other component, the time consumed for table lookup could severely damage the performance of our algorithm. Further on we will refer to the CPU load caused by the flow-handling routines as processing overhead or as overhead. Some measurement-based estimation of this overhead will be provided.

First we have measured the processing time of a packet when we have used our algorithm in the router, then based on these measurements, we compute the processing overhead introduced by our algorithm.

As we tried to measure the packet processing time we had to eliminate two problems. Due to the internal architecture of the NICs (Network Interface Card) and the load on the physical link the delays and losses on a NIC cannot be controlled. As a consequence during the overhead measurements the NICs or its drivers should not process the packets. The other issue was to force the packets to enter our algorithm. The algorithm is implemented in the IP Forwarding module of the IP layer. This module is not used if the packets are sent from or received by an upper layer. If we originate a flow from an upper layer and the destination is the same machine, the packets will be handled in the IP layer by the so-called 'loopback interface'. This is a virtual interface, which operates before the IP Forwarding module.

According to these requirements we have chosen a point in the Operating System, where the IP flow already left the input NIC, but did not reach yet the IP layer. It is the input pool of the Operating System, where the data is collected from every peripheral of the system before being distributed to the internal modules. This pool is called in the Linux terminology the 'backlog queue' and we will refer to it the same way. As the time needed for the IP layer to process a packet (us) is much smaller than the time resolution of the operating system (ms) we have measured the processing time for a large number of packets, then averaging them we got the packet processing time.

Thus, we have implemented a packet generator in the backlog queue. The packets were randomly assigned to different flows, the flows being differentiated based on the destination TCP port numbers. During our measurements, for a given system we did not change the number of generated packets, only the number of flows. The measured results are shown in Figure 11.

Without algorithm, the processing time of one packet in the Linux 2.0.32 kernel running on an Intel Pentium 166 based PC (iP166) is about t0=11.328125us (dashed line in Figure 11). Without our algorithm in the router, as expected, the number of generated flows does not alter the processing overhead. The processing time for Intel Pentium 333 based PC (iPII 333) is t1=3.533203us (dash-dotted line in Figure 11).

If we use the algorithm, then for different number of flows we have different time values. Figure 12 aggregates the measurement results. The graphs show that the packet processing time for the algorithm increases almost linearly with the number of flows generated and recorded in the memory of the router. It is also clearly seen the advantage of a fast

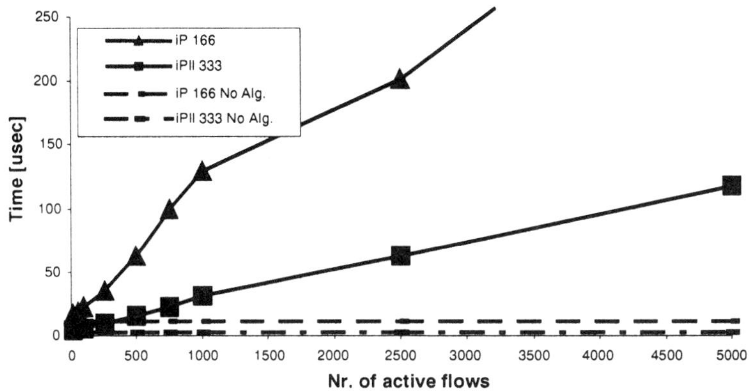

Figure 11 Packet proccessing time with and without the algorithm

processor. The delay introduced by the algorithm in the faster PC is less than one fourth than in the case of the iP166 PC. It means that the performance can be significantly improved if we use newer, thus more powerful processors. Also we have to mention that the table lookup procedure used is the simplest linear one. Better performance can also be achieved using an optimized search algorithm.

Having these measurements done, we calculate the overhead introduced by our algorithm in the router, due to the incapacity of the IP forwarding function to handle the packets as fast as they come from the NICs. If we consider NIC that has always a packet to the IP layer, the processing overhead depends on the link speed of NIC and packets size. If the packets are smaller, there are more packets on the same bandwidth, thus the router have to handle them faster. In other words the packet size is an important parameter in evaluating the processing overhead introduced by the algorithm.

In the followings we investigate two cases. In the first we have considered the packet size equals to the typical MTU in the Internet, 1500 bytes, while in the second we have considered the worst-case approach, where the packet size is set to a hypothetical average, 500 bytes. The link capacity was set to 100 Megabits/second (we conducted our measurements on a Fast Ethernet link). Table 1 and 2 show the processing overhead in [iP133 processor and iP333, respectively. For a given packet size we get the maximum number of generated packets per second: p0=100*10242/(1500*8) = 8738 packets/sec for the packets containing 1500 bytes and p1=26214 packets/sec for the 500 bytes long packets. Moreover, for a given number of flows in the lookup table, we also know from the measurements the maximum number of packets that can be transferred in a second if the algorithm is active. While the algorithm can handle more packets then allowed by the link capacity, the processing overhead of the router is considered 0the ratio of the maximum transferable packet rate and p0 (or p1).

Nr of flows	Processing overhead[%] with MTU=	
	1500 bytes	500 bytes
10	0	0
500	0	38
1000	12	70
2500	43	81
5000	71	90

Table 1 iP 166

Nr of flows	Processing overhead[%] with MTU=	
	1500 bytes	500 bytes
10	0	0
500	0	0
1000	0	0
2500	0	39
5000	3	68

Table 2 iP II 333

As a conclusion, the processing overhead is found to be acceptable for an Intel Pentium133 PC up to almost 900 flows if the packet has the MTU size, and 300 if the packet size is 500 byte. In the case of Intel PentiumII 333 the same value are 4900 flows and 1500 flows, respectively.

Appendix B

Figure 12 shows how the data packets are processed in a router. IP forwarding includes the selection of the output interface, the selection of the next hop, encapsulation, etc. Once all this is done, packets are queued on the respective output interface. The point where our algorithm is invoked is illustrated in figure 12 with dotted line. Note that each of the output queues of a router should have associated a Target window modifier and a Window marker.

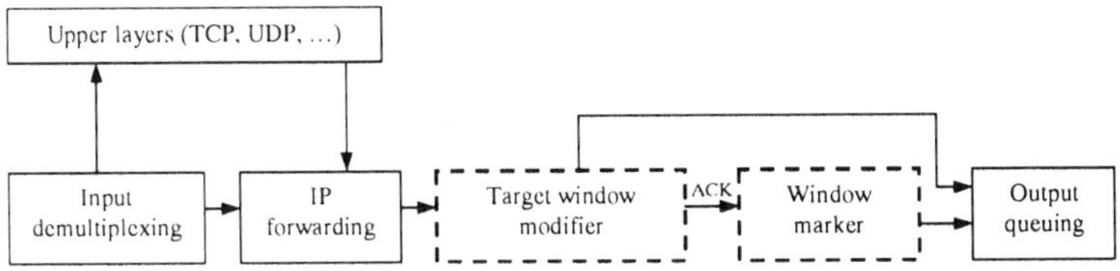

Figure 12 Router architecture

Target window modifier

Initialization:

count=0

for each packet arrival

if queue length > th_U

count=count+packet size

if count≥add_const

target window=target window/2

count=0

if queue length< th_L

target window=target window + packet size/div_const

Variable :

count - number of received bytes above th_U

Fixedparameters :

th_U - upper threshold of queue

th_L - lower threshold of queue

add_const - count maximum limit, then target window is divided by 2

div_const - fraction of the packet size added to the target window

Window Marker

for each ack packet arrival

connection lookup

if new connection

desired window = advertised window

else

if desired window>target window && queue length>th_U

desired window= desired window−(current ack sequence−last ack sequence)

if desired window<target window

desired window=target window

desired window=max (min (desired window, advertised window), MSS)

overwrite the advertised window of the
ack with desired window
last ack sequence=current ack sequence
release ack

Variable:

queue length - length of the queue of the output interface where the ack packet came from

Dynamic Programming Based Link Cost Function and Load Estimation Method for ATM Links

Applying Routing Intelligence in ATM Networks

M. Conte
Siemens AG Österreich
Boschstraße 10, 1190 Wien, Austria
marco.conte@siemens.at

Key words: Dynamic Routing, Dynamic Programming, ATM

Abstract: In connection-oriented networks dynamic routing schemes are designed to improve the call blocking performance by introducing network intelligence. An accurate model describing a networks statistical behaviour is the network Markov model. Based on this model the theory of dynamic programming for Markov processes (also called the Markov Decision Process theory, MDP) has in the past successfully been applied in the field of dynamic routing. The Forward Looking Routing (FLR) scheme [1] defines a dynamic link cost function for single-rate circuit-switched networks. In ATM networks carrying multirate traffic a more complex traffic theory is encountered than for single-rate traffic. Applications of the MDP concept for routing multirate traffic are made for example in [2], [3], [4] and [5]. In this work, a state dependent link cost function for multirate links is derived, based on the scalar link state model introduced in [5]. It is shown that for the single rate traffic model, this link cost function becomes equal to the FLR link cost function. An approximation method is proposed for determining the offered link load in weak mesh networks. Furthermore, a scheme for estimating the temporal evolution of the carried link load is developed. It is shown by simulation how the ATM link cost function and load estimation may be used within the frame of a dynamic routing scheme called DR/ATM. The performance of the DR/ATM scheme in simulation runs involving different network topologies is presented. Moreover the performance of the ATM link cost function in combination with link state flooding is shown. The simulation runs show that the blocking performance is substantially improved as compared with standard reference routing schemes.

1. INTRODUCTION

Routing in communication networks consists of finding paths, consisting of one or several links, through the network. The routing problem arises in connection-oriented telephony networks as well as in connection-less data networks. For telephony networks it is well known that the quality of the routing scheme employed significantly influences the call blocking performance, as shown e.g. in [1]. The Asynchronous Transfer Mode (ATM) supports connections with different and even variable bit rates and thus it allows to construct shared voice and data networks. Recent simulation results show that dynamic routing schemes generally reduce the blocking probability, or equivalently increase the throughput at a given blocking level, in multirate ATM networks, by introducing some additional network intelligence. For the design of a dynamic ATM routing scheme a trade-off between the benefits of routing optimisation and computational complexity must be carried out.

The present results were obtained as a part of a research study on dynamic routing which was carried out at Siemens Austria, under European Space Agency contract. The work started in 1996 with the scope of identifying promising dynamic routing scenarios for future ATM core networks and to analyse the possibility of involving satellite transmission to distribute routing related data. Some elements of dynamic routing for circuit switched telephony networks were reconsidered within an ATM environment. In particular the estimation of actual link loads from the frequency of call rejections was analysed. This technique was also used in a previous study [6]. A model dynamic routing scheme, called DR/ATM, was developed. DR/ATM was first developed for fully meshed networks, however, the scheme was later adapted to work within weak mesh networks as well. Finally, the ATM link cost functions were derived, which will mainly be discussed in the following.

2. ATM TRAFFIC THEORETICAL ASPECTS

2.1 Theoretical Background

ATM connections may specify individual bandwidth requirements in their traffic contract, depending of the connection type. Within an ATM network, an ATM node may receive a connection request to another node at any time. Consider short time intervals, where the number of existing connections is observed. Call requests arrive randomly, with the same constant probability for each time interval. This probability is independent of any previous call request. Existing connections may be terminated in each of the time intervals, with constant probability, independently of other calls. These assumptions are made, when call inter-arrival times, as well as call holding times of ATM connections are modelled as independent, exponentially distributed random values. This traffic model is also called Poisson-traffic model. The following derivations are based on the Poisson-traffic model. The carried link load of an ATM link carrying Poissonian ATM traffic is described as a vector $\underline{x} = (x_1, ..., x_K)$, where x_i is the number of established connections of type i, $i = 1, ..., K$. K is the number of connection request types which can occur. The vector $\underline{x}$ is also called the K-dimensional link state. The requested bandwidths of type i calls are denoted as c_i, $i = 1, ..., K$, $c_1 \leq c_2 \leq ... c_K$. The space of possible link states is delimited by the capacity of the link C as follows:

$$\underline{c} \cdot \underline{x} = \sum_{i=1}^{K} c_i x_i \leq C \text{, and } x_i \geq 0, \forall\, i = 1, ..., C. \tag{1}$$

The random process generated by the link states $\underline{x}$ during a period of time is a K-dimensional Markov process. I.e., at a given time, the transition probabilities from the actual state of the link $\underline{x}$ to other states do not depend on link states in the past, given the present state $\underline{x}$. More details about multidimensional traffic models can be found e.g. in [7] and [8].

In practical applications it may occur that the exact number of established calls is not known, whereas only the total occupied link capacity $u = \underline{c} \cdot \underline{x}$ is known. Possible values of u are 0, 1, ..., C, assuming that all requested bandwidth values c_i, $i = 1, ..., K$ are integers. The scalar u represents a random process in time, referred to as the scalar link state process. Depending on the requested bandwidth values, it may occur that some of the scalar states 0, 1, ..., K are never reached. For example, if all requested bandwidth values are even, no odd scalar link state can ever be reached. A recursive algorithm for the computation of the steady state probabilities q_i, $i = 1, ..., C$, of the scalar link state process is shown in [9]. The computation is as follows:

$$q_u = \frac{1}{u} \sum_{i=1}^{K} c_i a_i q_{u-i} \quad u = 1, ..., K, \text{ and } q_0 = \left(\sum_{i=1}^{C} q_i \right)^{-1}, \tag{2}$$

where $a_i = \lambda_i \cdot \tau_i$ is the offered load, λ_i is the arrival rate and τ_i is the mean duration for call type i, $i = 1, ..., K$. In contrast to the K-dimensional link state process, the scalar link state process is not a Markov Process, i.e. the transition rates from a state u to other states are not independent of the past given u. Nevertheless the process may be treated as a Markov process. This method is also applied in [5]. Another interpretation of this approximation is that a Markov process is defined, which has the same states and steady state probabilities q_i, $i = 1, ..., C$ and, on average, the same transition rates as the aggregated link state process. This process will be called the markovian aggregated link state process. The transition rates $a_{i,j}$ between states i and j of the markovian aggregated link state process are:

$$a_{u,u+c_i} = \lambda_i \text{ , if } u + c_i \leq C,\ i = 1, ..., K$$

$$a_{u,u-c_i} = \frac{\lambda_i q_{u-c_i}}{q_u} \text{ , if } u \geq c_i,\ q_u > 0,\ i = 1, ..., K, \tag{3}$$

$$a_{u,w} = 0 \text{ , for all other states u and w } (u \neq w).$$

By defining $a_{u,u} = -\sum_{\substack{i=0 \\ i \neq u}}^{C} a_{ui}$, $u=1, ..., C$,

the matrix of transition rates A of the Markovian aggregated link state process can be written as the $(C+1)\times(C+1)$ square matrix $A = (a_{ij})$. For the steady state probability vector $\underline{q} = (q_0, q_1, ..., q_C)$ the following system of equations is obtained:

$$\sum_{i=0}^{C} q_i a_{ik} = 0, \text{ for } k = 0, ..., C, \tag{4}$$

which can be written as a matrix equation: $\underline{q}\,A = \underline{0}$, where $\underline{0}$ is the C+1-dimensional zero vector. The loss rate L of the Markovian aggregated link state process is obtained as:

$$L = \sum_{i=1}^{K} \lambda_i b_i \text{ ,} \tag{5}$$

where b_i is the blocking probability for call type i. For a complete bandwidth sharing policy among the call types the blocking probabilities can be obtained from the steady state probabilities:

$$b_i = \sum_{u=C-c_i+1}^{C} q_u \tag{6}$$

Thus for complete bandwidth sharing, the blocking rates of the Markovian aggregate link state process are equal to the blocking rates of the original aggregate link state process. According to the theory of dynamic programming, [10], the following system of differential equations is obtained for the expected loss $v_u(t)$ at time t, given the initial state u:

$$\frac{d}{dt} v_u(t) = r_u + \sum_{j=0}^{C} a_{uj} v_j(t) \quad , u = 0, ..., C, \tag{7}$$

where r_u is the loss rate in state u. For long term considerations, $v_u(t)$ may be replaced by the asymptotic linear curve of the expected loss: $L \cdot t + z_u$, where the constant z_u depends on the initial state u of the link. The system of differential equations (7) becomes then a system of linear equations:

$$L = r_u + \sum_{j=0}^{C} a_{uj} z_j \quad , u = 0, ..., C, \tag{8}$$

which can be written as a matrix equation: $\underline{L} = \underline{r} + A\, \underline{z}$, where $\underline{L}$ is the C+1-dimensional column vector $(L, ..., L)^T$, $\underline{r} = (r_0, ..., r_C)^T$ and $\underline{z} = (z_0, ..., z_C)^T$. The values z_i, $i = 0, 1, ..., C$ are called the Howard relative costs. In [5] it is proposed that the system of equations (8) is solved, setting $z_0 = 0$. The reduced Howard relative cost $v_{ij} = z_j - z_i$ of j with respect to i is then used as link admission criterion.

2.2 Link Cost Function

After this overview of traffic theoretical methods we now proceed to the derivation of the multirate link cost function. Before introducing the cost function, we state a first lemma. Let $\underline{q}_i$ be the C+1-dimensional vector containing the value q_i at the i-th position and having all other entries equal to zero, i.e. $\underline{q}_i = (0, ..., 0, q_i, 0, ..., 0)$. The following equations are obtained using the definition of matrix A:

$$\underline{q}_n A\underline{z} = -\sum_{\substack{i=0\\ n-c_i\geq 0}}^{K} q_{n-c_i}\lambda_i\left(z_n - z_{n-c_i}\right) + \sum_{\substack{i=0\\ n+c_i\leq C}}^{K} q_n\lambda_i\left(z_{n+c_i} - z_n\right),\ n=0, ..., C. \quad (9)$$

In both summations the index i is taken from zero to K, only over those values of i, for which $n - c_i \geq 0$, respectively $n + c_i \leq C$.

Lemma 1:

$$\sum_{n=0}^{u} \underline{q}_n A\underline{z} = \sum_{\substack{n=u-c_K+1,\\ n\geq 0}}^{u} q_n \sum_{\substack{i=0,\\ u\leq n+c_i<C}}^{K} \lambda_i\left(z_{n+c_i} - z_n\right),\ n = 0, 1, ..., C. \quad (10)$$

Proof (by induction): for $u = 0$ lemma 1 becomes equal to equation (9), for $n = 0$. Assuming that equation (10) is proven for a certain value $u \geq 0$, it is obtained with equation (9) that:

$$\begin{aligned}\sum_{n=0}^{u} \underline{q}_n A\underline{z} + \underline{q}_{u+1} A\underline{z} = &\sum_{\substack{n=u-c_K+1,\\ n\geq 0}}^{u} q_n \sum_{\substack{i=0,\\ u\leq n+c_i<C}}^{K} \lambda_i\left(z_{n+c_i} - z_n\right) - \\ &- \sum_{\substack{i=0\\ u+1-c_i\geq 0}}^{K} q_{u+1-c_i}\lambda_i\left(z_{u+1} - z_{u+1-c_i}\right) + \sum_{\substack{i=0\\ u+1+c_i\leq C}}^{K} q_{u+1}\lambda_i\left(z_{u+1+c_i} - z_{u+1}\right)\end{aligned} \quad (11)$$

It can be seen that the second sum cancels with the indices n and i of the first (double) sum, for which $n + c_i = u + 1$, respectively $n = u + 1 - c_i$. Notice hereby that $c_K = \max\{c_1, ..., c_K\}$. The right hand side of (11) now becomes to:

$$\begin{aligned}&\sum_{\substack{n=u-c_K+2,\\ n\geq 0}}^{u} q_n \sum_{\substack{i=0,\\ u+1\leq n+c_i<C}}^{K} \lambda_i\left(z_{n+c_i} - z_n\right) + \sum_{\substack{i=0\\ u+1+c_i\leq C}}^{K} q_{u+1}\lambda_i\left(z_{u+1+c_i} - z_{u+1}\right) = \\ &= \sum_{\substack{n=(u+1)-c_K+1,\\ n\geq 0}}^{u+1} q_n \sum_{\substack{i=0,\\ u+1\leq n+c_i<C}}^{K} \lambda_i\left(z_{n+c_i} - z_n\right) = \sum_{\substack{n=(u+1)-c_K+1,\\ n\geq 0}}^{u+1} \underline{q}_n A\underline{z},\end{aligned} \quad (12)$$

which proofs the validity of the assumption (10). ❐ The value γ_{ik} is now defined as follows: if a call of type k can be established, starting at state i, γ_{ik} is equal to the reduced Howard relative cost of the transition. If starting at i a call of type k is blocked, γ_{ik} is equal to 1. In formulas:

$$\gamma_{ik} = \begin{cases} z_{i+c_k} - z_i & ,\text{if } i + c_k \le C \\ 1 & ,\text{if } i + c_k > C \end{cases} \quad \forall\, i = 0, \ldots, C,\, k = 1, \ldots, K \tag{13}$$

Moreover we let R_u be the loss rate of the link under complete bandwidth sharing policy, prohibiting states greater than u.

$$R_u = \sum_{\substack{i=u-c_K+1 \\ i \ge 0}}^{K} q_i \sum_{\substack{j=0 \\ i+c_j > u}}^{K} \lambda_j, \tag{14}$$

Lemma 2: The expected value of γ_{ik} which occur for transitions from a state i lower than or equal to the state u towards states above u, or for call blockings at state i is given by:

$$E[\gamma_{ik} \mid 0 \le i \le u,\, i + c_k > u] = \frac{\sum_{i=0}^{u} \underline{q}_i (A\underline{z} + \underline{r})}{R_u}, \quad u = 0, 1, \ldots, C. \tag{15}$$

Proof: by means of equation (10) and (14). The probability $P(m \to n \mid u)$ that, at a call arrival, a transition over the state u occurs from a state m (u-$c_K < m \le u$, $m \ge 0$) to a state $m + c_n$ ($1 \le n \le K$) is equal to $q_m\lambda_n/R_u$. For the probability $P(\text{loss} \mid m)$ for a call loss starting from m we obtain $P(\text{loss} \mid m) = q_m r_m / R_u$. By the definition of the conditional expectation value we obtain $E[\gamma_{ik} \mid 0 \le i \le u, i + c_k > u] = \Sigma\ P(m \to n \mid u) \cdot (z_n - z_m) + \Sigma\ P(\text{loss} \mid m) \cdot 1$ and, by inserting, equation (15) is obtained. ❐

We may now define the link cost function based on the actual aggregated state u, $\text{cost}(u) = E[\gamma_{ik} \mid 0 \le i \le u, i + c_k > u]$, and we define L_u as the rate of transitions from a state lower than or equal to u to states greater than u plus the blocking rate from states lower than or equal to u, under complete bandwidth at all states 0, 1, ..., C. More precisely L_u is defined as:

$$L_u = \left(\sum_{i=0}^{u} q_i \right)^{-1} R_u \tag{16}$$

We are now ready to state the following theorem:

$$\text{Theorem 1: } \text{cost}(u) = \frac{L}{L_u}, \; u = 0, 1, \ldots, C, \tag{17}$$

Proof: from equation (8) we know that $L = A\underset{\sim}{z} + \underline{r}$. Thus inserting into equation (15), we deduce cost $(u) = L \cdot \Sigma_i q_i / R_u = L/L_u$. ❐ If a single call type exists, the ATM link cost function cost(u) reduces to the well known link cost function for single rate circuit-switched traffic $B(N,A)/B(i, A)$, where B denotes the Erlang loss formula, N is the number of trunks, A is the offered load and i is the actual number of occupied trunks. We call equation (17) the "quotient formula" for multirate traffic and the cost function cost(u) the ATM-FLR cost function.

2.3 Determining the Offered Link Load

The call arrival rate for a link within an ATM network comprise the offered traffic between the start and the end node of the link, the directly carried traffic, as well as traffic which is routed over paths containing several links. For full and strong mesh networks we assumed that the first routing attempt is made on a pre-defined default path. The ATM link cost function is then computed for the default offered load. For weak mesh networks, where no default routes are used, the assumption is made that for a sufficient number of node pairs more than one equivalent minimum-hop paths exist, such that the traffic can be distributed among these paths. As a result, the traffic can be distributed among the links without increasing the sum of all link loads. Assuming optimal traffic distribution, the following optimisation problem is obtained:

$$\min \sum_{i=1}^{N} L_{(i)}(\lambda_{(i)}), \text{ given that } \sum_{i=1}^{N} \lambda_{(i)} = c, \tag{18}$$

where N is the number of links, $L_{(i)}(\lambda_{(i)})$ is the loss rate of link i, given the total call arrival rate $\lambda_{(i)}$ on link i, i = 1, 2, ..., N. The Lagrange function $\Phi(\lambda_{(1)}, \ldots, \lambda_{(N)}, w)$ for (18) is given as:

$$\Phi(\lambda_{(1)}, \ldots, \lambda_{(N)}, w) = \sum_{i=1}^{N} L_{(i)}(\lambda_{(i)}) + w \sum_{i=1}^{N} \lambda_{(i)}, \tag{19}$$

where w is a Lagrange multiplier. By differentiating with respect to λ_i, i = 1, ..., N the following optimality conditions are obtained:

$$\frac{d}{d\lambda_i} L_{(i)}(\lambda_{(i)}) = w, \forall\, i = 1,2,...,N \,. \tag{20}$$

Equation (20) states that if the assumption of optimal traffic distribution is fulfilled, which may be approximately the case in weak mesh networks, the derivatives of the link loss rates with respect to the total call arrival of the respective link must be equal. We use equation (20) to determine the offered link loads in weak mesh networks.

2.4 Link Load Estimation

If the aggregate link state u_0 at an initial time t_0 is known, the question arises as to, how the expected link load will evolve after t_0. As described in [10], for the Markovian aggregate link state process the following system of differential equations is obtained for the state probability distribution at time t, $\underline{\pi}(t) = (\pi_0(t), \pi_1(t), ..., \pi_C(t))$,

$$\frac{d}{dt}\pi_j(t) = \sum_{i=0}^{C} \pi_i(t) a_{ij} \quad, j = 0, 1, ..., C. \tag{18}$$

The system of differential equations (18) can be written as a matrix differential equation: $\dot{\pi} = \pi A$. Having an initial distribution $\underline{\pi}_0$, the state probability vector $\pi(t)$ for future time instants t, $t>0$, is obtained as: $\underline{\pi}(t) = \underline{\pi}_0\, e^{At}$, where e^{At} is the matrix exponential function of At, and it is defined as $(I + At + ½\,(At)^2 + 1/6\,(At)^3 + ...)$.

Equation (18) and its solution suggest that the expected actual link load may be roughly approximated by a scalar exponential function, decaying from the initial value u_0 to the asymptotic value u_∞. The asymptotic value can be obtained from the steady state distribution $\underline{q}$ as $u_\infty = \underline{q} \cdot (0, 1, ..., C)$. From the point of view of the well known theory of linear differential equations, the decay constant α of the exponential approximation should be in the range of the real component of the eigenvalue of A with the smallest absolute value, which is not equal to zero. For practical implementation, the value α was computed in analogy to the link load estimation in [6] as the sum of transition rates towards higher occupancy states u minus the transition rates toward lower states u, averaged over two points $u = C$ and $u = ½(C-u_\infty)$. The link load estimation $u(t)$ is then obtained as follows ($\alpha<0$):

$$u(t) = u_\infty + \left(u_0 - u_\infty\right) e^{\alpha t} \,. \tag{19}$$

3. ROUTING SCHEMES

3.1 DR/ATM

DR/ATM uses a routing architecture which was developed for the IN/DR scheme [6] and further adapted to ATM networks. It is based on regional routing servers, the routing control processors (RCPs). The ATM network is subdivided into routing domains. Each domain contains one RCP, and all ATM nodes within the domain are connected to the respective RCP. This concept of routing domains allows to achieve a scalability of the DR/ATM scheme to very large networks.

For each call, the first routing attempt is made on the default route, in strong mesh networks, or, in weak mesh networks, on a randomly chosen minimum-hop route. If the first routing attempt fails, the connection is cranked back to the source node, which sends a "routing query" to the RCP in its domain. The routing decision is computed within the RCP based on the ATM-FLR link cost function and on the link load estimations. The alternative route with the lowest path cost is chosen, if the path cost does not exceed a threshold, otherwise the call is rejected. The offered load value used for the link load estimations is determined from the frequency of routing queries. The routing decision is then sent from the RCP to the ATM node, which routes the call. In networks of large geographical extent, RCPs exchange data related to their domain with other RCPs periodically, via satellite broadcast.

3.2 LLP

The least loaded path routing (LLP) examines for all paths, the minimum free capacity of included links, i. e. the “bottleneck” of each path. The path with the largest bottleneck capacity is selected. Alternative paths are allowed only if the free capacity is greater than a certain threshold (trunk reservation). In the simulations shown in Section 4, LLP is implemented as follows:

- The first routing attempt is made on the default path.
- If the first attempt fails, the least loaded alternative path is selected
- Trunk reservation is applied

The actual link load is either derived from the link load estimation scheme or the exact link load is used.

3.3 Minimum-Hop Routing

Minimum-hop routing selects a route with the minimum number of hops, among the available routes, by means of the Dijkstra algorithm. If all minimum-hop routes are blocked, a route with the next higher number of hops is selected. If several equivalent minimum-hop routes are found, one of these routes is selected randomly. In Section 4, the link load for minimum-hop routing is known from link state flooding or no link state knowledge is assumed.

3.4 Minimum-Cost Routing

For minimum-cost routing, the route with the lowest total ATM-FLR path cost is chosen. The total path cost is computed as the sum of the actual link cost values for links belonging to the path. The actual links cost values are determined from the ATM-FLR link cost function and from the actual link load, as obtained from link state flooding. If the minimum path cost exceeds a path cost threshold, the connection is rejected.

4. SIMULATION

4.1 Network Models

Three model networks are considered for simulations, a full mesh, a strong mesh and a square mesh ATM network, each consisting of twenty nodes. The link capacities range from 50 to 100 Mbps, in the full and strong mesh network. In the square mesh network, all links have 100 Mbps capacity. The total numbers of (unidirectional) links are 380 in the full mesh, 280 in the strong mesh and 62 in the square mesh network. A total of 27 call types with sustainable cell rate (SCR) values ranging from 64 kbps to 4.5 Mbps, the average holding times were between 44 and 740 seconds. Complete fairness is applied, i.e. a call is accepted only if a call requesting the largest SCR value could be accepted instead. For all routing schemes only one alternative routing attempt is allowed at most for all connection attempts, except for non-alternate routing, where only default routes are allowed.

4.2 Simulation Results

Figure 1 shows blocking curves obtained in simulation runs for the 20 node square mesh network. The offered load is shown in percentage of the nominal traffic. The nominal traffic corresponds to the 1% blocking level for non-alternate routing. The lowest blocking probabilities are achieved using exact knowledge of the actual link load (dotted lines). It can be seen that in this case the use of the ATM-FLR cost function for minimum-cost routing improves the blocking performance slightly as compared with LLP.

Comparing DR/ATM with a single domain and LLP using the link load estimation scheme (solid lines), it is seen, that DR/ATM achieves an impressive reduction of the blocking probability, especially in the low load region. At 0.1% blocking probability, DR/ATM allows to carry over 5% more traffic than LLP.

Finally we consider the DR/ATM scheme with four routing domains (dashed line). The RCP-to-RCP update period is set to 30 seconds, i.e. each RCP sends its routing data, which is used for computing the link load estimation, every 30 seconds to the other three RCPs. It can be seen that only a moderate loss of blocking performance is obtained, as compared to DR/ATM, with a single routing domain. This shows that the concept of using routing domains may be successfully be applied in combination with the minimum cost routing and the link load estimation scheme.

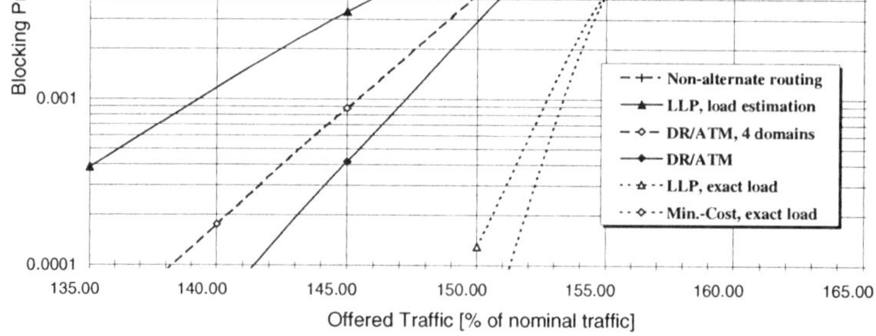

Figure 1: Blocking curves for the 20 node full mesh network.

The simulation results for the 20 node strong mesh network are shown in Figure 2. The uppermost line shows the blocking curve of non-alternate routing. The other two dotted lines represent the blocking curves of the LLP and ATM-FLR cost based routing with exact knowledge of link loads. It can be seen that by using the ATM-FLR cost function significantly lower blocking probabilities are obtained than with LLP.

As to the DR/ATM scheme and LLP using the link load estimation scheme, it can be seen that DR/ATM achieves lower blocking probabilities. However in the low load region, the improvement achieved by DR/ATM is less dramatic than for the full mesh network. A possible explanation for this effect is that in the link load estimation is less precise in the partial mesh case, particularly in the low load region. Nevertheless, at the 0.1% blocking level DR/ATM allows to carry 4.7% more traffic than LLP.

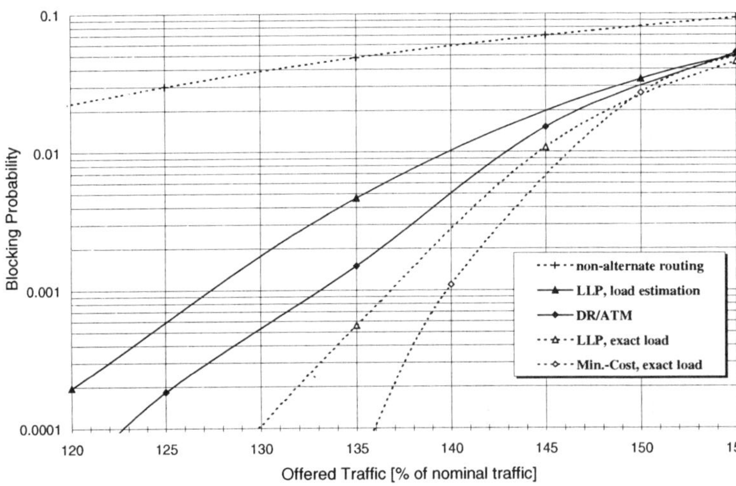

Figure 2: Blocking curves for the strong mesh 20 node network.

The blocking curves obtained for the 20 node square mesh network are represented in Figure 3. The highest blocking probabilities are achieved by minimum-hop routing without link state flooding (dotted curve). Lower blocking probabilities are obtained by minimum-hop routing with link state flooding and DR/ATM. The lowest blocking curve is obtained for minimum-cost routing using the ATM-FLR link cost and link state flooding.

At high load levels, the blocking curves of minimum-hop routing with and without flooding assume similar values. Moreover, the blocking probabilities for the two ATM-FLR cost based schemes DR/ATM and minimum-cost routing are close to each other. The reason for this is that the link load estimation scheme used for DR/ATM is more precise in the high load region.

As to the general routing performance, it is seen that at the 0.1% blocking level DR/ATM allows to carry 4.2% more traffic than minimum-hop routing with link state flooding. At the same blocking level minimum-cost routing with link state flooding allows to carry 8.3% more traffic than minimum-hop routing.

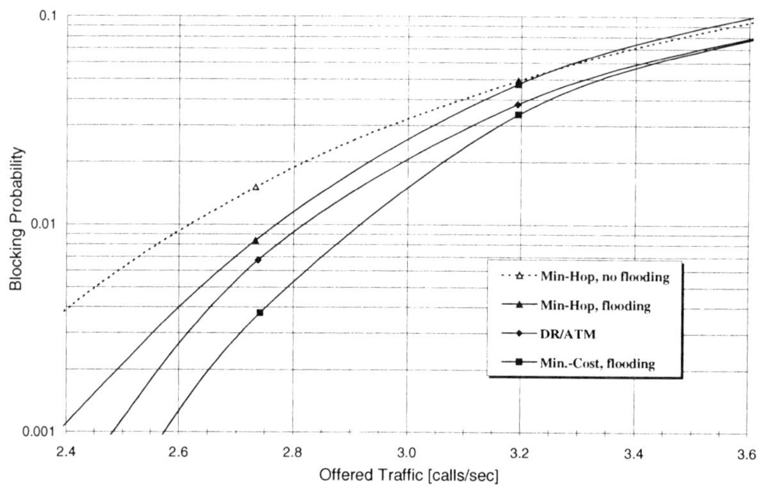

Figure 3: Blocking curves for the 20 node square mesh network.

5. CONCLUSION

Dynamic routing schemes have been shown in the past to reduce the call blocking probability in circuit-switched networks. In recent years several Markov decision theoretical approaches of various precision and complexity have been made to the routing problem for networks carrying multirate traffic. In this paper a new variant of defining a link cost function has been proposed. It has been shown that this link cost function can be computed by means of the so-called quotient formula. Moreover a method for obtaining offered link load values which can be used for the link cost computation in weak mesh networks has been introduced. The link cost function is independent of the type of the call to be routed. Thus possibly some routing performance is lost as compared to more precise routing schemes. However, it is seen as a benefit of the link cost function that only one actual link cost value per link needs to be stored and accessed in real time path computations.

The routing performance of the link cost function and the link load estimation method has been shown for a model routing scheme, DR/ATM. It is shown by simulation runs with different degrees of meshing that DR/ATM achieves an improved blocking performance, as compared with least loaded path routing and minimum-hop routing. Results regarding ATM-FLR link cost function and dynamic routing in networks running the Private Network Node Interface (PNNI) protocol can be found in [11]. A detailed description of the theoretical concepts and of the routing simulations can be found in [12].

6. REFERENCES

[1] K. R. Krishnan and T. J. Ott: „Forward-Looking Routing: A New State-Dependent Routing Scheme", ITC-12, pp. 1026-1031, 1989.

[2] R. Hwang, J. Kurose and D.Towsley: „State Dependent Routing for Multirate Loss Networks", IEEE Globecom '92, vol.1, pp. 565-570, Orlando, FL, USA, Dec. 1992.

[3] A. Kolarov, J. Hui: „On Computing Markov Decision Theory-Based Cost for Routing in Circuit-Switched Broadband Networks ", Journal of Network and Systems Management, Vol. 3, No. 4, 1995.

[4] Z. Dziong and L. Mason: "An Analysis of Near Optimal Call Admission and Routing Model for Multi-Service Loss Networks", IEEE Infocom '92, 1992, Vol. 2, pp. 142-152.

[5] K.R. Krishnan and F. Huebner-Szabo de Bucs:"Admission Control for Multirate Circuit-Switched Traffic", ITC-15, 1997.

[6] L. Bella , F. Chummun, M. Conte, G. Fischer, and J. Rammer,.: "Performance Evaluation of Dynamic Routing Based on the Use of Satellites and Intelligent Networks", in *Wireless Networks* 4, 167-180 (1998).

[7] H. Akimaru and K. Kawashima, Teletraffic Theory and Applications, Telecommunication Networks and Computer Systems, Springer-Verlag, 1993.

[8] R. B. Cooper, Introduction to Queueing Theory, North Holland, 1972.

[9] J. S. Kaufman, "Blocking in a Shared Resource Environment", IEEE Transaction on Communications, COM-29, Nr.10, pp. 1474-1481, October 1981.

[10] R. A. Howard: Dynamische Programmierung und Markov-Prozesse, (translation) 1965.

[11] M. Conte, J. Rammer, G.Fischer, N. Mersch, F. Chummun, L. Bella: "Simulation Study of Dynamic Cost-Based Routing in PNNI Networks, to appear in IEEE ATM Workshop 2000, June 2000.

[12] M. Conte: "Dynamic Routing Schemes for ATM Networks", Dissertation, TU Wien, to be published.

7. ACKNOWLEDGMENT

The work reported here was done under European Space Agency (ESA) contract. The contributions of F. Chummun, L. Bella, G. Fischer and J. Rammer are acknowledged. The author would like to thank J. Rammer for a critical reading of the manuscript.

Part Five

Services in Wireless Networks

A QOS PROVISIONING SCHEME FOR MULTIMEDIA WIRELESS NETWORKS

Mona El-Kadi, Stephan Olariu, Hussein Abdel-Wahab

Department of Computer Science

Old Dominion University

Norfolk, VA 23529-0162, U.S.A.

elkadi,olariu,wahab@cs.odu.edu

Abstract

It has been noticed that multimedia applications can tolerate and gracefully adapt to transient fluctuations in the quality-of-service (QoS) that they receive from the network. The management of such adaptive multimedia applications is becoming a new research area in wireless networks. As it turns out, the additional flexibility afforded by the ability of multimedia applications to tolerate and adapt to transient changes in QoS can be exploited by protocol designers to significantly improve the overall performance of wireless systems.

The main contribution of this paper is to propose a novel, rate-based, borrowing scheme for QoS provisioning in high-speed cellular networks carrying multimedia traffic. Our scheme attempts to allocate the desired bandwidth to every multimedia connection originating in a cell or being handed off to the cell. The novelty of our scheme is that in case of insufficient bandwidth, in order not to deny service to requesting connections (new or handoff), bandwidth will be borrowed, on a temporary basis, from existing connections in the cell. Our borrowing scheme guarantees that no connection gives up more than its "fair share" of bandwidth, in the sense that the amount of bandwidth borrowed from a connection is proportional to its tolerance to bandwidth loss. Importantly, our scheme ensures that the borrowed bandwidth is promptly returned to the connections.

Extensive simulation results show that our rate-based QoS provisioning scheme outperforms the best previously-known schemes in terms of call dropping probability, call blocking probability, and bandwidth utilization.

Keywords: Bandwidth allocation, cellular networks, QoS provisioning, multimedia traffic, reservation schemes, rate-based schemes.

1. INTRODUCTION

We are witnessing an unprecedented demand for wireless networks to support both data and real-time multimedia traffic. While best-effort service suffices for data traffic, the usability of real-time multimedia applications is vastly improved if the underlying network can provide adequate quality-of-service (QoS) guarantees. Admission control and bandwidth allocation schemes can offer wireline networks the ability to provide their users with such guarantees. Due to host mobility and notorious bandwidth limitations, the QoS provisioning problem is much more difficult in wireless networks. For example, a mobile host may be admitted into the network in a cell where its needs can easily be met, but the mobile host may eventually move to a cell that has little or no resources to offer. Since the user's itinerary and the availability of resources in various cells is usually not known in advance, global QoS guarantees are very hard to provide.

Admission control refers to the task of deciding if a connection should be admitted into, and supported by, the network. Admission control is necessary for real-time, continuous media connections since the amount of resources requested by these connections may not match the level of resources available at the time of connection setup. Admitting a connection into the network is tantamount to a contract between the network and the connection: on the one hand, the network guarantees that a certain level of resources will be maintained for the duration of the connection; on the other hand, the connection is expected not to request additional resources over and above those negotiated at connection setup. Traditional QoS parameters include bandwidth, end-to-end delay, and jitter. In addition, there are some QoS parameters that are specific to wireless networks.

It is typical in most admission schemes to deny service to a connection whose requests for resources cannot be met by the network. In such a case, the connection is said to be *blocked*. We will follow common practice and refer to connections as "calls". In cellular networks, an important QoS parameter is the *call blocking probability* (CBP), denoting the chance that a new connection request will be denied admission into the network. A similar situation arises when an established connection in one cell attempts to migrate into a neighboring cell (i.e. a handoff is attempted). If the new cell cannot support the level of resources required by the connection, the handoff is denied and the connection is dropped. The *call dropping probability* (CDP) expresses the chance that an existing connection will be forcibly terminated during a handoff between cells due to a lack of resources in the target cell. The CBP and

CDP together offer a good indication of a network's quality of service in the face of mobility.

The traditional admission control process outlined above is, in many cases, too conservative and pessimistic. Indeed, multimedia applications are known to be able to tolerate and adapt to transient fluctuations in QoS [3, 4, 8]. This adaptation is typically achieved by the use of an adjustable-rate codec or by employing hierarchical encoding of voice and/or video streams [3, 4, 11, 12]. The codec, along with appropriate buffering before play-out, can allow applications to gracefully adapt to temporary bandwidth fluctuations with little perceived degradation in overall quality. The graceful adaptation of applications to transient fluctuations in QoS is fundamental in wireless networks, where QoS provisioning is a very challenging task. As we shall demonstrate in this paper, the additional flexibility afforded by this ability to adapt can be exploited by protocol designers to significantly improve the overall performance of wireless systems.

As we briefly mentioned, once a connection is admitted into the network, resources must be *allocated*, at the negotiated level, for the duration of the connection. It is important to realize that in a cellular network where the user may move through the network traversing a sequence of cells, this commitment cannot be only local to the cell in which the connection originated. If the connection is to be maintained after the user crosses the boundary between neighboring cells (i.e. after a handoff), the network must guarantee an appropriate level of resources in each new cell that the user traverses [1, 5, 7, 9]. Without detailed knowledge about the intended destination of each connection, honoring this commitment is an extremely difficult task [5, 7, 9]. To address this problem, many QoS provisioning schemes *reserve* resources in cells on behalf of mobile hosts in anticipation of their arrival. Not surprisingly, the resource reservation problem has recently received well-deserved attention. We refer the reader to [7] and [9] for surveys of recent literature.

There are, essentially, two approaches to resource reservation:

- fixed reservation – where a certain percentage of the available resources in a cell are permanently reserved for handoff connections, and

- statistical reservation – where resources are reserved using a heuristic approach. These approaches range from allocating the maximum of the resource requirements of all connections in neighboring cells, to reserving only a fraction of this amount [5, 7].

In this paper, we propose a novel, rate-based, borrowing scheme for QoS provisioning in high-speed cellular networks carrying multimedia traffic. Our scheme includes a fixed reservation pool for handoffs. At call setup time, the connections are expected to specify (1) their desired amount of bandwidth, and (2) the minimum amount of bandwidth needed to ensure an adequate level of quality. Our scheme attempts to allocate the desired bandwidth to every multimedia connection originating in a cell or being handed off to that cell. The novelty of our scheme is that in case of insufficient bandwidth, in order not to deny service to a requesting connection (new or handoff), bandwidth will be borrowed, on a temporary basis, from existing connections in the cell. Our borrowing scheme guarantees that no connection will give up more than its "fair share" of bandwidth, in the sense that the amount of bandwidth borrowed from a connection is proportional to its tolerance to bandwidth loss.

There are four important points to note about our scheme: first, our scheme guarantees that the bandwidth allocated to a real-time connection never drops below the minimum bandwidth requirement specified by the connection at call setup time. This is very critical to ensuring that the corresponding application can still function at an acceptable level. Second, our scheme guarantees that if bandwidth is borrowed from a connection, it is borrowed in small increments, allowing time for application-level adaptation. Third, our borrowing scheme is *fair* in the sense that if bandwidth is borrowed from one connection, it is also borrowed from the existing connections. Specifically, if borrowing is necessary in order to accommodate a requesting connection (new or handoff), every existing connection will give up bandwidth in proportion to its tolerance to bandwidth loss. This motivated us to refer to our scheme as *rate-based.* Finally, the borrowed bandwidth is returned to the connections as soon as possible.

Extensive simulation results show that our rate-based QoS provisioning scheme outperforms the best previously-known schemes in terms of call dropping probability and call blocking probability. In addition, our scheme ensures a high bandwidth utilization in the cellular system.

The remainder of this work is organized as follows: Section 2 reviews relevant results from the literature; Section 3 discusses the details of our rate-based QoS provisioning scheme: Subsection 3.1 describes the assumed system parameters, Subsection 3.2 discusses the new call admission protocol, while Subsection 3.3 gives the details of the handoff management protocol. Section 4 gives a detailed description of our simulation model. The experimental results obtained from extensive simulations are presented in Section 5. Finally, Section 6 offers concluding remarks and points out directions for further work.

2. STATE OF THE ART

In order to set the stage for our rate-based QoS provisioning scheme, we now briefly review the bandwidth allocation and reservation schemes proposed in [7]. We chose these schemes as a benchmark since they are arguably better than other comparable bandwidth allocation and reservation schemes found in the literature [7].

The traffic offered to the cellular system is assumed to belong to two classes:

1 Class I traffic – real-time multimedia traffic, such as interactive voice and video applications,

2 Class II traffic – non real-time data traffic, such as email or ftp.

When a mobile host requests a new connection in a given cell, it provides the following parameters:

- the desired class of traffic (either I or II),
- the desired amount of bandwidth for the connection, and
- the minimum acceptable amount of bandwidth, that is the smallest amount of bandwidth that the source requires in order to maintain acceptable quality, e.g. the smallest encoding rate of its codec.

One of the significant features of the admission control and bandwidth reservation schemes in [7] is that in order to admit the connection, bandwidth must be allocated in the originating cell, and, at the same time, bandwidth must be reserved for the connection in all the neighboring cells. Specifically, for a new connection to be admitted in a cell, the cell must be able to allocate the connection its desired bandwidth. For Class I connections, the call will be blocked unless the desired bandwidth can be allocated to it in the original cell, and some bandwidth can be reserved for it in each of its six neighboring cells.

During a handoff, an established Class I connection is dropped if its minimum bandwidth requirement cannot be met in the new cell or if appropriate reservations cannot be made on its behalf in the new set of neighboring cells. However, Class II traffic has no minimum bandwidth requirement in the case of a handoff, and a call will be continued if there is any free bandwidth available in the new cell.

Numerous approaches for reserving bandwidth have been reported in the literature [1, 2, 4, 5, 6, 7, 9]. The schemes presented in [7] use statistical reservation techniques based on the number of connections in neighboring cells, the size of the connections in neighboring cells, the predicted movement of mobile hosts, and combinations of these factors.

It is worth noting that the reservation schemes in [7] keep the dropping probability for Class I connections very low, since the mobile host should find bandwidth reserved for it, regardless of the cell to which it moves. But bandwidth may be wasted in the neighboring cells (the host can only move into one neighbor), and the blocking probability in those cells may increase because unused bandwidth is being kept in reserve. In general, the schemes described in [7] favor minimizing the CDP at the expense of the CBP and give Class I traffic precedence over Class II traffic.

3. THE RATE-BASED BORROWING SCHEME

It is clear that keeping a small pool of bandwidth always reserved for handoffs, as in [7], yields low CDP. However, in our scheme, the size of the reserved pool is not determined by requests from neighboring cells, but is *fixed* at a certain percentage of the total amount of bandwidth available in the cell. We found that this produced results similar to the best results reported in [7], without the overhead of communication between neighboring base stations to request and release reservations. To further reduce the CDP in our scheme, we treat the reserved pool very carefully. We do not allow bandwidth from the reserved pool to be allocated to incoming handoffs unless the bandwidth is needed to meet the minimum bandwidth requirements of the connection. Like [7], our scheme gives precedence to Class I connections; Class II traffic does not make use of the reserved bandwidth. In order to lower the call blocking probability as well as the dropping probability, our scheme allows for borrowing resources (i.e. bandwidth) from existing connections. Our borrowing strategy has the following interesting features:

1 No Class I connection will ever have to give up bandwidth beyond the minimum level negotiated at call setup time.

2 If the cell does not have enough residual bandwidth to accommodate an incoming call, the existing connections will temporarily have to give up a certain amount of bandwidth (see Subsection 3.2 for details).

3 If bandwidth must be borrowed, it is borrowed gradually, in small increments, to allow time for application-level adaptation.

4 As soon as bandwidth becomes available due to a terminating call or to a mobile host leaving the cell, the borrowed bandwidth will be returned to the connections.

5 Our scheme is fair, in the sense that if bandwidth is borrowed, all connections will give up an amount of bandwidth proportional to their tolerance to bandwidth loss.

6 Our scheme requires minimal computational overhead and no communication overhead.

3.1. CELL AND CONNECTION PARAMETERS

Each cell maintains a pool of bandwidth reserved for Class I handoffs which, initially, represents r percent of the total bandwidth. At setup time, each connection specifies to the cell in which it originates a *maximum bandwidth* M (termed the *desired bandwidth*) and a *minimum bandwidth* m. The difference between these two values is the *bandwidth_loss tolerance* (BLT) of the connection. Thus, $BLT = M - m$. We note that for constant bit rate (CBR) connections $M = m$, indicating no bandwidth_loss tolerance and, thus, $BLT = 0$.

The cell maintains a parameter, f, $(0 \leq f \leq 1)$, which represents the fraction of the BLT that a connection may have to give up, in the worst case. This fraction is the *actual borrowable bandwidth* (ABB) of the connection. Thus,

$$ABB = f \times BLT = f(M - m).$$

By accepting a new call, the cell agrees that the supplied bandwidth will not fall below a certain level that we call the *minimum expected* (MEX) bandwidth. By definition, $MEX = M - ABB$. It is worth noting that $MEX \geq m$. Simple computation shows that

$$MEX = (1 - f)M + fm. \tag{1}$$

To prevent borrowing from producing noticeable changes in a connection's QoS, we introduce another network parameter, λ. The ABB is divided into λ *shares*, each share being equal to $\frac{M-m}{\lambda}$. This provides the basis for a method of borrowing bandwidth gradually from a set of connections whose allocated resources may be quite different.

A cell is said to be operating at level L, $(0 \leq L \leq \lambda)$, when all its ongoing connections have had L (or more) shares borrowed from them. Observe that for any connection, the ratio between the amount of bandwidth given up and its bandwidth_loss tolerance is a *constant*. Specifically, we have

$$\frac{\text{amount given up}}{\text{bandwidth\_loss tolerance}} = \frac{\frac{Lf}{\lambda} \times (M - m)}{M - m} = \frac{Lf}{\lambda}, \tag{2}$$

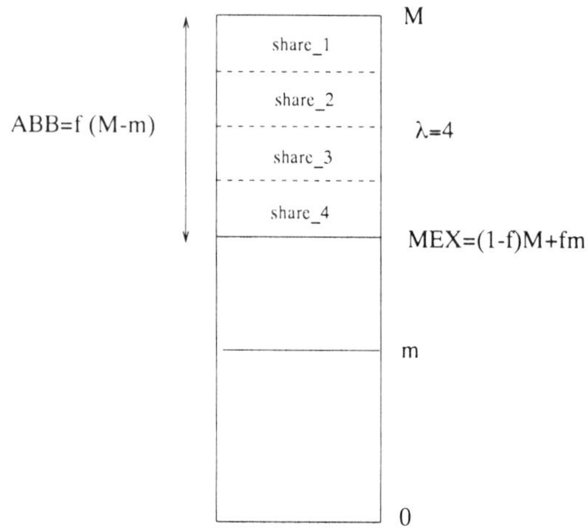

Figure 1 Illustrating the main connection parameters

which is a constant independent of the connection parameters. This is the sense in which we consider our borrowing scheme to be fair.

We note, however, that it is possible for a connection to be missing more than L shares after a handoff, due to the sacrifices made to prevent call dropping. However, our scheme attempts to restore bandwidth to handoff connections as soon as it becomes available. Figure 1 illustrates the concepts discussed above: this particular connection is in a cell with $f = \frac{2}{3}$ and $\lambda = 4$.

3.2. NEW CALL ADMISSION PROTOCOL

When a new call requests admission into the network in a cell operating at level L, the cell first attempts to provide the connection with an amount of bandwidth equal to its desired bandwidth minus L shares of its ABB, that is

$$M - \frac{ABB \times L}{\lambda} = \left(1 - \frac{Lf}{\lambda}\right) M + \frac{Lf}{\lambda} m. \tag{3}$$

If the amount of bandwidth specified in (3) exceeds the amount of bandwidth available, the cell tests to see if the call could be admitted if the cell progressed to level $L+1$. If transition to level $L+1$ will provide enough bandwidth to admit the call, the bandwidth is borrowed, the level is incremented, and the call is admitted; otherwise, the call is blocked. When the cell is operating at level $L = \lambda$, no more borrowing is allowed. It is important to note that our scheme never borrows from

CBR connections or from connections that have already lost more than L shares.

Every time bandwidth becomes available in a cell due to a connection releasing its bandwidth allocation, the cell will attempt to make a transition to the next lower level. As a result, the available bandwidth is returned to the connections that have lost bandwidth due to borrowing. All fluctuations in a connection's allocated bandwidth are gradual as only one share can be borrowed or returned at a time.

3.3. HANDOFF MANAGEMENT

The handoff admission policies differentiate between Class I and Class II connections. The reserved bandwidth is used only for Class I connections, which are admitted only if their minimum bandwidth needs can be met. When a Class I connection requests admission into a cell as a handoff, the cell checks to see if the minimum bandwidth requirement can be met with the sum of the available free and reserved bandwidth in the cell. If such is the case, the call is admitted into the cell and given bandwidth from the free bandwidth up to its desired level minus L shares. The connection is given bandwidth from the reserved bandwidth pool only if it is needed to reach its minimum requirement. If the minimum cannot be met using the free and reserved bandwidth, the cell tests to see if scaling to level $L + 1$ would free up enough bandwidth to admit the call. If so, the cell scales the other calls in the cell and provides the handoff call with bandwidth according to the guidelines described above.

On the other hand, Class II traffic will only be dropped if there is no free bandwidth left in the cell at all. The reserved pool is not available to these connections, because, as in [7], we assume that Class II traffic is able and willing to incur a possibly substantial fluctuation in service rather than be disconnected. Calls that have suffered a lowering of bandwidth due to a handoff will eventually be brought back to a reasonable level as their new cell has free bandwidth to give them. This is in contrast to the schemes presented in [7], which have no facility to improve connections which have been degraded due to a handoff.

4. SIMULATION MODEL

In order to evaluate the performance of our rate-based borrowing scheme, we implemented and simulated two other schemes for comparison. First, we implemented a request-based statistical reservation scheme from [7], termed the uniform and bandwidth-based model. According to this scheme, when reservations are made on behalf of a connection in neighboring cells, an equal amount of bandwidth is reserved in each neighboring cell, with no consideration of the most likely cell to which the host might travel. A cell does not reserve the sum of all the bandwidth it is asked to reserve, but just the largest of all the current requests.

We also simulated a simple scheme that reserves 5% of the total bandwidth in each cell for handoffs. New calls are admitted into the network if their desired bandwidth can be met, otherwise they are blocked. Class I handoffs are admitted if at least their minimum bandwidth requirements can be met. They are only given enough bandwidth from the reserved pool to meet their minimum, if there is too little free bandwidth available. Class II handoffs are admitted if there is any free bandwidth in the cell.

To simulate our rate-based borrowing scheme, we used a fixed reservation pool representing 5% of the total bandwidth. We set f to 0.5, thus permitting borrowing up to half of the bandwidth_loss tolerance. And we set λ to 10, so that each call had 10 shares to give.

To fairly contrast our scheme to the one in [7], we used the traffic types and characteristics given in [7], and modeled traffic behavior just as described there, with the exception of the handoffs. In [7], a handoff would occur during a connection with some given probability, and that probability would decrease exponentially with each successive handoff during the connection. We chose a different approach that seemed more realistic. We gave each mobile host a speed characteristic specifying the amount of time that will be spent in each cell during a call. Thus, longer calls are likely to experience more handoffs than shorter ones. Even with this minor change our results for the scheme from [7] correspond very closely to the results given there.

Table 1 shows the exact characteristics of the traffic used in our model. Each of the six types occurs with equal probability. For the results discussed in the following section, the speed was set to a host spending from 1 to 15 minutes in a cell, with an average of 5 minutes per cell. Each cell has 30Mbps of bandwidth. The network is a hexagonal grid of size 6×6 consisting of 36 cells. Traffic is provided to each cell at the level being measured. If a host moves out of the 6×6 grid, it is as though the connection ended normally – hosts do not "bounce" back into the network.

5. EXPERIMENTAL RESULTS

Figure 2 compares the values of bandwidth utilization for the request-based reservation scheme from [7], for a fixed reservation scheme with

CLASS	AVG BPS	MIN BPS	MAX BPS	AVG	MIN	MAX
Class I	30Kbps	30Kbps	30Kbps	180s	60s	600s
Class I	256Kbps	256Kbps	256Kbps	300s	60s	1800s
Class I	3000Kbps	1000Kbps	6000Kbps	600s	300s	18000s
Class II	10Kbps	5Kbps	20Kbps	30s	10s	120s
Class II	256Kbps	64Kbps	512Kbps	180s	30s	36000s
Class II	5000Kbps	1000Kbps	10000Kbps	120s	30s	1200s

Table 1 Traffic characteristics for our simulation model

$r = 5\%$, and for our rate-based borrowing scheme with $r = 5\%$, $\lambda = 10$ and $f = 0.5$, so that at most half of a call's bandwidth_loss tolerance can be borrowed. For the fixed reservation scheme and the rate-based borrowing scheme, at the maximum connection rate, the bandwidth utilization comes close to equaling the bandwidth outside of the reserved pool. The results for the request-based reservation scheme are worse than for the other two, because we did not implement a cap on the size of the reserved pool.

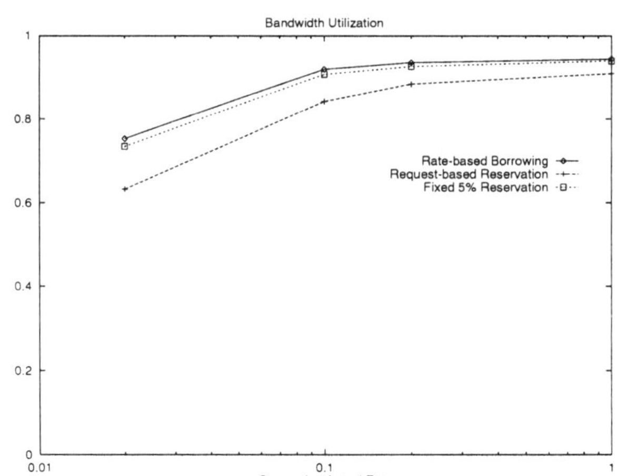

Figure 2 A comparison of bandwidth utilization by the three schemes

Figures 3 and 4 show, respectively, the CDP for Class I traffic alone and for Class I and II traffic combined. The borrowing scheme outperforms the other two schemes in both cases. In fact, the dropping probability for Class I connections is very close to zero. The motivation, of course, for favoring Class I connections by giving them exclusive use of the handoff reserves, is that real-time connections would suffer an actual

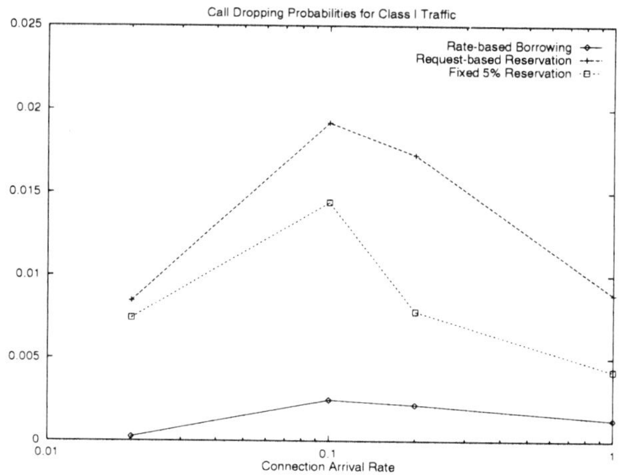

Figure 3 Illustrating call dropping probabilities for Class I traffic

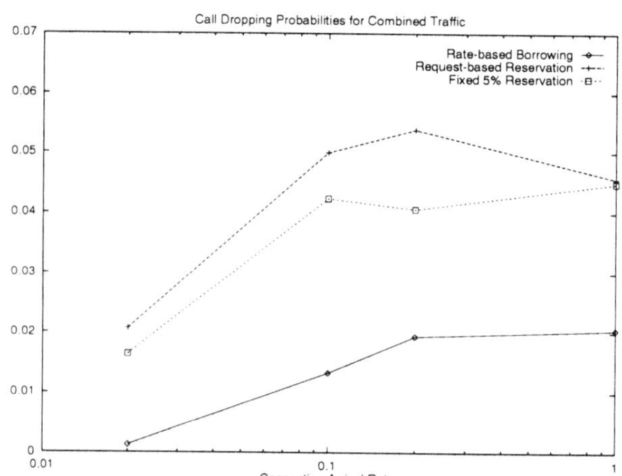

Figure 4 Illustrating call dropping probabilities for Class I and Class II traffic combined

loss by being dropped. We assume that a Class II application, although inconvenienced by being dropped, would be able to resume its transmission at a later time, without significant loss. Despite this, Class II traffic fares significantly better under our rate-based borrowing scheme than under the others; it is especially important that our scheme returns bandwidth to connections who have suffered cuts during a handoff.

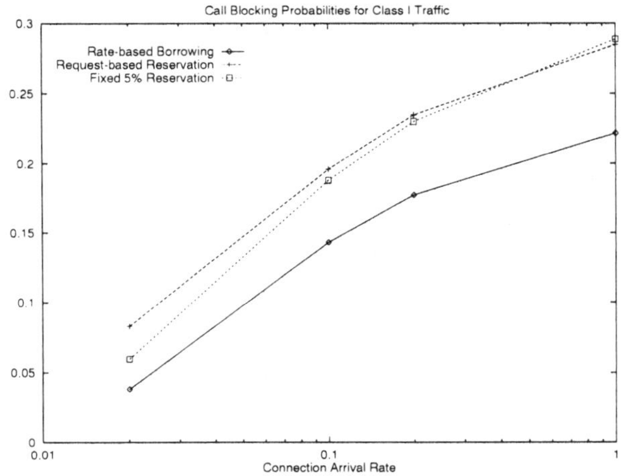

Figure 5 Illustrating call blocking probabilities for Class I traffic

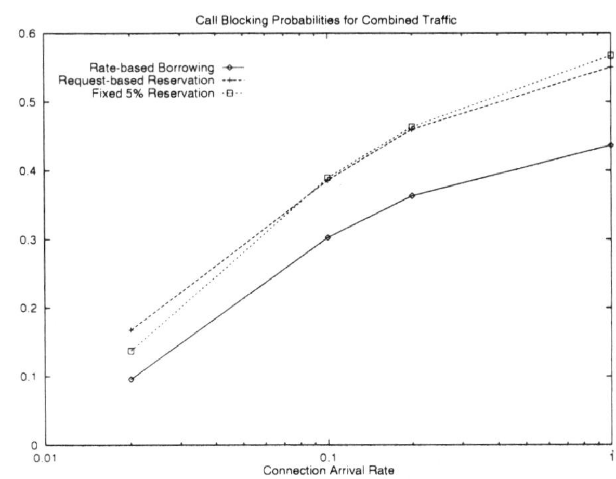

Figure 6 Illustrating call blocking probabilities for Class I and Class II traffic combined

Next, Figures 5 and 6 illustrate, respectively, the call blocking probabilities for Class I traffic alone and for Class I and II traffic combined. They demonstrate how borrowing allows a significant improvement in the CBP while also improving the dropping probability. As with CDP, the combined traffic also fares worse than Class I traffic alone in terms of CBP. However this is not due to any bias in the algorithms, but rather to the characteristics of the traffic being simulated. The Class II traffic requires more bandwidth on average.

6. CONCLUDING REMARKS AND DIRECTIONS FOR FUTURE WORK

Since multimedia traffic is intended mainly for human consumption [8], and since human senses are most forgiving, multimedia applications can tolerate and gracefully adapt to transient fluctuations in the QoS that they receive from the network. We have demonstrated that the additional flexibility afforded by the ability of multimedia applications to tolerate and adapt to transient changes in the QoS parameters can be exploited by protocol designers to significantly improve the overall performance of wireless systems.

Our main contribution is a novel, rate-based, borrowing scheme for QoS provisioning in high-speed cellular networks carrying multimedia traffic. To the largest extent possible, our scheme attempts to allocate the desired bandwidth to every multimedia connection originating in a cell or being handed off to that cell. The novelty of our scheme resides in the fact that, in case of insufficient bandwidth, in order not to deny service to requesting connections (new or handoff), bandwidth is borrowed, on a temporary basis, from existing connections.

An important characteristic of our rate-based borrowing scheme is that no connection gives up more than its "fair share" of bandwidth, in the sense that the amount of bandwidth borrowed is proportional to the connection's tolerance to bandwidth loss. Importantly, our scheme ensures that the borrowed bandwidth is returned promptly to the connections.

Extensive simulation results reveal that our scheme features very low call dropping probability, low call blocking probability, good bandwidth utilization, and reasonable success with keeping both classes of connections operating steadily near their desired bandwidth.

Several problems are open, though. First, it would be nice to abandon the idea of a fixed reservation of bandwidth in favor of an adaptive reservation strategy. One possible avenue in this direction is to use recent traffic history as a predictor of future demands. Second, it is important to perform a sensitivity analysis of the overall performance relative to changes in the values of the system parameters r, f and λ.

Bandwidth borrowing subjects connections to possibly frequent fluctuations in the amount of bandwidth they are provided. In the two comparison schemes, the only fluctuation in the bandwidth provided would occur due to a handoff. In a simulation run at a rate of one connection per second, we found that the bandwidth supplied to a connection fluctuated an average of once every 10 seconds. It is clear that this issue requires more research.

Bandwidth borrowing decreases the probability that calls will always be provided their desired amount of bandwidth. In simulation we noticed that the calls that lost bandwidth during a handoff were always steadily replenished; however, the network wide average bandwidth provided to each call was about 85% of the desired amount. In the case of gracefully adaptive multimedia applications it makes sense to introduce a new QoS parameter that specifies, for each connection, the fraction of time that it can tolerate operating below its expected bandwidth level. We are working on incorporating such a QoS parameter into cellular networks. This promises to be an exciting area for further research.

Acknowledgments

Part of this work was supported by ONR under grant N00014-97-1-0562.

References

[1] A. Acampora and M. Naghshineh, Control and Quality-of-Service provisioning in high-speed microcellular networks, *IEEE Personal Communications*, Vol. 1, Second Quarter 1994.

[2] P. Agrawal, D. K. Anvekar, and B. Narendran, Channel management policies for handovers in cellular networks, *Bell Labs Technical Journal*, 1 (1996), 96–109.

[3] S. Chen and K. Nahrstedt, Distributed Quality-of-Service routing in ad-hoc networks, *IEEE J. Select. Areas in Communications*, 17, (1999), 1488–1505.

[4] H. Kanakia, P. P. Mishra, and A. Reibman, An adaptive congestion control scheme for real-time video packet transport, *IEEE/ACM Transactions on Networking*, 3, (1996), 224–239.

[5] D. Levine, I. Akyildiz and M. Naghshineh, A resource estimation and call admission algorithm for wireless multimedia networks using the shadow cluster concept, *IEEE/ACM Transactions on Networking*, 5, (1997), 1–12.

[6] M. Naghshineh and M. Schwartz, Distributed call admission control in mobile/wireless networks, *IEEE J. Select. Areas in Communications*, 14, (1996), 711–717.

[7] C. Oliviera, J. Kim and T. Suda, An Adaptive Bandwidth Reservation Scheme for High Speed Multimedia Wireless Networks, *IEEE J. Select. Areas in Communications*, 16, (1998), 858–874.

[8] S. V. Raghavan and S. K. Tripathy, *Networked Multimedia Systems*, Prentice-Hall, 1998.

[9] H. G. Perros, K. M. Elsayyed, Call admission control schemes: A review, *IEEE Communications Magazine*, November 1996, 82–91.

[10] L. Trajković and A. Neidhardt, Effect of traffic knowledge on the efficiency of admission-control policies, *ACM Computer Communication Review*, 29 (1999), 5–34.

[11] N. Tran and K. Nahrstedt, Adaptive adaptation by program delegation in VOD, *Proc. Int. Conf. Multimedia Computing and Systems*, 1998, 96–107.

[12] B. J. Vickers, M. Lee, and T. Suda, Feedback control mechanism for real-time multipoint video services, *IEEE J. Select. Areas in Communications*, 15, (1997). 512–530.

Profile-based IN Service Provision for Third Generation Mobile Communication Networks

Hoang Nguyen-Minh, Harmen R. van As
Vienna University of Technology, Institute of Communication Networks
Favoritenstrasse 9/388, A-1040, Vienna, Austria
Tel.: + 43-1-58801-38815, Fax: + 43-1-58801-38898
Email: Hoang.Nguyen-Minh@tuwien.ac.at

Abstract: The number of mobile users requiring high-quality and mobility-related services is increasing ever so rapidly. An important issue is efficient management of service provisioning, locating mobile users as well as operating network service databases. In this paper, we introduce a system model that integrates the intelligent network (IN) and mobile communication networks, thus providing fast, flexible IN service design, creation, and provision independent of the infrastructure of the diverse communication networks. This paper also examines a profile-based replication strategy, a *user profile replication* (UPR) that can reduce the update costs. We also present a method to distribute network databases and service control management of the VLRs and MSC/SSPs by taking into account *user mobility behaviour.* A *most frequent location area* (MFLA) is introduced which also reduces the number of location updates in such a way that the change of the location information and the service profile database are managed by the VLR/MFLA of the user instead by the HLR/SCP. With the MFLA concept, the currently centralized database and the service control management becomes distributed. This reduces the communication (signaling messages) and computational cost (database accesses) as compared to the *distributed VLR scheme* and a *reference scheme* for UMTS/IMT-2000 regarding IN service provision and location management.

Keywords: IMT-2000, UMTS, Intelligent Network (IN), HLR, VLR, location update, mobility management.

1. INTRODUCTION

The third generation mobile communication systems (UMTS/IMT-2000) are designed to provide global operation and an enhanced set of services capabilities while at the same time improving network performance significantly. These networks will evolve from the existing wireless and wired networks by adding the necessary capabilities to support intelligent and mobility-related services and features. For that propose, the evolution will be based on increased IN capabilities for rapid introduction of services, efficient service control and provision, incorporating database and signaling capabilities necessary to support terminal, personal, and service mobility functions.

Many research works have been carried out to develop IN concepts for third-generation mobile communication networks. In [1], a network model to integrate the common functionalities of the mobile network and a method to provide the IN service efficiently have been proposed. This method downloads the IN service profile to the Visitor Location Register (VLR) registered by the mobile terminal (MT) during location registration. It also provides the IN service to the VLR when an IN service call arrives from the MT. In this paper, this scheme is referred to as the Lee's scheme. In [2], a first step towards a user-profile management and a virtual home environment (VHE) by employing IN capabilities set 3 (CS-3) has been proposed. A vertical architecture with full independence of the user-profile database and the service management can be used to facilitate the provision of multiple services, allowing services to be active in parallel.

In this paper, an integration of the IN and mobile communication networks is studied. We also propose a profile-based replication that can be applied to distribute service provision and location management for future mobile communications. A new location strategy called *user-profile replication* (UPR) is proposed. UPR reduces the communication (signaling messages) and computational costs (database accesses) as compared to Lee's scheme [1] and a reference scheme used for UMTS/IMT-2000 [5,6]. A *most frequent location area* (MFLA) is then introduced which also reduces the number of location update messages between the Home Location Register (HLR) and the Visitor Location Register (VLR) in such a way that the location information change and service profile downloading are managed by the MFLA of the user instead of the HLR and the Service Control Point (SCP). The UMTS/IMT-2000 scheme is considered as reference architecture for comparison. We give the protocols of location updating/registration and IN service call procedures. We compare the three algorithms and show the effectiveness of the UPR scheme by an analytical method.

The rest of the paper is organized as follows. In Sections 2 and 3, we describe the IN capabilities set (CS) and the integrated system model for mobile communication systems. In Section 4, we introduce the procedures for location update and IN service call setup of those schemes. We present an analysis and performance comparison of these algorithms in Section 5. Finally, some concluding remarks are given in Section 6. A list of abbreviations is given at the end of the paper.

2. IN CAPABILITIES SETS (IN CS) IN MOBILE NETWORKS

UMTS/IMT-2000 are third-generation mobile communication systems that aim to support global service portability with a QoS that is at least comparable to that offered by the wired networks in various wired and wireless environments. UMTS/IMT-2000 should take care of the interest of developing and deploying personal communication services (PCSs) in the sense that it could use functional model on IN functional entities (FEs) to develop new services in a modular way.

IN is a high-level network architecture that facilitates and speeds up service implementation and provisioning in a cost-effective manner. In addition, IN can be applied to several types of networks such as PSTN, ISDN, and mobile networks. The IN architecture based on CS-1/2/3 capabilities can be used to support mobile communication networks and the interest of integrating them in IMT-2000/UMTS has increased [2, 5, 6]. ITU-T has proposed the IN capability sets (CS) such as CS-1, CS-2, and CS-3 [3, 4]. CS-1 advocates a centralised architecture in the sense that it does not support the distribution of control and data storage functionalities. CS-1 specifies the IN services at the wired network level and generally supports only personal mobility. In CS-2, not only personal and terminal mobility is provided, but also service mobility in mobile networks. CS-3 is advised to provide integrated services of wired and wireless networks, as well as the first set of capabilities for *service cohabitation*.

ITU-T also specifies the IN Conceptual Model (INCM) that is the framework to describe and design the IN architecture. This model provides four views of telecommunication networks through which the services could be modelled. It comprises the Service Plane, the Global Functional Plane, the Distributed Functional Plane, and the Physical Plane. These four planes show the capability of IN from each different side [3, 4].

3. SYSTEM MODEL FOR INTEGRATING OF IN AND MOBILE NETWORKS

In this section, we analyse the integrated model and study how IN services are provisioned in the third generation mobile communication networks such as UMTS/IMT-2000. The proposed system model for third mobile communication networks based on IN is shown in Figure 1. In this configuration, MSC/SSPs (service switching points in the mobile switching centers) and the service control points (SCPs) are interconnected over a common channel-signaling network (SS7 network).

In IN, the SSP recognizes the request for the IN service and then sends the query to the SCP, which processes the IN service profile and returns it to the SSP. In mobile communication networks, the Home Location Register (HLR) and the Visitor Location Registers (VLRs) have been used to manage the location of the MT and the subscription profile of the users. Thus, an SCP constitutes a distributed transaction system in which functions are distributed among the *service control modules* (SCMs), the *service data module* (SDM), and the HLR. The HLR in the model only stores the home user profiles and their current location. The SDM can be divided into many modules, which store a common set of data for all the services supported by that domain. With this integration model it is possible to efficiently implement various services as well as to ensure mobility. This model also supports the separation of the mobility-related data managed at the *transport level* from the user-related service data managed at the *service level*.

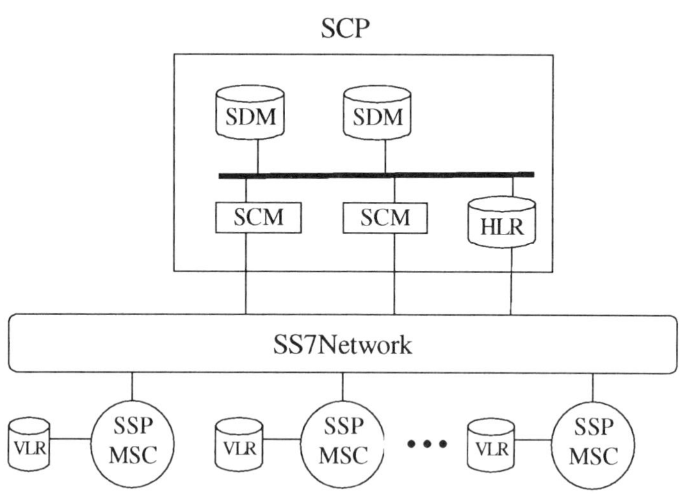

Figure 1. Reference architecture

3.1 Lee's Scheme

Lee et al. have proposed a distributed VLRs scheme [1], which can reduce the signaling costs and the load of the SCP. In this strategy, the SDM is distributed to the VLR where the user is currently located. User and service profiles are downloaded to the VLR when an MT performs the location registration. Another aim of the VLR is to response to IN service requests from the corresponding MT during the call setup procedure. The MSC/SSP recognizes the IN calls of the MT and sends the IN service request to the VLR instead of the SCP.

3.2 User profile replication (UPR) with MFLA strategy

A. Strategy Definition

In the UPR, the system handles a per-user mobility profile recording his most probable mobility patterns, $\{(A_1, p_{A1}), (A_2, p_{A2})\ldots(A_k, p_{Ak})\}$, whereby p_{Ai} denotes the probability that an MT is in location A_i, and $p_{A1} > p_{A2} > \ldots > p_{Ak}$. The probability κ to find an MT in $\{A_i\}_{i=1}^{k}$, whereby k is number of location areas, is given by

$$\kappa = \sum_{i=1}^{k} p_{Ai} \tag{1}$$

The profile for each user is stored in the SDM/SCP and the VLRs by $\{A_i\}_{i=1}^{k} = \{A_1, A_2, \ldots, A_k\}$. Each user must store an ordered set of locations $\{A_i\}$ considered as a location area for him in its registered MT. The first entry in $\{A_i\}$, A_1, is called the most frequent location area (MFLA). Within $\{A_i\}$, the MT does not generate a location update. For incoming calls, the MT has to be paged in all the locations in the $\{A_i\}$. When the MT moves outside $\{A_i\}$, it has to register the new location to the VLR/MFLA instead of the HLR/SCP. If an IN service is requested from an MT, the IN service profile and the service-related database can be provided locally by the replicated VLR where the MT is currently located. It has then not to be provided by the SCP.

This scheme can reduce the number of unnecessary location updates by taking into account the user mobility patterns, but it increases paging delays, required memory, as well as profile updating cost.

B. User Profile List (UPL) updating and maintaining

Replicating user profiles can reduce the location update time. This is because the location of the called MT is more likely to be obtained by a single query from the local VLR, rather than by a high-latency remote query. However, the associated cost of replication with caching is the update cost incurred in processing the user profile to form the *user profile list* (UPL), replicating and maintaining consistent replicas to the locations where it is stored. Denote this cost as $C_m(k)$, whereby k is the user profile size. We assume that the frequency at which the system invokes the list maintenance function is proportional to the user call arrival rate λ_c and the number of calls originated and terminated by the user, N, for updating the UPLs. Thus, the average cost per time unit of maintain the UPL is

$$M = (\lambda_c/N)C_m(k) \quad (2)$$

Clearly, if a user neither calls nor receives calls, the UPLs should not be changed. If the user makes many calls, the system should be prepared to update the list every M calls, if necessary.

4. INTELLIGENT NETWORK SERVICE IN UMTS/IMT-2000

4.1 Location update protocol

Figures 2a and 2b show the conceptual location registration in IMT-2000/UMTS [7,8], further on referred to as a reference scheme, and in Lee's scheme [1].

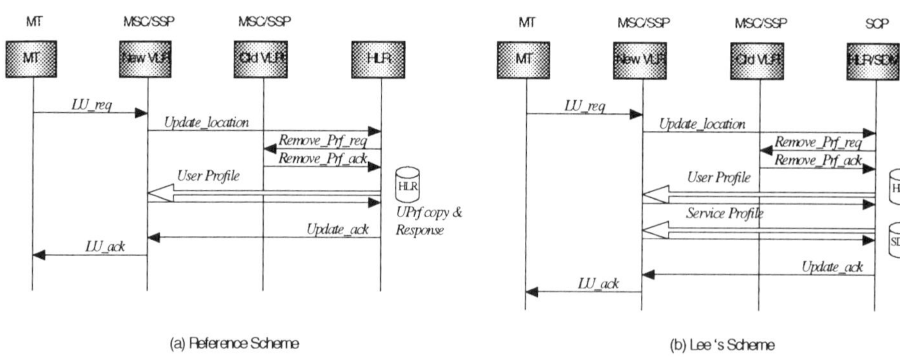

Figure 2. Location update procedures

If an mobile terminal (MT) roams to another mobile switching center (MSC) but still remains inside $\{A_i\}$, the URP scheme does not generate an location update request (*LU_req*) and hence does not involve costs (*case 1* in Figure 3). The MT notifies the system only when it leaves the area covered by *{A_i}*. For the first registration outside {A_i} (*case* 2 in Figure 3), the new MSC gets the *user profile* and the *service profile* from its MFLA, whereby the new location of the MT is forwarded to the MFLA by a forwarded pointer. If the MT moves to another MSC outside of {A_i}, the message sequences are identical to the *case 2* procedures except for a request for a redirection pointer from the old MSC to the new one, as can be seen in *case 3* of Figure 3. The redirection pointer in the location database of the old MSC helps to reduce the miss rate during the call delivery procedure. When an MT re-enters its *{A_i}*, the above-mentioned messages are exchanged and in addition the MFLA is notified that the MT is inside *{A_i}* and the forwarded pointer is deleted as shown in *case 4* of Figure 3. In this case, all profiles related to the user still remain in *{A_i}*, thus the copy of those profile is unnecessary. This scheme reduces the access to the HLR/SCP as much as possible. The HLR/SCP keeps all the user profiles, service profiles and their pointed MFLAs, which will be updated and maintained *offline* periodically depending on the number of calls originated and terminated by the users.

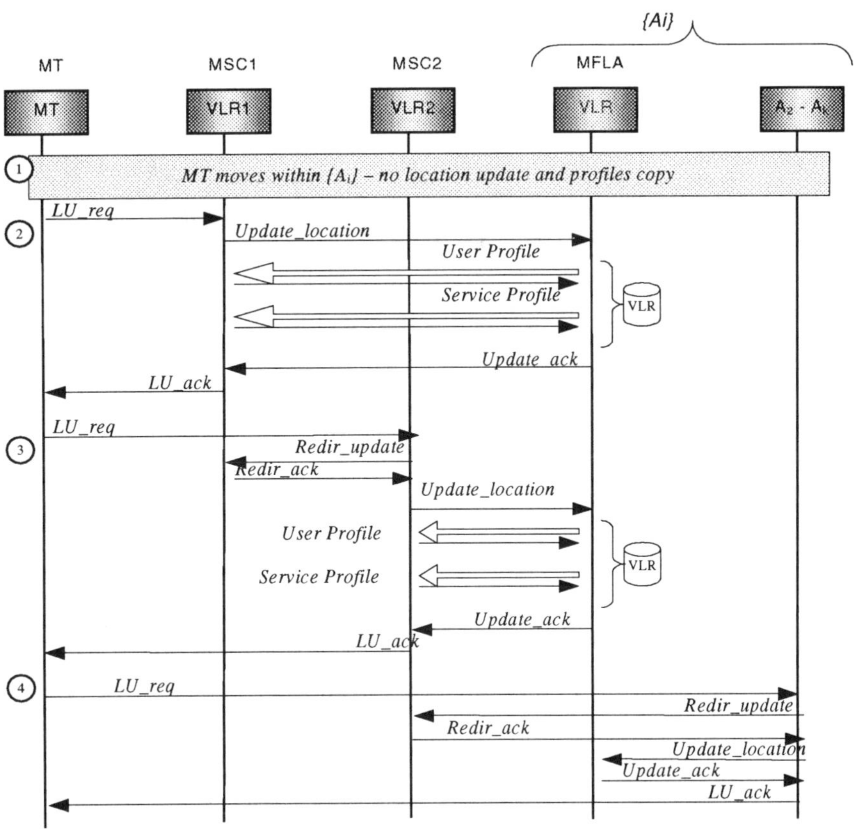

Figure 3. Location update procedures for UPR scheme

4.2 IN service call setup

Figure 4a illustrates a scenario for an IN service call setup for the reference scheme. The call setup request is first routed to the nearest MSC/SSP. The SSP recognizes the IN call and triggers an IN user checkpoint service, which is responsible to check the calling user's right at the SCP. The SCP processes the IN service profile and returns the result to the SSP. Many requests from many MSC/SSPs may result in an overload at the SCP as well as a bottleneck at the HLR.

In the Lee's and UPR schemes, both user profile and service profile are downloaded into the VLR/SSP when the user performs the location registration. Thus, the processes for the IN service and queries of related profiles take place at the local VLR within the SSP in a cost resulting reduction. The IN service call setup procedure is shown in Figure 4b.

In this procedure, we assume that the SSP is responsible for chaining the elementary services and applies its basic triggering capabilities of the IN CS-3 [2,4], which enables two or more services to be active during the call on the same SSP.

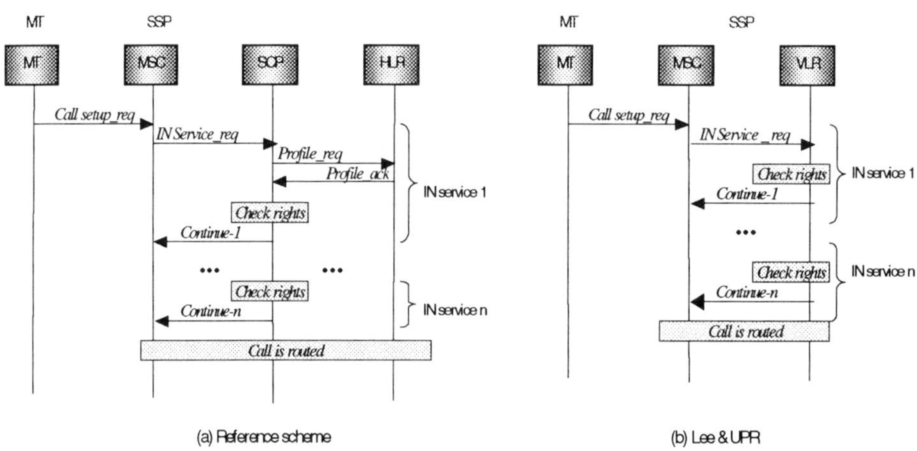

Figure 4. Scenario for IN service call setup

5. PERFORMANCE ANALYSIS

5.1 System Analysis

In this study, we assume that the density of users is uniform throughout the area, that the direction of motion with respect to the border is uniform on $[0, 2\pi)$, and that all the cells are of the same shape and size as well as that they form together a contiguous area. Let t_c and t_m be independent and identically distributed random variables representing the call inter-arrival time and the location area (LA) residence time. We assume t_c and t_m to be exponentially distributed with the rate of λ_c and λ_m, respectively. To simplify the analysis, we assume that all MTs have same values for κ; λ_c, and λ_m.

We assume that every VLR serves a location area (LA) that can be connected to a corresponding MSC/SSP. For simplicity, we also make in this contribution the assumption that all VLRs are connected to only one SCP/HLR through a common channel-signaling network No. 7 (SS7 network). Let p_{IN} be the probability that call is an IN service call.

In the UPR with the MFLA strategy, the system maintains a record of each user's most itineraries. Thus we can assume that the probability distribution of a user's location is known exactly. In practice, it will be necessary either to ask the user to provide it or to estimate this using information such as the user's past calling history. The MT keeps track of the list of the location areas $\{A_i\}$. When the system changes the contents of the set, it will notify the MT.

5.2 Cost Analysis

To compare the different schemes in terms of cost, each activity is given a cost. The frequency of occurrence for each of these events for the different schemes is computed from the message sequence diagrams. The IN service management cost is divided into two components: (1) the *updating cost, U*, which incurs in completing the location update procedure and service profile download cost, (2) the *IN call processing cost, I,* which incurs in completing the IN call setup procedure. We consider communication (routing, sending, receiving messages) and database processing as our basic measures of cost. The location update and IN call setup procedures involve the setup of dedicated channels, thus we ignore this cost for all calculations.

Some costs and parameters used to evaluate the analytic model are summarized in Table 1.

Table 1. List of cost parameters

Parameters	Meaning
D_h	The cost for a query or an update of the HLR
D_v	The cost for a query or an update of the VLR
D_s	The cost for a query or an update of the SDM/SCP
N_{lo}	The cost for a signaling message through the local link within the MSC
N_{re}	The cost for a signaling message through the remote link to another MSC

Table 2. Updating activities, probabilities and costs

Scheme	Case	Description	Probability	Cost
IMT-2000	--	--	--	$U_{IMT}=6N_{re}+2D_v+D_h$
Lee	--	--	--	$U_{Lee}=8N_{re}+3D_v+D_h$
UPR	1	Moves inside $\{A_i\}$	$p_{u1}=\kappa^2$	$U_1=0$
	2	Exit from $\{A_i\}$	$p_{u2}=\kappa(1-\kappa)$	$U_2=6N_{re}+3D_v$
	3	Moves outside $\{A_i\}$	$p_{u3}=(1-\kappa)^2$	$U_3=8N_{re}+4D_v$
	4	Entry into $\{A_i\}$	$p_{u4}=\kappa(1-\kappa)$	$U_4=4N_{re}+2D_v$

Table 3. IN call processing costs

Scheme	Cost
IMT-2000	$I_{IMT}=np_{IN}(4N_{re}+2D_s+D_h)$
Lee & UPR	$I_{Lee,UPR}=np_{IN}(4N_{lo}+2D_v)$

Tables 2 and 3 summarise all possible events, their corresponding probabilities, and the costs for updating and IN call processing procedures of the reference model, Lee's, and UPR scheme. The average cost per unit time for the UPR scheme can be expressed as the product of the cost and the rate of their occurrence as

$$U_{UPR}=\lambda_m\sum_{i=1}^{4}p_{Ui}U_i+\frac{\lambda_c}{N}C_m(k)=\lambda_m C_u+\frac{\lambda_c}{N}C_m(k) \tag{3}$$

The total cost per unit time is the combined cost of location updating and IN service management, denoted as T, which is the sum of the two costs, given as

$$T=U+I \tag{4}$$

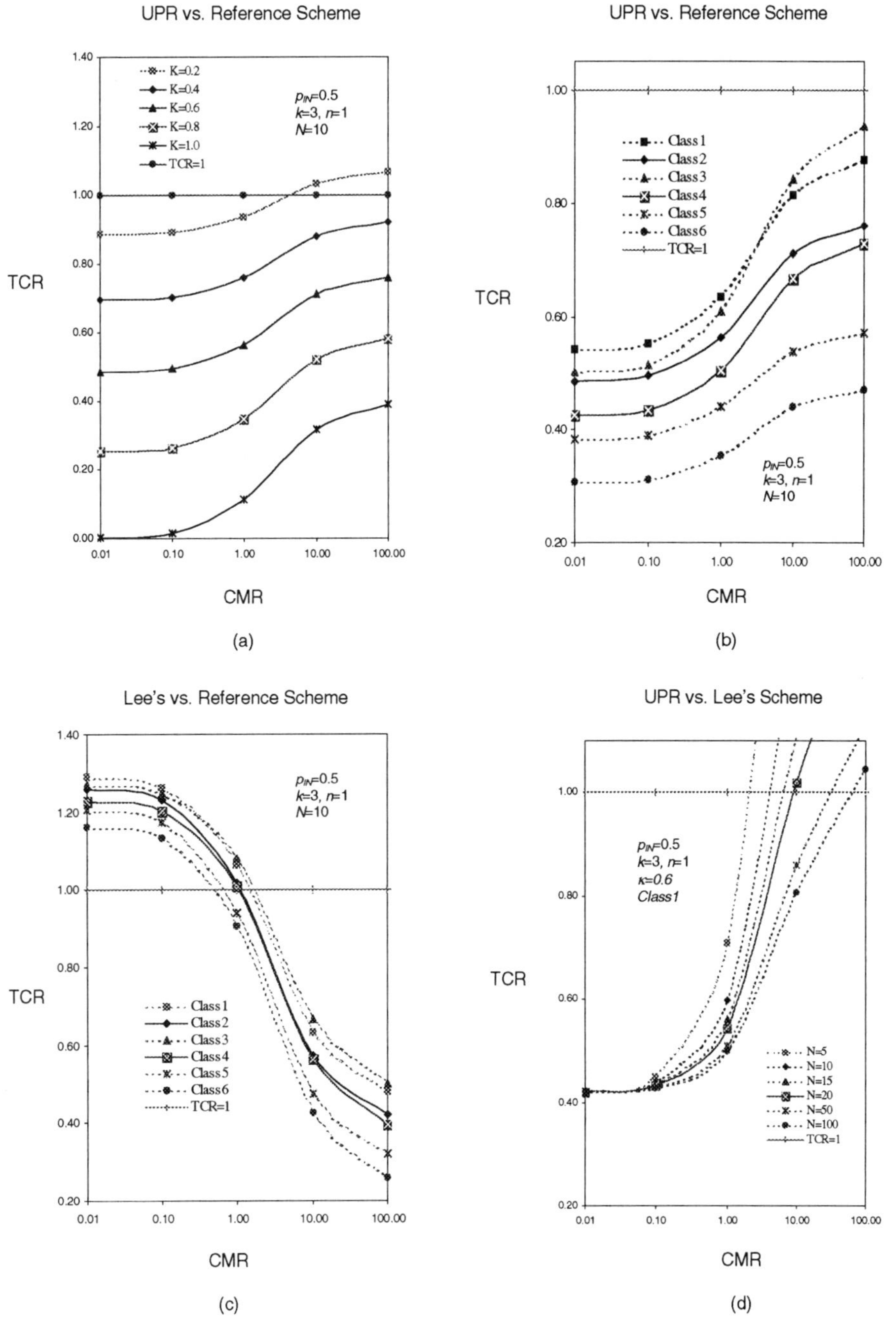

Figure 5. Total Cost Ratio (TCR)

5.3 Numerical Results

In evaluating and comparing the cost of those strategies, we consider the call to mobility ratio (*CMR*) as the ratio of the call arrival rate to the mobility rate such as

$$CMR = \frac{\lambda_c}{\lambda_m} \tag{5}$$

In order to be able to estimate the total cost per unit time, the call arrival rate λ_c, and the rate at which the MT moves between LA, λ_m, are needed. We use the call to mobility ratio (*CMR*) to study the performance of these schemes and compare the cost of those strategies by the ratio of the total costs (*TCR*) described as follows

$$TCR = \frac{T}{T_{ref}} = \frac{C_u + CMR * C_i}{C_{u-ref} + CMR * C_{i-ref}} \tag{6}$$

Table 4. Coefficient parameters

Class	N_{lo}	N_{re}	D_v	D_s	D_h
1	1	2	1	2	3
2	1	2	1	2	5
3	1	3	3	2	9
4	1	3	3	2	15
5	1	2	1	2	10
6	1	3	3	2	30

The methodology of evaluation used is to establish a common unit of measure for all costs terms, for example time delay. We use the six sets of values for the cost parameters N_{lo}, N_{re}, D_v, D_s, and D_h as shown in Table 4. The value of N_{lo} is normalized to one since it can be seen as the lowest among the other costs. The optimum value of k, the number of location areas where user profiles are replicated, is limited by the incremental cost of maintaining the list. It could be chosen as large as the cost of the list maintaining permits. According to [9], $k \in \{3,4,5\}$ are good choices for the practical systems, to meet the call set-up delay requirements. The number of calls that the system *offline* updates (or maintains) the user profile lists (UPLs), denoted by *N*, can be chosen as 10. The cost for updating and maintaining UPLs per user is supposed to be the product of *k* and its average updating cost C_u. The number of service checkpoints, *n*, selected from Figure 4 and Table 3, is one for the analysis. It's clear that if n is greater than one the cost of IN service management reduces due to the access and processing of other IN services still remaining at the local VLR/SSP.

Figure 5a plots the variation of the normalized total cost between the UPR and reference model schemes with a wide range of CMR (from 0.01 to 100). As κ increases, the total cost for UPR falls rapidly. Figure 5b shows the ratio of the total cost between the UPR scheme and the reference scheme for the six classes in Table 4. For the proposed strategy, UPR always performs extremely well compared to the reference scheme. When the cost of querying/updating of the HLR is much higher compared to the VLR, the UPR shows increased performance. Lee's scheme (distributed VLR scheme) reduces the IN service management cost significantly only if the user call to mobility ratio is greater than one (CMR $\geq$ 1) as shown in Figure 5c. Additionally, as the frequency of *offline* updating the user profile lists (UPLs) decreases (N varies from 5 to 100), our proposed scheme of UPR gives a better performance than Lee's scheme (Figure 5d).

6. CONCLUSION

In this paper, we propose an integration model of IN in future mobile communication networks and a profile-based replication method to efficiently provide IN service. In the proposed network model, two mechanisms have been applied: the distribution of information of the HLR to VLRs and the user profile replication (UPR) strategy for handling the location management and service provisioning. Detailed procedures for the message sequence for location updating and IN service call setup protocols have been presented. Performance analysis shows that the UPR scheme can give the best solution when the replication of user profiles is optimal if the user has a highly localised mobility pattern. We use the most frequent location area (MFLA) to reduce network signalling traffic and database updating as well as the querying delay during location registration and call delivery procedures. The proposed MFLA introduces a new concept for distributed location database access and service control management dependent on per-user mobility behaviour. The numerical results show that the UPR strategy can significantly reduce the total cost compared to other schemes. This UPR strategy could be a very good candidate for distributed location management of future mobile communication networks. In these strategies, user profiles have been copied to a visited network to improve performance but security, integrity, and confidentially must be considered and maintained.

ABBREVIATIONS

FE	: Functional Entity
IN	: Intelligent Network
IN CS	: IN Capabilities Set
IMT-2000	: International Mobile Telecommunications 2000
ISDN	: Integrated Services Digital Network
HLR/VLR	: Home/Visitor Location Register
LA	: Location Area
LU	: Location Update
MFLA	: Most Frequency Location Area
MSC	: Mobile Switching Center
MT	: Mobile Terminal
PCS	: Personal Communication Service
PSTN	: Public Switched Telephone Network
SCM	: Service Control Module
SCP	: Service Control Point
SDM	: Service Data Module
SS7	: Signaling System No.7
SSP	: Service Switching Point
UMTS	: Universal Mobile Telecommunication System
UPL	: User Profile List
UPR	: User Profile Replication
VHE	: Virtual Home Environment

References

[1] Y. Lee, H. S. Kim, and J. S. Song, "*IN service provision using VLR in IMT-2000 network*", Computer Communications, Vol. 23, No.11, June 2000, pp. 1027-1033.

[2] C. Diquelou, "*IN CS-3: The first step towards User Profile Management and the VHE concept*", Proceedings of ICIN'2000 Conference, Bordeaux, France, Jan. 2000, pp. 145-150.

[3] T. Magedanz, R. Popescu-Zeletin, "*Intelligent Networks*", ITP Press, 1996.

[4] ITU-T, Recommendations [Q12xx] of Intelligent Network, Sept. 1999.

[5] M. Laitinen, J. Rantala, "Integration of IN services into future GSM network", IEEE Comm. Mag., June 1995, Vol. 33, No. 6, pp. 76-85.

[6] N. Faggion, T. Hua, "*PCSs through the evolution of fixed and mobile comms. and the IN concept*", IEEE Network, July/Aug. 1998, Vol. 12, No. 4, pp. 11-18.

[7] ITU-T, Draft Recommendation Q.FIF, Sept. 1997.

[8] ETSI, 3G, TS [22.xxx], Documentations of Technical Specification Group Services and System Aspects, 1999.

[9] G. P. Pollini, C-L. I, "*A Profile-based Location Strategy and its Performance*", IEEE JSAC, Vol.15, No. 8, Oct. 1997, pp. 1415-1424.

Agents for Support of WAP-based Services

Mihhail Matskin and Runar Bell
Department of Computer and Information Science, Norwegian University of Science and Technology,N-7491 Trondheim, Norway
e-mails: {Mihhail.Matskin,Runar.Bell}@idi.ntnu.no

Key words: Intelligent agents, mobile communication services, WAP, WML

Abstract: We consider an application of agent-based methods to development of WAP-based services. Special attention is given to supporting off-line autonomous processing and personalizing WAP-based services. A developed Java-based WML-Agent system is used as an example of utilization of agent technology in WAP-based communications.

1. INTRODUCTION

WAP (Wireless Application Protocol) is a new rapidly growing area of mobile communications. Because of its promising as a standard the current tendency that more and more companies support WML–pages for their customers will be continued in future. This tendency is also a basis for fast growing number of available WAP-based services. At the moment most of efforts are directed to transferring information from HTML to WML-oriented view and this is a practical need because of without available information there is no basis for advanced services. The problem is that specific features of WAP-based communications often are not enough taken into account in new implemented services. The WAP/WML is a new field and there should be a new original way of utilization of its features. Of course there is a similarity between utilization of HTML-pages and WML-pages. However the traditional solutions for WWW cannot be cloned for WAP/WML directly. The basic differences as we see them are as follows:

- The users of WAP will not surf through pages not knowing what they are looking for but rather they will surf in order to cover their needs
- It may also be assumed that WAP users will not want to surf the Internet at all but just have a direct access to a specified information

The last assumption is, maybe, too limited from our point of view because of anyway users may need search of desired information. However a definitely important requirement from the WAP perspective is a high level of personalization in the WAP-based services. We see the following reasons for that:

- The first reason is a high cost of mobile communications
- The second reason is that most of WAP users assumed to be non-experienced Internet users (which is, maybe, too limited view because of knowledge of Internet is increased very fast)
- The third reason is a limited expressive capability of WAP devices

It is possible that the last limitation will be relaxed in the future but at the moment we should take it into consideration. A consequence of the first reason is that the mobile user connection time to the network for requesting information should be minimized and that as much as possible work for surfing should be done off-line. Another consequence from the first reason is requirement to precision of delivered information – scrolling of a huge amount of information usually given by WWW search engines may require a lot of time (and money as a duration of mobile connection can be quite long).

In order to be able to get information from WWW the user can make use of several methods:

- It is possible to make search using some of the Internet search engines (AltaVista, HotBot, Infoseek etc.)
- It is possible to browse through one of many catalogue listings (Yahoo etc.)
- It is possible to guess a hyperlink to a place where the user may find the information s/he is looking for or just go browsing through hyperlinks

These ways of obtaining information can be applied for WAP-based services only partly because of the above-mentioned reasons. However in the case of WAP the user may better know what s/he is looking for and where to find it and this can be a basis for efficient solutions.

Some support for customers of WAP-devices is provided by operators of mobile networks who specify available services on special homepages. Definitely this branch will be developed in future. However our goal is to allow user to choose an information s/he is interested in, to make the chosen information available, updateable and adaptable and to make such service flexible and dynamic.

In order to reach the goal we apply an agent-based approach. In particular we would like to give the WAP user an agent-based tool for selection, personalization, finding and updating information to be accessed via WAP devices.

The rest of the paper is organized as follows. First we consider briefly WAP features and intelligent agent characteristics. Then we present our vision of utilization of agents in WAP-based communications. Next we describe a Java-based WML-Agent system which we develop for providing agent-based services in WAP-based communication. Then we consider some related works. In the conclusions we consider a summary of results and future works.

2. WAP – WIRELESS APPLICATION PROTOCOL

Here we consider very briefly only basic features of the WAP.

The basic idea of WAP [7,18] as a standard for wireless communications is to extend communication ability of wireless telephony. In other words cellular phones are no longer just phones but rather communication devices capable of running applications and capable to communicate with other devices and applications over a wireless network. The standard takes into account the limitations of wireless data network compare to wired networks (for instance, less bandwidth, more latency, less connection stability and less predictability) and limitations of mobile handsets compare to personal computers (for instance, small screen size, complicated text input, limited memory, slow CPU).

The WAP standard specifies end-to-end application protocol and an application environment based on a browser as two essential elements of a wireless communication. The application protocol is a layered communication protocol which is embedded in each WAP-enabled user component. The network side includes a server component which implements the other end of the protocol that is capable of communicating with any WAP component. The server component often takes the role of a gateway for routing requests from user component to an application server and the gateway can be physically located in a telecom network building a bridge between wireless and computer networks.

A “microbrowser” integrated into a handset allows to display information to a user and to accept requests from the user. The basis for information representation is WML – a tag-based document language which is an XML-based language. The basic unit of WML is a card which specifies a single user interaction. Navigation occurs between cards which are grouped into decks. A deck is the top-most element of a WML document and it presents possible alternatives for user interactions.

It is assumed that when the user logs on to a WAP gateway s/he supplies the address of a starting page (WML deck), a portal, and then performs interactions according to the deck description. It is desirable for user to be able to find all the information s/he would ever need linked directly or with as few links as possible from the starting deck. It is also preferable that favorite decks/cards which the user visits more often can be reached through a fewer set of links than decks/cards which are used once for a while.

3. INTELLIGENT AGENTS

We apply an agent-based approach to support of WAP-based communication. Here we give only a brief description of the basic notions related to agents (for more detailed overview of basic agent properties we refer, for example, to [1,2,3,12,20]).

“Agents” is a new approach to building software in a distributed decentralized environment. In spite of different views to agents in different research societies there increases an agreement of what are basic agent characteristics. Here we summarize some essential characteristics by adopting a definition from [20] as the most representative one from our point of view. According to that we consider an agent as a software entity which represents somebody (a human or another agent) in computer environment and enjoys the properties of autonomy, pro-activity, reactivity and social ability (they are referred as weak agent properties in [20]). By autonomy an ability to operate without intervention from outside is assumed. By pro-activity a goal-oriented behavior and ability to take initiative is understood. Reactivity assumes perception and affecting the environment and social ability assumes communication with other agents in order to achieve own goals.

Among other possible agent characteristics we should also point out mobility [20] and we consider mobility as a very useful agent feature. However, we treat mobility rather as an optimization of agent communication than a necessary agent property. This is why agent mobility was left outside the scope of the paper.

In this work we mostly employ autonomy, pro-activity and reactivity of agents while for supporting social ability we refer to [8,9] where architecture for agents cooperative work support is developed. The above-mentioned agent characteristics are very suitable for achieving our goals because of they allow us to support personalization of behavior, off-line activity and environment monitoring as specific requirements of WAP-based communication which we referred in the Section 1.

4. WHAT DO AGENTS HAVE TO DO IN WAP-BASED APPLICATIONS?

As we mentioned in the Section 1, our concern is to provide user with a means for description of required services and for making such services available via WAP-device (WAP-phone) in a flexible way.

The way of how WAP-services are now supported by service providers is proposing some predefined set of services and making these services available via WAP-handset menu. A typical set of services includes weather forecast, news, banking and finances, newspapers, travel information, currency exchange rates, access to catalogues etc. These services are quite reasonable selection but they can't satisfy all customers of course and, what is more important for us, they are not enough personalized. By personalization we understand not only supporting static profiles but also dynamic adaptation. How to personalize the set of services and select/add new services?

Let us consider some possible scenarios of user WAP-based communication.

The ideal starting point for user communication with WAP-based device is a WML deck which always reflects what the user wants to see at that precise moment. In this way it should be as easy as possible to retrieve the information without typing advanced queries or long hypertext links in order to get to a certain destination. For example, if the user likes to read the international news or weather forecast on his way to work in the morning, then the best way would be to get an overview of all the latest, and still not read, news from the favorite news agencies as either a starting card or as a card close to the starting card in the hierarchy. Of course, during the day user preferences or requirements for information could be changed to, for example, more business-oriented news in daytime or TV-programs in the evening on the way home. Taking this into account the starting card should reflect such shift of focus of interest and it should allow to get access to preferable information as easy as possible. At the same time the previous focus of interests could be not important anymore and it should not correlate with the current focus.

Some another scenario may reflect a permanent need for awareness of important events and newly available information of great importance for the customer. As a typical example we can consider an information from stock exchange market, especially, awareness of changes in particular stocks quotas beyond a predefined threshold. Another scenario may require customer to search for some important information which can't be presented as an element in his card before hand (or can't be clearly defined and may require some filtering) and to notify him about its availability.

We consider only few different possible scenarios for a user. Each of these scenarios could be specialized for different customer. We think that it would be impossible to create a WAP portal which is accommodated for all the different needs a user might have. It also may be very hard and time consuming work to manually create lots of portals which are meant to be updated regularly and reflect different user needs. From the user point of view, the best way to reach the documents of interest as easy as possible could be make them available through own personalized portal. Since the users are interested in calibrating their portal not very often somebody is needed to keep it constantly up to date. This is a job which is just perfect for a software agent.

Software agent can be an autonomous constantly running piece of software. At period when it is not necessary to do anything, it will hibernate but when a change in the source for portal or some event occurs, the agent awakes and updates the information on the WAP portal. When the user connects to personal portal s/he will get a pre-generated and up-to-date documents without waiting for completion of queries at servers around the world. As we refer to agent features in the Section 3 autonomy, reactivity and pro-activity should play a key role in implementing such agents. Pro-active behavior should rely on a customer profile presented in a system. In a more advanced case agents learning ability can be developed and monitoring of customer's actions may be a source for deciding about a preferred time for availability of particular information. In the same way a pro-active reasoning using available sources about customer (for example, from user profile, from electronic notebook etc) can be quite advanced with using logical rules and deductive methods.

5. A JAVA-BASED WML-AGENT SYSTEM

As an example of utilization of agent technology in WAP communication we developed a Java-based WML-Agent system.

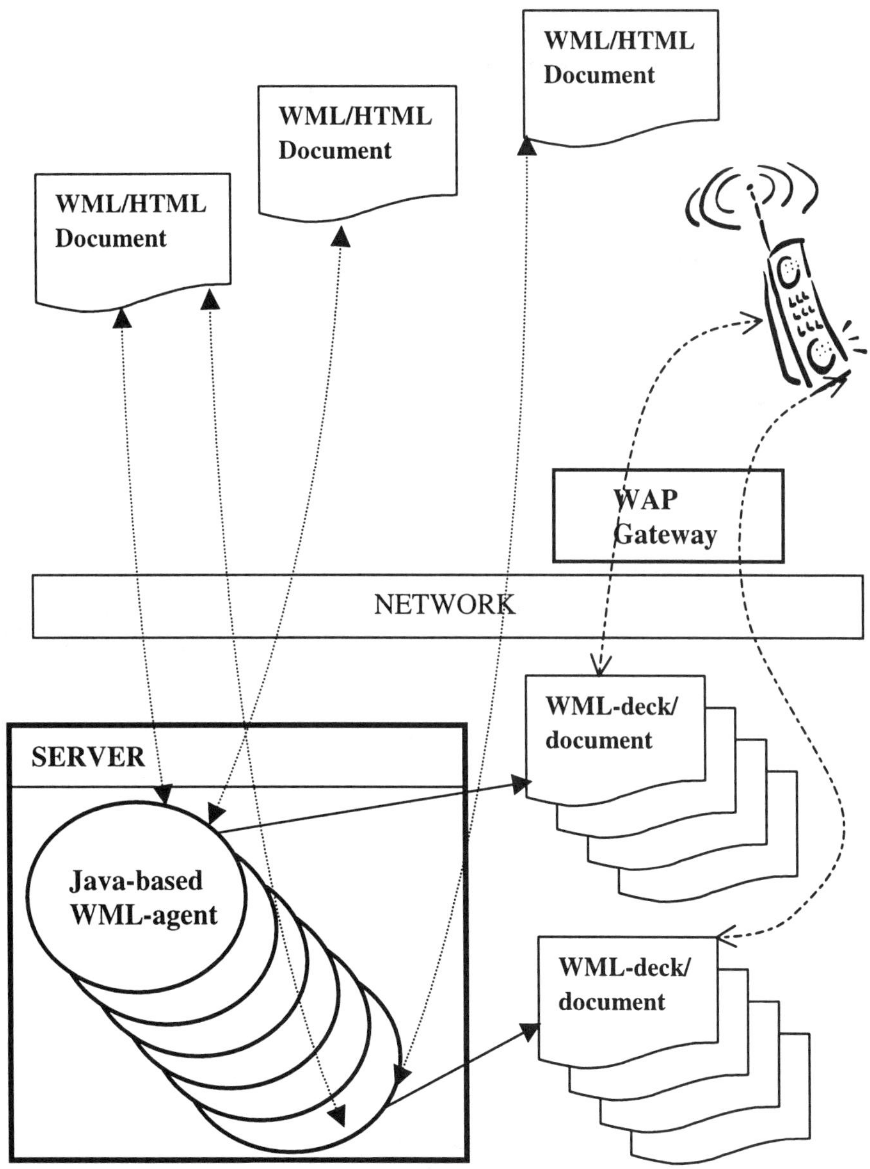

Figure 1: Java-based WML-Agents

The main idea behind Java-based WML-Agent is creating an intelligent agent which is

- highly adaptive to extracting interesting information from various WML/HTML-pages
- able to locate, collect and present information in a WAP device oriented view
- able to gather information from pages which are commonly updated without having customer to do that himself
- able to reorganize an order and priority of information to be presented to the customer
- able to employ a simple pro-active behavior
- able to process user requests off-line

The Java-based WML-Agent should inhabit the Internet, collect information from different sources in the Internet and provide WML-decks which the customer can access via WAP gateway as it shown in Figure 1 (this figure also shows that there can be many agents corresponding to different customers or profiles).

Each user should be able to design his or her starting page telling Java-based WML-Agent what information the user would like to have and, possibly, where to find it. Since most non-experienced customers probably may find this a hard and time consuming task, it should be possible use Java-based WML-Agent by a service provider in order to create certain starting pages targeted to large group of users. This could be, for example, a creation of profile with the target group being customers interested in sports and fashion in the age of 18 to 24 years. Some other profiles would cover target group of peoples interested, for instance, in technology, economy and politics. To combine these two options (different individual and group profiles) it should be possible to let the user create his own profile by combining several profiles together. In our case the user could, for example, indicate that s/he is interested in sports and technology but not in fashion, economy and politics.

In a general case of Java-based WML-Agent it should be possible for the customer to make registration (profile presentation) for a service both using computer-based and WAP-handset based interfaces as it shown in Figure 2 (however in the current version of Java-based WML-Agent only a computer-based interface for registration for a service is implemented).

The result of registration is a record in customer database and a generated personal WML-agent who may start running on a server of a service provider.

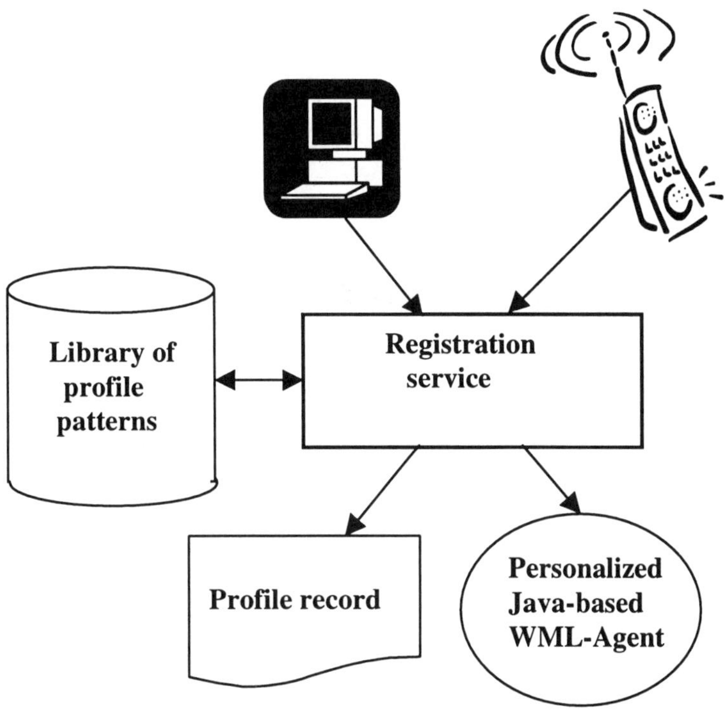

Figure 2. Registration service

As a summary of the scenario requirements from the Section 4 we can say that ability to dynamic shift of focus of required information and dynamic reorganizing the WML-decks (as a basic means for WAP-based interaction) are of a great interest. There are actually two parts of the problem:

- Description of customer's needs and keeping updated focus of customer's interests
- Organizing visualization of required information

These parts were main focus of the work for implementation of the Java-based WML-Agent system.

6. A BRIEF OVERVIEW OF THE JAVA-BASED WML-AGENT DESIGN

Here we briefly describe some basic features and functionality of the Java-based WML-Agent system.

The Java-based WML-Agent system may be divided into two parts:

- One which takes care of the Graphical User Interface (GUI) and
- One which lies behind the GUI and does all the processing

The GUI part provides means for creating, updating and customizing user cards and requirements and the processing part does search, adaptation and generation of WML decks and cards.

An example of GUI for Java-based WML-Agent is presented in Figure 3.

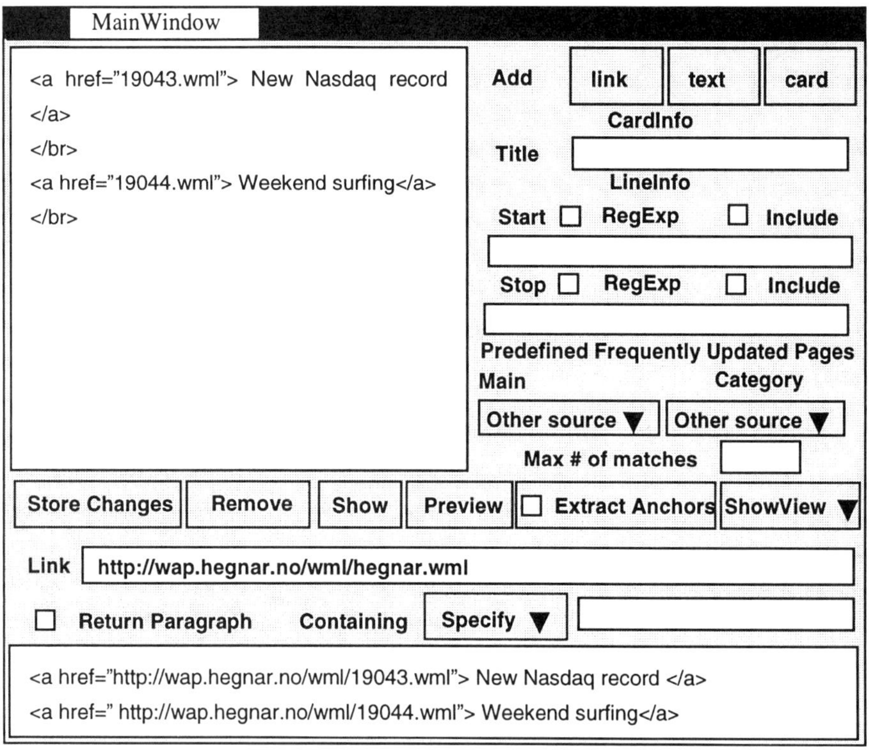

Figure 3. System GUI

Some examples of functionality provided by the GUI are as follows:

- Creating new cards, new fields in already existing cards, new links and text
- Visualizing control of input/output to/from the application
- Presentation and selection of a set of predefined resources. These resources may be easily updated and customized
- Displaying information about the current card
- Defining frequently updated pages. This allows to define WML-pages of special interest which should contain up-to-date information

- Defining limitation for the amount of deliverable information
- Defining filters for information filtering from predefined pages

The processing part of Java-based WML-Agent allows to display the required information according to customers needs, to process queries from the customer off-line and to present results of processing in the form of WML-cards.

A basic loop of an agent processing part can be very generally described as follows:

1. Receive an event from the customer, environment or other agent components
2. Process the event by event handler and call query processing, filtering or pro-activity handling components. All the above components use customer profile record.
3. After processing the event by query processing, filtering or pro-activity handling components generate or update corresponding WML-deck
4. If there are more events left in the queue then go to the first step, otherwise go to "hibernate" state

Events can be generated externally (for example, by clock or customer) or internally (for example, by pro-activity handling component).

Another information retrieval feature that is provided by the Java-based WML-Agent is information filtering. User can describe what particular kind of information from predefined WML-decks and HTML-pages (usually they are some pages provided by newspapers or news agencies) s/he is interested in. This can be described by a regular expression and the agent will filter updated information in the pages according to the expression. Such filtering allows more personalized and precise search that may decrease the amount of deliverable text. This is especially useful with a limited size of screen of a WAP-device.

The Java-based WML-agent system employs basic agent features like autonomy, reactivity and a simple pro-activity. In the current version we restricted pro-activity to ability to a simple reorganizing WML-decks according to results of processing by putting the hottest news on the top. However further development of this feature will be our prioritized work in a new Java-based WML-Agent version.

7. RELATED WORKS

To the best of our knowledge we didn't find yet works on intelligent agents for WAP-based services. However there are works on development of News Agents for the Internet and Web-pages. Some of them are as follows.

Reference.com [13] searches for any information in over 150,000 Usenet newsgroups and publicly accessible lists. The results are either emailed daily to the customer or available at personal WWW site.

RoboBear [14] performs automated searching, downloading and decoding the Usenet newsgroup articles. RoboBear is an application that must be downloaded to the user computer before it is of any use. Then it must be set up to search the newsgroup articles, using the news server's built-in search capabilities.

MyCNN (CNN Custom News) [15] is a site where the user is able to highly personalize his starting page at CNN. This includes having constantly updated weather forecast for cities of interest, stock quotas, sports, showbiz, world news etc.

Infoseek News [16] is a personalized news service that allows to express personal news interests and to create own customized news page.

My Yahoo! [11,19] delivers customized headline news, sports scores, stock quote, weather reports, web resources, personalized version of the Yahoo! directory and many other things. It allows users to find website determined by their interests and people with the similar interests using Automated Collaborative Filtering (ACF) technology in partnerships with Firefly Network, Inc. (for more information about ACF see, for example, [4]).

Paperboy [17] is a free news ticker which compiles a personal newspaper from daily newspapers based on topics. The result is a list of web links which can be delivered daily by email. Customer can also specify which articles s/he wants by giving the system some keywords. Complex queries are also possible with syntax similar to well-known search engines.

Newsweeder II [5] is a service that regularly searches for Web pages and Netnews articles that are interesting to the customer. It applies advanced machine learning techniques to pages and articles which the customer liked in the past.

NewT [6] is a USENET news filter that can be "trained" by giving it a series of examples that show in which kind of articles the user is interested in. From this the NewT agent can search all news articles trying to find other articles which are similar to those initially indicated by the user. When the agent presents the other articles that it has found the user gives feedback according to their applicability; thus the NewT agent can widen or restrict its searching next time.

All the above news services are very interesting and inspiring but they do not exploit pro-activity in the sense we considered it previously and, what is more important for us, they are not designed to be used in WAP-based communication. In other words they don't put enough attention to the special WAP communication features we mentioned in the Section 1.

Several mobile communication suppliers provide WAP start-portals for their customers. However some of them, at the moment, lack the ability to personalize the content of the portal while even those who allow describe personal WML-decks do not support their dynamical changing and pro-active updating.

8. CONCLUSIONS AND FUTURE WORKS

The main goal of this work is to demonstrate advantages of application of agent-based methods to WAP-based communication support. The general benefits of agents in WAP supported communication services are based on their ability to

- off-line processing via autonomous behavior
- personalization of services via pro-activity and adaptation
- relaxation of limitations of current WAP-based devices via precision of delivered information

Another important advantage of the agent-based approach is possible decentralization of the agent-based services and, as a result, a better scalability of service providing in a case of very large number of subscribers.

In this paper we don't consider the issues related to the costs of providing the agent-based services. However we understand that these issues become important in developing future versions of the system.

With a new technology such as WAP their are expected a lot of expansions in the near future both regarding usability due to better equipped terminals and higher transfer rate due to improvements in the telecommunication infrastructure. In addition to these general improvements we see the following extensions to our Java-based WML-Agent as our next works:

- Taking into account a context (for example, time) of request. This may increase pro-activity of the services and it may not require special support from a mobile network
- Push-services. The WAP specification is going to define a "push" mechanism which will allow any Internet server to send information to the user.

Another future work we see in increasing a social ability of Java-based WML-Agents by allowing them to communicate and negotiate to each other for better satisfaction of the customer requests. In this part we refer to our work on co-operative work support by Agora platform [8,9,10].

ACKNOWLEDGEMENTS

This work is partially supported by the Norwegian Research Foundation in the framework of the Distributed Information Technology Systems (DITS) program and the ElComAg project.

REFERENCES

[1] J. .M. Bradshaw (Editor), *Software Agents*, AAAI Press/The MIT Press: Menlo Park, CA. 1997, p. 291-316.

[2] M. N. Huhns, M. P. Singh, L. Gasser (Editors), *Readings in Agents,* Morgan Kaufmann Publishers, 1998, 523 p.

[3] N. R. Jennings and M. J. Wooldridge (Eds.), *Agent Technology: Foundations, Applications and Markets.* Springer Verlag, 1998, 323 p.

[4] J. A. Konstan, B. N. Miller, D. Maltz, J. L. Herlocker, L. R. Gordon and J. Riedl, *GroupLens: applying collaborative filtering to Usenet news.* Communications of the ACM, 40(3), 1997, pp. 77-87

[5] K. Lang, *NewsWeeder: Learning to Filter Netnews.* Armand Prieditis and Stuart Russell (eds.). Machine Learning, Proceedings of the Twelfth International Conference on Machine Learning, Tahoe City, California, July 9-12, 1995, San Francisco, CA : Morgan Kaufmann Publishers, 1995.

[6] P. Maes, *Agents that Reduce Work and Information overload.* Communications of the ACM, 37(7), 1994, pp. 31-40

[7] S. Mann, *Programming Applications with the Wireless Application Protocol: The Complete Developer's Guide.* John Wiley & Sons, 2000, 256 p.

[8] M. Matskin, M. Divitini and S. A. Petersen. *An Architecture for Multi-Agent Support in a Distributed Information Technology Application.* International Workshop on Intelligent Agents in Information and Process Management on the 22nd German Annual Conference

on Artificial Intelligence in Bremen (KI-98), TZI-Bericht Nr. 9, Germany, September 15-17, 1998, pp.47-58.

[9] M. Matskin, *Multi-Agent Support for Modeling Co-operative Work.* In: T. Yongchareon, F. A. Aagesen, V. Wuwongse (Eds.) Intelligence in Networks. The Fifth International Conference SMARTNET'99, 22-26 November 1999, Thailand, Kluwer Academic Publishers, 1999, pp. 419-432.

[10] M. Matskin, O. J. Kirkeluten, S. B. Krossnes and Øystein Sæle. *Agora: A Multi-Agent Platform and its Implementation.* Proceedings of the 2000 International Conference on Artificial Intelligence (IC-AI'2000), June 26-29, Las Vegas, Nevada, USA, CSREA press, 2000, 7 p.

[11] M. Miller, *The Yahoo!* Publisher: Que, 2000, 339 p.

[12] H. S. Nwana, *Software Agents: An Overview*, Knowledge Engineering Review, 11(3), 1996, 40 p.

[13] URL: http://www2.reference.com/

[14] URL: http://www.flashpoint.com.au/

[15] URL: http://www.mycnn.com/

[16] URL: http://infoseek.go.com/

[17] URL: http://www.paperboy.de/

[18] URL: http://www.wapforum.org/

[19] URL: http://my.yahoo.com/

[20] M. Wooldridge and N. Jennings, *Intelligent Agents: Theory and Practice.* The Knowledge Engineering Review, 10(2), 1995 p. 115-152.

H^3M - A Rapidly Deployable Architecture with QOS Provisioning forWireless Networks

Edwin C. Foudriat, Research Professor, foudr_e@cs.odu.edu
Kurt Maly, Head, maly@cs.odu.edu
Stephan Olariu, Professor, olariu@cs.odu.edu
Department of Computer Science, Old Dominion University
Norfolk, VA 23529
Telephone 757-683-3915
Fax 757-683-4900

Abstract

Future wireless networks will have to be based on highly mobile systems that are self-organizing, rapidly deployable, heterogeneous, and will not rely on expensive infrastructure. Since these networks will be called upon to support real-time interactive multimedia traffic, they must be able to provide their users with adequate quality of service (QoS) guarantees. In addition, we want to consider networks with highly dynamic resource requirements, such as QoS, by multiple, dynamically changing groups.

The architecture and protocol is based upon a protocol called DRAMA.[4 - 8]. We have taken the kernel of DRAMA - operating a framed combination of TDMA and CSMA/CD in a metropolitan area with multichannel load balancing - and created a new architecture suitable to highly mobile environments. We refer to it as the Hierarchical Heterogeneous Highly Mobile network (H^3M, for short). As the name suggests, the H^3M network consists of a hierarchy of heterogeneous hosts distributed over a geographical area and linked together in a wireless communication system. At the bottom level of the hierarchy we have a cluster architecture whose connectivity and management activities are assumed by a mobile base station (MBS). In turn, the MBSs are organized into a virtual network, essentially emulating a local area network like structure. The protocol for arbitrating as to who sends what on what frequency at what time is based on a combination of TDMA and CSMA/CD subframes and frequency allocations to a cluster of nodes that allows for dynamic bandwidth allocation and which support multicasting.

Simulation and analysis have shown that H^3M is robust, scales well and provides much higher efficiency throughput than other protocols, while supporting the same degree of host mobility. Importantly, H^3M turns out to be well suited for handling on-demand multimedia communications in a heterogeneous, highly mobile environment. Multicasting is supported with minimal overhead.

1.0 Introduction

In recent years, wireless and mobile communications have seen an explosive growth both in terms of the number of services provided and the types of technologies that have become available. Indeed, cellular telephony, radio paging, cellular data, and even rudimentary multimedia services have become commonplace and the demand for enhanced capabilities will continue to grow into the foreseeable future [13, 14, 15, 16, 17, 18, 1, 19, 20, 21, 22]. It is anticipated that in the not-to-distant future mobile users will be able to access, while on the move, data and other services such as electronic mail, video telephony, stock market news, map services, electronic banking [17, 18, 1, 19, 20, 22].

However, the well-known cellular networks are too rigid and inflexible to adapt to situations in which rapid deployment is critical. Such is the case in disaster relief, search and rescue, law-enforcement, collaborative computing, multimedia classroom, miliary operations, interactive mission planning, and similar special purpose applications. To address the need for self-sufficiency and rapid deployment, a number of designs have been suggested including packet radio networks (PRN) [13, 24, 15] and mobile ad-hoc networks (MANET) [17, 1, 23, 24, 19]. While these architectures feature a number of desirable characteristics [1, 17], many assume that all hosts in the network are identical in processing and communication capabilities. This, in turn, imposes numerous limitations of these networks both in terms of robustness and overhead in routing messages end-to-end.

Our main contribution is motivated by the observation that in practical situation it is very unlikely that all the hosts are identical. It is very often the case that some of the hosts have much higher computational and communication capabilities than others and hence they could serve as mobile base stations (MBS). In this capacity they can serve as cluster heads and can handle the management activities inherent to the efficient operation of the network.

Our architecture, which we refer to as the Hierarchical Heterogeneous Highly Mobile network (H^3M), consists of a hierarchy of heterogeneous hosts distributed over a geographical area and linked together in a wireless communication system. At the bottom level, we have a cluster architecture whose connectivity and housekeeping is assumed by a mobile base station (MBS). In turn, the MBSs are organized into a virtual network, essentially emulating a local area network-like structure. This is a virtual structure - the actual implementation depends on the availability of resources in the network.

In section 2, we discuss the basic architecture and operations of H^3M including intra- and inter-cluster communication, framing for multimedia traffic, call setup, bandwidth management and load balancing. In section 3, we handle the cluster organization and operations and both node and cluster mobility. In section 4, we discuss performance studies for both synchronous and asynchronous traffic. Section 5 compares H^3M with other mobile protocols based upon important features such as delay, jitter, routing, bandwidth management, multicasting and load management.

2.0 H³M Architecture and Operational Protocols

Each cluster operates as a quasi-independent LAN whose protocol is patterned upon the concepts first developed in DRAMA - Dynamic Resource Allocation in a Metropolitan Area network [4 - 8]. DRAMA's objectives were:

- to deliver a mix of real-time and datagram traffic to a metropolitan sized network using multiple LANs with high efficiency and robustness; and
- to provide LAN sharing of the multiple bands so that load balancing is effectively achieved over an extremely wide range of load conditions.

For H³M, we retain the basics of these objectives but modify them to support mobility and heterogeneity in the nodes:

- to deliver a mix of real-time and datagram traffic with high efficiency and robustness to metropolitan size networks using multiple clusters ;
- to provide clusters sharing the frequency and bandwidth range of the network so that load balancing is effectively achieved over an extremely wide range of load conditions; and
- to provide QoS provisioning and multicasting in real time over wide ranges of operational parameters such as receiving and transmission capabilities, traffic types, load conditions, and node mobility.

2.1 H³M Architectural Assumptions

We consider a number of nodes distributed over an area linked together in a wireless communication system. To do so we assume neighborhoods of nodes which form local clusters. Each cluster supports both inter- and intra-cluster communication. A typical configuration is shown in Figure 1. We make a number of assumptions about the nodes and clusters.

Power capability: Each node has a transmitter and at least one receiver.

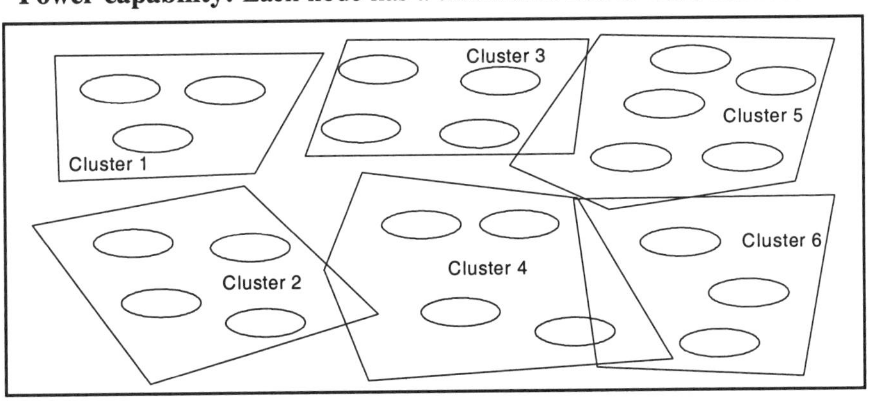

Figure 1 Nodes partitioned into a network of clusters

Transmitters may have different power capabilities. Some nodes, like those which are vehicle based, may have strong transmitter signals which carry across all clusters; others may be hand-held devices which can only reach nodes within their own cluster and possibly nodes in adjacent clusters. Each receiver is capable at any one time of tuning to any cluster's transmissions. Each has good signal-to-noise so that it can receive signals from distant clusters assuming sufficient transmitter power.

Frequency scheme: Each cluster operates on its own CDMA code (non-interfering frequency) in such a way that inter-cluster interferences are kept at an acceptable level.

Cluster basic operations: Each cluster integrates the its nodes into a local area network (LAN). It is assumed that *at least one node* in a cluster has sufficient transmitter power to be heard by all clusters in the network. This node will be designated as the *cluster sender transmitter* (CST) and will most likely be the MBS. Each node has *at least one receiver* tuned to listen to its cluster's transmissions and their exist sufficient *additional receivers* in the cluster so that the cluster is able to receive and translate transmissions from all other clusters. Nodes with additional receivers assigned to monitor outside clusters are *cluster monitor receivers* (CMR).

Mobility: Mobility is the norm rather than the exception. We make no prior assumptions about mobility. In fact, our approach supports a large range of mobility patterns.

2.2 Intra-cluster operational protocol

A frame, shown in Figure 2, has three subframes - sync header, a TDMA and a CSMA/CD subframe - which are shown in Figures 3a, 3b and 3c, respectively. Frame time is assumed to be of the order of 10 msec. The header subframe, Figure 3a, sent by the MBS, provides cluster control information needed by nodes and other clusters and possibly information about subsequent frames if a change is in order. The TDMA subframe is time slotted. It provides support for real-time information using the fact that each call (*circuit*) has a reserved slot. A major feature concerning *circuits* is that much of the wasted call capacity is recovered for those which are silent since a silent call packets contains only 1 or 2 bytes needed to maintain the resource. The subframe boundary between the TDMA and CSMA/CD subframes is dynamic. At the end of the TDMA frame, the remainder of the frame is devoted to asynchronous data using a CSMA/CD protocol similar in operation to the well known Ethernet.

As the core of the H^3M architecture, like DRAMA [4 - 8], the cluster protocol is based upon framing. To work for the entire network that might span an area of many kilometers, each cluster has a non-interfering frequency and bandwidth. (see details in Section 2.4). Nodes use this frequency for all intra-cluster communications and, as discussed in Section 2.3, to send information over the whole network. The bandwidth at which the cluster sends is allocated dynamically in order to provide load balancing QoS (see Section 2.4). In both its TDMA and CSMA/CD subframes, signal overlap from multiple senders will corrupt the information. During the TDMA subframe, the next node must not start sending until the prior node's signal has passed. During the CSMA/CD subframe, a collision event can occur up to the cluster's round trip transit.

2.3 Inter-cluster operational protocol - Communication between clusters

Based upon the architectural features in section 2.1 and the intra-cluster protocol above, we derive procedures which allow a node to communicate with any other node in any other cluster. Three situations occur.

Single receiver case: A receiving node does not have sufficient receivers to be able to tune to the sending node's frequency. In this case, the receiving cluster's *cluster monitor receiver* (CMR) which monitors the sender's cluster transmissions must receive the information and retransmit it in a packet in the cluster's frequency band in a succeeding frame so that it is forwarded to the receiver.

Low-power-sender case: A sending node does not have sufficient power to be heard by its

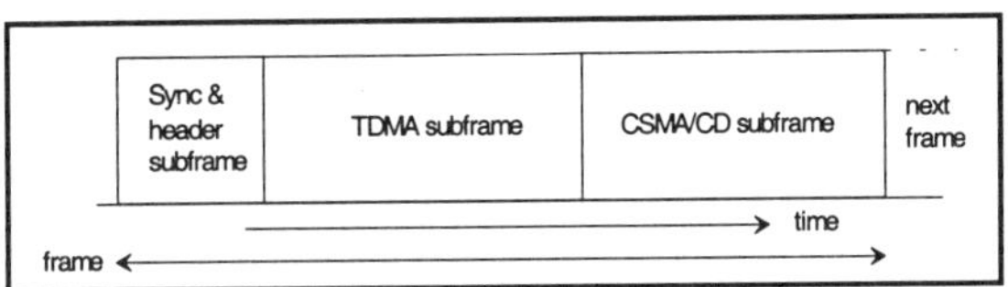

Figure 2 H3M frame structure

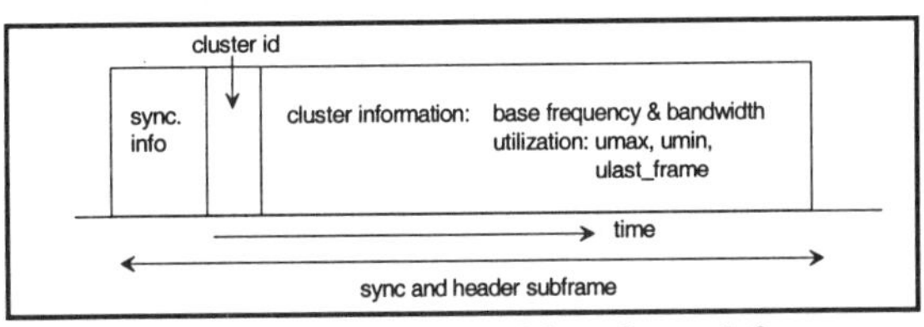

Figure 3a. Sync and header subframe structure - provides cluster operational informa-

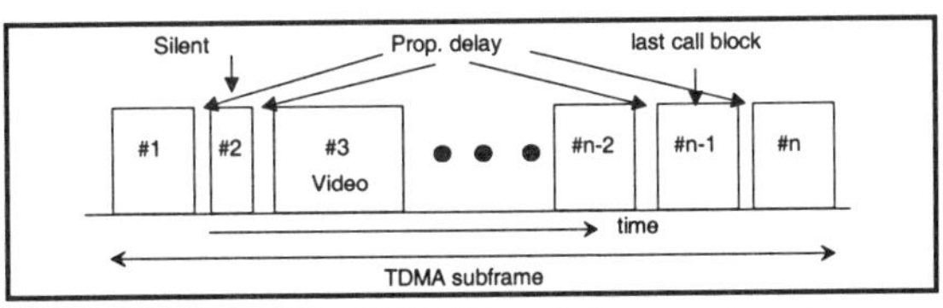

Figure 3b. TDMA subframe structure

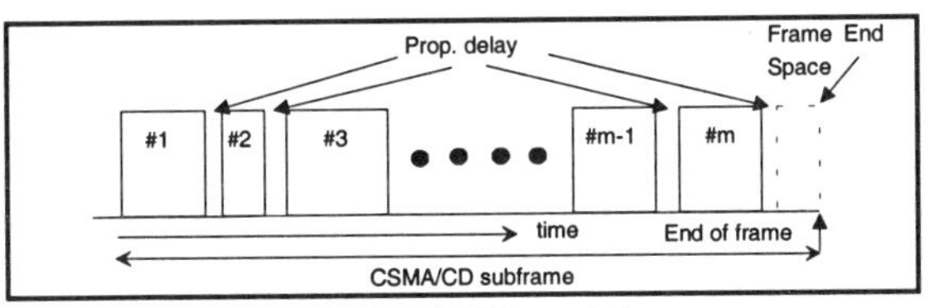

Figure 3c. CDMA/CD subframe structure

receiving node. In this case, the sending cluster's *cluster sending transmitter* (CST) must copy the sender node's information and retransmit it at the cluster's frequency during a subsequent frame so that it can be heard by the receiver node or receiver cluster's CMR.

Best case: The communicating node has sufficient transmitting power so that information sent can be received directly and the receiving node has sufficient receivers that it can tune to the other's cluster frequency. In this case, no forwarding protocol is needed as is for the first two cases.

Note that the first two cases can be additive, i.e., both situations can exist simultaneously or each can occur separately. In a full duplex circuit for each direction different cases can occur. Note that the data transfers rates between sender and receiver are not seriously delayed by operating under other than the best case scenario. For example in the low-power-sender case using the TDMA subframe, the sending node transmits its information in one packet and the CST then retransmits this information in a subsequent frame. The minimum forwarding delay is the one frame in the best case and 3 frames under the condition where both sender and receiver cases exist. It is also true that the potential information communication rate of a cluster is reduced in situations other than the best case. We will discuss and compare performance features of H^3M in Section 4 of the paper.

Establishing and assigning connections - call setup protocols

With both TDMA and CSMA/CD subframes available, H^3M is able to provide a wide range of services. It is necessary to establish procedures and commit resources so that connections operate efficiently. Three connection types are reasonable. The first are *circuits (calls)* which use the TDMA subframe. They enable time-critical - typically voice, video and control - data transfer. The second are datagrams which consist of multiple packets. They use CSMA/CD subframes but in order to conserve resources may use setup to enable decisions based upon transmitter and receiver cases availability as described above. We designate multiple-packet datagram handling as *datagram circuits.* The final connection type is single-packet *datagrams*. It uses the CSMA/CD subframe also. *Datagrams* are used as the protocol to establish the others.

The protocol steps for *datagrams*:

1. The sender places the packet.
2. It is copied and retransmitted by the sender's leader in a subsequent CSMA/CD subframe.
3. The packet is received by the receiving node's CMR and delivered in a subsequent receiving cluster's CSMA/CD subframe.

Steps 1 - 3 are followed for all *datagrams.*

4.Based upon the cases above, two alternative possibilities exist:
a. Not low power sender - the sender's transmitter power is sufficient for the monitoring node in receiving cluster to hear. In this case, the CST does not transmit, i.e., skips step 2 above and completes step 3.
b. Not single receiver - the receiving node may have a receiver tuned to the sending cluster's frequency. In this case it is able to receive the packet directly from the sender's signal as a result of step 1 or 2 so the CMR can ignore in step 3.

While not part of the *datagram* delivery procedure, the receiver is now aware of the situation under which he is capable of receiving additional information from the sender. If he receives the information as a result of steps 1 - 3 and cannot dedicate a receiver to monitor the sender's subsequent transmissions, he knows both sender's CST and 'CMR must participate. If 4a, only the sender's CST need not participate; if 4b, then the CMR need not participate. This information is vital for all participating in the setting up of *circuits* and *datagram-circuits*.

Establishing *datagram circuits* and *circuits* requires the commitment of resources over an extended period of time. Based on the *datagram* delivery procedure above *datagram circuits* and *circuits* that do not have the best case scenario require the CST and/or the CMR to participate for subsequent delivery. In addition, the *circuit* requires the commitment of TDMA subframe block(s) in one or both clusters.

The protocol steps for both circuit types, using the single packet datagram as the handler, are:

1.Steps 1 - 4 above are accomplished. In the call setup, packet information is placed by the CMR whether he was able to read the sender's packet directly or its CST retransmission.
2.The receiver must now decide:
a. whether to accept or reject the call; and
b. whether he can commit a receiver.
c. if he accepts in a and not b when establishing a *circuit*, he must also reserve a TDMA subframe block in his cluster.
3.With these decisions made, he sends a *datagram* packet acknowledge in return. It contains information about the receiver and his cluster's commitment, and the sender's CST requirement.
4.Return packet follows the same procedure as steps 1 above. Note, the return packet may be used to establish the return half of a duplex circuit, thus creating a full-duplex circuit.
5.As in step 2 above, the sender must now reserve and commit his necessary resources based upon how the return packet is to be handled. In addition, based upon information in the return packet, the potential

participants are able to establish how subsequent packets must be handled base upon the commitment of CSTs and CMRs and that the necessary resources have been committed.

The steps provide *datagram circuits* and *circuits* with the optimum commitment of resources and hence provide maximum efficiency under heterogeneous conditions. Note, it is not necessary to use a full duplex connection in either *datagram circuits* or *circuits* and each direction may use different resources.

2.4 Bandwidth management and load balancing

Over an extended period, some clusters are very busy while others may under utilize their capacity. In DRAMA [4 - 8], a mechanism existed where capacity can be reassigned by allocating additional frequency bands as needed. This system required each node to have as many transmitters and receivers as bands available. For heterogeneous mobile wireless-based systems using multiple bands does not appear to be practical. Some nodes may have a wealth of electronic gear connected with their mission; others may have a simple digital-based T/R which provides only simple communication.

A solution whereby a single T/R unit is used while enabling capacity redistribution is suggested here. Both transmitters and receivers may be tuned to any frequency within the overall network frequency range. Further, each transmitter and receiver is built whereby its bandwidth size can be changed at least in steps. For example, a receiver would have an itermediate frequency bandwidth capable of change and a transmitter could modulate over a smaller or larger bandwidth depending upon its band pass filters. With this frequency structure, the system becomes a variable-bandwidth, variable-frequency wireless system (VB/VFWS).

Assuming VB/VFWS units can be built, the major operational problem is to partition the total frequency range so that capacity changes have the minimum impact on other clusters, and to provide a load balancing system able to allocate bandwidth as needed to the cluster as needed and resolve priorities when bandwidth is scarce. To change capacity in increments, a cluster most likely needs a free frequency range either directly above or below its present range or a total reassignment and readjustment of all frequency bands without interference and signal corruption.. This, in turn, requires cooperation between clusters as to what frequency and bandwidth each uses. Methods for the allocation of frequency and bandwidth and their performance they provide will be discussed in a later paper, but the basic load balancing techniques were proven in reference [7-8]. We assume that frequency and bandwidth changes can be made in *O(1 u*sec*)*, so that changes can be made rapidly and successfully between frames.

We pose the question of how each receiver listening to the cluster may be

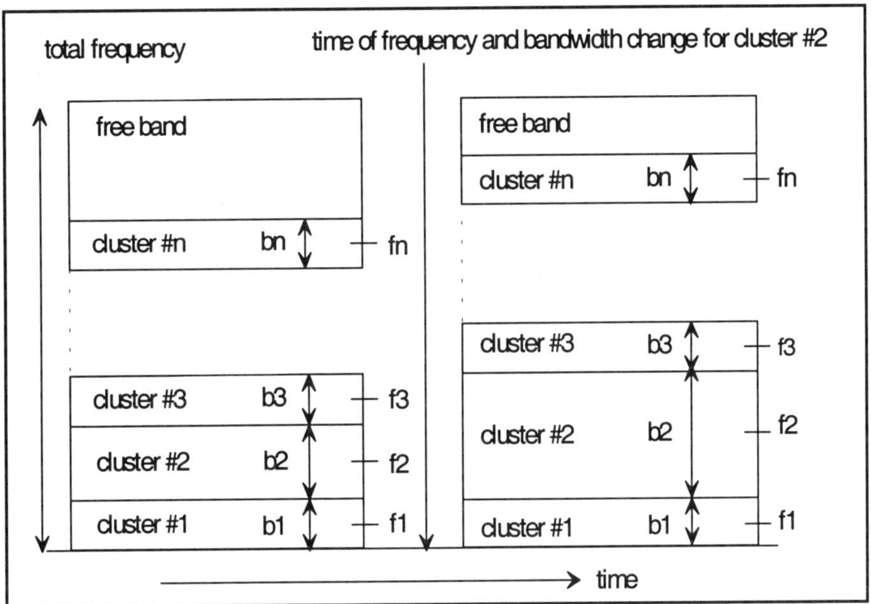

Figure 4 Typical bandwidth change for cluster 2

made aware of change, since the change must be *non-interferingly* implemented for all receivers in the network. One solution appears to be signaling during the synchronizing signal subframe block at the beginning of each frame as noted in Figure 3a. The MBS would indicate changes in frequency and bandwidth for some succeeding frame. Each receiver and transmitter, after detecting a cluster signal change in the sync subframe, would retune at the appropriate frame time. We illustrate this change in bandwidth in Figure 4. Assuming load changes would occur in *O(1 min.),* the VB/VFWS should easily be able to maintain clusters with well balanced performance over a wide range of loading conditions.

3.0 Cluster and network operations

In this section, we discuss how clusters and the network operate to provide high mobility and dynamic load balancing. Although we discuss mechanisms for cluster and network operations, in many cases there may be additional operational requirements based in the actual utilization which may dictate and/or override the procedures we discuss. Therefore, the discussion should be used to understand what features exist, how they are implemented and how they may be used in H^3M when these type activities take place in a real-world environment.

3.1 Cluster organization

The only low-level requirement for cluster organization is based upon the affect of transit time over the distance between and spanned by its nodes and possibly maximum transmitter signal strength of a node. Cluster dimensions should be *O(1 km)*. The actual decision on cluster organization should be done in-situ based upon distance, frequency, bandwidth, terrain, the number of nodes accommodated, and intra- and inter-node higher level activity, etc. Note that cluster can overlap in span since they operate at non-interfering frequencies. It is assumed that the network will operate over in line-of-sight frequency ranges. Network dimensions are *O(10 km)*.

As described in DRAMA, the TDMA subframe supports circuit-type data. Each member of the cluster is aware that *n* calls exist and know that after the nth call, the CSMA/CD subframe begins. The CSMA/CD subframe is similar to Ethernet operations. Major exceptions are that due to framing, packets arriving during the TDMA frame are delayed until the start of the CSMA/CD subframe and after the mth packet has been transmitted there may be wasted capacity if no packet is able to fit into the space before the frame ends.

3.2 Cluster formation

The operational features needed to form a new cluster are:

1. an MBS node for control and a CST node with sufficient transmitter power to carry across the network;
2. nodes with sufficient additional receivers to act as CMRs to support intercluster reception;
3. cluster dimensions should be *O(1 km)*.
4. an unused frequency and minimal bandwidth within the overall network environment;
5. very probably, a need for some activity unrelated to H^3M's architecture or protocol; and
6. a mechanism for interested nodes compare theses activities so as to decide to join.

Because of 5, we provide only a initialization protocol based upon the apriori fact that nodes[1] have used 6 to determine that they need to disconnect from their present cluster and form a new one. To simplify the protocol, we assume that the potential MBS, CST and CMRs, as in 1 and 2 are identified and that they have successfully terminated any *circuits*. We assume nodes satisfy 3. The first step is to select identify the MBS and the resources for 4. The MBS then notifies all existing clusters concerning the new cluster. Upon

[1]Closely spaced nodes do not have to be in the same cluster since clusters can physically overlap without interference.

[2]We do not treat the condition of a new node joining the network because of the diversity of situations under which networks may operate and different solutions exist for different environments. One example is to use low powered beacon signals where clusters advertise their frequencies.

completion of 4 and notice that all clusters are prepared to receive the new cluster, the MBS terminates is residence in his present node and begins frames in the new cluster. Upon receiving this message, all CST and CMR nodes terminate and join the new cluster. At this point, the new cluster is fully operational such that nodes with existing calls can now join using the protocol for node movement (section 3.3). The need for the new MBS, CST and CMRs to terminate all calls should not seriously hamper critical communications since the new cluster can be up and running in very few frames after termination. It begins as a lightly loaded cluster such that resurrecting terminated calls should require minimal delay. All other nodes joining can now do so without any call interruptions.

3.3 Node movement[2]

The situation involving mobile nodes can be very dynamic. Nodes may move to a location where it can no longer work effectively within its present cluster or a group of nodes may form a cluster as above. Note, clusters can overlap physically, since cluster organization is primarily dictated by distance. We have not included reorganization due to cluster size or load since an H^3M cluster can operate under an extremely wide range of load using load balancing technique discussed in section 2.4. These situations must be resolved without serious cluster and network performance degradation. Let us take up the situations seriatim.

Single node movement: Where a node moves, the node needs to join another cluster in which it can operate efficiently. The first step is to determine into which cluster the node should move.[3] The node can then determine its new cluster operational features such as node id and band by using the information available in the cluster's header frame, Figure 3a. Cluster activity unrelated to protocol or architecture is outside H^3M's scope but needs to be done based upon step 7 of Section 3.2. The node should then transfer its active calls as follows:

1. For outgoing *circuits*, once a new cluster has been selected, the node should reserve resources in the new cluster before the switch is completed. Again, the exact protocol, including when, is a matter of further study. However, since any resource reservation in the new cluster can be maintained by silent call block, the capacity cost of reserving resources should be minimal. Further, the cluster to which the node is moving has the option of obtaining additional capacity so it should be able to anticipate future load changes very effectively. At the point of

[3]New cluster selection requires knowledge of the location of clusters in direction the node is headed. Since our protocol does not use routing, alternative source of location information is needed. Since location is needed by many operations we assume that it can be made available without addition to the protocol.

the actual switch, the node can issue "final" call blocks in the TDMA frame for the cluster from which it is departing, retuned it transmitter and receiver and joins the new cluster at its next frame.

2. For incoming and outgoing *datagram circuits*, no resource reservation is necessary. It can inform the new cluster of its present CST and CMR activity; that is which outgoing packets need to be retransmitted and which incoming packets need to be copied, translated and retransmitted because the node does not have receivers to commit.

3. The node then needs to transfer and multicast an "*I am here*". Since the actual transition should be *O(10 msec.)*, the likelihood of missing critical data is minimal. Further, reliability of information transfer should be handled at the transport level so duplicate packets can be send for any information destroyed or lost. If additional time is needed for the node to get established, then for critical messages, the node could use full-duplex circuits and an XON-XOFF protocol to span the transfer period.

Cluster breakup: The breakup of a cooperative-based cluster is accomplished in much the same manner as the movement of a single node. However, the first step for those nodes which are to break off is a decision whether to form a new cluster or to join an existing cluster already covering the area. Once this decision is made, the nodes can use the procedures outlined above to make connections.

4.0 Performance of H^3M

In the past, extensive analysis and simulation have demonstrated the feasibility and performance capability of DRAMA [4-8]. As a first step to be assured that H^3M can be provide similar performance in an extended environment, it is necessary to extend the analysis of DRAMA into the areas where differences occur. One major difference is that additional loads occur in H^3M due to repetition of TDMA and CSMA/CD packets needed to forward information to other clusters. Secondly, H^3M uses VB/VFWS to obtain load balancing whereas DRAMA allocated or released channels with fixed bandwidth. Factors affecting capacity and delay for both the TDMA and the CSMA/CD subframes under varying bandwidth conditions were performed. Details of the analysis are presented in Appendix A and B.

TDMA subframe performance for a fixed set of circuit conditions is completely predictable. The time for each *circuit's* information to be delivered from sender to receiver depends only on the repetitive transfers needed, as seen in Figures 2 and 3. Load changes occur due to different mix of circuit

---

[4] Actually, video circuits should be more efficient than voice especially for larger bandwidths and diameters since the propagation delay is needed only after the *circuit* frame transmission is complete.

Voice circuits in TDMA sub-frame				
Distance 1 km		Number of voice circuits in TDMA		
Bit rate, x 10+6	Round trip prop, bits	40% of frame	60 % of frame	80% of frame
1	10	11	18	24
5	50	59	88	118
10	100	112	169	226
20	200	206	310	413
Distance 2 km		Number of voice circuits in TDMA		
1	20	11	17	24
5	100	56	84	112
10	200	103	154	206
20	400	176	264	352
Distance 5 km		Number of voice circuits in TDMA		
1	50	11	17	23
5	250	49	74	98
10	500	81	123	164
20	1,000	122	183	244

Table 1. Voice circuits for differing frame utilization, different bit rates and cluster distances

types (voice, video, real-time control, etc.) and include factors such as intra- and inter-cluster calls, call silent periods for each type, maximum and minimum bandwidth needed for each type, and the propagation time needed for each call block to pass the next call location. For example, video *circuits* do not have silent periods as do voice *circuits*. The subframe termination boundary occurs after all nodes have had a chance to send their *circuit* calls without overlap. All *circuits* (including repeated) are numbered and are sent in order; when a *circuit* terminates all call numbers above are decremented by 1.

The ability of the system to handle node *circuit* load can be studied using the analysis in Appendix A. Maximum allocated TDMA load is considered to

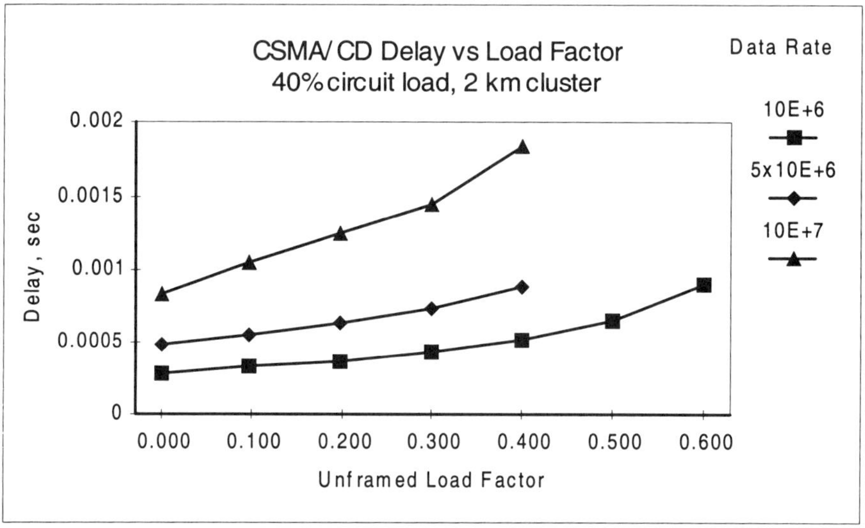

Figure 5 H³M CSMA/CD sub-frame performance for a wide range of bandwidths

be 95% of frame time. Assuming 50% of the calls are silent, this provides an minimum CSMA/CD subframe of approximately 50 - 55% of frame time depending upon the sync subframe size. This still accommodates a reasonable number of single *datagrams* and *datagram-circuits*.

We have used equation A1) to determine the number of voice circuits available under various conditions of bandwidth and cluster diameter. We assume one call per node. These results are shown in Table 1. They demonstrate that H³M can support a wide range of nodes over differing cluster distances.. Further noting that each video circuit is equivalent to ~7 voice circuits for the assumptions made[4], we can provide an estimate of the number of video circuits which can be adequately accommodated in a single cluster. By adjusting bandwidth, H³M can support a node count in a single cluster from very small to extremely large with out operational or performance deterioration. Hence, it scales well with regard to both node size and cluster diameter.

Considerable analytical and simulation work was conducted to document the CSMA/CD performance for DRAMA [4-5]. The early work [4] plus work by others [9, 25] examined the effect of framing on CSMA/CD while the later work [5] showed that when a LAN uses multiple channels, it is more effective to segregate the synchronous and asynchronous data onto separate channels. H³M does not have this option available, so we needed to extend the original work to verify that changing bandwidth and sending frequencies is a feasible alternative method for load balancing. In Appendix B, we describe a study that extrapolates the original work to H^3M.

Figure 5, the result of that study, shows the delay performance for CSMA/CD subframe packets for a typical system with 2 km cluster diameter, 40% TDMA *circuit* load and a bandwidth range factor of 10. Figure 5 uses the unframed load for its load factor, i.e., the load occurring over the 10^{-2} sec. frame length assuming no TDMA *circuit* load. An effective CSMA/CD load, λ_{eff}, can be considered. It is:

$$\lambda_{eff} = (1 - t_{TDMA}/T)\lambda \qquad \text{B3)}$$

where the variables are defined using equation A1). For the situation shown in Figure 5, the effective load is 1.667 of the unframed load. Thus, at 40% CSMA/CD load factor, the system is supporting a 68% load on the CSMA/CD subframe.

One additional feature for the CSMA/CD subframe performance should be noted. We suspect that by working with the delay factors to reduce the effect of collisions at the beginning of the CSMA/CD subframe, it should be possible to support larger load factors more effectively. This should widen the range of bandwidth and cluster diameter conditions which H^3M provides.

5.0 Comparison of H^3M with other Mobile Communication Protocols (MCP)

To put H^3M in perspective, we need to compare it to other MCPs [1, 10-12, 17, 19, 22] . We do so by discussing areas where features differ significantly..

Routing - H^3M does not need routing tables. It is only necessary that cluster CMRs to be aware of all members residing in that cluster and making the necessary copies, translations and retransmissions. Copy, reformat and retransmit requirements are established at *circuit* setup time. In comparison, for most MCPs, routing tables are required and must be maintained at either global or quasi-global (hierarchically) level and must be periodically updated to enable inter-cluster communications [10]. For example, in reference [10], routing overhead can be as high as 500 kbps for in-cluster updating and 18.3 kbps for out-of-cluster updating. During large node movements, both route updates and data packet loss may occur. This significantly reduces throughput and it may take a significant time to recover.

Delay and jitter - These are hard to quantify in MCPs because of the various ways in which packet access and routing are provided, the effects of hopping and secondary factors relates to transmission lost and retransmission. Further, the importance of delay and jitter depends upon the type of information being transferred.

Delay - in H^3M *circuit* hop count ranges from 0 to a maximum of 3 regardless of cluster or distance. Delays are predictable within *O(10 msec.)* at setup time and do not change with mobility. For many MCPs, hop count is unknown until call setup and may change due to loading. In reference [10], for example, the average hop count is 10 and ranges from 4 to as much as 11. This creates average delays ranging from 100 msec. to 1 sec. depending upon routing. Further conditions such as mobility, varying loads, changed circuit bandwidth needs, etc. over the *circuit's* life span may change, causing extreme different delay in predicting delay in many MCPs.

Jitter - for *circuits* in H^3M, this factor occurs mainly due to call active/silent activity since dynamic subframing is used to recover capacity. While this causes more jitter than when the fixed circuit time blocks are used as in many MCPs, the jitter is still within subframe times and its nature is highly regular because it is manly due to call silent/active changes and call admission and terminations.

Multicasting - multicasting causes no difficulty in H^3M. Since a cluster's CMR monitors all packets from a sender's cluster, it only needs to recognize that a particular packet is destined for a node in its cluster; then make the necessary copy, translation and retransmission. For multicasting, special addressing is needed but can easily be resolved during protocol setup or dynamically during cast setup. Further, there is no additional capacity costs and routing issues found in other MCPs.

Bandwidth and capacity utilization - H^3M has a number of these features which makes it superior to other MCPs. First, with dynamic framing, recovery of silent *circuit* activity is easily accomplished. Second, without routing (see above) there is no need to use capacity to propagate routing tables. Further as noted above, routing problems during high node movement may cause packet loss further diminishing utilization.. Third, hopping in H^3M is minimized such that clusters are not required to transfer packets in which they have no interest. Fourth, multicasting does not impact capacity as it does in MCPs where clusters must propagate packets as in hopping. Fifth, node movement is accomplished with only a small increase in reserved but unused resources. Finally, load balancing (see below) provides an excellent and effective method of sharing capacity between clusters so that loss due to loading effects is significantly ameliorated.

Load balancing and fairness - dynamic load changes are probably the most difficult situation to handle in many MCPs. The problem's causes are varied and diverse. In most cases, they must be handled by alternative routing; this complicates call handling, network control and management significantly. In H^3M, dynamic load balancing is a fundamental feature of its design and can be readily implemented using VB/VFWS. Further, as investigated in past DRAMA studies, it is very effective and can be implemented in a time frame that alleviates most situations under which load changes normally occur.

6.0 Summary of results

The items in section 5 demonstrate that H^{3}M provides a QoS which is superior to many MCPs. The features provide support for real-time, multimedia traffic in a wireless network that is highly mobile, rapidly deployable and effectively integrates heterogeneous nodes without an expensive infrastructure.

We have adapted the best features developed in DRAMA to wireless networks. We have devised techniques for inter-cluster data transfer which are both efficient and easily implemented and have significantly extended the DRAMA concepts by using variable bandwidth/variable frequency (VB/VFWS) equipment. These techniques significantly reduce the number of receivers which nodes must have and make it feasible for simple nodes with only a single T/R to participate in the network; they support highly predictable QoS features that support multimedia in data transfer environment that is most suited for each media data type.

VB/VFWS enables adoption of multiple channel concepts first developed for DRAMA to be adapted for load balancing in H^{3}M. Load balancing enables fairness for data traffic when vastly different loading and loading types occur in clusters over the network. This feature easily extends QoS to a type of traffic which has been difficult to service in many networks, wireless or not. Finally, we have extended the analysis and simulation work in DRAMA to H^{3}M to show that the latter system can be effective and to reinforce the understanding of its limitations.

References

1. Lin, R.; Gerla, M.: "Adaptive Clustering for Mobile Wireless Networks"; *IEEE Jour. On Selected Areas in Comm.*; Vol 15; No. 7; Sept 1997; pp. 1265 - 1275.
2. Nakamura, H.; Tsuboya, M.N.; Nakajima, A.: "Applying ATM to mobile Infrastructure Networks"; *IEEE Comm. Mag.*; Jan 1998; pp. 66-73.
3. Zhou, S.; Senevirtne, A.; Percival, T.: "An Efficient Location Management Scheme for Hybrid Wireless"; Faculty of Engineering, Un. Of Tech, Sidney.
4. Sharrock, S.M.; Maly, K.J.; Du, H.C.; Ghanta, S.: " A CSMA/CD-Based, Integrated Voice/Data Protocol with Dynamic Channel Allocation," Tech. Rpt.# TR-86-009; Computer Science Dept.; Un. of Minn.; Minneapolis, MN; July 22, 1986
5. Sharrock, S.M.; Maly, K.J.; Ghanta, S.; Du, H.C.: "A Framed, Moveable-Boundary Protocol for Integrated Voice/Data in a LAN," *Proc. Of SIGCOM ' 86*; August 1986, #9.
6. Sharrock, S.M.; Maly, K.J.; Ghanta, S.; Du, H.C.: "A Broadband Integrated Voice/Data/Video Network of Multiple LANS with Dynamic Bandwidth Partitioning," *Proc. INFOCOM '87*; March 1987; pp 417-425.
7. Maly, K.; Overstreet, C.M.; Qui, X.; Tang, D.: "Dynamic Resource Allocation in a Metropolitan Are Network," Tech. Rpt.# TR-88-03; Computer Science Dept.; Old Dominion University; Norfolk, VA 23529-0162; Feb. 16, 1988.

8. Maly, K.; Overstreet, C.M.; Qui, X.; Tang, D.: "Dynamic Bandwidth Allocation in a Network," *Proc. of SIGCOM'88*;
9. Metcalfe, R.M.; Boggs, D.R.: "Ethernet: Distributed Packet Switching of Local Computer Networks," *Communications of the ACM*; vol 19; July 1976.
10. Iwara, A.; Chiang, C.; Pei, G.; Gerla, M.; Chen, T.: "Scalable Routing Strategies for Ad Hoc Wireless Networks," *IEEE Jour. On Selected Areas in Communications*; Vol 17, No. 8; August 1999; pp 1369 - 1379.
11. Jubin, J.; Tornow, J.D.: "The DARPA packet radion network protocols," *Proce. IEEE*; Jam 1987
12. Shacham, N.; Craighill, E.J.; Poggio, A.A.: "Speech transport in packet-radio networks with mobile nodes," *IEEE J. Select. Areas of Commun.*; Dec 1`983; pp 1084-1097.
13. N. Abramson, Ed., *Multiple Access Communications: Foundations for Emerging Technologies,* IEEE Press, New York, 1993.
14 D. J. Baker, "Data/voice communication over a multihop, mobile, high frequency network", *Proc. MILCOM'97*; Monterey, CA, 1997, 339-343.
15. D.Bertzekas and R. Gallager, *Data Networks,* Second Edition, Prentice-Hall, 1992.
16. W. C. Fifer and F. J. Bruno, "Low cost packet radio," *Proceedings of the IEEE,* 75, (1987), 33-42.
17. M. Gerla and T.-C. Tsai, "Multicluster, mobile, multimedia radio network," *Wireless Networks,* 1, (1995), 255-265.
18. E. P. Harris and K. W. Warren, "Low power technologies: a system perspective," *Proc. 3-rd International Workshop on Multimedia Communications,* Princeton, 1996.
19. R. Ramanathan and M. Steenstrup, "Hierarchically-organized, multihop wireless networks for quality-of-service support," *Mobile Networks and Applications,* 3, (1998), 101-119.
20. D. Raychaudhuri and N. D. Wilson, "ATM-based transport architecture for multiservice wireless PCN," *IEEE Journal of Selected Areas in Communications,* 12, (1994), 1401-1414.
21. R. Sanchez, J. Evans, and G. Minden, "Networking on the battlefield: challenges in highly dynamic multihop wireless networks," *Proc. of IEEE MILCOM'99,* Atlantic City, NJ, October 1999.
22. J. E. Wieseithier, G. D. Nguyen, and A. Ephremides, "Multicasting in energy-limited ad-hoc wireless networks," *Proc. MIL COM'98,* 1998.
23. W. Mangione-Smith and P. S. Ghang, "A low power medium access control protocol for portable multimedia devices," Proc. Third International Workshop on Mobile Multimedia Communications, Princeton, NJ, September 1996.
24. M. Joa-Ng and I.-T. Lu, "A peer-to-peer zone-based two-level link state routing for mobile ad-hoc networks," *IEEE Journal of Selected Areas in Communications,* 17, (1999), 1415-1425.
25. Kleinrock, L and Tobagi, F.A.; "Packet Switching in Radio Channels: Part 1," *IEEE Transactions on Communications*, COM-23; 12, Dec. 75; 1400-1416.

Appendix A: H^3M TDMA analysis

TDMA subframe performance is predictable. The time for each *circuit's* information to be delivered from sender to receiver depends to a large extent on the repetitive transfers used. The TDMA block for the *circuit* is based its count number which decrements by 1 for each prior count *circuit* that terminates. The *circuit* access time it determined by it count and the number of bytes each prior *circuit* has to send and the propagation delay between each prior *circuit* node. While there are many variables which can be random over time, most will not vary greatly between frames. What is more important than frame delay and jitter is how many *circuits* can be supported. This relates to the number of nodes a cluster can support assuming each node is talking on a single *circuit* at any one time.

The frame structure is shown in Figure 2 and the TDMA subframe structure is shown in Figure 3b. The subframe termination boundary occurs after all nodes have had a chance to send their *circuit* calls without overlap. Let:

η_υ = number of voice circuits including repeated calls
η_τ = number of video circuits including repeated calls
p = percentage of silent voice circuits
d = cluster diameter, km
β = data rate, bps
b_υ = circuit voice bit rate per frame (32K bps) = 320 bpf
b_τ = circuit video bit rate per frame (200K bps) = 2K bpf
b_s = circuit voice silent bit rate = 16 bits
δ = round trip propagation = $2*5d\beta$ - signal travel time is assumed as 5μsec/km
T = frame time = 10 msec.
b_f = frame sync block size = 128 bits

Then for a TDMA sub-frame:

δ_d = prop.delay in bits = $(\eta_\upsilon + \eta_\tau)\delta/3$ – assume average node separation 1/3 d
δ_d = voice circuit delay in bits = $RU[(1\text{-}p)\eta_\upsilon]b_\upsilon + RD[p\eta_\upsilon]b_s$
δ_d = video circuit delay in bits = $\eta_\tau b_\tau$
and

$$\tau_{\text{TMDA}} = \text{TDMA subframe time} = (b_f + \delta_d + \delta_d + \delta_d) \qquad \text{A1)}$$

For the analysis, *RU* and *RD* designate round upper and lower respectively since partial voice circuits are not realistic. Repeated calls, both incoming/outgoing and voice/video, use the same load resources as to local cluster calls so they can be treated in an identical manner. Further, it is assumed video circuits do not have silent periods as do voice circuits.

By solving for η_υ assuming $\eta_\tau = 0$ and $p = 0\%$, we can calculate the number of voice calls per unit of bandwidth assuming that the TDMA subframe take up (40% - 80%) for a reasonable percentage of the TDMA load. The results in Table 1 are conservative since allowance has been made for all voice *circuits* to be active. With p = 50% the space allotted for *datagrams* and *datagram-circuits* is 80% to 40% of the frame time. It is just that there is no assurance that this condition would exist at all times. Further noting that each video circuit is equivalent to approximately 7 voice circuits, we can provide an estimate of the number of nodes which can be adequately accommodated in a single cluster. This analysis can result in strategies for acquiring or releasing bandwidth. These results are shown in Table 1

Tables 1 illustrate the wide range of capacity available for TDMA subframe voice circuits. Clusters of widely varying count are feasible so that nodes which frequently communicate with other nodes and are within a few kilometers can be easily accommodated within a single cluster. Further, in many situations a cluster can support a number of video circuits.

Table 1 also illustrates that at higher bit rates and larger diameters, propagation delay creates some reduction in performance. Some ways to regain some of this lost capacity would be to change the *circuit* protocol slightly so that each node would complete all its established *circuits* before the next node starts, increase the size of *circuits* blocks and have them sent in alternative frames and/or other mechanisms which would cut down on the need for inter-nodal propagation delays.

Appendix B H^3M CSMA/CD analysis

With the help of the performance analysis for DRAMA [4 - 5], we are able to predict with good accuracy the performance of H^3M for the CSMA/CD subframe. The CSMA/CD subframe operates similar to Ethernet. However, two effects influenced by framing can alter the performance significantly. They are:

1. packets arriving at the end of the frame which are too large to fit into the space available must be held until the start of the next CSMA/CD frame; and
2. packets arriving during the TDMA subframe are held until the start of the CSMA/CD subframe.

For unframed CSMA/CD, Metcalfe and Boggs [9] have shown that delay is dependent upon the ratio of packet transmission time to propagation delay in the network; i.e., the relative performance measures are the same whether packet size is 2000 bits and the propagation delay is 10 μsec., or the packet size is 50,000 bits and the propagation time is 250 μsec. For framed CSMA/CD, larger packet size leads to increased normalized delays; i.e., for a given frame length, F, and packet data transmission time, P, the delay increases as P/F increases. As shown in reference 4, for ratios ranging between 2% - 5%, framing causes little significant increase for the major portion of the load range, but for ratios of 20%, increases become significant at loads of 30% and higher. For large packets of information, it is better to schedule them either as *circuit* blocks for the TDMA subframe or to break the packet up into small packets to be transmitted as a *datagram circuit*.

Reference [4] also studies CSMA/CD packet delay due to split framing using an approximate analytical model and simulation to extend the results. Four factors influence the packet delay:

1. those arriving during the voice region;
2. those arriving during the data region but delayed due to transmission of those in 1;
3. those arriving during the remainder of the data region; and
4. those arriving within one packet time of the end of the frame which are delayed until the beginning of the subsequent data region.

Taking all four contributions [4]:

$$\text{Delay} = (V^2/2F)(\lambda^2+\lambda+1)+(P/2F)(P+2V-V \text{ if } (\lambda V/P)>1 \quad \text{B1)}$$
$$= (V^2/2F)(\lambda+1)+(P/F)(P/2+V)+P \quad \text{otherwise}$$

where

V = voice length region,
λ = data packet arrival rate over entire frame,
F = frame length, and
P = packet transmission time.

The statistical value of number of voice calls based upon fluctuating silent periods can be estimated as:

$$\underline{V} = N(sp+d)+c \quad \text{and} \quad \text{B2)}$$
$$\underline{V}^2 = N^2d^2+c^2+2Ndc+s^2Np+s^2p^2N(N-1)+2Nsp(Nd+c)$$

where

N = number of voice calls,
s = voice sample packet transmission time,
d = voice slot overhead time (control bits and propagation time),
p = fraction of time call is talking, and
c = time for framing information.

Reference [4] shows that by substituting equations B2) into B1), it accurately predicts the delay, D, for small packet arrival rates, λ, for a wide range of voice calls, N, and fraction of calls talking, p. However, equation B1) does not include the effect of collision resolution and so, as the arrival rate increases, the delay predicted by equation B2) is too low. Reference [4] simulated DRAMA for a wide range TDMA subframe loads (15% - 75%) and load ranges from 0% - 70%[5] obtained curves which show that actual delay which occurs. It is easy to fit these curves very accurately using poly-

[5]At the higher TDMA sub-frame loads, the higher load ranges tend to overload the system so finite delay, D, does not exist.

nomial regression and to interpolate between curves for other loads. Since data rates, λ, can be considered relative, it is possible to use the curves of reference [4] for a wide range of conditions as long as packet size/frame size condition are similar to those used in the simulation - 5%. We use this technique to demonstrate that H^3M is capable of supporting a wide range of bandwidth conditions efficiently. Figure 5 shows[5] the delay performance for CSMA/CD subframe typical system with 2 km cluster diameter, 40% circuit load and a bandwidth range factor of 10. This is important since H^3M supports varying loads by changing bandwidth, where as original DRAMA does so by changing bands. By working with the delay factors to reduce the effect of the collision problem at the beginning of the CSMA/CD subframe it should be possible to support a wide range of bandwidth and cluster diameter conditions. One additional feature for the CSMA/CD subframe performance should be noted. Figure 5 shows performance for the unframed load factor, i.e., the load occurring over the 10^{-2} sec. frame length assuming no circuit load. An effective CSMA/CD load, λ_{eff}, considered. It is:

$$\lambda_{eff} = (1 - t_{TDMA}/T)^{-1}\lambda \qquad \text{B3)}$$

where the variables are defined for equation A1). For the situation shown in Figure 5, the effective load is 1.667 of the unframed load. Thus, at 40% load factor, the system is supporting a 68% load on the CSMA/CD subframe.

One can conclude, as a result of the above analysis and discussion, that the CSMA/CD subframe performance of H^3M should be similar to that expected nominal CSMA/CD and that it should adequately support packet data transport. As we have noted in section 2.4, when cluster bandwidth is inadequate, additional bandwidth can be allocated in order to relieve congestion.

Part Six

Infrastructure for Flexible Services

Provision of Signalling for Programmable Services

M. Banfield, S. Simpson, D. Hutchison

Distributed Multimedia Research Group,
Computing Department,
Lancaster University,
Lancaster, LA1 4YR,
United Kingdom
e-mail: [banfield, ss, dh]@comp.lancs.ac.uk

Key words: Programmable and Active Networks, Caspian, ReTINA, IEEE P.1520, SOAP

Abstract: Additional network complexity driven by the demand for new broadband services increases the need for network control and management signalling. This paper takes stock of this trend and suggests an approach within the context of IEEE P.1520 to separate signalling and associated broadband intelligent services from multimedia data transport so that each may be developed to their full potential. The authors draw on their experience of development of two signalling systems, one a TINA NRA inspired Connection Management System, and the second based on the P.1520.3 Programmability Architecture, in proposing a new Signalling Transport Service Provider rôle.

1. INTRODUCTION

The emergence of Intelligent Networks (IN) and other advances in telecommunications services such as the Telecommunications Management Network (TMN) have been built on extensions of the once simple call control signalling models. Even the Internet, whose connection-less IP protocol lacks an identifiable unique call set-up signalling stage, is adopting signalling through protocols such as RSVP and SIP.

This paper anticipates new initiatives within the rapidly expanding telecommunications, networking and information technologies industries. The trend is to introduce additional new functionality (programmable services) into the network while consolidating the move towards deregulation and imposition of open free market conditions in the provision of communication services. It is well understood that, coupled with these market condition changes, the next generation of multimedia applications will generate new demands for Quality of Service (QoS), multicast and broadband services, which will in turn produce both economic and technological challenges.

The separate demands of future signalling and multimedia transport are analysed. The evolution of the Internet to provide a broadband multi-service transport medium will require the adoption of a sophisticated signalling environment. As opposed to traditional data services, multimedia traffic requires network resource management. The various uses of signalling include, end to end QoS management, and gathering of IP-POTS gateway resources. Far from replacing signalling, emerging technologies such as MPLS and DiffServ are introducing more control messaging into the Internet.

In heterogeneous services networks, where the current approach of dedicated service provision signalling protocols are inappropriate, a common signalling framework needs to be developed. Common strategies need to be adopted to provide consistent and reliable control messaging. The paper considers new open signalling and open market models (such as TINA and IEEE P.1520), and a new rôle, namely the Signalling Transport Service Provider (STSP), is proposed.

In making a preliminary evaluation of STSP we show results obtained on next generation signalling performance, with measurements made of two Open Signalling solutions. These are placed into context against alternative active technology signalling approaches.

2. THE EMERGENCE OF INTELLIGENT BROADBAND NETWORKS

In recent years, demand for advanced network services coupled with moves to deregulate the telecommunication market place have led to a number of important initiatives within the domain of open signalling. Currently network operators are forced to use multiple network control and management applications each tied to a particular technology (or worse still to a specific manufacturer's equipment). Such an approach is costly since it requires the high overhead of installing, training, running, and maintaining multiple management systems and reduces competition by tying an operator to a particular vendor's equipment.

This has resulted in a number of initiatives typified by TINA-C and IEEE P.1520 to develop architectures based on the specification of open interfaces. This should allow a single control and management system to be deployed that is independent of the underlying network technologies and the selection of vendor's switching equipment. Furthermore, the establishment of open interfaces enables the further introduction of competition into the telecommunications marketplace, enabling third party access to the underlying network resources. Of these many programmes, IEEE P.1520 is considered further in the paper; it is one of the most significant activities covering a broad spectrum of technologies including ATM switch control, SS7, MPLS, and Differentiated Services.

2.1 IEEE P.1520

P.1520 is an IEEE standards development project created in early 1998 by the OPENSIG community. Its aim [PIN-api] is to establish an open architecture for network control and define the interface between network control and management functions. Rather than specify static protocols these interfaces are designed to provide a powerful programmable API for the network infrastructure just as operating system APIs (e.g. Microsoft's Win32) currently provide to the application programmer.

The concept of programmable interfaces in P.1520 is an extension of the use of reference points in TINA; however, P.1520 does not prescribe the use of RM-ODP as a means of specification. Currently, interfaces have been described using a combination of OMG IDL (e.g. the 'L' interface for a DiffServ Router [PIN-IP007]) and in traditional protocols (e.g. the qGSMP representation of the CCM interface for ATM switches [PIN-ATM019]).

Central to P.1520 is the development of the four-layer P.1520 Reference Model inspired by the principle of opening the telecommunications infrastructure to the free market (Fig 1).

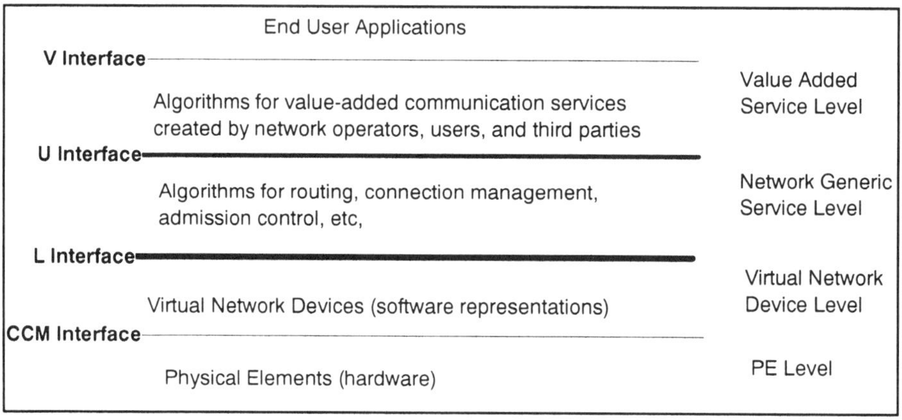

Figure 1. The P.1520 Reference Model

The model introduces four entities and four network API interfaces. The differentiation between the VASL (value-added service level), NGSL (network generic service level), and VNDL (virtual network device level) are made to separate the functionality into the classic three roles of hardware owner, network operator, and value-added-service provider.

3. DEFINITION OF THE SIGNALLING TRANSPORT SERVICE PROVIDER

The next generation of advanced telecommunications service architectures, such as TINA and P.1520 as well as the current IN approach, share in common a high degree of utilisation of inter-component signalling independent of the multimedia transport. This identifies a separation between the transport of multimedia data and the signalling data. The separation may be only logical, where signalling and multimedia data share a common physical network, or may be physical with the provision of an isolated network dedicated to transport of signalling messages.

The TINA approach for example, relies upon a distributed processing environment (DPE) to provide an object-oriented model that enables inter-component communication independent of location. This is likely to be closely related to the OMG's CORBA standards with extensions to suit the specific requirements posed within the telecommunications domain. A Kernel Transport Network (kTN) is provided to interconnect DPE nodes and provide inter-node communications. No assumptions are made as to the type of service provided by the kTN, which could be connection-less or connection-oriented. The kTN communication is based upon a specialisation of GIOP, such as IIOP for an IP based kTN or an SS7 ESIOP (Environment Specific Interoperability Protocol) for a Signalling System 7 based kTN.

While much work has been performed on identifying the main business rôles and interfaces, little thought has been given to the provision of signalling functionality or the relationship established between this and the communicating entities. We add the concept of a Signalling Transport Service Provider (STSP) whose function is to provide the transport of signalling messages (Fig 2). This applies equally to TINA, P.1520 and IN architectures but, in this paper, P.1520 is used as the reference model for further exploration.

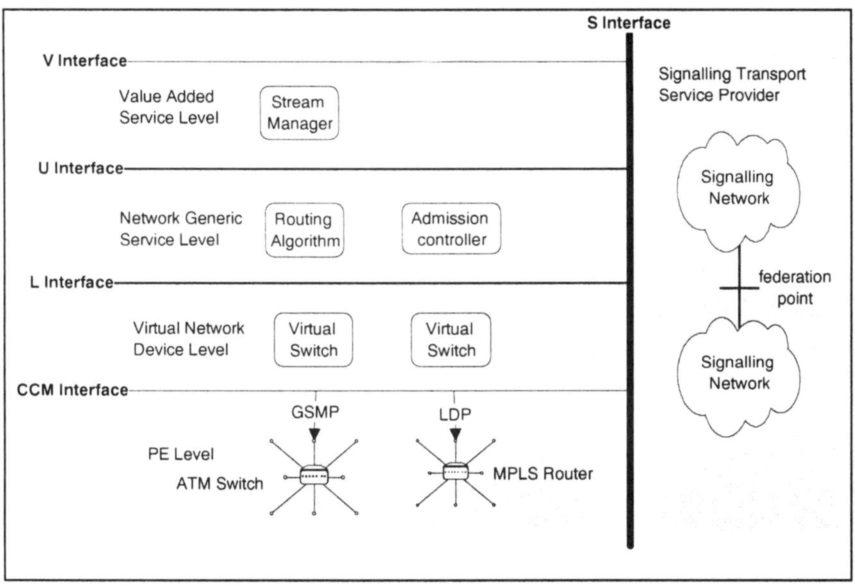

Figure 2. Introduction of the STSP and 'S' Interface into the IEEE P.1520 Reference Model

The isolation of this rôle extends the open market principles embraced in Open Signalling architectures. It provides for the outsourcing of signalling transport by the connectivity provider. This thereby allowing both the connectivity provider and the STSP to concentrate on building networks for efficient provision of their particular type of connectivity. This separation is, of course, a logical one. While in some circumstances the separation of STSP and multimedia transport network provider is made, in many cases it will be more appropriate in terms of cost for a single provider to combine both functionality sets. This detail is hidden from the control and management software, which should be unaware of how the signalling is transported, or of the ownership of the transport network.

Nevertheless, there are circumstances where this separation may be realised physically. The connectivity provider may not own a network appropriate for the transport of signalling (e.g. a wireless provider), or perhaps such high demands on network service quality are made that increased reliability through the use of an isolated signalling network provider may be appropriate. Another very important issue is that there can be no guarantee that value added service providers (located at the VASL) are physically connected to the multimedia network on which they compose their "value-added services". For example, a US based value added service provider may retail a VPN service over a network owned by British Telecommunications plc and located in the United Kingdom. A third party STSP, perhaps an ISP, can act as a communication enabler between these geographically disconnected providers.

4. REQUIREMENTS ON THE SIGNALLING TRANSPORT REFERENCE POINT

The IEEE P.1520 model is based upon a hierarchy of horizontal interfaces. In order to achieve our goal, we add to the model a new vertical interface (the 'S' interface) through which the VASL, NGSL, and VNDL establish a relationship with the signalling transport service provider (STSP). Two or more entities that wish to communicate must both form a binding with an STSP to enable them to do so. Since communication may be both intra-level (between entities located at the same service level) and inter-level (entities located at different service levels), the specification of the interface may be extended to service specific requirements noted for each signalling relationship.

The 'S' interface is used to negotiate the service requirements of the communicating entities much in the same way as a Service Level Agreement (SLA) is negotiated between an end customer and network operator over the UNI. This paper does not attempt to fully specify the interface down to IDL; this is a matter for further work and the authors believe there is much to be gained from utilising the techniques from RM-ODP[ODPspec]. Instead, we recognise five key areas of functionality that should be exposed over the 'S' interface:

1) Reservation of Service Quality – an agreed binding level of service quality must be negotiated between the parties. This could include high-level criteria such as "mean time to failure" and "percentage packet loss" or more detailed QoS characteristics such as bandwidth, delay, error rate and jitter.

2) Composition of Virtual Signalling Groups – although the signalling network provider may have many concurrent users of the signalling network, each user should be kept isolated from one another. Virtual signalling groups should be supported in a similar way to the provisioning of virtual private networks (VPN) to end-users. These groups may be used to constrain both intra and inter service level communication but should also enable federation between co-operating service providers.

3) Timescale and priority of interaction – a distinction is often drawn between the priority and timescale of connection control signalling compared to FCAPS functionality [ITU-T-M.3400] (Fault, Configuration, Accounting, Performance, and Security Management).

4) Quality of Protection – signalling messages may have very high security requirements, since the interception or spoofing of signalling may provide a loophole to compromise the integrity of the multimedia data.

5) Charging – given that the rôle of the Signalling Transport Service Provider may be outsourced to an alternative operator, provision should be made for a transparent charging system. Even where both signalling and multimedia transport are kept in-house it may be useful to know the associated cost of the signalling created by new intelligent and value-added services in order to charge correctly.

The next point to consider is the set of communications primitives provided to the users of the signalling network. Here, four possible options have been considered:

1) Network layer access – an OSI layer 3 service is provided to the customer (i.e. the Connectivity Provider of the multimedia network). The user can select their own protocols by either using those pre-provided (e.g. TCP installed in the DPE workstation's Operating System protocol stack) or by building customised protocols to their own needs.
2) Network and protocol selection – many high level protocols (OSI layer 4 and above) are designed solely for specific lower layer network infrastructures (OSI layer 3 and below). An example of this is IIOP, a specialisation of GIOP that applies only to IP networks. It is therefore not always appropriate to provide simple layer 3 access to the signalling network; instead the full protocol stack must be provided. This could be in the form of TCP/IP, IIOP/TCP/IP, or SAAL/ATM.
3) Universal communications primitives – given the previous option, it is indeed possible for the selection of a number of different protocols to be in concurrent use on a given signalling network (e.g. IIOP and RMI). It remains the job of the applications to agree, prior to communications, on choice of signalling protocols. Instead this option, the network exposes a set of basic universal communication primitives that can be used to

usefully express any communication requirement. The signalling network then maps these primitives down on to the most suited protocol for that particular communication session.

4) Full DPE Services – the final option is for the STSP to provide a full range of DPE functionality that the users can utilise to communicate. This would include RPC services, messaging services, naming services, streaming services, etc.

The third option of providing universal communication primitives seems the most appropriate option given the heterogeneity of types of network and service required. It hides the network provision from the customer while enabling the customers to construct a DPE environment suited to their own specific requirements. This could mean mapping the universal primitives to a DPE personality such as CORBA or RMI, or perhaps a Microsoft Message Queue (MSMQ) [MQref].

5. REALISATION OF OPEN SIGNALLING SYSTEMS

To investigate further the distinct requirements of signalling traffic, we have made an analysis of next-generation open signalling protocols. Recently, there have been a number of research projects creating consortiums large enough to build platforms to demonstrate the potential future developments in the management of advanced telecommunications services. Lancaster University has participated in a number of these, with the resulting platforms being established in our Distributed Multimedia Research Group (DMRG) laboratories.

5.1 The ReTINA Connectivity Service

ReTINA was until its completion during 1999 a TINA auxiliary project working to validate and develop the ideas encompassed in the TINA architecture. The goal of ReTINA was to develop and demonstrate an industrial-strength open distributed processing environment (DPE). This DPE supports distributed real-time multimedia applications over emerging broadband networks. The project embraced many areas of distributed system research including end system control, network management, DPE services, and software engineering tools. In this paper, we consider the results obtained from the evaluation of the Connectivity Service platform [ReTINA-cm] developed by Alcatel, Siemens, Broadcom, and Lancaster University.

The ReTINA Connectivity Service platform is based upon the principles of the TINA Network Resource Architecture [TINA-nra], and the Network Resource Information Model [TINA-nrim]. The platform is composed of two components, the TINA Connection Manger and an optional additional QoS Manager.

The network is represented by the Connection Manager through the composition of a hierarchy of Connection Performers (CPs) acting over child subnetworks (Fig 3) until, at the lowest level, the open interfaces to switch resources is reached. A separate signalling network known as a kTN is used to transport this inter-component DPE signalling.

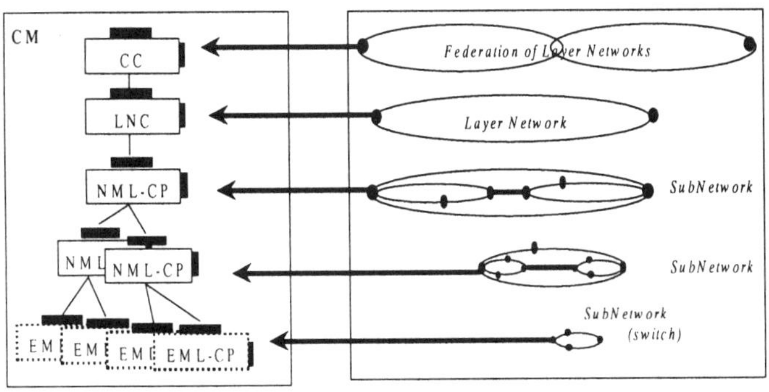

Figure 3. TINA Connection Management Architecture

Only basic call and QoS management functionality is provided by the Connection Manager. A QoS Manager can be optionally installed to provide advanced levels of QoS monitoring and estimation. However, this feature is not considered further as is a complex subject worth a publishable paper in its own right, and provides a more advanced QoS model than that exposed by other approaches such as ATM-F UNI v3.1.

Each Connection Performer (CP) exposes a CORBA interface to allow others to invoke requests to establish and release connections at the given subnetwork level. In addition, there are other more complex CP functions, many specific to the functionality appropriate for that particular hierarchical layer.

5.1.1 Implementation and results of the Connectivity Service

An implementation of this architecture was made on a Sun SPARC Solaris 2.5 platform using the Orbix v2.3 CORBA 2 compliant ORB from Iona. The Connection Manager was deployed over a cluster of three Sun Workstations all connected via an IP (over Ethernet) based kTN, so to simulate the distributed component nature of the TINA DPE. Element Layer mapping agents (using TINA terminology, the EML_CP) were then developed to enable the control of AAL5/ATM encoded video streams across two ATM switches (Fore ASX100 and ASX200).

Measurements were made of the bandwidth used on the IP based signalling network by "snooping" and analysing packets in transit across the signalling network. The implementation is based upon Orbix generated IIOP v1.0 messages; this is a specialisation of GIOP v1.0 designed to work specifically over IP-based networks. The IIOP protocol operates over TCP/IP, which gives a low error rate but higher delay characteristics. These properties mean that the IIOP protocol is well suited to the QoS requirements in this signalling domain.

Obviously, the amount of signalling bandwidth consumed is dependent upon the number of switches used in a route and the height of the hierarchical decomposition performed. Indeed, it is unlikely that there will be a one to one mapping between DPE node and Connection Performer (CP) element (e.g. NML_CP); many CP's may be co-located thereby constraining communication to use of the operating system IPC mechanisms. Given this factor, it is therefore appropriate to measure the bandwidth consumed between two dislocated CP components. Since both NML_CP and EML_CP expose the same generic interface, there is no differentiation between the signalling bandwidth requirements of either component. Analysis of the signalling network traffic shows that the bandwidth used to request the set-up/release of a connection between a parent and child node in the hierarchy is dependent on the type of connection requests (e.g. uni/multicast). It is also dependent on whether the operation was successful or if it generated an exception. Table 1 shows the size of IIOP messages used to request the successful set-up/release of a unicast, unidirectional connection between two intermediate nodes (NML_CP).

Operation	Request	Reply	Total
setup_snc	287 bytes	198 bytes	485 bytes
release_snc	248 bytes	12 bytes	260 bytes

Table 1. Size of IIOP messages for connection establishment and release

These figures only express the bandwidth consumption point to point between two CPs. To obtain a realistic figure as to the total bandwidth used to establish/release a connection it is necessary to sum the multitude of individual CP to CP signalling. This is dependant on height and breadth of the hierarchical topology, and for the remainder of the paper we simplify this by assuming the construction of hierarchies with a regular branching factor (an example of this being a binary tree).

Given this hierarchical signalling relationship it is obvious that kTN bandwidth consumption may be reduced by the construction of broad flat hierarchies. This infers a high branching factor, increased complexity routing performed at a given CP, and limited potential for computation distribution (at least at the granularity of the CP). Such properties restrict the potential scalability for concurrent use by introducing the bottleneck of nodes of high computation. This can be overcome by utilising the total combined power of the DPE nodes instead through the deployment of a less flat hierarchy of distributed CPs. Inevitably, there is a trade-off to be made against limiting kTN usage and exploiting DPE distributed concurrent processing. This is a complex problem that must take into account the requirements posed by the particular network deployment. However, for our purposes we consider the trade-off made on a simple network of 8 switches.

A network of 8 switches can be constructed in four regular hierarchies, 1x8, 1x2x4, 1x4x2, and 1x2x2x2. For each of these four hierarchies, measurements were made of kTN usage and processing time to establish a connection. Processing time is given as two values; average processing time per DPE node, and total DPE computational time. Per DPE node-processing time is a measure of the time from the reception by the CP (at level n) of a *setup_snc* operation to the time the CP invokes the set-up operation on the next CP at level n+1. Consequently the time includes both marshalling/de-marshalling computational time, plus the running of the routing algorithm on the topology modelled by the CP. Clearly this time is a factor of the DPE node platform (Orbix 2.3 on Solaris v2.5) and the efficiency of the ReTINA routing algorithm. However this is meant to illustrate the kTN and DPE processing trade-off and the figures should not be read as precise figures for all TINA Connection Management systems.

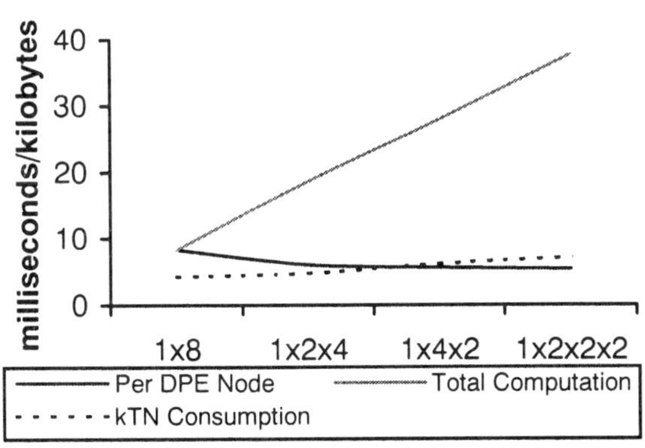

Figure 4. Trade-off between DPE computation and kTN usage

Observing the result (Fig 4), there is visibly a trade-off between per node processing, and kTN usage when choosing the appropriate hierarchy, the choice depending upon power of DPE nodes and scalability requirements of the system. If a flat hierarchy (1x8) is chosen, all connection establishment requests must pass through a relatively slow root CP. When the 1x2x2x2 topology is used, processing time in the root CP per setup/release is less, enabling more connection requests to be processed by this CP node.

Total Computational time appears though to suggest that broad hierarchies are considerably more efficient in terms of DPE usage. This is true, but the figure is simply a summation of the per node DPE processing time for all nodes within the environment. Much of this computation is the marshalling/de-marshalling of DPE requests between nodes, and will be performed in parallel between DPE nodes, and does not represent latency of connection establishment to the customer. The Connectivity Provider when deploying their topology must, therefore, make a judgement between the costs of deploying many low powered parallel DPE nodes or a limited number high-powered nodes.

Extra bandwidth will, of course, be consumed on the kTN due to the overhead of the network and transport protocols. This must be coupled with any necessary fragmentation of the IIOP message, which is dependent upon the underlying kTN MTU (Maximum Transmission Unit) size. The scenario established within our laboratory utilised a kTN provided by an STSP based upon a 10Mbit/s Ethernet infrastructure.

Since IIOP utilises TCP as its transport protocol, it was possible to discover the likelihood of IIOP message fragmentation through "snooping" the TCP MSS (Maximum Segment Size) advertised by the TCP stack. Analysis with *tcpdump* showed the MSS to be 1460 bytes, consistent with the 40 bytes of TCP/IP header (both IP and TCP headers are by default 20 bytes) and the 1500 bytes maximum capacity of an Ethernet frame. In many legacy systems, X.25 is another likely kTN medium. Here too, fragmentation is unlikely to be necessary, given the 576 byte X.25 MTU.

In addition to this, there may be further signalling through the use of legacy switch control protocols. An example of this within early ReTINA demonstrators was the deployment of the EML_CP component on a Solaris workstation on a different kTN subnet than the switch. The ATM switch was not itself a DPE node, and set-up/release, QoS monitoring, and configuration requests where passed between the EML_CP and switch using the SNMP protocol. Later ReTINA Connection Managers incorporated the switch as a full DPE node, and part of the EML_CP was deployed on the switch itself, therefore not requiring extra SNMP communication over the kTN signalling network.

5.2 The PINE Approach

An alternative approach to heavily structured solutions typified by ReTINA is PINE (Programmable Interfaces for Network Elements), a programmable network framework currently under development by Lancaster University within the EURESCOM funded Caspian (P926) project. Within this paper, PINE is chosen to highlight an alternative signalling model.

The PINE framework builds upon the wealth of background in Open Signalling architectures combined with the developments within the Internet community to provide a multiservice environment for dynamic service provision. PINE combines the IEEE P.1520 programmable architecture with dynamic injection of active code into network elements, two previously opposing approaches to advanced service provision.

The IEEE P.1520 IP Subworking group (now established as a separate project denoted P.1520.3) has identified the provision of the 'L' interface to be inadequate for the modelling of the control characteristics of IP network elements (e.g. routers). A two-layer abstraction was defined [PIN-IWAN] distinguishing between the "L+" service-specific interface and the "L-" service-independent interface, the latter being an abstract representation of an IP network element independent of whether the device in question is for example a DiffServ Router, or an MPLS switch. This interface, however, opens up the router's internals, something that many vendors may not be keen to do, and provides a very complex abstraction to program. The "L+" reference point is a service-specific interface (e.g. DiffServ Router Control Interface) providing an easy to program API that hides the underlying router complexity from the programmer. The result being that there is no single L+ interface, but a set of interfaces providing the programming of specialised services. Recently this model has been extended [PIN-IP013] to a three-tier abstraction model; "L+" interfaces now referred to as the "service-specific building block", the "L-" interface being the "resource building block". Additionally a "base building block" layer has been added that has no service or resource significance from a behavioural or packet-processing perspective. This division is currently under review and is unstable, and therefore PINE is built upon the preceding two-tier abstraction.

Our implementation of PINE is based upon the deployment of LANode (Lancaster Active Node) [LARA] routers at the Element Layer. These router nodes allow code providing control interfaces (at the "L+" reference point) to dynamically be uploaded and instantiated. The router can therefore be easily configured to provide different service abstractions of the routing resources, and specific service interfaces can quickly be deployed, tied to individual customers' requirements.

LANode offers a programmable API providing base abstractions of router resources (e.g. the routing table) on which the dynamically uploaded code can provision its service-specific control interfaces. As such, this

LANode programmable API corresponds to the "L-" reference point in the IEEE P.1520.3 model.

At the Network Management Layer (NML) subnetworks have a peer-to-peer relationship rather than the client/server relationship of the ReTINA Connection Management Architecture. Subnetworks are opaque in that they hide their underlying routing element infrastructure. From the SML viewpoint, each subnetwork is a virtual router, and just as with the physical routers, the virtual NML router supports the dynamic instantiation of new control interfaces. These control interfaces exposed by the virtual NML routers are equivalent to the "U" interface reference point (Fig 5).

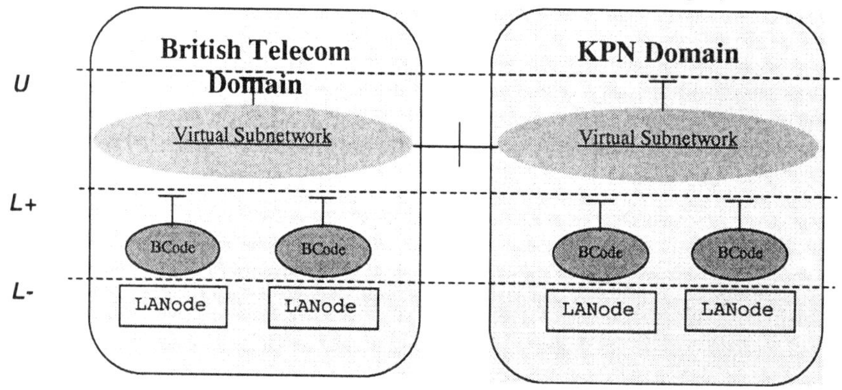

Figure 5. The PINE Architecture

5.2.1 PINE Signalling

Signalling within the PINE framework follows a distinctly different model to the ReTINA Connection Manager, with the use of the same physical IP network for carriage of both signalling and multimedia data. However, the distinction between signalling control data and the multimedia data are still maintained; this contrasts with most "active network" solutions which utilise in-band signalling: data and control are combined into the same stream, an example of such an approach being SwitchWare [Sware98].

Within PINE, two distinct types of control signalling exist. In the most coarse-grained approach, new algorithms can be uploaded to the active node that drastically alter the characteristics of the network element. For example, new packet scheduling algorithms could alter the type of Quality of Service (QoS) available, perhaps changing from Differentiated Service QoS to per flow provision. Coarse-grained signalling is therefore performed through the uploading of Java 2 bytecode that accesses the platforms "L-" API. The transport of this bytecode signalling data is carried over the conventional HTTP protocol. This has considerable advantages over the in-band approaches used by many conventional "active network" platforms and typified by the ANEP protocol. Firstly, the HTTP protocol is well accepted today for transport of HTML documents; all firewalls already support the conveyance of HTTP data. PINE therefore supports the control of nodes over multiple security domains as our bytecode signalling traffic can pass freely through today's firewalls. Alternative protocols such as ANEP [ANEP97] cannot boast this advantage. Secondly, as HTTP is built upon TCP/IP which is supported by most modern OSes, many nodes can easily be extended to support PINE, either as a network node or as a control node inaugurating the uploading of bytecode. Alternative protocols (e.g. ANEP) are not supported by current protocol stacks, and require the addition of extra modules into the OS to capture, process and deliver (to corresponding process) the packet contents.

A finer-grained control is granted through the bytecode algorithms exposing L+ control interfaces. It is not appropriate to attempt to specify the method by which interfaces are instantiated since this can be tailored to the customer's requirements. It is, however, likely that a common RPC (remote procedural call) standard would be adopted, possibilities including CORBA/IIOP, DCOM, and RMI. In our implementation we have chosen a further alternative through the development of components exposing interfaces using the SOAP (Simple Object Access Protocol) [SOAPspec] protocol.

SOAP is a distributed RPC protocol based upon the common HTTP and XML standards. Existing RPC protocols (e.g. DCOM and IIOP) are poorly suited to the Internet environment. Wide-scale firewall deployment is often incompatible with current RPC mechanisms, many firewalls supporting only a few well established services (HTTP, SMTP, etc). Organisations have already invested great resources into HTTP security mechanisms (e.g. SSL) that can be readily utilised by SOAP, whereas by contrast other secure distributed mechanisms (e.g. DCE) require considerable investment.

Existing approaches (including DCOM and CORBA/IIOP) require hefty runtime support code in order to implement their rich service sets. Many of these features are rarely used, and inappropriate for this domain. Few network elements have built in support for DCOM or CORBA, where as HTTP support is now commonplace.

The SOAP protocol has the following advantages which make it appropriate for dynamic L+ interface provision:

a) HTTP supported by current firewalls – as explained earlier in regard to coarse-grained signalling, this enables support for inter-security domain signalling.
b) Lightweight – SOAP has cut down RPC functionality. Additional unused complexity is minimised so to keep interface implementations compact; important since this is carried within the signalling bytecode.

c) Object model agnostic – is independent of programming language, object model, processor, platform, and OS consistent with the heterogeneous network element infrastructure.

Figure 6 shows the signalling relationships within the element layer (LANode) of the PINE framework.

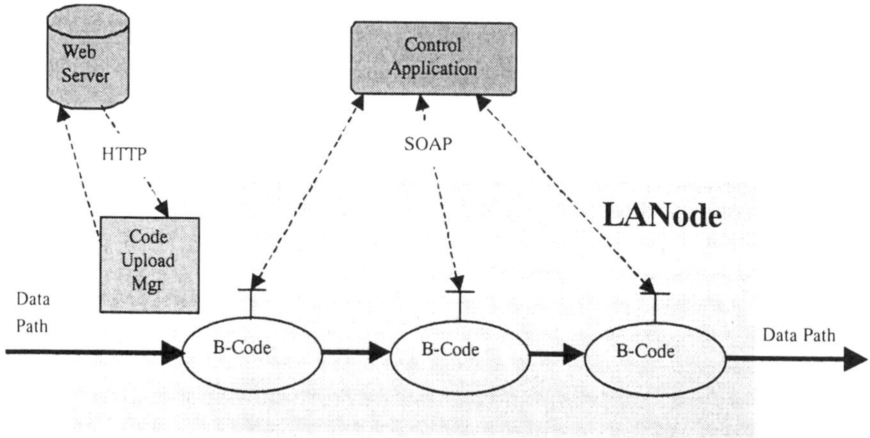

Figure 6. HTTP and SOAP signalling

Although SOAP is a new protocol it has evolved from several years of XML research [XML-RPC]. An internet draft has been submitted to the IETF, and SOAP has support from major distributed object technology vendors including Iona and Rogue Wave. Indeed Microsoft have added their voice to the SOAP community, adopting SOAP as an integral part of the Windows DNA 2000 Architecture [DNA2000].

As an example of a service deployed within the PINE framework, a simple application has been developed to expose an L+ interface to enable influence of routing tables by algorithms and software situated off-router. An example of such a scenario is the establishment of Virtual Private Networks over which the customer has direct control. The L+ interface exposes a view of the router resources tailored to the permissions of the customer.

An excerpt of this interface (defined in CORBA IDL for simplicity) is provided (Fig 7) to depict a typical L+ interface.

```
module routing {
  struct RouteId { string dest, mask; };
  struct RouteParams { string gateway; };
  struct Route { RouteId id; RouteParams params;};
```

```
    typedef sequence<Route> Routes;

    /* Accesses an L+ routing table. */
    interface RoutingTable {
      /* Get all routes. */
      Routes getRoutes();

      /* Get the parameters for one route. */
      RouteParams getParams(in RouteId id);

      /* Create a new route (or alter parameters of existing route). */
      void addRoute(in RouteId id, in RouteParams params);

      /* Delete a route.*/
      void delRoute(in RouteId id);
    };
  };
```

Figure 7. A Route Control L+ interface

Implementations have been made exposing both SOAP and IIOP implementations of this interface to run on LANode routers. This has enabled some comparison of the relative merits of both solutions and conclusions to be drawn. Three factors were considered for detailed analysis: size of SOAP/IIOP messages, time taken to process SOAP/IIOP request and response, and size of bytecode used to implement each solution.

The following series of results was obtained instantiating the routing interface on LANode router based on an Intel Pentium II 400MHz processor. The LANode software utilised the Linux 2.2.13 kernel, and a JVM based on the Blackdown JDK 1.2.2RC4 using native thread support. The IIOP instantiation of the L+ interface exploited Orbacus 3.2 (Java binding) a CORBA 2 implementation from Object Orientated Concepts [OOC]. The SOAP implementation used the SOAP/Java (version 0.3) message parser from DevelopMentor [DevM]. This also required the use of a SAX XML parser, for these testes the Xerces/Java v1.0.0 parser from Apache.

Measurements were made of SOAP and IIOP message sizes and round trip time for request/response. The "getParams()", "addRoute()", "delRoute()" methods detailed in the IDL in figure 7 were taken as example L+ interface calls. It was necessary to factor out the network traversal time, and any work the interface implementation may make on receipt of the call in order to achieve a true representation of the call processing time.

Table 2 shows the round trip request/response time

Method	IIOP	SOAP
getParams()	1.95ms	41.27ms
addRoute()	2.03ms	40.50ms
delRoute()	1.72ms	38.16ms

Table 2. Round trip request/response times for IIOP and SOAP interface implementations

These figures show SOAP to be considerably slower and require more processing than equivalent IIOP invocations. The "on the wire" size of the invocation and response message for SOAP was also found to be considerably larger.

Table 3 provides a comparison of the relative size of IIOP and SOAP request and response messages. Since SOAP is still relatively new and unknown, an example HTTP/SOAP message is illustrated in Fig 8.

RPC protocol	Request	Response	Total
IIOP	100 bytes	39 bytes	139 bytes
SOAP	609 bytes	520 bytes	1129 bytes

Table 3. SOAP and IIOP message sizes for getParams() method

```
------------------------------------------------------------------
POST /L+interface/routing HTTP/1.0
Content-Type:  text/xml
SOAPMethodName:  urn:schemas-lancs.ac.uk:caspian#delRoute
User-Agent:  Java1.2.2
Host:  wand:8888
Accept:  text/html, image/gif, image/jpeg, *; q=.2, */*; q=.2
Content-length:  362

<?xml version='1.0' ?>
<s:Envelope xmlns:ns2='urn:schemas-lancs.ac.uk:caspian'
xmlns:xsi='http://www.w3.org/1999/XMLSchema/instance'
xmlns:s='urn:schemas-xmlsoap-org:soap.v1'>
     <s:Body>
              <ns2:delRoute>
                       <id s:href='#sid1' />
              </ns2:delRoute>
              <ns2:RouteId s:id='sid1'>
                       <dest>148.88.153.50</dest>
                       <mask>255.255.0.0</mask>
              </ns2:RouteId>
```

```
</s:Body>
</s:Envelope>
```

Figure 8. Sample HTTP/SOAP message to invoke delRoute() method

Clearly the figures obtained show that SOAP performs poorly when compared to the IIOP protocol, this despite SOAP lacking the rich set of RPC features found in IIOP, and without the overhead of the CORBA object model (which could indeed by built over SOAP) in our Orbacus experiments.

However, within our chosen usage domain these issues are not the highest priority. The Java bytecode for these interface implementations are considerably larger than SOAP/IIOP messages. Given the rapid interface dynamic instantiation feature that requires the upload of bytecode onto the LANode router, it is important that the implementation bytecode be as compact as possible.

A comparison of the total bytecode size of the interface implementations is made in table 4. Class files that account for these totals have been selected based on their need to be present in code that exposes the interface of an object implementing RoutingTable. The ORB and the HTTP server have been excluded from these, since they would likely be provided locally on the system.

For both CORBA/IIOP and SOAP, this includes similarly sized parts such as classes representing types defined in the IDL, a skeletal interface implementation, and a main program to instantiate the implementation, and bind it to the relevant communication services.

In the IIOP version, a selection of other IDL-generated classes (stubs, helpers and holders) constitutes the bulk of the extra weight, since little of their function is needed for SOAP. What is needed includes a method dispatcher, SOAP body classes representing method requests and responses, and the registration of Java types to SOAP types, all of which have been hand-written.

The code-size advantage of SOAP has come from eliminating unnecessary code for simple distributed services.

RPC protocol in implementation	Size of implementation
IIOP	26129 bytes
SOAP	11241 bytes

Table 4. SOAP and IIOP implementation

This shows that although SOAP has a considerably higher overhead in terms of message size and required processing, the actual implementation is approximately only 43% of the size of the equivalent IIOP implementation. Therefore, the choice as to which protocol to use to provide L+ interfaces is not specified within our PINE framework; the most appropriate protocol should be chosen given the usage requirements posed. This should include consideration of interface lifetime, likely interface usage, type of clients, and security restrictions.

6. RELATED WORK

The approaches to signalling identified within this paper represent only two possible solutions. In tandem with our "Open Signalling" inspired solutions, other frameworks have been developed with similar goals. Of particular interest is the Parlay Group, the Darwin project conducted at Carnegie Mellon University, and the "Active Reservation Protocol" being developed by USC/ISI.

The Parlay Group is a closed consortium formed in April 1998 by 5 partners (British Telecom, Ulticom, Microsoft, Nortel Networks, and Siemens). In 1999, the consortium was expanded further to include AT&T, Cegetel, Cisco, Ericsson, IBM and Lucent. Its rôle is to define an API [ParlayAPI] to expose network capabilities while ensuring network security and integrity. This enables the introduction of a new telecommunications business model [Parlay99], separating network ownership and service provision, similar to that of P.1520. However, the Parlay API only represents the "U" interface reference point and does not consider how these interfaces map onto the "L" interfaces (known within Parlay as Resource Interfaces) of the underlying network equipment. This is considered a matter for implementation, and therefore outside the remit of Parlay. Our PINE framework is compatible with Parlay; "U" interfaces supporting the Parlay API can be instantiated to meet customer demands.

The Darwin Project [DARWIN98][DARWIN99] aims to develop a framework for customisable resource management for the support of value-added services. Similar to our LANode routers used within PINE, limited active processing is supported on the control plane through "delegates", active code installed on network nodes. Delegates are however more autonomous when compared to PINE, where the purpose of active code is to provide control interfaces tailored to the specific customers services and the intelligence resides within the NML layer.

Naturally, delegates pose significant security concerns, which are addressed through both runtime (e.g. Java 2 JVM sandbox) and compile time mechanisms. Additionally, the concept of delegate owner is introduced to control those permitted to introduce delegates onto the router.

Other signalling is supported through Darwin's own Beagle signalling protocol. Unlike traditional signalling protocols (RSVP, and PNNI), Beagle signalling affects virtual network meshes rather than individual streams.

The "Active Reservation Protocol" (ARP) project [ARP99] at ISI is developing "a framework for implementing and deploying complex network control functions using an active network approach". It is argued that rather than the need for a long protocol standardisation process, only a single implementation is required to form a new standard (the standard is the implementation code itself). Similar to the LANode routers within PINE, a programmable interface at the "L-" reference point is defined known as the "Protocol Programming Interface" (PPI). Although the approach shares a similar design, the goals are different; ARP aims to enable rapid deployment of new (network element to network element) protocols as opposed to exposure of programming interfaces within PINE.

7. DISCUSSION AND CONCLUSION

This paper has discussed the context and the increasing rôle of signalling in the provision of telecommunication services. The signalling generated by increased broadband IN functionality spawns Quality of Service requirements quite separate and distinct from those of multimedia data.

Many of the emerging models such as TINA and IEEE P.1520 choose to push the rôle of provision of signalling traffic into the functionality provided by the Distributed Processing Environment (DPE). However, the current DPE architectures, including CORBA, DCOM, and RMI, are designed as generic DPE platforms. They do not tackle the issue of provision of a signalling network, aiming to be independent of the underlying network infrastructure. This paper therefore highlights signalling network provision, and suggests the isolation of signalling provision as a specific rôle to consider when designing an advanced-services network. A context for this rôle of signalling provision has been provided through the explanation of its relevance to the TINA and IEEE P.1520 architectures.

These next-generation approaches to the provision of advanced network services are realistic and realisable. Prototype realisations have been established in research labs such as ours. As has been shown, although signalling has distinct characteristics from multimedia data, signalling itself is quite diverse. ReTINA and PINE characterise two distinct approaches to the provision of signalling transport; ReTINA adopting a separate signalling network (known as the kTN), while PINE uses the same network for signalling as for data transport.

While minimising signalling bandwidth is obviously an important consideration, it is not the only aim of an intelligent broadband signalling protocol; other properties such as extensibility and computation distribution may be considered. For example, although considerably more efficient, the ATM Forum's UNI3.1 protocol lacks the extensibility of the open signalling approach. The efficient location of signalling nodes need not be made so as to minimise signalling but to maximise distribution of processing. Such a decision is unique to each particular situation, taking account of costs of signalling bandwidth against cost of computational nodes.

Despite the claims made by some about the Internet's lightweight approach to signalling, the provision of adequate signalling functionality is of increasing importance. Recently there has been considerable interest on network security following attacks on prominent Internet sites. Governments fear terrorist attacks on our networking infrastructure at a time when we are increasingly reliant on an on-line society. Much of the interest has gone into securing end-to-end data communications on an individual stream through adoption of encryption protocols. Yet security of signalling provision is of even greater importance since, if compromised, it has the potential to bring down or violate the entire network. Recently the important rôle of intelligent signalling technology was highlighted in the UK with the outage of several IN services. While we do not know if this was simply a software fault or a malicious attack, the unavailability of 0800 (freephone) numbers to key services caused significant disruption to many people.

Signalling security is of great importance and Quality of Protection (QoP) has been identified as a key goal of the identification of the separate rôle of signalling provision. Both ReTINA and PINE attempt to address this issue, ReTINA with the separate signalling network, and PINE by using regular secure HTTP technologies (e.g. SSL). However, this issue is far from solved, and we expect ever more advanced approaches to be developed to match the increasing sophistication of network hackers.

In conclusion, those involved in the development of advanced network services are concerned with the construction of ever more complex architectures. It is clear that the signalling network is the foundation of an advanced services network, and without adequate provision of the signalling network, the whole architecture cannot be effectively realised.

ACKNOWLEDGEMENTS

We gratefully acknowledge the EURESCOM funded Caspian project (P926) and ACTS programme ReTINA (AC048) project for the sponsorship of M. Banfield and S. Simpson.

REFERENCES

[ANEP97] Alexander, D.S., Braden, B., et al., "Active Network Encapsulation Protocol (ANEP)", July 1997, http://www.cis.upenn.edu/~switchware/ANEP/

[ANON99] Tschudin, C., "An Active Networks Overlay Network (ANON)", IWAN'99, Berlin, June/July 1999.

[ARP99] Braden, B., "ASP EE: An Active Execution Environment for Network Control Protocols", http://www.isi.edu/div7/ARP, December 1999.

[DARWIN98] Chandra. P., et al. "Darwin: Resource Management for Value-Added Customisable Network Service", IEEE ICNP, Austin (USA), October 1998.

[DARWIN99] Takahashi, E., et al. "A Programmable Interface for Network Resource Management", IEEE OPENARCH, New York (USA), March 1999.

[DevM] SOAP/Java version 0.3 message parser, DevelopMentor Inc., http://www.develop.com/soap/

[DNA2000] "Windows DNA 2000", Microsoft Developer Network Journal, Issue 16, January/February 2000, Pages 18-22

[LARA] Cardoe. R., et al. "LARA: A Prototype System for Supporting High Performance Active Networking", IWAN'99, Berlin, June/July 1999.

[MQref] Dellinger, C., "Codename Falcon: The Microsoft Messaging Queue", Microsoft Interactive Developer Journal, March 1998.

[ODPspec] ISO/IEC, "Open Distributed Processing", ITU Recommendation X.901 – X.904, 1995

[OOC] jOrbacus v3.2, Object Orientated Concepts Inc., http://www.ooc.com

[ParlayAPI] Parlay Group, "Parlay API Specification 1.2", The Parlay Technical Team, September 1999, http://www.parlay.org/

[Parlay99] Parlay Group, "Parlay API Business Benefits White Paper", Issue 1.0, June 1999, http://www.parlay.org/

[PIN-api] Biswas, J., et al. "Application Programming Interfaces for Networks", IEEE P.1520 Whitepaper, January 1999.

[PIN-ATM019] Adam, C., Huard, J-F., Lazar, A., Nandikesan, M., "The qGSMP Protocol", [P1520/TS/ATM-019], IEEE-P.1520 ATM subworking group, July 1999

[PIN-IP007] Raguparan, M., Biswas, J., Wang, W., "P1520-L Interface requirements for IP routers that support differentiated services" [P1520/TS/IP007], IEEE-P.1520 IP Subworking group, May 1999.

[PIN-IP013] Vicente, J., et al. "Proposed L- Interface Architecture", [P1520/TS/IP013], IEEE-P.1520.3 IP Subworking Group, January 2000

[ReTINA-cm] Banfield, M., Edwards, C., Charton, N., Hutchison, D., "Providing Scaleable QoS-based Connectivity Services", TINA'99 Conference, April 1999.

[SOAPspec] Box, D., et al. "SOAP: Simple Object Access Protocol", INTERNET-DRAFT <draft-box-http-soap-01.txt>, Internet Engineering Task Force, November 1999.

[Sware98] Alexander, D.S., et al. "The SwitchWare active network architecture", IEEE Network, 12(3):P29-36 May/June 1998.

[TINA-bm] TINA-C, "Business Model and Reference Points Version 4.0", May 1997

[TINA-nra] TINA-C, "Network Resource Architecture v3.0", Deliverable no: NRA_v3.0_97_02_10, February 1997.

[TINA-nrim] TINA-C, "Network Resource Information Model Version 3.0", Deliverable no: NRIM_v3.0_97_12_17, December 1997.

[PIN-IWAN] Denazis, S., et al. "Designing Interfaces for Open Programmable Routers", IWAN'99, Berlin, June/July 1999.

[XML-RPC] UserLand Software Inc, http://www.xml-rpc.com/

A VERSATILE NETWORK INDEPENDENT SERVER ARCHITECTURE FOR MULTIMEDIA INFORMATION AND COMMMUNICATION SERVICES

Wolfgang Kellerer
Munich University of Technology, Institute of Communication Networks
Arcisstr. 21, 80333 Munich, Germany. Email: kellerer@ei.tum.de

Abstract The success of new service architectures will substantially depend on how they cope with existing technologies. Information and communication technology is characterized by a multitude of heterogeneous transmission platforms on which a multitude of different systems provide services to the users. The latter dispose of various terminals and varying network access points. Despite of one global communication standard the heterogeneity is even supposed to increase in the near future. For flexible service provisioning and to support various heterogeneous technologies, new concepts have to be worked out addressing especially the rising customer requirements and providing network access to new market entrants.

This paper presents a new component based server architecture that realizes convergence of heterogeneous systems on a service level, abstracting from the networks. The control of information and communication services is strictly separated from the control of communication bindings in the networks. Network adaptor units enable the connection to any network type including packet oriented networks and broadcast networks. The server architecture is targeted to high-level services rather than basic services that are already provided in the networks. Especially, the combination of previously separated services allows a variety of new service offers. The structure of the service architecture components according to a session concept in access, service, and communication session supports an easy service control and management. The signaling between the components with simple messages bases on the principles of the IETF SIP concept. Service and user mobility features based on a directory server concept and a service discovery mechanism for adaptive selection of resources support flexible service provisioning. The functionality of the server architecture is illustrated by a detailed example.

1. INTRODUCTION

The market for information and communication services is presently characterized by quick changes. Permanent technical improvements and the rising demand for electronic communication in general lead to a competition situation, where providers come up with new services in shorter and shorter life cycles. In addition, the progressing liberalization in telecommunications fastens these changes in service provisioning. New market entrants show up and have to differentiate themselves from the incumbent operators in prices and (even more important) by offering unique services. Since most parts of networks are still under control of former monopolists, new approaches are needed to open networks to provide new services in a flexibe and quick way to the users.

Important improvements have also been made in the underlying network techniques, which contribute to the rapid changes in the services market. New and advanced transmission techniques (photonics, wireless) and improvements in data processing (processors, memory) provide increasing bandwidth and lead in general to enhanced capabilities of communication networks. Increasing service offers and decreasing prices also raise the demand of users for more quality and for more supplementary features in their information and communication services. Competing providers have to address this demand: Existing services are enhanced with new features (e.g. supplementary services in PSTN and mobile networks by the Intelligent Network), new services are being developed (e.g. GPRS for wireless data communication) or new services are being created by the combination of previously separated services. This has led to a rising variety of generally available services and service systems (e.g. ISDN telephony/fax, wireless telephony, SMS, GPRS, WAP, WWW) as well as personalized services (e.g. unified messaging, one number services).

We can also observe an increasing variety in the control of communication networks. Unlike the attempts to build a homogeneous global network infrastructure (as with B-ISDN/ATM in the 90ies), currently even more and differentiated network technologies and protocols come up (e.g. H.323, HSCSD, GPRS). In the same way it cannot be predicted, whether the next generation system for wireless communication (UMTS) in fact is able to integrate all future services. In addition, packet oriented IP-based networks move in the focus of interest, even though they lack in providing reliable quality for telecommunication services. Beneath that, digital broadcast networks gain importance. In this way not the attempt to *one* global infrastructure but to an infrastructure consisting of multiple heterogeneous parts can be observed.

Similar to the networks, the systems enabling information and communication services – the service architectures – are also very heterogeneous. Network specific architectures like POTS exist beneath service integrating architectures like ISDN. Application specific architectures like T.120 or DSM-CC complement the variety.

The user requests services that are easy to handle, allow personalization, can be provided in short time, and fullfil all their complex demands. Customized services should be accessible independently of specific networks and terminals.

One approach that supports the necessary convergence is provided by the IP technology. With the IP protocol, different transmission networks converge in one superior transmission protocol on the network layer. However, IP based services are terminal based. This means that all service software has to be integrated in the terminals and kept consistent. The IP approach is not sufficient for the above stated demands for this reason. In addition, IP based networks lack in providing features that are indispensable for telecommunication services. Guaranteed quality of service, security mechanisms, regulatory aspects, and billing concepts are missing.

To find a more suitable solution, we have to consider on the one hand the requirements and on the other hand the heterogeneity of the standards. A feasible approach would be to support the convergence of the different networks and systems on a *service level* superior to all the existing architectures. Service architectures that realize this concept would bring together existing networks and systems to provide best usage for the customers. According to the antagonistic principles of the location of network intelligence for service control (intelligent networks vs. intelligent terminals) we can call this "Intelligence on top of the networks".

Along with this more network centric approach this paper describes a component based server architecture. The control of information and communication services is strictly separated from the control of communication relationships within the networks. Adaptation units realize the connection to any heterogeneous network. The server architecture is targeted to high-level services rather than basic services that are already provided in the networks. Especially the combination of previously separated services allows a variety of new service offers. Beneath well-known telecommunication services and their enhancements by supplementary services especially broadcast services and asynchronous services are focused for service combinations. The structuring of the service architecture components according to a session concept in access, service, and communication sessions reduces complexity for service control and management.

There are other approaches that as well address the problem of infrastructure independent service provisioning. [1] describes an architecture for service provisioning for hybrid services based on programmable networks. Service control components are distributed among network nodes and terminals. This programmable network approach does not fit with our requirements, since it affects infrastructure and user equipment. Other approaches are based on migration issues towards high-level architectures like TINA and propose the integration of heterogeneous networks therein e.g. [2, 3], but only consider legacy telecommunication networks.

This paper is structured as follows. First we give an overview over the evolving techniques for multimedia service architectures and point out the requirements for future architectures. Section 3 describes our approach that deals with the mentioned requirements by introducing a novel network independent server architecture. The key concepts of this architecture are presented in Section 4, whereas Section 5 describes some personalization and mobility features. Finally Section 6 illustrates the functionality with a service example. A summary and an outlook on further work are given in the conclusion.

2. MULTIMEDIA SERVICE ARCHITECTURES

To convey further understanding of the basic principles of our architecture and to elaborate the motivation, we will first discuss the characteristics of existing service architectures regarding their suitability according to our requirements (see also [4]).

Telecommunication Service Architectures. Telecommunication service architectures are evolving from many service specific networks (e.g. POTS, X.25) to multi-service systems. Narrowband ISDN has been the first step towards an integrated services network. The service control of ISDN is tightly integrated within the switching protocols and therefore very network specific. Broadband ISDN is based on the same principles, however.

The Intelligent Network (IN) enhances the public ISDN (PSTN or mobile networks) with a centralized control of supplementary services, which is separated from the switching system. Neither user terminals nor the user network interface need to be touched for the provisioning of supplementary services. But the development of services is restricted to a limited call model resulting from the PSTN.

IP Based Service Architectures. The internet considered as a service architecture provides a platform for flexible service provisioning by the common open interface of the IP protocol. However services e.g. email, ftp, WWW, are running on the end systems. That means the appropriate service software has to be present in all involved terminals and has to be integrated there for any new service. Looking beyond simple information services, we will see in addition that the IP architecture is lacking important features indispensable for telecommunication services (e.g. QoS, security, and billing mechanisms are missing).

Even architectures for IP telephony, like H.323 of the ITU-T [5] or SIP of the IETF [6], are not fulfilling all requirements for flexible multimedia service provisioning yet. H.323 is in accordance to ISDN quite protocol oriented and allows only a very limited implementation of non-standardized services, since it is limited to the ISDN call model. Based on simple textual messages the SIP IP telephony is more suited for open service development. In addition SIP is transaction oriented similarly to HTTP, i.e. call states are not stored longer than the call establishment phase. Therefore services are not restricted to an unflexible call model. But on the other hand no influence from within the network on the services is possible after their initiation.

Furthermore both IP telephony architectures outline the need for a central network intelligence, located e.g. in a network server, for reliable service control. H.323 proposes the use of a Gatekeeper whereas SIP uses SIP servers (Proxy Server, Redirect Server).

TINA. The internet concept provides a network layer platform for all entities interworking for a service. High-level service architectures like TINA [7] instead are based on a complex middleware for the interconnection of components on application layer. In TINA the middleware (the TINA consortium proposes CORBA) carries the signaling between all components of the service and network architecture. Hence a great amount of flexibility for the provision of new services can be provided. However the middleware has to be present in every terminal and in each switching node. This of course restricts a wide-spread usage. Moreover TINA is focused on an ATM/B-ISDN infrastructure. Nevertheless the service architecture concepts very well support decomposition into separated functional areas (access session, service session, communication session). The TINA session concept based on a network centric service session manager allows a much more generic control of services than the telephony call models.

To summarize, we can see that none of the existing architectures as they are is qualified for an overall architecture to realize the convergence of networks on the service level. Especially the correlation of network characteristics within the service control is too tight. Other concepts are not suited for a wide spread usage yet since they affect the software in the terminals. A centralized approach seems to be advantageous for the control of complex services since it provides the necessary overview independent from the current status of a certain terminal.

Towards APIs. In addition to the service architectures discussed above, a number of approaches are emerging that define programming interfaces (API) for services. The IETF standardizes for example an access point from the internet to IN Service Control Points (SCP): PINT [8]. The PARLAY Group defines an open interface to allow third party providers to control services over a PARLAY API [9]. Both approaches illustrate the trend towards open interfaces for previously closed networks and are very well suited as APIs for superior network independent architectures.

Terminals and Customer Requirements. As we have seen internet services for example require powerful PCs as universal internet terminals, on which the service software is running. Existing terminals for telecommunication services, e.g. telephones or mobiles, instead are application specific. Also in future, specific terminals will exist and even be preferred since they are more adapted to certain use cases, e.g. PDAs. To provide access to personalized services we will have to improve on the service server side.

New Requirements. As we have worked out, an effective way to cope with the existing heterogeneity of terminals, networks, and service architectures will be an overlay service provisioning of all infrastructures together. This would address the communication demand and reflect the various customer access points. A service architecture is supposed to support the following features to provide this kind of convergence.

- support heterogeneity of networks and terminals in a simple way, transparent to the service control
- independency of networks, but possibility to use network specific features as far as needed
- control of services spanning multiple networks
- rapid and easy service development

- easy extensibility for additional networks

3. THE NOVEL APPROACH: NETWORK INDEPENDENT SERVER ARCHITECTURE

To enable network infrastructure independent service definition and control we propose a new service architecture. The architecture defines the components of a control server that resides within several communication networks or that has got access to those. Fig. 1 shows the general concept.

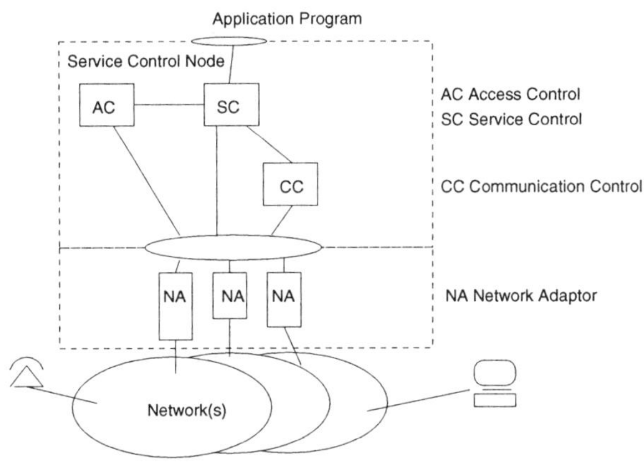

Figure 1 Server architecture general concept

Within the server architecture we distinguish two areas. One part (the lower part in Figure 1) is responsible for the strict separation of the network control and the service control. This is achieved by Network Adaptors (NA). The NAs convert the service control signaling into network specific signaling.

The part that performs the control of the services uses a well-defined interface for the communication with the network adaptors. In particular the service control consists of three main functional blocks. The Access Control components (AC) perform the user access to the server and manage all user specific data. The execution of a specific service is controlled by the service specific Service Control components (SC). These make use of the Communication Control components (CC) to establish communication relationships in the different networks.

The server architecture is open to application service providers, which could provide application specific data for the service execution or which could start a service from within an application specific context.

This architectural concept provides an overall use of heterogeneous networks for the provisioning and control of services. Thereby the operators of such a server need not to be the owners of the networks, but have a point of access to them. Open APIs like an advanced PARLAY API will support the realization of such an architecture. The server could be located at any place within the networks, e.g. close to an IN SCP or at a CATV head end. The more networks to be reachable the better it is. The components of the architecture could also be distributed over multiple servers using a middleware platform.

Services. The server provides its services to all customers in the connected networks. However it is not aimed to offer services that are already present in the different networks (e.g. simple telephony, conference, data transfer), but high-level services that go beyond the functionality of a specific network. These are

- services spanning several networks
- combination of basic services (e.g. conference plus video retrieval)
- separation of access network and execution network (e.g. for VoD)
- supplementary services
- personalized services (e.g. UPT-like one number service, unified messaging)

Section 6 presents a detailed example for typical services.

Networks. The users have access to the server via heterogeneous communication networks. The server accepts requests via signaling channels as well as usage channels; the latter means that the user explicitly addresses the server, connected to the network adaptors. From a network's perspective the server looks like a well-known element. For example in the Intelligent Network we use an SCP based network adaptor. In IP based networks the network adaptor acts as a server (web server, mail server) and could be addressed in this way. Therefore access networks may be e.g. (interfaces in parentheses) POTS (UNI), ISDN / B-ISDN (D-channel or B-channel), IN (SCP), IN-SSP (INAP), Internet (via web server: HTML, WAP, email, etc.), GSM, UMTS, GPRS.

For the establishment of connections or connection-less communication relationships, the server makes use of communication networks. When we separate access networks from communication networks, we have to point out that of course one network may serve for both. In the same way the service control information, which flows between a user and the particular Service Control (SC) during an active service, could be carried on the access network or on other connections that have been set up by the Service Control. In addition to the networks mentioned for access networks further examples for communication networks are DVB, DAB, satellite networks, and CATV. The networks are controlled via standardized interfaces like PINT, PARLAY, JTAPI, JAIN, CGI-scripts, or Gatekeeper, if available or via the user network interfaces.

4. SERVICE CONTROL ARCHITECTURE

So far we have presented the overall concepts of our server architecture. In the following we will describe some more details about the main mechanisms providing the characteristics of the architecture.

4.1. PLATFORM AND INTERFACES

The interconnection of the architecture's components could be set up via any common high-performance signaling network. In general, any distribution of the components is possible then depending on the signaling performance. Since we do not proclaim a new internal signaling network, a subnetwork of existing networks could be used ("virtual signaling network", e.g. high-speed LAN).

The separation of the service control components into the three blocks is based on the session concept of TINA [7] and other similar approaches. Like there we distinguish access session, service session, and communication session. However, the further structuring and functionality differ from the TINA recommendations as can be seen in the following. The main defined interfaces of our architecture result from this separation by sessions. Additional interfaces have been set up for adaptation and supporting components.

The interfaces (see Fig. 2) are defined by the set of messages that can be exchanged with them. All in all we have defined 12 messages to keep the complexity to a minimum. The messages are text based following the SIP concept [6], to support easy message processing.

In particular the messages are:

`ACCESS` starts an access session and delivers information about the current configuration of the user access (terminal, network, NA).

`START` transmits the user's selection to start a service.

`CREATE` The creation of a service session is requested by `CREATE`.

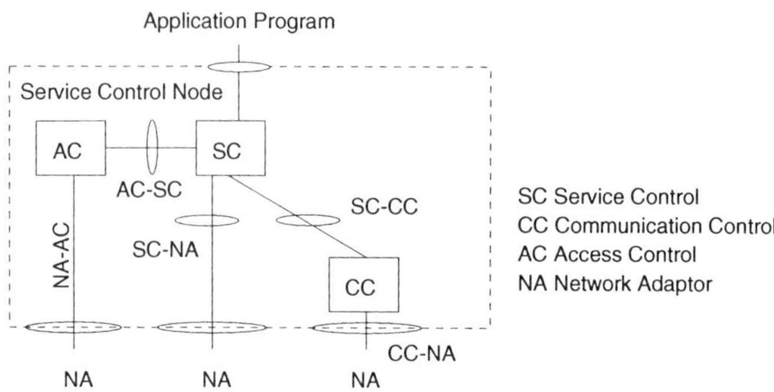

Figure 2 Main interfaces of the service architecture

`INFO` is used for requesting further information exchange between the components e.g. address of invited user. `INFO` requires an answer that is also delivered in an `INFO` message.

`STATUS` In contrast to `INFO`, `STATUS` delivers information without preceeding request, e.g. to store the session status in an user agent.

`END/BYE` Whereas `END` terminates a whole session, `BYE` is used to end the participation of one single user in a session.

`INVITE/SETUP` Both are used for the addition of a new user, a resource, or a connection to a session. `INVITE` is used by the Service Control whereas `SETUP` is a Communication Control message. Both messages are the first of a three way handshake protocol.

`OK` acknowledges a requested communication relationship (`INVITE/SETUP`) or if the requested configuration is not possible it suggests an alternative.

`ACK` is the answer on `OK` and confirms the establishment of the previously reserved configurations.

`REGISTER` is used for the registration of resources and network adaptors.

Infrastructure Independent Addressing. User addressing is performed in two steps. On the one hand each user has got one or more addresses that identify him in each of the different networks connected to the server. At least one address is known to the server from the access network. Further addresses (in different networks) could be obtained when the user connects from different network adaptors or registers addresses in an access session. All addresses are stored with the user profile to be able to contact the user over different networks. The user could configure one main contact address.

For the assignment of the addresses to the corresponding network, we use the addressing scheme: `user@network`. Whereas for `user` the address of the customer in a specific network is used, e.g. a telephone number or an email alias. `network` labels the network or network domain in that `user` is a clear identifier. Valid addresses may be `08928923505@dtag.tel` or `01721234567@D2.gsm` or `kellerer@ei.tum.de`. We have defined this addressing scheme in respect with the addressing of the IETF IP telephony in the SIP protocol [6]. So we achieve compatibility to existing addresing schemes in addition.

For a clear, network independent identification of a user within the server architecture, i.e. among the components, for the identification in different sessions, we use another well-known user indication scheme. Each user is assigned with a `login` and a `password`. In particular both are necessary when a registered user logs on from a previously not used network. The user profile is identified by the login and all user-network addresses are stored there for retrieval within the communication session.

4.2. COMPONENTS

The architecture consists of the area for network adaptation and the area for network independent service control. Fig. 3 shows the decomposition of these two areas in the most important components. Beneath the Network Adaptors (NA) and the components of the three main blocks that represent the session concept, further components are presented that support the latter in fulfilling their functions. These Service Support components (SS) are necessary for adaptation and registration of service and network resources. They will as well be described in the context of the three sessions explained below.

Service Access. The components of the block Access Control (AC) realize the access session and support the user's access to the server (`ACCESS`). The service access fulfils two functions. First to support the user in selecting and starting an information and communication service (`START`). This corresponds to the TINA definition. Second to support user registration and user profile customization (`ACCESS` and `INFO`). In contrast to the TINA definition an access session only lasts until the access connection of a user exists independent from service execution. This means an access session could also be the receipt of an email by the server containing user information (including encrypted login and password).

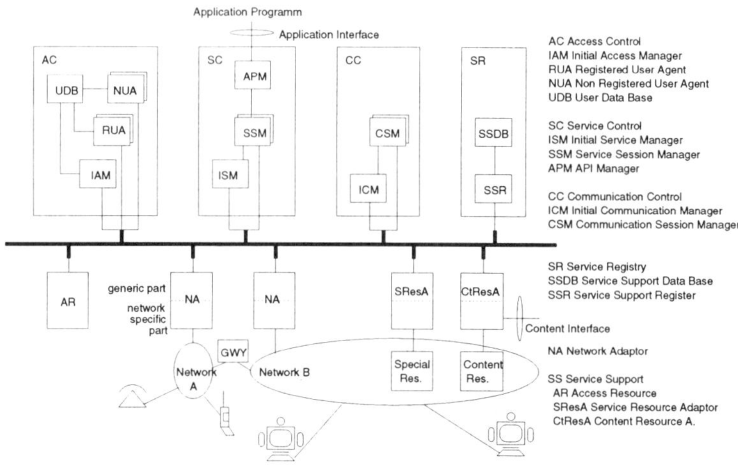

Figure 3 Component decomposition of the server architecture

For the management of user specific data the AC block contains a data base (UDB), which stores a user profile for every registered user with his login. The UDB contains all user data that is relevant for service control, in particular these are

- networks and corresponding addresses, by which the user could be contacted
- terminals the customer uses in combination with the addresses
- the terminal configurations
- the current or, if the user does not permanently update, the last user configuration (network, terminal)
- services for which the user has registered
- parameters for personalized services, e.g. call forwarding configuration
- service independent preferences and configurations of the user regarding QoS, billing

In this way the access session supports *user* as well as *service mobility*, since it supports the user to access services from any network or any terminal.

The User Profile Data Base (UDB) is realized as a directory server using LDAP. Thus a customer could also use his profile for other applications outside the server, e.g. for a browser independent storage of personal bookmarks or for terminal configurations (user specific look and feel).

A so called User Agent (UA) controls the access session for one user on the server and represents the user in all phases of service initiation and execution even beyond the access session (e.g. billing). The UA is selected and instantiated by the Initial Access Manager (IAM) for registered users (RUA). For non-registered users (NUA) a user agent is created for temporary storage and management of the user data.

Access Resource. The Network Adaptors enable the user connection from any network by the transformation of the network specific signaling into the generic messages of the server architecture. For the realization of access functions like authorization and authentication hypertext servers seem to be very useful to set up a convenient front end to the customer. Most terminals are already equipped with a browser. Scripts (e.g. CGI) or applets could control the user interaction with the server components. However there are quite a number of different hypertext servers dedicated to specific situations, e.g.

- HTML server, the most common information server in the internet,
- WML server for WAP clients in wireless communications, and
- VoxML or VXML server for speech based browsing [11].

In addition, many other XML based servers are imaginable in future. Each hypertext language could be transmitted over any network using e.g. HTTP. So it would be a great effort to integrate all servers in each of the network adaptors.

Instead we have separated the hypertext servers in our architecture from the NAs and concentrated them in an Access Resource (AR). The AR provides a centralized hypertext server functionality to all forms of access. This means, the same HTML server, equal if the user logs on via ISDN dial-up (NA-ISDN/PPP) or LAN (NA-LAN) will support every HTML/HTTP based request to the server architecture.

Service Control. When the user has started a service via the access session, a service session is activated (`CREATE`). The central Element of each service session is in analogy to the TINA concept a Service Session Manager (SSM). The SSM controls one instance of a selected service, which has been instantiated with user specific parameters. Created by a central Initial Service Manager (ISM) the SSM is able to communicate directly with the user via a NA to request further information (`INFO`) or exchange control information (`INVITE, BYE, END`). Therefore the SSM uses the NA of the access session or establishes a new connection via the Communication Control (CC).

Service Context. The service session manages a central view of the status of a service, which is represented by the service context. The service context contains user specific information of the access session as well as service specific parts from the service definition. For description of communication relationships defined for a service only qualitative descriptors are used. We have defined for example three QoS classes: asynchronous, relative QoS, and guaranteed QoS. These descriptors are necessary for the Communication Control to select a suitable infrastructure for the execution of the service.

Service Support Components. There are many services who's content is not created by the customers themselves (e.g. telephone dialog) but is delivered by media servers (e.g. video server). These resources are represented in our architecture by the class of Content Resources (CtRes) as a subclass of the Service Support Components (SS). Content Resources could be integrated directly by the SSM into a service (`INVITE`). Beneath that, media server could be treated as normal users if the appropriate control interfaces are not present.

Communication Control. The setup and control of all communication relationships for one service are performed by one Communication Session Manager (CSM) for each service. The CSM maps the qualitative description onto concrete connections and selects the appropriate networks according to the communication part of the session context, which is received from the SSM in the `SETUP` message. The mapping also depends on the user configuration data, which could be obtained from the UDB or the UA (`INFO`). All this information is used by the CSM to decide, which NA to choose to make the best selection in respect with the given situation.

The connection setup is done according to a three way handshake protocol that is repeated in several steps (cf. atomic action protocol of the ITU-T [12, 13]). First the required configuration is requested from the SSM to the CSM (`SETUP`). The CSM selects the NAs and additional service support components and forwards a `SETUP` with concrete data. Each component has to answer the request with an `OK` message either acknowledging the configuration or proposing an alternative. When the CSM has received all `OK`s the configurations are confirmed and activated with `ACK`. If alternative suggestions from the components do not fit the

CSM it starts the procedure again with SETUP or has to report to the SSM with an OK including alternative solutions.

Special Resources (SRes) e.g. converters or gateways support the Communication Control to provide network independent service execution. The SRes residing in the networks could be controlled similar to the networks by SRes adaptors.

5. PERSONALIZATION AND MOBILITY

Beneath the basic features, which we have described above, the server architecture includes some additional features to support a flexible provisioning of high-level information and communication services.

Mobility: User Agent and Directory Server. Mobility is supported by the centralized management of the user profile including her or his current "location" in a data base (user mobility). Moreover the server architecture allows the access of services from different terminals over heterogeneous networks providing the requirements (QoS, UI) could be met (service mobility).

XML based Service Definition. Each service definition in the server architecture consists of a static and a dynamic part. The dynamic part determines the sequential execution of a session (service logic) in respect with events and the service context. The service context is the service specific part of a session context (see previous section), and contains the static part of a service definition.

Whereas the static part of the service definition, the service parameters, is stored conventionally in data structures, we have chosen a new approach for the service logic description. In contrast to techniques like in the IN, where services only could be defined according to a limited set of building blocks (SIBs), or ISDN/H.323, which have a strict protocol based definition, we propose an approach using a scripting language based on the eXtensible Markup Language (XML). In particular we refer to the Call Processing Language CPL of the IETF IP Telephony Working Group [10] for the definition of our tags. XML based scripting languages provide an easily readable and verifiable way of service definition to support a flexible service development [11]. A simple service definition allows users as well as external providers to use the service architecture for customized services.

Service Discovery: Network Adaptors Registry. An advanced service discovery mechanism supports an online reaction of the server to changes in its environment. For this purpose the architecture includes a Service Registry (SR) in that analogous to the User Profile Data Base available characteristics of network adaptors and adaptors of service support components could be stored.

Along with the approaches of service discovery (e.g. JINI, SLP) components could register with the Service Registry (REGISTER) and offer their services. The CSM, when mapping the qualitative description, could retrieve the data from the SR and select appropriate components. Hereby we could also think of selections made upon pricing information and even upon the current traffic situation, which could be registered, too.

6. A MULTIMEDIA SERVICE EXAMPLE

To convey further understanding of the control flow and procedures within the proposed service architecture, a detailed example will be given. We illustrate the universal application of the server architecture with an example based on an advanced voice telephony service that is extended with an information service. The customer Mr. Joe Black is currently on a trip in a car and is able to communicate only via GSM. He wants to contact Mrs. Jane Green, which is working in her office and connected to ISDN and to the company's LAN. We assume both users being registered with the server. The requested services are executed in the following way. Fig. 4 illustrates the steps.

1 Customer Black dials the number of the server and the GSM network adaptor (NA-GSM) accepts the incoming voice call and forwards it to the voice browser Access Resource. The AR begins an access session (ACCESS).

2 The user is logging in via DTMF and his RUA (RUAb) is started. Automatically his current configuration is stored.

3 The RUAb offers various services, which he could use with the current connection. The possible selections are presented via speech synthesis by the voice browser.

4 Black chooses *Call "Jane"* and starts his service (START).

5 The access session creates the service instance *Call "Jane"* (CREATE), which is automatically enhanced with the internal logins of Mr. Black und Mrs. Green. The personal alias "Jane" is transferred for Black into the login of Jane Green. The access session ends.

6 Since the SSM has got all data necessary for the service execution, it requests the establishment of a connection between Black and Green with SETUP.

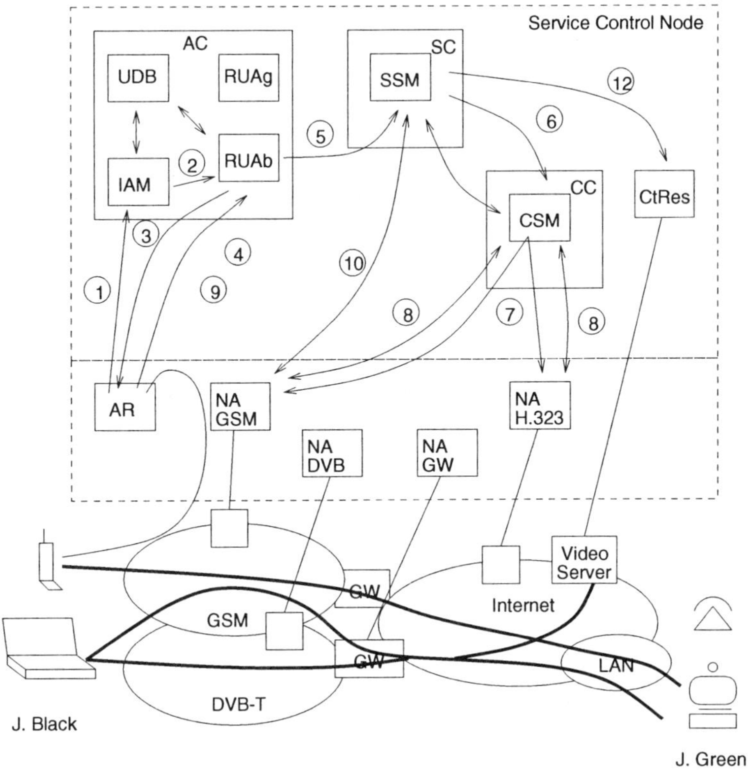

Figure 4 Example

7 The CSM retrieves information about the location of Black and Green and about available networks. The logins are mapped by the UDB to the appropriate network addresses then. At the same time a UA for Green is created to store the current configuration (RUAg). The CSM selects GSM for the real-time connection (guaranteed QoS) for Black and H.323 for Green, since Black prefers an IP based call for lower costs. A H.323/GSM Gateway is available. The control for GSM is done via an advanced PARLAY interface in GSM and for H.323 via a Gatekeeper interface.

8 After all NAs have acknowledged a successful reservation of the connections with `OK`, the connections are finally set up after the `ACK` from the CSM.

After a short voice discussion the users want to consult a document, a video documentation about a product, to get more information about a business decision. Since the GSM connection is not sufficient in bandwidth for Black, he activates his portable PC, which has in addition to another GSM module a DVB-T (Digital Video Broadcast) receiver.

9 User Black starts a new access session; this time originating from his PC via GSM. The access session is now processed via a HTML server (AR). It is identical to the one before (same RUAb) except for the presentation (screen instead of voice). After the login procedures the RUA automatically stores the DVB-T receiver.

10 After Black has activated the service *"Video Retrieval Within Call"* (`START`), the UA checks if this service could be combined with the existing voice dialog. After that the service is integrated within the active service session. The service session takes over the HTML server of the access session and the GSM NA/AR for further message display and the presentation of the video. The SSM requests the input of the video server's address and the chosen video (`INFO`).

11 The further proceeding is similar to the steps described before. But the connections are set up partly in different networks (DVB-T, GSM and Internet, in which the video server is located). For user Green the internet connection remains in use, since the video server itself is connected to the internet.
For both users it is transparent, that the exchanged data is transmitted over different networks and even the interaction channel and the downstream channel are separated. This illustrates the independence of the service architecture from the control of communication relationships within the networks.

12 The video server is known to the server as a content resource (CtRes). Thus the SSM could activate the video server access for Black and Green. The control of the video source is performed by a specific protocol of the video server via the channels that are established by the CSM.

7. CONCLUSION

It has been shown that service provisioning has to consider the increasing heterogeneity of network technologies. Current service architectures are not enough independent from network infrastructure to provide the demanded flexibility for multimedia information and communication services. A feasible solution is the convergence on the service level.

A network independent server architecture has been proposed. Network adaptors in combination with an adaptive selection mechanism in the communication control allow the complete network independent control of services over heterogeneous networks. Networks include telecommunication networks and information networks as well as broadcast networks. The most important advantages of the server architecture are:

- flexible application of the service control and of deployed services in heterogeneous networks with heterogeneous terminals without further changes
- independency from network providers
- adaptive selection of communication networks for service execution
- services can be deployed in networks they were originally not designed for

To complete the architecture some points are still open for further research. For example the integration of pricing within the access session is under investigation. In addition the interaction with some kind of Network adaptors has to be detailed. We have already realized components for an integration of broadcast networks, especially DVB-T [14].

A very critical subject is of course service interactions. We have studied interaction issues for multimedia services and found many new causes for interactions not mentioned in literature yet that will greatly influence the correct behavior of services [15]. Some more strategies for an early avoidance of service interactions will have to be integrated into the server architecture. For the validation a prototype realization of the server architecture is currently under development using UML and SDL in combination with tools.

References

[1] C. Gbgaguidi, J.-P. Hubaux, M. Hamdi, and A. Tantawi. A Programmable Architecture for the Provision of Hybrid Services. IEEE Communications Magazine, Vol. 37, No.7, July 1999, pp. 110-116.

[2] E. Holz, O. Kath, M. Geipl, G. Lin, and V. Vogel. The CAMOUFLAGE Project – Introduction of TINA into Telecommunication Legacy Systems. In Proc. of TINA'97, Novemver 1997, pp. 206-215.

[3] D. Guy. TINA-IN-Workgroup: Intelligent Networks Migration Towards TINA. 1999. Http://www.tinac.com/wg_sig/in/Main.htm

[4] W. Kellerer and B. Quendt. Multimedia Service Architectures. In Proc. of EUNICE'98, Munich, Sept. 1998, pp. 67-76.

[5] M. Korpi and V. Kumar. Supplementary Services in the H.323 IP Telephony Network. IEEE Communications Magazine, Vol. 37, No. 7, July 1999, pp. 118-125.

[6] H. Schulzrinne and J. Rosenberg. Internet Telephony: architecture and protocols – an IETF perspective. Computer Networks, Vol.31, 1999, pp. 237-256.

[7] TINA Service Architecture, Telecommunications Information Networking Architecture Consortium, Http://www.tinac.com, 1997.

[8] PINT – PSTN To Internet Integration. IETF. Http://www.ietf.org /html.charters/pint-charter.html, 1999.

[9] Parlay - the API for Secure and Open Access to Networking Functionality for Third Party Applications outside the Network. In Proc. of: ICIN'2000, 6th International Conference on Intelligence in Networks, Bordeaux, January, 2000, pp. 17 - 21.

[10] J. Lennox, H. Schulzrinne. CPL: A Language for User Control of Internet Telephony Services. Internet Draft, Internet Engineering Task Force, IPTEL WG, February 1999. Work in progress.

[11] B. Guedhami, W. Kellerer and C. Klein. Web enabled Telecommunication Service Control using VoxML. In Proc. of SmartNet 2000, Vienna, Sept. 2000.

[12] H. Müller. Flexible Signalisierungsarchitektur für Multimediadienste mit heterogenen Endgeräten. Dissertation, Technische Universität München, 1996.

[13] ITU-T Recommendations X.851/2. Information Technology – Open Systems Interconnection - Protocol For The Commitment, Concurrency And Recovery Service Element. ITU-T, 1993.

[14] P. Sties and W. Kellerer. Radio Broadcast Networks Enable Broadband Internet Access For Mobile Users. In Proc. of EUNICE'99, ISBN 84-7653-717-4, Barcelona, September 1999.

[15] W. Kellerer, P. Sties, and P. Moritz. Service Interactions beyond IN: The new Challenge for Multimedia and Convergence. In Proc. of ICIN'2000, 6th International Conference on Intelligence in Networks, Bordeaux, January 2000, pp. 277-282.

Part Seven

Mobile Agents

MobiliTools: An OMG Standards-Based Toolbox for Agent Mobility and Interoperability.

Bruno Dillenseger
France Télécom R&D (formerly known as Cnet), BP98, F-38243 Meylan Cedex, France

Abstract: One of the keys to success for applications of mobile and/or intelligent agents in large-scale open systems such as Internet is the ability of heterogeneous agents to cooperate and negotiate, and meet if they are mobile. This heterogeneity support requires the adoption of standards at the underlying distributed system level to support interoperability in agent management, mobile agent transport, and agent communication transport. This paper shows how both OMG standards and a modular architecture based on three kinds of component — agent mobility kernel, agent communication tools, and agent activity kernel — makes it possible to build a variety of heterogeneous mobile agent platforms with ad hoc features while preserving interoperability.

1 YET ANOTHER JAVA MOBILE AGENT PLATFORM?

1.1 A new paradigm for distributed systems

Classical techniques for distributed systems are based on client/server, code on demand, and remote evaluation paradigms, which finally result in moving code, and/or data, and/or control, as described in [14]. Now, mobile agents bring everything together into a new paradigm.

This paradigm has been introduced by Telescript [15] throw the *remote programming* concept, to reduce network load and latency, and to suit temporary network connectivity. As underlined in [9], there is little chance to find a "killer application" of mobile agents, but the paradigm is nice for any distributed application spread in a large-scale dynamic open system, where adaptation capability, through dynamic re-distribution of a set of cooperating agents, is a key to coping with changing hosts and network conditions, or to optimize the execution of distributed services.

But this nice anthropomorphic paradigm may not be so easy to handle practically. Besides security issues, which are critical to real large-scale applications, transparency, reliability, scalability and interoperability are other key challenges.

1.2 Limitations of today's mobile agent platforms

1.2.1 Transparency

Today's typical mobile agent platforms are built on a centralized programming language, enhanced with remote communication capabilities, and finally completed with mobility features (e.g. Java-based platforms). This final add-on of mobility deeply changes the behaviour of the original programming framework. For instance, many useful JDK packages are not designed for mobility, and transparency to mobility issues arise for any access to resources such as threads, files, sockets...

This is the reason why Java-based frameworks include specific models and tools for agent activity, communication and mobility, and specify programming restrictions. For instance, creating threads is discouraged (or forbidden) by Voyager [20] and Grasshopper [17], because the platform needs to tightly manage the agent activity. In some platforms, insufficient or disregarded restrictions can result in unspecified and indeterminist behaviour if an agent moves while it is involved in communication. As a matter of fact, communication has an impact on agent activity, and mobility has consequences on both communication and activity.

Full transparency would consist in having strong mobility as defined in [6], maintaining not only the agent state, but also the state of its activity and of its bindings to resources, including on-going communications.

1.2.2 Scalability

Both activity and communication models are of great importance for scalability. Java-based platforms that create (at least) one thread of activity per agent are examples of non-scalability if one imagines hundreds or thousands of agents needing to meet in one place.

Communication tools are also determining in scalability. Agents need to communicate locally, to take advantage of the remote programming paradigm, but also remotely, as explained in [13]. Remote communication may be implemented in a number of ways, with more or less state-of-the-art properties in terms of persistence, reliability, guaranty of delivery and causality ([2], [13]). Unfortunately, these outstanding properties typically rely on distributed algorithms introducing scalability limitations.

1.2.3 Interoperability

Last, but not least, it must also be considered that mobile agents' specific properties are dedicated to large-scale, dynamic, open distributed systems (e.g.

Internet). In such a context, heterogeneous mobile agents need a common high-level communication language to understand each other, as well as a standardised distributed execution and communication infrastructure to interoperate.

FIPA's [16] and KQML-based Agent Communication Languages are emerging standards for making agents understand each other, negotiate and cooperate. But high-level communication also requires a lower level of interoperability, on the communication transport level. Unfortunately, no standard communication infrastructure actually emerges to transport messages between heterogeneous agents. Mobile agents also need to move around in a standardised infrastructure, dealing with a common conceptual framework.

Today's mobile agent platforms typically come with specific integrated frameworks making it difficult to introduce interoperability support. Nevertheless, Voyager's CORBA support and Grasshopper's MASIF compliance are encouraging effort examples towards interoperability.

1.3 MobiliTools' specific approach

MobiliTools is a set of CORBA-based Java tools for mobility that can be used separately. The specific architecture relies on two main principles:

1. a clear separation between object mobility support, communication tools, and activity management;
2. use of standard middleware for agent and communication transport.

Principle (1) is motivated by the idea that there is no universal mobile agent framework. It is preferable, instead, to create a number of interoperable agent frameworks by choosing and combining different communication tools and agent activity schemes, on top of a mobility kernel. For instance, if at least one of the communication tools is independent from the mobility kernel, it can be used by any other agent platform or software to interoperate.

Principle (2) enforces interoperability by choosing a standard communication layer, not only between agents, and between agent platforms, but also between agents and legacy applications. Moreover, communication middleware comes with useful generic services and tools meeting typical distributed systems' needs.

Mixing these two principles results in the architecture shown by Figure 1. Any component may be replaced or reused to build a variety of agent frameworks with a common support for agent and/or communication transport.

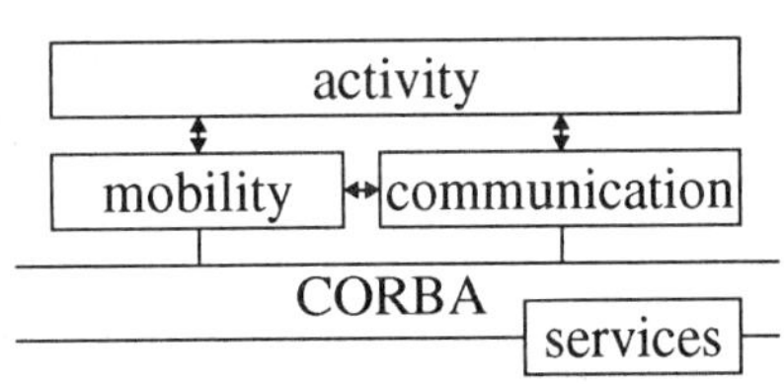

Figure 1: MobiliTools architecture

2 OMG STANDARDS AND AGENT TECHNOLOGY

2.1 Corba

OMG's Common Object Request Broker Architecture makes it possible for distributed programmes to perform remote calls on each other, regardless of their programming languages, in an object-oriented manner, while hiding network layers and operating systems heterogeneity. This standard is the result of a consortium grouping the major companies in information technology, and has several commercial and free implementations. CORBA support in some web browsers and in Java 2 is a sign of maturity.

CORBA comes with common services for distributed systems such as localisation (naming service, trader), and event-oriented communication (event service). Persistence, transactions, and security are also addressed. All these topics are of great interest for mobile agents, and everything can be re-used (as is, or as implementation "templates"), without enforcing any programming language (provided that the mapping exists from the Interface Definition Language to the target language), while relying on a well known, specified and widely available standard.

CORBA is an opportunity for interoperable basic management of agents, transport of mobile agents, and transport of agent communication. [3] describes several agent platforms developed on top of middleware such as CORBA. These platforms show in particular how several programming languages may co-exist to allow several programming levels, and how the middleware can be fully hidden to the agent programmer.

CORBA implementations do not actually support object mobility, but they can be used for every stationary component in a system of agents: execution environments hosting agents, infrastructure for agent communication, directory service...

2.2 Mobile Agent System Interoperability Facilities

OMG's first contribution to agent technology is the MASIF specification [10], dedicated to the interoperable management of agents and agent platforms. MASIF's framework is based on the following concepts: *Agents* autonomously act on behalf of a person or an organization called an *authority*. Agents are executed in *places*, hosted by *agent systems* (see Figure 2). *Mobile agents* have the ability to move from place to place, between agent systems, provided that their *agent system type* is recognized by the destination agent system. Agent systems are also bound to an authority, and may be grouped into a *region* if they are bound to the same authority. Agents are given a globally unique *name* resulting from the triplet {authority, agent identity, agent system type}.

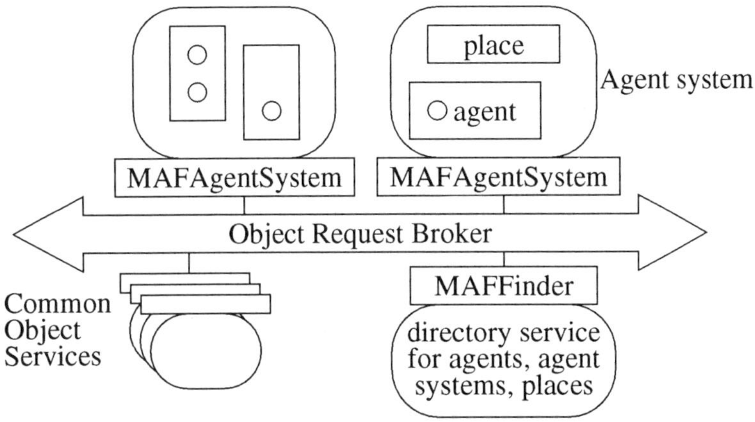

Figure 2: MASIF conceptual framework, with MAFFinder and MAFAgentSystem interfaces.

This framework is managed via two CORBA interfaces. Interface `MAFAgentSystem` must be implemented by agent systems to manage agents (create, suspend, resume, terminate), to receive migrating mobile agents, and to transfer agent classes. Interface `MAFFinder` is dedicated to registration and lookup of agents, places and agent systems.

2.3 CORBA 2.3, OMG Agent Working Group

OMG's interest in mobility and agent technology is growing. CORBA 2.3 specifications are contributing to object mobility support by including an object-by-value feature that makes it possible to pass programming language objects as invocation parameters.

As far as agent technology itself is concerned, MASIF is only a preliminary step in OMG's work. The Agent Working Group (AWG) [19] was created at the end of 1998, in order to open a forum for educating OMG in agent technology, and develop an architectural framework supporting agent technology in a compatible and complementary way with OMG's specifications. The AWG is also interested in coordinating standardisation work with other consortia in the agent field, such as FIPA.

The AWG started to write an "Agent technology green paper" [11], issued a Request For Information on "Agent technology and Object Management Architecture" in 1999, and is currently working on an "Agent Technology White Paper and RFP Roadmap" [12]. RFPs will focus on interoperability, agent communication language, security, mobility, as well as distribution, robustness and scalability.

3 THE AGENT COMMUNICATION TRANSPORT SERVICE

3.1 Overview

The Agent Communication Transport Service (ACTS) is a CORBA service for transporting messages between heterogeneous agents, whatever mobile or not, CORBA objects or not. Accordingly to the decomposition suggested in Section 1.3, the ACTS is a communication tool, independent from both the mobility kernel and the activity model. Although it is independent from MASIF, the ACTS may be considered as a complement enabling interoperability between agents for remote communication, through the definition of extra interfaces. A detailed description of the ACTS can be found in [4]; we present the basics below, and then compare the ACTS with other related work.

3.1.1 How it works

The ACTS is based on one or several servers, playing the role of *message port factory*. Basically, *message ports* are stationary FIFO buffers where agents can add and retrieve messages of CORBA "Any" type. Note that agents need not be CORBA objects. A message port can be switched from this default *store mode* to *forward mode*, by declaring a *message port listener*. A listener is a CORBA object that receives pending and incoming messages. This listener may be invalidated, either explicitly, or as soon as a CORBA communication failure occurs with this object. Such a communication failure may spring from a loss of network connectivity with the listener, or may be caused by an obsolete CORBA object reference due to the listener mobility. No message is lost, and the FIFO order is maintained anyway.

3.1.2 Typical ACTS usage

The ACTS may be distributed on a number of servers running on well connected nodes (ACTS servers can be considered as e-mail servers). An agent may have one or several message ports in different network areas in order to improve communication performance and/or reliability. According to its specific constraints, an agent may choose either a pure asynchronous communication model, where it polls its message port (store mode), or a more "reactive" model where it gets incoming messages on the fly (forward mode). In the latter case, the new reference of the listener has to be registered after each move in order to keep the "reactive" behaviour. Note that the forward mode should be handled very carefully, since each forwarded message creates a thread of activity in the listener.

3.1.3 Customization: ACTS personalities

The ACTS *personalities* hide the CORBA infrastructure and the ACTS interfaces, while providing easy-to-use communication utilities for Java. ACTS personalities also come with enhanced transparency support, advanced communication features, and higher level addressing.

The ACTS *Mailbox personality* wraps message ports into Mailbox Java objects. Mailboxes are designated with high-level addresses, consistent with MASIF's region concept (`agent_name@region_name`). Multicast and unicast features are supported by addresses transparently targeting a group of mailboxes in a given region (`group_name@region_name`). The CORBA naming service is used to register and find the ACTS servers and the mailboxes' message ports:

- name "/MAF/*region_name*/acts/factory" for ACTS servers;
- name "/MAF/*region_name*/acts/mailbox/*mailbox_name*" for message ports bound to ordinary mailboxes, or arbitrary unique names in naming context "/MAF/*region_name*/acts/mailbox/*mailbox_name*/" for message ports bound to group mailboxes.

Section 4.3 details the specific naming service usage for scalability.

The ACTS *Logged Mailbox personality* is a Mailbox extension providing the programmer with communication tracing tools and event ordering based on a Lamport Clock mechanism [8]. The ACTS *FIPA personality* is a FIPA-oriented use of the ACTS, compliant with FIPA'98 specifications for Agent Management and agent-agent interactions [5].

3.2 Further discussion on the ACTS and communication issues

3.2.1 Agent communication schemes

[1] identifies two communication schemes in agent-based systems: agent-to-agent communication where partners are addressed by globally unique identifiers, and anonymous communications where partners do not know each other (event model). Through the Mailbox personality and its multicast/unicast enhancement, we see that the ACTS supports both schemes, both in forward and store mode. Another way to achieve this is to mix the message ports with the CORBA event service, but the event service can't be used directly by agents because of their mobility.

3.2.2 Communication delivery

Three basic techniques can be used (and mixed) to reach a moving destination:
1. use a directory which binds constant names to changing locations;
2. broadcast;
3. replace the mobile agent by a forwarding "ghost" on each move;

Technique (1) is often criticized for it relies on a centralized service. Nevertheless, this technique is currently of common use in mobile phone. Applicability domain of technique (2) is typically the LAN, where broadcasting does not necessarily generate extra messages (e.g. Ethernet). Larger-scale broadcast is a problem since it typically consumes too much network bandwidth and processing time in all the recipients (and/or in any intermediate communication element). Technique (3) comes with risks of reference chain breaking and forwarders proliferation. Moreover, it can not be applied when the reason for mobility is a node or network link shutdown.

All these techniques can be defeated in the case of highly mobile agents because messages may be routed permanently and never reach their destination. [13] presents a solution derived from the distributed snapshot algorithm. It is based on a synchronisation between message propagation and moving agents on communication links. However, this work needs to be continued in order to take network and node faults into account, and scalability is likely to be a problem.

The ACTS approach is different: an agent is always addressed by other agents through a single reference that never changes (the message port). The only reference that needs to be updated is the reference to the listener when a message port is operated in forward mode. Doing this update is of the agent's responsibility. In the special case of a highly mobile agent, it is recommended not to use the forward mode, not because messages could be lost, but because messages might never reach the moving listener. The store mode seems to be the right communication model in this case.

4 THE SIMPLE MASIF IMPLEMENTATION

4.1 SMI overview

Accordingly to the decomposition of agent platforms given in Section 1.3, SMI implements a mobility kernel in Java. Starting from MASIF specification, SMI aims at providing a generic, light-weight and well-specified environment for mobile Java objects.

4.1.1 Agencies

An agency is an execution environment for mobile agents, called agent system in MASIF's terminology. Basically, they are instances of class `Agency` running in a Java Virtual Machine. Each agency belongs to a region, has a name (unique in the given region), and is bound to an authority. An agency is also a CORBA server implementing MASIF's `MAFAgentSystem` interface. Its CORBA object reference is registered in the naming service (see Section 4.3).

Agents can be managed through the `MAFAgentSystem` interface and methods of class `Agency`. Operations include creating and terminating an agent, suspending and resuming an agent activity, moving an agent, listing the names of hosted agents, and getting information on a local agent.

4.1.2 Mobile objects/agents

Agencies have methods for creating and managing any Java object implementing the `MobileObject` interface. This interface mainly consists of call-backs related to the lifecycle of mobile agents (see Section 4.2). `MobileObject` implementations also have to implement the `java.io.Serializable` interface since Java serialization is used to generate mobile agents' states. As specified by MASIF, an agent resides in a place, and has a unique name combining an identity, an authority and an Agent System Type identifier.

4.2 MobileObject lifecycle

The design of interface `MobileObject` is a straightforward mapping of the MASIF framework: agents may be created, moved, suspended, resumed and terminated. Agents have to be informed when such lifecycle events start, succeed or fail (see Table 1), not only to properly react, but also to be able to deny permission: an agent can refuse creation, mobility, or reinstallation after a move, by throwing an exception in the corresponding call-back.

Table 1: Agent lifecycle management and `MobileObject` interface.

Agency method	involved MobileObject call-back(s)
`createAgent`	`afterBirth`
`resumeAgent`	`resume`
`suspendAgent`	`suspend`
`moveAgent`	`beforeMove afterMove afterMoveFailed`
`terminateAgent`	`beforeDeath`

For instance, method `moveAgent()` of class `Agency` involves a number of steps which can fail for various reasons: the specified agent or the destination agency doesn't exist, the destination agency can't be reached because of a communication problem (network, CORBA, naming service), or agent de/serialization has failed. But the agent may also abort the move by throwing an exception before (in `beforeMove()`) or during serialization, during or after (in `afterMove()`) deserialization. If the move is aborted after the serialisation step, the `afterMoveFailed()` call-back is invoked.

4.3 Naming service distributed exploitation

SMI agencies are bound to unique names in the CORBA naming service, according to a naming scheme extending MASIF's concept of region: "/MAF/*region_name*/agency/*agency_name*". As a result, agencies (like mailboxes' message ports and ACTS servers, see Section 3.1) can be found via high-level deterministic names, helping region interconnection.

A specific naming service administration is required to avoid a bottleneck effect. The first idea is to distribute the naming service on several servers, with one name server per region. Each name server contains the name bindings for its own region, and is federated with the other name servers in the "/MAF/" naming context. As a result, resolution of name "/MAF/regionA/..." with region B's name server is transparently forwarded to region A's name server. To go further on distribution, region names may contain sub-regions (e.g. "regionA/sub-region1/..."). In this case, one name server can be responsible for each sub-region. Note that this distribution also applies for the ACTS servers.

4.4 Back to MASIF and interoperability

MASIF specifications practically supports interoperability for basic agent management tasks, through the definition of:

- a common framework of places, agent systems, region, etc.;
- a service for agent, place and agent system registration and lookup;
- an external interface for agent lifecycle.

All these points don't require a smart interpretation, and their implementation is quite straightforward. But interoperability is not fully specified for agent mobility, and is not addressed at all for agent communication. Since the latter issue is explicitly not in the scope of MASIF (the ACTS described in Section 3 suggests one solution), let's focus on the former issue. MASIF's mobility support is based on two operations:

- `receive_agent()` is invoked on the destination agency to transfer an agent — parameters include the agent profile, the agent state, the agent class

name, and a CORBA object reference to the agent system providing the agent's classes;
- `fetch_class()` is invoked by the destination agency on the class provider to get the incoming agent's locally undefined classes.

4.4.1 Agent profile

Heterogeneity management is based on the provisioning of an *agent profile*. A profile contains a set of identifiers specifying the agent programming language, the agent system type, versioning information, and serialization format. Identifiers are already defined for Java, Tcl, Scheme, Perl, Aglets, MOA, AgentTcl and Java object serialization.

SMI naturally gets the Java language and the Java object serialization identifiers, and is given a free identifier for "SMI agent system type". SMI's policy is to reject agents of any other agent system type trying to move in. Since a dedicated exception is missing, the generic `MAFExtendedException` exception is thrown. It could be imagined that hosting an agent of a different agent system type but of the same programming language could be easy, especially in the case of Java. But several implementation choices remain about de/serialization, class loading and agent lifecycle hooks. Let's discuss the interoperability issues in the case of Java as a common programming language (in the case of heterogeneous programming languages, we imagine a pseudo-agent system switching agents on to the right agent system).

4.4.2 Agent deserialization and classloader

Using standard Java object serialization does not mean that a standard `ObjectInputStream` can be used for deserialization. A specific classloader must be provided for each agent in order to fetch missing classes from the specified class provider, using the specified codebase, for the specified agent profile. This classloader must be supplied by a specific `ObjectInputStream` deserializing the agent state.

There are several other implementation choices about class loading issues, which may lead to non-interoperability. For instance, the classloader used for agent deserialization may be quite different if it assumes that classes are transferred as a whole as a parameter of `receive_agent()`, or downloaded on the fly from the class provider if they are locally undefined.

Missing classes in the destination agent system may be fetched either always from the same agent system, or from the source agent system. The former technique introduces a serious bottleneck, and may prevent an agent from moving from agency B to agency C if the class providing agency A is unreachable. The latter technique may cause a proliferation of classes, since it requires that the agencies keep byte code for hosted agents' classes. The main issue is scalability, since the amount of byte code stored in each agency may rapidly grow. SMI uses this technique however, because it results in a much more fault tolerant overall distribution. This has to be tuned and refined, but detecting and discarding useless classes is complicated by Java's reflective features.

4.4.3 "Internal" interfaces

Finally, the main difficulty for interoperability within a given programming language, is that standard hooks must be specified to tell the agent it is going to move or die, or it has just moved, or it has just been born... A common lifecycle interface such as SMI's `MobileObject` (see Section 4.2) should be defined for each language.

Local interactions with the agency and the other agents also need to be specified. For instance, an agent willing to move must be given a standard way to request the move from the agent system it is residing in. Then, supporting the remote programming paradigm for heterogeneous agents requires a standard mechanism to initiate and handle a local communication tool through a standard interface.

4.5 To be added: agent activity models

For the sake of genericity, SMI does not enforce any agent execution model. Agents are responsible for starting, suspending, resuming and terminating their activity accordingly to the corresponding lifecycle call-backs. Agents may launch a thread of activity, or share a pool of threads. The former approach fully supports autonomous agent activity, but is not scalable, while the latter approach is essentially dedicated to event-driven agents, like in [2].

Event-driven activity may be implemented using the reactive programming model. Such a model consists in splitting execution into logical time slices, or *instants*. Reactive objects react to events, combinations of events, or absence of events, and generate events that are consumed in the same instant. An instant ends when all events are consumed, and a new instant starts when new external events appear.

The benefit of such an approach is that between two instants, the state of an agent is stable and very well defined. Then, move requests can be transparently executed after the end of each instant, without affecting the programming model. Moreover, work described in [7] has produced a Java prototype able to run thousands of reactive objects, which is a promising performance regarding scalability concerns.

5 CONCLUSION

Through the presentation of MobiliTools, this paper practically explores:

- the applicability of OMG standards for making interoperable mobile agent platforms;
- how a mobile agent platform can be built as a combination of a mobility kernel, communication tools, and agent activity support.

Although MASIF brings limited interoperability support, mainly because of the "internal interfaces" issue, it is an interesting starting point for the architecture of mobile agent platforms. CORBA is convenient to implement the stationary parts of the global infrastructure, responsible for transporting and managing agents and messages. The naming service, used in an appropriate manner, provides a scalable directory for high-level location-independent references.

At last, the approach based on the assembly of independent components improves comprehensibility of transparency issues, and leads to a variety of interoperable combinations suited to various needs. For instance, the ACTS may be used in any mobile agent platform without any other MobiliTools. In the same way, SMI may host any agent activity and communication framework while managing mobility through the `MobileObject` interface.

Next steps include tuning and completion in order to fully implement MASIF, enhance the communication support, and offer a couple of agent activity models. Strong mobility support is on the way, on the basis of the reactive programming model.

ACKNOWLEDGEMENTS

The Agent Communication Transport Service has been developed for the MIAMI project [18], with design contribution from Stefan Covaci and Alexander Yip. The FIPA personality has been made by Huan Tran Viet. The logged mailbox personality has been made by Anne-Marie Tagant.

REFERENCES

1. J. Baumann, F. Hohl, N. Radouniklis, K. Rothermel, M. Strasser: Communication concepts for Mobile Agent Systems. In mobile Agents: 1st International Workshop MA'97, Lecture Notes in Computer Science, April 1997, Springer, pp. 123-135.
2. L. Bellissard, N. De Palma, A. Freyssinet, M. Herrmann, S. Lacourte: An Agent Platform for Reliable Asynchronous Distributed Programming. Symposium on Reliable Distributed Systems (SRDS'99), Lausanne (Switzerland), 20-22 October 1999.

3. B. Dillenseger: From Interoperability to Cooperation: Building Intelligent Agents on Middleware. Lecture Notes in Artificial Intelligence 1437 (Proc. of IATA'98), Sahin Albayrak, Francisco J. Garijo Eds. Springer 1998, pp. 220-232.
4. B. Dillenseger, Huan Tran Viet: Towards full agent interoperability. In Proc. of 2nd International ACTS Workshop on Advanced Services in Fixed and Mobile Telecommunications Networks. 9-10 September 1999, Center for Wireless Communications, Singapore.
5. FIPA 98 Specification. Foundation for Intelligent Physical Agents (Geneva, Switzerland),1998. see [16]
6. A. Fugetta, G. P. Picco, G Vigna: Understanding code mobility. IEEE Transactions on Software Engineering, vol. 24, No 5 (1998), pp. 342-361.
7. L. Hazard, J.-F. Susini, F. Boussinot: The Junior Reactive Kernel. Rapport de recherche No 3732, July 1999, INRIA Sophia Antipolis (France).
8. L. Lamport: Time, clocks, and the ordering of events in a distributed system. Communications of the ACM, July 1978, Vol. 21, No 7., pp. 558-565.
9. Danny B. Lange, Mitsuro Oshima: Seven good reasons for mobile agents. Communications of the ACM, Vol.42, No 3, March 1999, pp. 88-89.
10. Mobile Agent System Interoperability Facilities Specification. Joint submision: GMD Fokus & IBM Corp., supported by Crystaliz Inc., General Magic Inc., The Open Group. OMG TC document orbos/97-10-05 (1997).
11. OMG Agent Working Group: Agent Technology Green Paper. Document ec/99-12-02, 24 December 1999. see [19]
12. OMG Agent Working Group: Agent Technology White Paper and RFP Roadmap. Ref. internet/99-11-01, draft .02, 29 November 1999. see [19]
13. Amy L. Murphy, Gian Pietro Picco: Reliable communication for highly mobile agents. In proc. 1st International Symposium on Agent Systems and Applications, 3rd International Symposium on Mobile Agents, Palm Springs (USA), D.S. Milojicic ed., october 1999, IEEE Computer Society, pp. 141-150.
14. J. Vitek: New paradigms for distributed programming. In proceedings European Research Seminar in Advanced Distributed Systems, Zinal (Switzerland), march 17-21, 1997.
15. J. White: Telescript technology: the foundation for the electronic market place. General Magic White Paper, General Magic, 1994.

WEB REFERENCES

16. FIPA - http://www.fipa.org/
17. Grasshopper - http://www.ikv.de/
18. MIAMI - http://www.fokus.gmd.de/research/cc/ecco/miami/
19. OMG Agent Working Group - http://www.objs.com/isig/agents.html
20. Voyager - http://www.objectspace.com/

Software Agent Constrained Mobility for Network Performance Monitoring

C. Bohoris, A. Liotta, G. Pavlou
Center for Communication Systems Research
School of Electronic Engineering and Information Technology
University of Surrey, Guildford, Surrey GU2 5XH, UK
{C.Bohoris, A.Liotta, G.Pavlou}@eim.surrey.ac.uk

ABSTRACT

During the recent years of research on mobile agents, significant effort has been directed towards the identification of models of agent mobility suitable for network management applications. Also, a lot of research work is currently being carried out trying to provide an assessment of mobile agent frameworks used to build agent-based network management systems. In this paper we clarify three different models of agent mobility, present a mobile agent-based performance monitoring system that exhibits the "constrained mobility" model, and discuss its practical use for dynamically programming network elements. The implementation of this system is presented and compared with static object approaches. Furthermore we provide a performance evaluation of the mobile agent based system as it compares with Java-RMI and CORBA distributed frameworks, in order to assess the advantages, along with the overheads, of agent solutions.

KEYWORDS

Mobile Agents, Constrained Mobility, Performance Monitoring.

1. INTRODUCTION AND BACKGROUND

Network management has been the subject of intense research over the last decade, with the relevant progress being twofold: on the one hand, architectures and algorithms for solving management problems have been devised; and on the other hand, different management technologies have been proposed and standardized. From the protocol-based approaches of the early 90's, exemplified by the Simple Network Management Protocol (SNMP) [1] and OSI Systems Management (OSI-SM) [2], the focus moved to distributed object-based approaches in the mid to late 90's, exemplified first by the Common Object Request Broker Architecture (CORBA) [3] and later by Java's Remote Method Invocation (Java-RMI). More recently, the focus seems to be shifting back to protocol-based approaches the emerging Directory Enabled Networks (DEN) framework.

The paradigm of moving management logic close to the data it requires is a technique that has been conceived early in the evolution of management architectures, the relevant framework known as "management by delegation" [4]. Subsequent research showed the applicability of this concept in the context of OSI-SM [5] while a similar approach was subsequently standardized, the Command Sequencer Systems Management Function (SMF). More recently, the same concept has been proposed in the context of SNMP through the Scripting MIB. While such approaches are specific to the respective management frameworks, delegation in the context of general distributed object frameworks is achieved through *object mobility*. Mobile objects are usually termed *mobile agents* and when they act through emerging behavior in the Artificial Intelligence (AI) sense, they become *intelligent agents*. Mobility and intelligence are though orthogonal properties.

The emergence of mobile agent frameworks has led many researchers to examine their applicability to network management and control environments. [6] considered first code mobility and presented a taxonomy of the relevant aspects. [7] considered mobile agents in the context of the Intelligent Network (IN) and proposed an agent-based architecture for "active" IN service control. [8] discussed the general issues of using mobile agents for network management while a number of other researchers have attempted to use mobile agents in specific network management case studies. It is believed that mobile agents can provide better solutions at least to performance and fault management problems, given the large amount of data that needs to be moved around in respective solutions based on traditional approaches.

Mobile agents may move around the network from node to node and clone / destroy themselves according to their intelligence. We term this situation "strong mobility" and it is this property that has not yet been exploited to good effect in network management. An alternative possibility for mobile agents is to move from node A to B, typically guided by a "parent" stationary agent, and stay there until their task is accomplished. We term this situation "constrained mobility" and we believe it is this approach that can be readily exploited in management environments. In this case, instead of predicting the required functionality, standardizing and providing it through static objects in network elements, mobile agents could support it in a dynamic, customizable fashion. The key advantage in this case is that the target node needs only to provide the required "bare-bones" capability which could be dynamically augmented through mobile agents, with the mobile agent logic able to change to reflect evolving requirements over time. Such a possibility would obviate the use of functionality such as the OSI-SM

Systems Management Functions (SMFs) and similar capabilities provided in SNMP.

In the work described in this paper we are trying to evaluate the use of mobile agents for network performance monitoring, assuming a constrained mobility paradigm in which a mobile agent is sent to execute and monitor information within a network element. The latter can be managed through a collection of static agents that offer similar capabilities to a OSI-SM or SNMP Management Information Base (MIB). The evaluation is twofold: first, we are interested in assessing the usability of a mobile agent platform as opposed to a static object platform such as CORBA or Java-RMI, and in particular the agent customization aspects; and second, we would like to examine the performance implications of using mobile agents in order to assess if the provided flexibility is potentially outweighed by the additional performance overhead. This work has been partly carried out in the context of the MIAMI ACTS project (Mobile Intelligent Agents in the Management of the Information infrastructure) [9], which examines the impact and possibilities of using mobile agent technology for network and service management.

The rest of this paper has the following structure. In Section 2 we summarize briefly the way in which performance monitoring is supported through generic but predefined functionality in the context of OSI-SM, SNMP and CORBA-based management systems. In Section 3 we examine three models of agent mobility that can be applied in network management applications. In Section 4 we concentrate on the performance monitoring system and present our agent implementation based on constrained mobility. In Section 5 we present an evaluation and assessment of the performance monitoring system and in Section 6 we present a summary and conclusions.

2. STATIC PERFORMANCE MONITORING

Performance management is one of the management functional areas identified initially in OSI Systems Management (OSI-SM) [2]. It addresses the availability of management information in order to be able to determine network load under both natural and artificial conditions. It supports the collection of performance information periodically in order to provide statistics and allow for capacity planning activities. Performance management needs access to a large quantity of dynamic network information. An important issue is to provide this information to management applications with a small impact on the managed network. A key requirement is the ability to convert raw traffic information to traffic rates with thresholds applied to them so that Quality of Service (QoS) alarms can be generated. An additional requirement is the periodic summarization of a variety of performance information for trend identification and capacity planning purposes.

QoS management applications monitor performance aspects both "within" the network and at "edge" nodes where customer services are offered, trying to identify potential performance degradations. They may subsequently trigger the reconfiguration of parts of the network in order to alleviate congestion e.g. by changing the routing strategy, re-allocating resources such as bandwidth to trails, etc. Monitored aspects of services may include the service availability, the supported capacity in terms of available bandwidth and the end-to-end delay. In the case of a "leased line" service with guaranteed QoS, e.g. as part of a Virtual Private Network (VPN) service, its availability may be affected by faults while the available capacity and delay may be affected by network congestion when the provided bandwidth is multiplexed. In general, performance management is coupled with both fault and configuration management.

A simplistic approach for collecting the required performance information is through periodic polling. In this case, the collected raw data is processed either at a centralized Network Management Station (NMS) or at Element Managers (EMs) which may form part of a hierarchical management system e.g. following Telecommunications Management Network (TMN) [10] principles. The problem with this approach is that it generates substantial management traffic and, subsequently, does not scale (it should be mentioned though that the generated traffic is smaller in the hierarchical compared to the centralized case). An alternative approach is to delegate monitoring activities to the network elements, reporting only QoS alarms or summarized reports to higher-level managers. The OSI-SM Metric Monitoring [11] and Summarization [12] systems management functions have addressed this requirement through generic Support Managed Objects (SMOs), which need to be provided in managed network elements. Facilities similar to metric monitoring have been subsequently provided in SNMP environments, initially in the Remote Network Monitoring (RMON) [13] specification. In addition, similar facilities could be provided in CORBA-based network elements.

The problem with such generic functionality is that it needs to be first researched, standardized, implemented and eventually deployed in network elements; this process typically takes a long time. In addition, any modification, e.g. for providing more sophisticated features that were not thought out in advance, needs to go through the full research, standardization and deployment cycle. For example, [14] identified additional functionality that combines the capabilities of metric monitoring and summarization objects in a powerful fashion but such additions need to go through a new standardization cycle. The specification of the OSI-SM systems management functions took a long time and is partly responsible for the perceived complexity and lateness in the deployment of OSI-SM-based network elements.

Mobile agents could provide similar facilities in a dynamic fashion, allowing for network elements with bare-bones *real resource* only and no *support* management information. Additional generic capabilities could be provided through mobile agents which would be sent to execute within a network element. The behavior of those agents could be altered dynamically at any time. The flexibility provided to management applications would be enormous, since they could now "customize" network elements for performance and other management activities according to their requirements and would not be restricted by the available standardized facilities. On the other hand, network elements should be able to host mobile agents through suitable platform infrastructure and real resource managed objects should be realized as static agents. The performance implications of using mobile agents universally are addressed later in this paper.

3. AGENT MOBILITY IN MANAGEMENT

The problem of identifying the features that distinguish an agent from a common computational entity has raised controversial arguments for nearly a decade and only recently those features have been identified. Theoretical studies on agents and artificial intelligence have come to the conclusion that a computational entity can be regarded as an agent if it exhibits some of the following properties: social ability, autonomy, reactivity, proactivity, adaptability, persistency, and ability to learn, communicate, co-operate, and move [19].

Other research work focusing on the practical application of agent theories tends to characterize agents with a subset of the above properties. For instance, Mobile Agents are commonly defined as computational entities that act on behalf of some other software entity, exhibit some degree of autonomy, and are particularly featured with migration capability.

The chief benefits that agent mobility can bring into the network management arena, for each of the five management functional areas, are highlighted in [8]. Some of those benefits include reduction in network traffic, efficient utilization of computational resources, support for heterogeneous environments, and increased flexibility. Nevertheless, the use of mobile agents does not come without costs. In particular, code migration incurs additional traffic into the network, absorbs considerable resources from the agent hosts, and is associated with migration delays of the order of seconds or even tens of seconds, depending on the agent configuration and functionality [20] (see also Sec.5.2 below). In some cases the agent migration overheads outweight their benefits and make this approach inconvenient. It is therefore important to grasp the various aspects of agent mobility and to relate them to network management in order to identify those aspects that are particularly beneficial.

In the following subsections we define three different types of agent mobility, ranging from the simplest, light-weight form of mobility to the most heavy-weight one. For each case we elaborate on its benefits and limitations in relation to network management, identifying advantageous scenarios.

3.1 Constrained Mobility

One of the most elementary forms of code mobility is defined in [6] as Remote Evaluation (REV), after the pioneering work described in [21]. In REV, an application in the client role can dynamically enhance the server capability by sending code to the server. Subsequently, clients can remotely initiate the execution of this code that is allowed to access the resources collocated within the server. Therefore, this approach can be seen as an extension of the client-server paradigm whereby a client in addition to the name of the service requested and the input parameters can also send code implementing new services. Hence the client owns the code needed to perform a service, while the server offers both the computational resources required to execute the service and the access to its local resources.

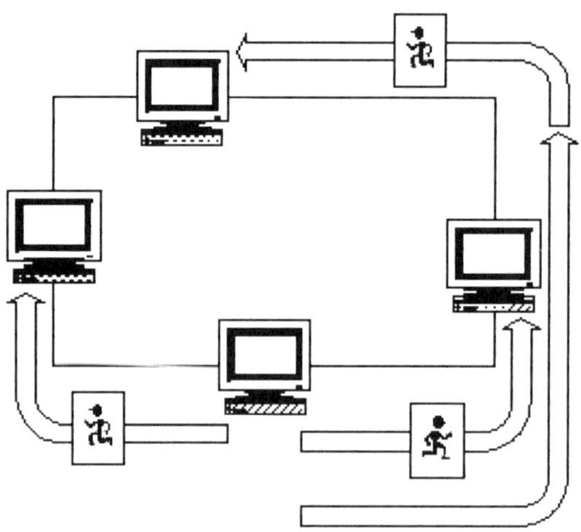

Figure 1 – Constrained mobility. The agent is created and initialised by a client application and is then shipped to an agent host. The agent execution is then confined to this host.

A natural evolution of the REV model involves sending code possessing one or more of the above mentioned agent features – e.g. mobility and autonomy. This type of agent mobility can be defined as *constrained mobility* since the agent, upon its creation in a client site, is only allowed to migrate to a remote server where its execution will be confined.

When constrained mobility is adopted in management the agent is created by a client acting in the manager role and is, then, dispatched to a target network element acting in the server role (Figure 1).

This approach is particularly suited to dynamically programming or upgrading network devices. In this case, agents do not need to be particularly sophisticated. In a simple scenario these agents do not even need to have any sort of autonomy – e.g. they do not need to have the ability to select their target network element, since the manager can provide this information – and could be as simple as collections of objects that can be executed in a remote virtual machine. Therefore, mobility degenerates into a simple dynamic mechanism to efficiently deploy or upgrade network protocols or services.

Agent deployment overheads, namely *deployment traffic* and *delay*, represent the drawbacks of general MA approaches. In constrained mobility the agent does not need to incorporate complex migration features and, as a result, its size and the incurred network traffic are minimal if compared to the other forms of agent mobility (see sections below). Similarly, it will not be necessary to use general purpose MA platforms – usually associated with heavy-weight migration mechanisms [20] – and, thus, the agent migration time can be considerably reduced.

In conclusion, constrained mobility is a particularly well-suited mechanism to dynamically program network elements. It can outperform traditional centralised management for data-intensive tasks and when high degrees of semantic data compression need to be achieved – e.g., through data aggregation or analysis. Constrained mobility is typically advantageous to perform off-line analysis of bulk data and, more generally, to implement tasks whose duration is at least comparable with the overall agent deployment time.

3.2 Weak Mobility

Similarly to constrained mobility, in weak mobility the agent is created and initialised by a client application and is, then, shipped to an agent host. However, 'weak' MAs are not confined to this host since they are meant to perform the same task in more than one location (Figure 2). Weak MAs do not retain any knowledge of the data processed or of the actions performed in previously visited hosts and, consequently, they can only implement tasks in which this information is not required.

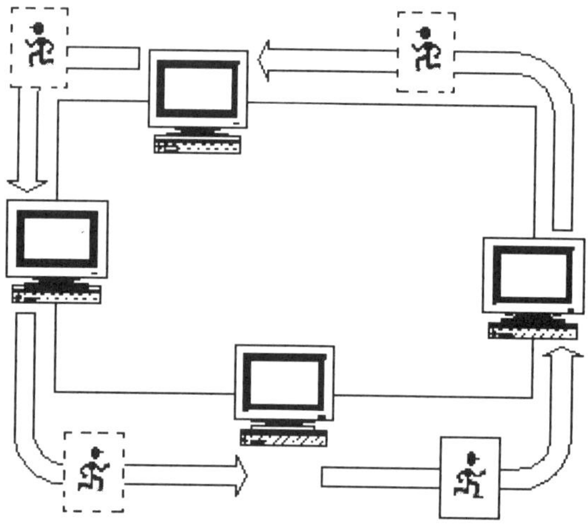

Figure 2 – Weak mobility. The agent task involves the visitation of many hosts, but no information gathered in previous visits is preserved.

A convenient use of weak mobility is for dynamic decentralisation of management tasks that are otherwise performed in a centralised fashion. The agent is delegated part of the management responsibility and will incorporate functionality such as procedures aimed at data semantic compression or aggregation.

A trivial example showing the main advantages of weak mobility is the case in which the management station has to search for a single value in a table, a data structure typically used to store information inside devices. In SNMP management the whole table has to be transferred from the remote element to the management station, where the table rows are searched for the value. Hence, large tables will incur heavy unnecessary traffic into the network and will result in computational overload on the management station.

A more efficient approach is adopted by OSI management which supports remote scope and filtering operations. Thus, the searching routine is executed in the device and, consequently, only the retrieved value is transmitted to the manager. The drawback of this approach is that routines such as the one implementing scope and filtering have to be hard coded into the network elements which tend do become complex since a large number of routines need to be implemented. Even worse the introduction of new routines requires a cumbersome standardisation process and their deployment needs complex software upgrade.

If constrained mobility was to be used the agent incorporating the search routine would be shipped to the network device, where it would retrieve the requested value from the local table and issues this value to the manager.

This solution addresses the shortcomings of the OSI approach, retaining its performance benefits. However, constrained mobility is not suitable in the more general case in which tasks analogous to the searching routine are to be run on multiple network elements. In fact, as the number of network elements to be searched grows, the management station will be overloaded and the network capacity around it will be saturated by the simultaneous generation and transmission of the agents.

Tasks characterised by a relatively short duration and involving multiple network elements are more efficiently implemented according to the weak mobility model. Since only a single MA is issued by the manager, both network and computing resources around the manager station are preserved. Weak mobility is, thus, suitable to collect on-line data and to perform simple control and configuration tasks from several network elements.

Therefore, weak mobility can lead not only to a reduction in network load but also to a more even utilisation of processing resources through a dynamic serialised distribution of simple management tasks.

3.3 Strong Mobility

With strong mobility agents in addition to being able to access and process data from network elements can also accumulate information and preserve it upon migration (Figure 3). This feature allows for the implementation of more elaborate tasks in which the agent operations depend on data gathered in previously visited hosts. In other words, the agent operation can be altered by the data.

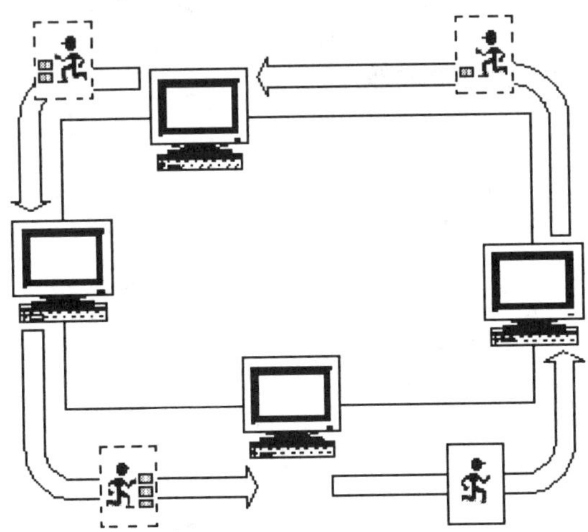

Figure 3 – Strong mobility. The agent task involves the visitation of many hosts and the preservation of information gathered in previous visits.

In management, strong mobility is particularly well suited to configuration tasks and to data-intensive tasks involving data aggregation from highly distributed network elements and on-line data analysis. A simple example is a task involving the collection of utilisation information from a relatively large number of network elements. In a traditional SNMP-based system the management station has to poll every single element in order to collect the required raw performance information before it can produce a useful utilisation rate. This may incur heavy traffic and may even not be acceptable if the number of the elements is too large. OSI management offers a more efficient mechanism for obtaining the utilisation rates, as this is done locally at the network element, but still requires further aggregation of this information at the management station.

Constrained mobility does not suit this task since it would require the deployment of a number of MAs equal to the number of network elements. Each MA would typically be executing for a time negligible with respect to its deployment time and, then, the agent deployment overheads would be unacceptable.

On the other hand, weak mobility would represent an inefficient imitation of the OSI approach. In fact, the agent would have to visit sequentially all the elements and report the local utilisation rates back to the manager before migrating to the next element. Then, we still would not achieve a total decoupling between network element logic and network management logic, since the manager would be still concerned with the collection of data from each element and with its aggregation.

Such decoupling can be achieved through strong mobility, in which case the agent will preserve the utilisation rates of all previously visited elements and will then be able to perform a further level of data aggregation independently from the manager.

The main drawback associated with strong mobility is the agents' size. 'Strong' agents need to incorporate more intelligence that others and tend to be larger in size. More critically, the agents' size can vary significantly depending on the amount of information that has to be preserved during migration. It is, therefore, important to design the agents in such a way to limit their size variations – e.g., by allowing only semantically compressed information to be stored.

In conclusion, strong mobility can be employed to implement data-intensive tasks requiring aggregation of information from different network elements.

4. CONSTRAINED MOBILITY IN A PERFORMANCE MONITORING SYSTEM

In this section we elaborate on the model of constrained mobility by describing its application in a mobile agent based performance monitoring system. As mentioned earlier the constrained mobility model is particularly suited to network management tasks that require a relatively long period of time to execute, when dynamic programming of network devices is required, or when a large number of data collected is intended for off-line analysis. An effective performance monitoring system typically possesses all of the above characteristics making the constrained mobility agent-based approach particularly suitable to it. A user of such a system typically wishes to gather performance information from a number of different machines in the network, receive performance reports on a scheduled basis and on-the-fly notifications when a performance threshold is triggered. The user can analyse the performance reports and notifications obtained to determine utilization trends, isolate performance problems, and possibly solve them before they adversely impact network performance. In this way performance monitoring can also aid in capacity planning and in the provisioning of a consistent level of service to all users of a network.

4.1 System Realisation Based on Constrained Mobility

A typical scenario of operation for an agent-based system is actually very similar to a scenario followed by a system based on static distributed objects. Initially, in the client side of both systems, a static "master" entity is responsible for accepting a user request and initiating the management service. This entity corresponds to the Network and Element Management Layer (NML and EML) functionality in terms of the TMN model.

In the distributed objects system an object with that role would send a remote request for the initiation of a management service to an object located in a remote server machine. These objects located in the server side of the system are responsible for fulfilling the service request using the required management logic, which pre-exists there in a fixed manner. The server objects can also remotely communicate with the "master" object in the client side in order to report important information.

In the case of the agent-based system (Figure 4) using the model of constrained mobility, a "master" agent in the client side will initially create a mobile agent that owns the required management logic to fulfil the service request. This mobile agent will then migrate to the remote server machine where it can have local access to the resources required to perform its task. A static "target" agent that pre-exists in the server side is responsible for providing access to those resources to such a mobile agent on request. The functionality provided by this "target" agent at this level corresponds to the TMN Network Element (NE) level.

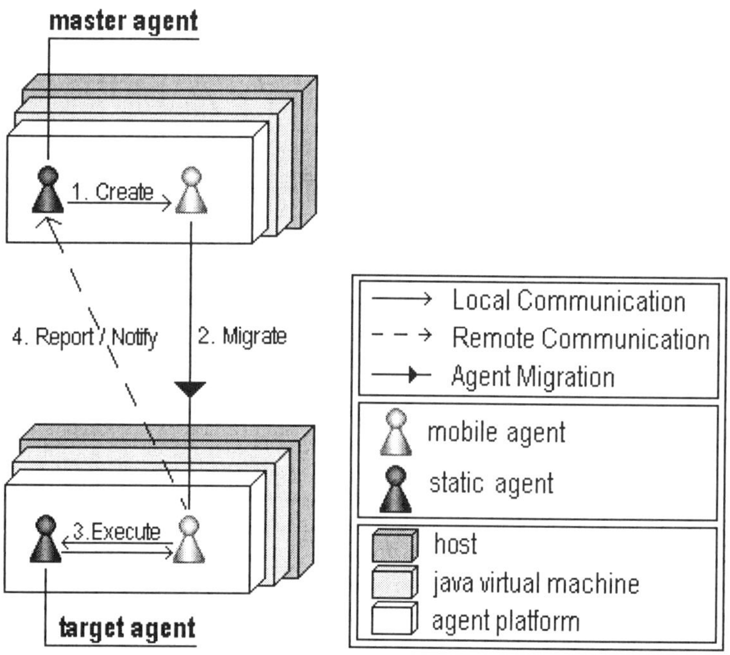

Figure 4 – A mobile agent based network management system using the model of constraint mobility.

The performance monitoring system we developed is based exactly on the above design scenario of an agent based network management system using constrained mobility. A user request for performance monitoring of a remote machine will initially be passed to the "master" agent, which will create a mobile performance monitor agent. This mobile agent will be provided with the specific monitoring parameters as set by the user and will then migrate to the remote machine. Upon reaching its destination the MA will contact a "target" static agent that pre-exists there and is responsible for providing "raw" performance information on request. The performance monitor agent will process this information to obtain rates of utilisation and loss. The performance monitor agent will remotely send reports of the results gathered back to the "master" agent in the client side, on a scheduled basis. It can also send notifications in real time when an applied performance threshold is triggered. The performance monitor agent provides the functionality of a metric monitor and a summarisation object as specified in the X.739/X.738 standards.

4.2 System Implementation

For the development of our mobile agent-based performance monitoring system we used purely the Java programming language, with all system classes built using Sun's JDK version 1.1.7b. The Grasshopper agent platform version 1.2 was also used, providing a simple execution environment for agents, and an API covering all the required basic agent functionality.

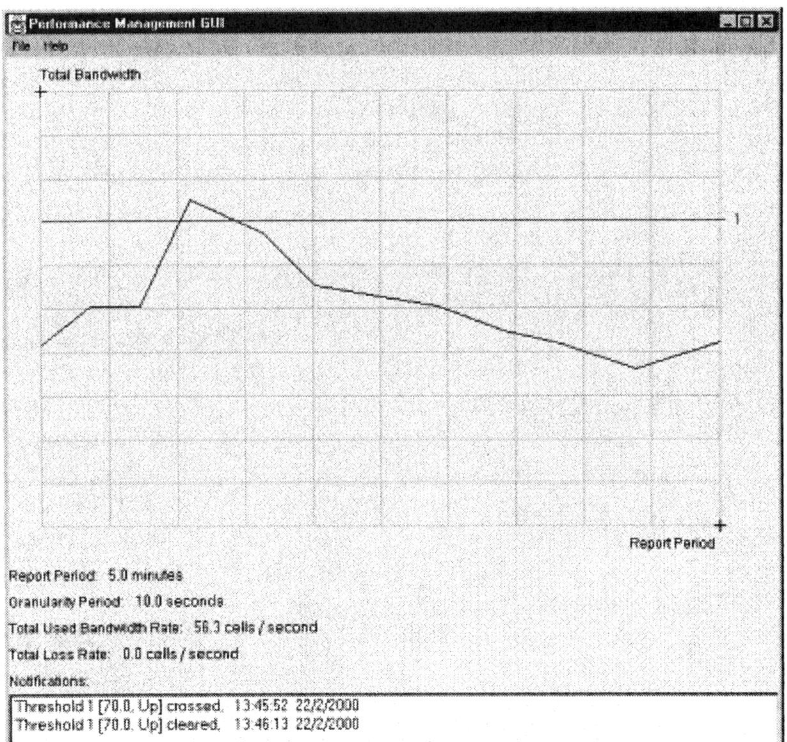

Figure 5 – The performance monitoring system graphical user interface. The graph of utilisation rates gathered goes over the threshold line for a while, indicating that this threshold was exceeded as it can also be seen from the notifications displayed in the status window, located in the lower part of the picture.

The whole development work was done under Sun's Solaris version 7 of the UNIX operating system. For the development of the "target" agent, the AdventNet SNMP version 2.0 libraries were used in order to obtain raw performance information by querying an SNMP agent. The system was built to operate in two different modes for the monitoring of IP and ATM traffic, respectively. All information gathered during the performance monitoring process appears on a graphical user interface in the client's machine, as shown in figure 5.

5. EVALUATION AND ASSESSMENT

While in the previous section we showed how mobile agent technology can be used instead of a static distributed object technology for building hierarchical management systems with the additional advantage of dynamic customisation and object migration, in this section we look at the performance implications of using mobile agent technology. As such, we decided to design and build two additional versions of the system, using Java-RMI and CORBA respectively as static distributed object platforms. The reason we chose Java-RMI is that the Grasshopper platform also uses Java-RMI's JRMP protocol (as well as a proprietary protocol), so we are able to see the precise overheads incurred by the mobile agent support infrastructure. In addition, the comparison with CORBA allows us to draw conclusions on the overheads of mobile agents platforms compared to an emerging distributed object technology for network/service management.

In the case of the Java-RMI/CORBA based performance monitoring system, a "master" object located in one machine sends a request for performance monitoring to a "factory" object located at a network element. When a request arrives to the "factory" object it is responsible to locally create a new instance of a performance monitor object that will perform performance monitoring and summarization functions. A "target" object is also located at the network element and provides raw performance information. The functionality and algorithms in all systems were identical so that we could directly compare the different approaches. It should be noted though that in the case of a static distributed object approach the functionality of the performance monitor object is static and cannot be altered, in a similar way to OSI-SM and SNMP support object facilities. Finally, we chose to use CORBA with the Java mapping for reasons of uniformity and we used the Sun Microsystems openly available version of CORBA with the Java mapping.

5.1 Environment and Methodology

Three performance monitoring systems were used, all containing the same functionality, written using Grasshopper mobile agents, Java-RMI and CORBA. For the evaluation we considered four different cases, one for every system running over its standard communication protocol, and an additional case for the mobile agent Grasshopper system running over the JRMP protocol. We were interested in measuring the following aspects for each case. First, we measured remote invocation response times. Timestamps were taken using the *currentTimeMillis* method of the *java.lang.System* class. A list of 25 elements was remotely transferred 100 times between two objects located in different machines, each time measuring the total time and finally calculating the average and standard deviation of these measurements. The same procedure was repeated while increasing the number of elements in the list to 50, 75, and 100. This operation in fact models the periodic summarization reports generated and remotely sent by the performance monitor mobile agent.

For the same experimental cases, we measured the TCP packet sizes using the tcpdump program that originated at the Lawrence Berkeley laboratory, reporting the total payloads at the TCP level. All these measurements where taken using two different machines over a lightly loaded 100 Mbit/sec Ethernet in the role of the management network with the following specification: Sun Microsystems Ultra-10, 256MB of memory, Sun's Solaris 2.5.1 version of UNIX. Finally, we measured the additional overheads incurred during MA deployment, namely the agent deployment time and the total amount of bytes incurred during the agent transmission from the manager to the network element.

5.2 Response Times

The response times of management operations for each of the three performance monitoring systems have been considered. We examined the performance aspects of remotely invoking operations between two objects located in different machines. An array of objects (class "java.util.Vector") containing 25, 50, 75 and 100 "Double" numbers respectively was passed as a parameter in the Mobile Agent, RMI and CORBA systems.

The total time required to complete the objects transfer, for each of the four different solutions described above, is reported in Figure 6, which depicts the resulting measurements in the form of statistical boxes.

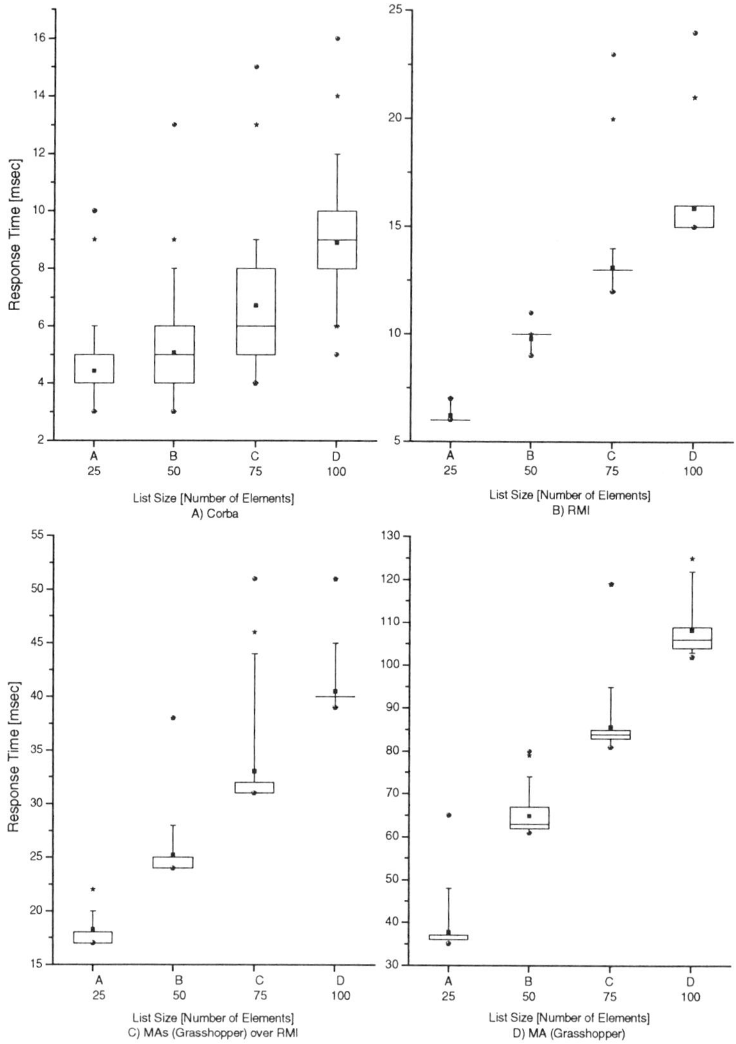

Figure 6 – Statistical Box Charts showing response times for each of the four experimented cases. The boxes include the 25-75% boundaries, the mean values (a black square) and the median values (a line). The 5-95% range boundaries are delimited by whiskers. The outliers are depicted with black circles and stars.

Figure 7 combines the same results in a single graph depicting only mean values and best linear fit, for an easier comparison.

Some important results can be drawn from Figure 6 and Figure 7. First, there is a significant performance penalty to pay for remote method invocations in the context of a mobile agent platform compared to Java-RMI and CORBA. Second, Grasshopper performs much better over RMI in comparison to the default proprietary protocol. The Grasshopper approach is at least three times slower than the Java-RMI and CORBA ones. Finally, by observing the difference in slopes from Figure 7, we can conclude that these two former approaches, along with the MA over RMI implementation, tend to scale much better than the plain MA approach.

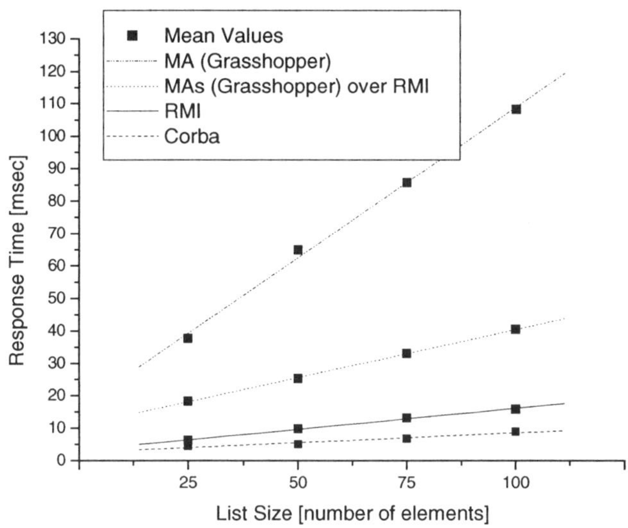

Figure 7 – Mean values and best linear fit of response times.

An additional performance overhead in Grasshopper is the initial time for the performance monitor mobile agent to migrate to the network element. The mobile agent needed an average of 1505 milliseconds to migrate, a performance overhead much larger than the time required to create a performance monitor object through its factory in the static RMI/CORBA approaches, which is less than 15 msec. In other words there is additional overhead of two orders of magnitude. In cases of constrained mobility, which is the approach used in this paper, this is a one-off overhead and can be tolerated. On the other hand, this measurement shows that agent mobility has relatively high performance overheads and this should be born in mind when designing systems exhibiting full mobility.

5.3 Total Packet Sizes

We also measured the packet sizes in all four cases. An array of objects (class *java.util.Vector*) containing 25, 50, 75 and 100 *double* numbers respectively was remotely sent using remote invocations in the Mobile Agent, RMI and CORBA systems. Each time, the payload of the TCP packets was measured. A graph of the results gathered can be seen in Figure 8. It is interesting to observe that Grasshopper configured with its proprietary protocol incurs levels of traffic comparable to those incurred by the two distributed objects systems. When Grasshopper was configured to operate over Java-RMI's JRMP it clearly incurred the biggest amount of traffic.

We also measured the packet overhead of migrating the performance monitor mobile agent to the network element. The required data was 2854 bytes, which again is much higher than the amount of data required to create a performance monitor object through its factory in the static approaches, which is around 500 bytes. This again will incur a substantial traffic overhead in full mobility environments, but can be tolerated in the case of constrained mobility.

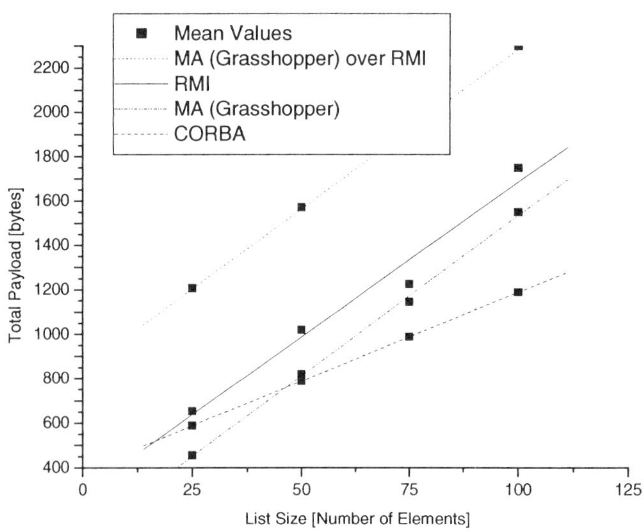

Figure 8 – Mean and best linear fit of total incurred TCP payloads, measured as the sum of all the bytes incurred in the network to complete the given task.

6. SUMMARY AND CONCLUSIONS

In this paper we present three different models of mobility that can be used in management applications. We also describe a performance monitoring system that uses the model of constrained mobility, and use it to evaluate the performance overheads of an agent based system compared with distributed ones based on Java-RMI and CORBA.

Constrained mobility involves the migration of an agent to a remote machine, where it executes a task and terminates upon completion. This is a particularly suitable model for tasks requiring a long period of time to complete. Also in scenarios where information intended for off-line analysis is collected by the agent in the remote machine. Finally programmability of network elements can be achieved in a way that the functionality at the network element level can be extended or customised, as we also demonstrated through the description of our agent-based performance monitoring system.

Weak mobility involves the migration of a mobile agent to a number of machines without preserving information gathered from previous visits. This is a suitable model for performing a short-term task repetitively in a number of machines. Also in scenarios where information intended for on-line analysis is collected by the agent in the remote machine.

Strong mobility involves the migration of a mobile agent to a number of machines while it preserves its state formed during previous visits. It is best suited for scenarios where the information collected from previous visits can affect the current or future behaviour of the agent. The task the agent has to complete in every machine should be a short term one, and therefore this model can be applied when information is collected for on-line analysis.

The nature of mobile agents does not allow a general mobility model of deployment in management applications. A suitable model can be selected by examining the requirements of a specific application.

In our performance monitoring system, mobile agents are created at the network management level according to user requests and then migrate to network elements to perform monitoring functions in a local manner. The behaviour of the monitoring algorithms can be customized, enabling dynamic programmable functionality to be provided directly in the managed network elements.

One of the key targets in embarking in this exercise was to evaluate the use of mobile agent technology in comparison to static object approaches in network management environments. The design and implementation presented in section 4 show that mobile agent platforms exhibit the same programmability characteristics to static object platforms. In addition, both remote method invocations and local invocations are possible. The same object-oriented principles and similar Application Programming Interfaces (APIs) can be used in mobile agent environments. A key advantage of mobile agents is the provision of dynamic services in network elements that have not been pre-programmed with such facilities. The customisation of mobile agent behaviour can provide a powerful mechanism for "intelligence on demand".

On the other hand, while design and programmability aspects are similar to static object approaches, there is a performance overhead to pay when using mobile agents. Remote method invocations are at least three times slower than those in Java-RMI / CORBA and this difference could be more pronounced when comparing performance to the protocol-based OSI-SM and SNMP approaches. In addition, agent migration incurs a substantial overhead in terms of both latency and required data to be transported across the network. This is less of an issue in constrained mobility environments but could lead to performance and scalability problems in environments where a large number of mobile agents migrate relatively often – i.e., in weak and strong mobility systems.

While initially mobile agent frameworks were thought as rivals to static distributed object frameworks, a view also stated in [7], the two approaches need to coexist. Static object approaches can provide superior performance characteristics. Real synergy could thus be achieved if stationary agents could be provided using static objects, with method invocations being possible between mobile agents and static objects in both directions. Such an environment would combine the best of both worlds. We are currently working on enhancing Java-RMI with an environment supporting constrained mobility.

ACKNOWLEDGMENTS

This work was partly undertaken in the context of the ACTS MIAMI project, which is funded by the Commission of the European Union.

REFERENCES

[1] J.Case, M.Fedor, M.Schoffstall, J.Davin, *A Simple Network Management Protocol (SNMP)*, IETF RFC 1157, 1990.

[2] ITU-T Rec. X.701, Information Technology - Open Systems Interconnection, *Systems Management Overview*, 1992.

[3] Object Management Group, *The Common Object Request Broker: Architecture and Specification (CORBA)*, Version 2.0, 1995.

[4] Y. Yemini, G. Goldszmidt, S. Yemini, *Network Management by Delegation*, in Integrated Network Management II, Krishnan, Zimmer, eds., pp. 95-107, Elsevier, 1991.

[5] N. Vassila, G. Pavlou, G. Knight, *Active Objects in TMN*, in Integrated Network Management V, Lazar, Saracco, Stadler, eds., pp. 139-150, Chapman & Hall, 1997.

[6] M. Baldi, S. Gai, G.P. Picco, *Exploiting Code Mobility in Decentralised and Flexible Network Management*, Proc. of the 1st International Workshop on Mobile Agents, Berlin, Germany, April 1997.

[7] Breugst, M., Magedanz, T., *Mobile Agents - Enabling Technology for Active Intelligent Network Implementation*, IEEE Network, Vol. 12, No. 3, May/June 1998.

[8] A. Bieszczad, B. Pagurek, T. White, *Mobile Agents for Network Management*, IEEE Communications Surveys, Vol. 1, No. 1, http://www.comsoc.org/pubs/ surveys, 4Q1998.

[9] Mobile Intelligent Agents for the Management of the Information Infrastructure (MIAMI) ACTS project, project page: http://www.fokus.gmd.de/research/cc/ima/ miami/, page at Univ. of Surrey: http://www.ee.surrey.ac.uk/CCSR/ACTS/Miami/

[10] ITU-T Rec. M.3010, *Principles for a Telecommunications Management Network (TMN)*, Study Group IV,1996.

[11] ITU-T Rec. X.739, Information Technology - Open Systems Interconnection, *Systems Management Functions - Metric Objects and Attributes,* 1992.

[12] ITU-T Rec. X.738, Information Technology - Open Systems Interconnection, *Systems Management Functions - Summarization Function,* 1993.

[13] S. Walbusser, *Remote Network Monitoring (RMON) Management Information Base*, IETF RFC 1271, 1991.

[14] G. Pavlou, G. Mykoniatis, J. Sanchez, *Distributed Intelligent Monitoring and Reporting Facilities*, IEE Distributed Systems Engineering Journal (DSEJ), Special Issue on Management, Vol. 3, No. 2, pp. 124-135, IOP Publishing, 1996.

[15] Foundation for Intelligent Physical Agents, web page: http://www.fipa.org/.

[16] Object Management Group, *Mobile Agent System Interoperability Facilities Specification*, orbos/97-10-05, 1997, ftp://ftp.omg.org/pub/docs/orbos/97-10-05.pdf

[17] The Grasshopper Agent Platform http://www.ikv.de/products/grasshopper/index.html.

[18] D. Griffin, G. Pavlou, P. Georgatsos, *Providing Customisable Network Management Services Through Mobile Agents*, Proc. of the 7th International Conference on Intelligence in Services and Networks (IS&N'00), Athens, Greece, February 2000.

[19] M. Wooldridge, N.R. Jennings, *Intelligent Agents: Theory and Practice,* The Knowledge Engineering Review, Vol.10, N.2, pp.115-152, 1995.

[20] G. Knight, R. Hazemi, *Mobile Agents based Management in the INSERT Project,* Journal of Network and Systems Management, Vol.7, N.3, September 1999.

[21] J.W. Stamos, D.K. Gifford, *Remote Evaluation,* ACM Transactions on Programming Languages and Systems. 12(4), pp.537-565, October 1990.

PACKET ROUTING WITH GENETICALLY PROGRAMMED MOBILE AGENTS

Jon Schuringa
Günter Remsak
Institute of Communication Networks
Technical University Vienna, Austria
{jon.schuringa,guenter.remsak}@tuwien.ac.at

Abstract This paper presents novel mobile agent routing techniques. We use genetic programming to build mobile agents that monitor the network status and set the routing tables in the network nodes in such a way that it maximizes network throughput and minimizes the overall packet delay. Performance is measured by simulation using a realistic network and traffic model that is capable of generating fractal traffic. The result is a high performance, self-configuring routing method, irrespective of the network topology and traffic.

Keywords: Routing, Mobile Agents, Genetic Programming

1. INTRODUCTION

The increasing complexity and diversity of communication networks make them hard to manage, managing them efficiently when traffic is strongly fluctuating is even harder. It is widely known that the mobile agents concept can be used to provide the basis for a self-configuring network (e.g. [2, 3]).

We describe in this paper a self-configuring routing system, which is based on "AntNet" [1]. The idea used has its origins in the behavior of real ants. Real ants are capable of finding shortest paths by using information (pheromones) deposited by other ants [4, 5].

Apart from significant improvements in the AntNet algorithm, we use genetic programming techniques to build mobile agents that monitor the network status and set the routing tables in the network nodes in such a way that it maximizes network throughput and minimizes the overall packet delay.

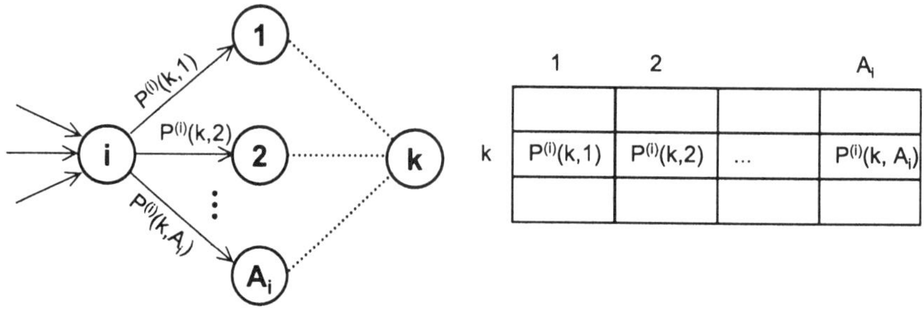

Figure 1 Routing table $P^{(i)}$ at node i

The routing table (Fig. 1) indicates how the packets arriving at that node must be routed on the outgoing links, depending on the final destination of the packets. The routing table for node i is a matrix $P^{(i)}$ with dimension $N \times A_i$, where N is the number of nodes in the network and A_i is the number of neighbors of node i. $P^{(i)}(k,j)$ is the fraction of traffic with destination k, that is routed through neighbor j at node i. It is the task of the routing agents, described in the following section, to set the routing table.

2. AGENT ROUTING

This section will first give a summary of the agent routing algorithm and then discusses the differences with the AntNet system.

Agents are generated concurrently with normal packets, but far less frequently. The main task of the agent on the path to its destination is to monitor and collect the network condition on the route between source and destination. The agent takes the same route as normal packets, according to the probability in the routing table inside the visited node, however for agents the probability is adjusted: we increase the probability for outputs with small queue sizes. This mechanism helps the agent in finding new, better routes. When the agent arrives at its destination, it takes the same route backwards and updates, according to the collected information and locally available information in the node, the routing tables at every visited node. The agent dies when it returns back to its source node.

Below are the main differences between our system and the AntNet system:

- High priority agents

  In AntNet the backward agents have higher priority than normal data packets, to quickly propagate the information the forward agent collected. Forward agents have the same priority as data packets, so they also experience the same network conditions. The fact that agents experience these delays is used by the algorithm (it is one of the main principles).

The disadvantage is however that it could a long time before an agent can react to specific situations like congestion. Forward routing systems [7, 8] do not inhabit the slow round trip delay.

In our system we use backward routing, however our agents always have a higher priority than normal data packets. They no longer experience the normal packet delay; they calculate it by using the current queue size and the link capacity. On the return path the calculated delays are used as if they were the real experienced delays. The advantage is that the same information is propagated much faster in the network. The effect on the network performance will be discussed later.

- Collecting neighboring data

  Directly related to the previous point, we also know the transmission queue sizes of all neighboring nodes along the visited path. This information is collected by our agent also, but only if the delay is small enough. To be precise, we only use the information if the calculated delay to the neighbor is smaller than the mean delay to that neighbor. Calculating the extra delays to neighboring nodes complicates the algorithm, because it may happen that more than one possible path between two nodes occur. Section 2.1 gives a detailed explanation of the problem and its solution.

- Inforcement rule

  For each node that the backward agent visits, we update the probability routing table. The amount should be proportional to the goodness of the trip. We use genetic programming to find a way to measure the goodness of the trip, it should find the best trade-off between adaptivity and stability.

  Different than in the AntNet system, we allow negative inforcements also. Very bad scoring trip times could be negatively awarded. The inforcement rule gives a measure of how good the agent's trip was. This measure, which is a number between -1 (bad) and 1 (good), is used to update the routing table. An inforcement of 0 will leave the probability unchanged. Let r denote the inforcement, then the following formula defines how to update the probability: let P be the short form of $P^{(i)}(k, o)$ (i.e., the current probability

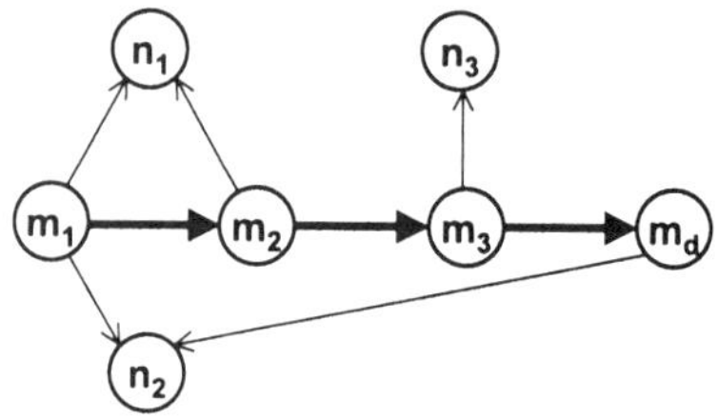

Figure 2 Example of the delay graph; The nodes on the main path m_i are connected by bold arrows, the other nodes are neighboring nodes n_j of which we have the delay info as well.

to use output o for packets with destination k at node i). The new probability P' is defined by

$$R_{plus} = r\left[\frac{1}{2} + \left(\frac{1}{2} - P\right) sign(r)\right] \tag{1}$$

$$P' = P + R_{plus} \tag{2}$$

By normalization, the other probabilities P_j (all outputs not used by the agent) should be adjusted:

$$P'_j = P_j\left(1 - \frac{R_{plus}}{1-P}\right) \tag{3}$$

The main difference in the algorithm between our system and AntNet is that our agents inspect the transmission queues, allowing lower agent delays and more network information per agent. We call it the "Queue Inspecting Agents" (QIA) system.

2.1. EFFICIENT ALL PAIRS SHORTEST PATH ALGORITHM

The method we applied for collecting delay information along the visited path, together with information about neighboring nodes could lead to the existence of multiple paths between two node pairs. We are only interested in the shortest path, so we have to apply some sort of shortest path algorithm. We could use Floyd's or Dijkstra's algorithm, however, due to their computational complexity ($O(N^3)$ and $O(NE\log N)$ respectively), not well suited for our purposes. If we make a graph of the delay information (Fig. 2), we can see that we have a graph with special properties for which we developed a simple and efficient algorithm based on theorem (4). To compute the shortest path between all pairs has computational complexity $O(N^2)$. Note however that the algorithm is processed in different nodes and each node on the return path only needs to know the delay from itself to the other nodes, which can be computed in $O(N)$ time and therefore the algorithm is scalable.

The main idea is based on two properties:

- The vertices in the graph that represent neighboring nodes, have no outgoing edges.

- Let $m_1, m_2, ..., m_k$ represent the nodes the agent visited on the forward path. There exists one and only one path from m_i to m_j for $1 \leqslant i < j \leqslant k$, while in the other direction (from m_j to m_i) no paths exists.

Let $\mathcal{G} = (\mathcal{V}, \mathcal{E}, w)$ a weighted, directed graph. $\mathcal{V}$ consists of the sum of the visited vertices $\mathcal{V}_m$ (the main agent track) and $\mathcal{V}_n$ (the neighboring vertices), respectively. Note that the outgoing degree of the vertices in $\mathcal{V}_n$ equals 0. Now let $\mathcal{V}_m = \{m_1, m_2, ..., m_k\}$, i.e., the agent started in vertex (node) m_1 and visited all m_i until m_k. The edges between two vertices in $\mathcal{V}_m$ are directed only from $m_i \to m_{i+1}$. We claim that the shortest path $\mathcal{D}$ between vertices from $\mathcal{V}_m$ and all other vertices in $\mathcal{V}$ equals

$$\mathcal{D}(m_i, y) = \min[\mathcal{D}(m_{i+1}, y) + w(m_i, m_{i+1}), w(m_i, y)] \tag{4}$$

Proof by induction on $\mathcal{V}_t$:

Basis: Let $i = k$, from the vertex m_k only edges to $\mathcal{V}_n$ exists, therefore the minimum distance to all other vertices equals $w(m_k, y)$ which is satisfied in (4): $\mathcal{D}(m_k, y) = \min[\infty, w(m_k, y)]$.

Induction step: Assume $\mathcal{D}(m_{i+1}, y)$ is the minimum distance between m_{i+1} and y. Note that there are two possible ways to go from m_i to y. From all the outgoing edges in m_i, there is one going to m_{i+1}, all others go to a vertex in $\mathcal{V}_n$. Since the outgoing degree of the edges of the vertices in $\mathcal{V}_n$ equals 0, there are only two edges from m_i that may lead to y. That is either directly with cost $w(m_i, y)$, or via m_{i+1} with cost $\mathcal{D}(m_{i+1}, y) + w(m_i, m_{i+1})$, from these two possibilities we have to take the minimum, which is exactly equation 4. □

We must take care that the properties of the graph $\mathcal{G}$ remain valid during the forward trip of the agent. There are cases that need special attention:

- Do not go to an already visited node. This is fulfilled by how the agent moves in the network. If the agent has no other choice,

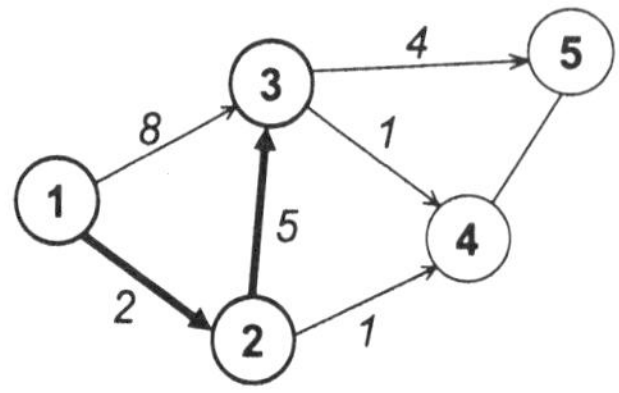

Figure 3 Forward agent arrives at node 3.

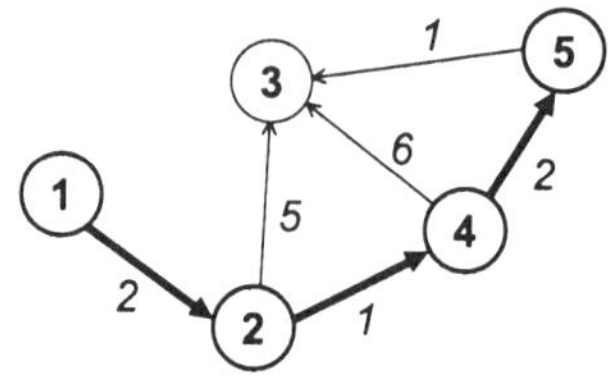

Figure 4 Forward agent arrives at destination node 5.

i.e., all other nodes already have been visited, the agent does not continue on its forward path. It will start with its backward trip to update the routing tables.

- Do not include delay info to an already visited node. This is a small restriction that could loose some information. Let m_a denote the node that we visited in the past and is a neighbor of our current node m_b. Including the delay info $m_b \rightarrow m_a$ would bring no information to nodes we visited: (1) before m_a and, (2) nodes we will visit after m_b, but it could for nodes between m_a and m_b.

- If the agent enters a node that was previously a neighbor node, we have to distinguish between two cases:

  1 Imagine the case depicted in Figure 3; the agent arrives at node 3, of which we have delay information from node 1. In this case, the total delay of going from 1 to 3 via 2 is less than going directly from 1 to 3, and therefore we *delete* the direct delay info from 1 to 3.

  2 When the direct link is shorter, we *change* the visited path by including the shorter link in the visited path, and deleting the delays that are no longer part of the visited path.

  Let's assume that after the agent arrived in node 3 in Fig. 3, the agent continues to node 4. Node 4 is an again a neighbor of node 2, but this time the direct link is shorter. Node 3 is deleted from the visited list. If the agent's next hop is to node 5, then we will have the graph as shown in Fig. 4.

As an example, the minimum distance from node 4 to 3 in Fig. 4 is the direct link $w(4,3) = 6$ or via node 5. The cost of going via node 5 equals $2 + \mathcal{D}(5,4) = 2 + 1 = 3$. So, the minimum distance from node 4 to 3 equals 3.

The algorithm to compute the shortest path in a node consist of a single loop over all destination nodes (thus, complexity $O(N)$) and selecting the minimum distance according to equation 4. Note that the shortest path information calculated in one node is used by other nodes on the return path.

3. THE SIMULATION MODEL

In this section we will discuss the simulation model of the nodes in our network (Section 3.1) and the model of the traffic generator (Section 3.2). We implemented these models in a (C++) simulation tool we wrote ourselves.

3.1. NODE MODEL

All incoming packets are directed to the router (Fig. 5), which has a deterministic service time per packet. We choose this to be sufficient small compared to the speed of the incoming packets. The function of the router is to inspect the destination address in the packet header and perform a routing table lookup. Whenever an agent arrives and it has to perform certain functions in the node, the router forwards the agent to the processor. It may send the agent to another node by setting the destination and forward the agent to the node's router. We found out that the processor queue-server system has a very small influence on the total network performance, because: a) the number of concurrent agents in one node is very small and, b) the algorithm is fast (Section 2.1). Compared to the propagation delays on the links, the processor time is negligible. Nevertheless, we used a deterministic service time of 1 msec. per agent.

If the router has to forward the packet (or agent), packet transmission is started immediately on the output, or is queued if the output is not idle. All output queue space is dynamically shared among all outputs.

3.2. TRAFFIC MODEL

It is important that the generated traffic in our simulation represents real traffic closely, because the agents that will be produced are best suited to deal with the traffic that was used in the simulations. What we use are $N(N-1)$ independent traffic sources (one for each source - destination pair) that can generate Poisson as well as fractal traffic. Ignoring the long-range dependence typically results in overly optimistic performance predictions and inadequately network resource allocation [9, 10]. The method used is called "Random Midpoint Displacement"

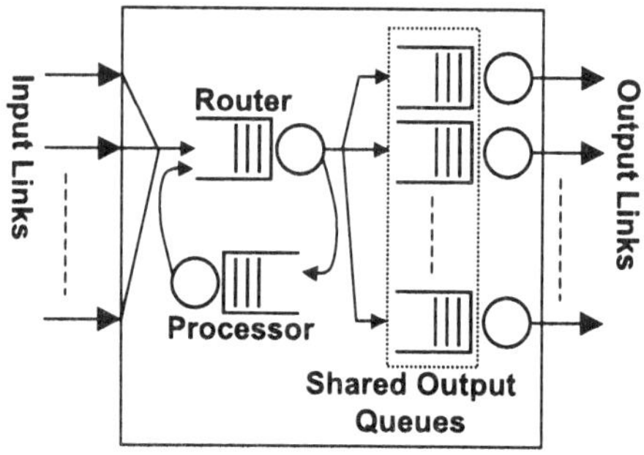

Figure 5 Network node model

(RMD), which is fast, simple, efficient and adequate for qualitative studies.

4. THE EVALUATION METHOD

The main problem of our system is how to update the routing table with the information the agents collect. We want to find a formula for this non-trivial problem by using genetic programming techniques.

The process starts with completely random expressions, for each we run a simulation and we evaluate it using a fitness formula:

$$score = c * \frac{B_{success}}{B_{offered}} + (1 - c) * \frac{max(d_{lim} - d, 0)}{d_{lim}} \quad (5)$$

Which yields a number between 0 (bad) and 1 (good). $B_{success}$ denotes the total number of bytes that were successfully delivered, $B_{offered}$ is the total number of bytes that were offered to the system. With parameter c we can control the importance of throughput versus packet delay, we used $c = 0.9$. A delay d greater than d_{lim} is set to d_{lim}, we used 10 seconds for this limit. We now use Darwinian principles [6] like natural selection, mutation and crossover to build new expressions which we also evaluate, whenever this gives us a better expression, the better one replaces a less good one.

The process continues until we reach a converged state (i.e., no progress during the last x tries). The result is a set of expressions, which are best suited in solving the problem from all expressions we tried. However, we only looked at one specific topology and traffic characteristic so far. To see if the expression we found is more general, we evaluate the expression in about 200 different network configurations. For this purpose we use a set of powerful computers, including a supercomputer. The results are stored in a database.

The expressions may contain constants, variables and operators. The simulation tool represents expressions by a tree structure, which allows easy manipulation (e.g. mutation and crossover) of the expressions while preserving a correct syntax. The variables that may be used by the genetic programming process are listed below. Assume the agent is on its backward path and enters a node, then the following information is now available:

- d_{no} - The new delay info delivered by the agent to go from the current node to node n using output o.
- $\overline{d_n}$ - The mean delay to go from the current node to node n.
- σ_n - The standard deviation of the collected delays from the current node to node n.
- $P(n, o)$ - The current routing table probability to use output o for packets with destination node n.
- h_{no} - The number of hops the last agent visited on the way to node n using output o.
- tq - A number between 0 and 1 indicating the ratio between the total used queue space and the total available queue space in the current node.
- oq_o - A number between 0 and 1 indicating the ratio between the used queue space by output o and the total used queue space in the current node.
- B_n - The best delay seen by the last w agents to go from the current node to node n.

4.1. OTHER ROUTING PROTOCOLS

Similar like in the evaluation of the AntNet system, we compare the agent based routing system with other routing methods. We implemented in our simulation tool the following routing protocols:[1]

- OSPF: At the simulation start the routing tables in the node are filled according to a shortest path algorithm with a cost function based on the link capacity.

[1]Note that we only implemented the functionality of the routing protocols that is relevant to this study.

- OSPF with load balancing: If multiple paths with equal costs to a destination exist, OSPF may use load balancing to spread the load over multiple paths.

- Daemon: This is the same routing protocol as used in AntNet. It is a routing protocol that knows at any time and without a delay the complete state of the network. Based on this information, a shortest path for each packet is calculated at each hop. The main purpose of this routing protocol is to obtain an empirical upper bound of the achievable performance.

- Daemon Source Routing: Under certain circumstances, OSPF and the agent routing achieve a better performance than the daemon discussed above. Packet delays can be higher, which was caused by oscillating packets. To resolve this we implemented another daemon, which uses source routing. At the generation of the packet a shortest path is calculated to the destination, and this path remains fixed for this packet.

5. RESULTS

This section presents the best agent we found with the genetic programming method (Section 5.1), and discusses the performance of that agent for two scenarios (Section 5.2).

5.1. THE BEST AGENT

Recall that all agents we evaluated are the same, except for their inforcement rule. In this section we present the agent that has the best overall characteristics. We evaluated the agent with all kinds of topologies, network sizes (up to 500 nodes), link capacities, traffic load, etc.

$$r = \begin{cases} \max\left[0.0068, P(n,o)^{\left(\frac{0.01*{d_{do}}^2}{{B_d}^2}\right)^{0.98}}\right] & \text{for } tq \leq 0.9 \\ \frac{0.76}{2.02h\exp(h*oq_k)-0.95} & \text{for } tq > 0.9 \end{cases} \tag{6}$$

The agent consists of two parts; the one that is taken depends on the node's buffer fill ratio. Let us first consider the first part, i.e., buffer fill ratio is less than 90%. The term $\frac{{d_{do}}^2}{{B_d}^2}$ is a relative measure of how good the trip was. The fact that it is relative is important because otherwise the agent would not perform well in different scenarios, i.e., it will most likely not work equally well in small and large networks.

Furthermore, the formula uses the current routing table probability. For probabilities close to 1, r will be close to 1, so the new probability will remain close to 1, even if the new trip delay time d_{do} is relatively bad. If the probability is close to 0, r will strongly depend on the term $\frac{{d_{do}}^2}{{B_d}^2}$. Assume that $P(k,j) = 1$ for a certain output j and destination k, so $P(k,x) = 0$ for $x \neq j$. Without the "max" function in (6) the probabilities $P(k,\cdot)$ would never change again, the small constant 0.0068 is used to prevent this. This is a way to prevent oscillations in the network, because it takes at least one other agent (who takes the same low probability route) to make a significant routing table probability change.

The first part of (6) balances the network load, so if the traffic increases until congestion arises, it is likely that the queues of multiple nodes will quickly be more than 90% filled, and for these nodes, the second part of (6) will be used by the agents. The inforcement r will be higher for routes that have a low hop count. This has a positive impact on the throughput, because the result is that the agent now tries to use routes with a minimal number of hops and therefore also a minimum chance of being discarded by a full buffer. The second variable is oq_k (which has a smaller influence on r than the hop count) it encourages to use a small local output queue. Recall that the output buffer space is dynamically shared by all outputs.

5.2. SIMULATION RESULTS

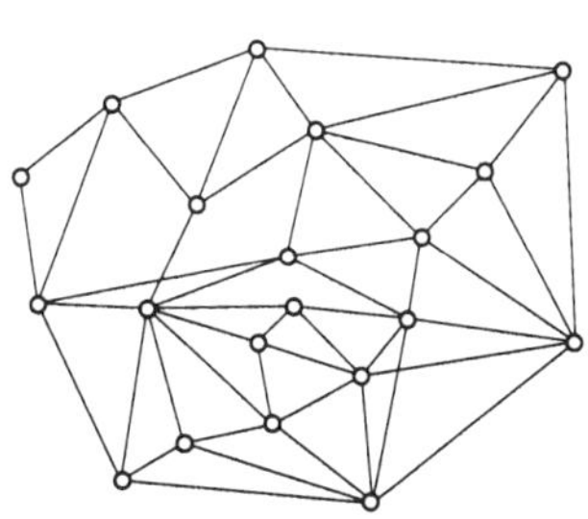

Figure 6 Random topology network

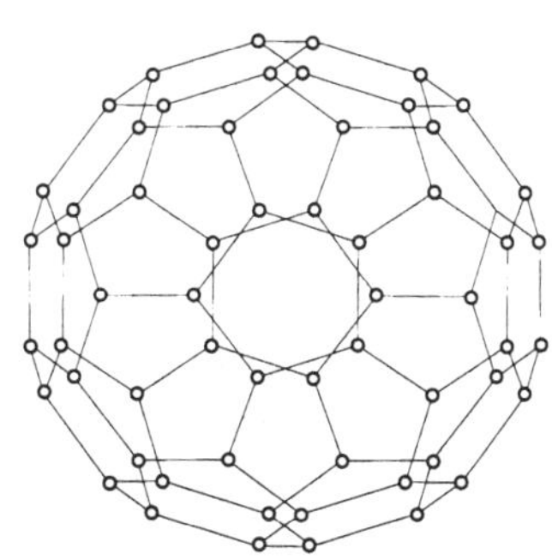

Figure 7 Bucky ball topology

We show the performance of the agent for two different networks: a randomly generated network (Fig. 6) and a network with a regular structure, the so-called Bucky ball (Fig. 7). We chose the Bucky ball because we expected this topology to be relative hard for our agent system. There might be a high number of more or less equal cost paths, which has a negative influence on the convergence time.

Table 1 Network properties

	Bucky Ball	*Random*
Topology	Regular (Carbon 60 molecule)	Irregular (random)
Nodes	60	20
Bidirectional Links	90	44
Link capacity	5 and 10 Mbit/sec uniform distributed	0.25, 0.5, 0.75 and 1 Mbit/sec uniform distributed
Propagation delay	25 msec.	U(5,10) msec.
Source rate	U(0,Max) packets/sec	U(0.1Max,Max) packets/sec
Source type	Fractal (H=0.7)	Poisson
Packet Size	U(100,1500) bytes	U(100,1500) bytes

The properties of the two simulated networks are shown in table 1.[2][3] Both simulations use an output queue space of 1 Mbyte per node and the size of the observation window w (used for B_n) equals 200. All simulation runs simulated 1000 seconds, with the routing tables initially set according to the shortest paths.

OSPF using load balancing did not show a significant improvement on OSPF without load balancing in our networks, the same holds for both daemons, the daemon with source routing performed equally well, or slightly better than the other daemon. That is why OSPF with load balancing and the normal daemon are not shown in the graphs in this section.

Bucky ball: Figure 8 shows the bucky ball network throughput for different network loads. All routing methods achieve the same throughput for low and moderate load. OSPF is the first routing method that starts dropping packets, followed by AntNet. Our agent system (QIA) that uses: high priority agents, information from neighboring nodes and an inforcement rule obtained with genetic programming shows a significant improvement on AntNet.

[2]The parameter *Max* in the table is varied in the simulation to get the different traffic load scenarios.

[3]With the notation $U(a,b)$ we mean a number uniformly distributed on the interval $[a,b]$.

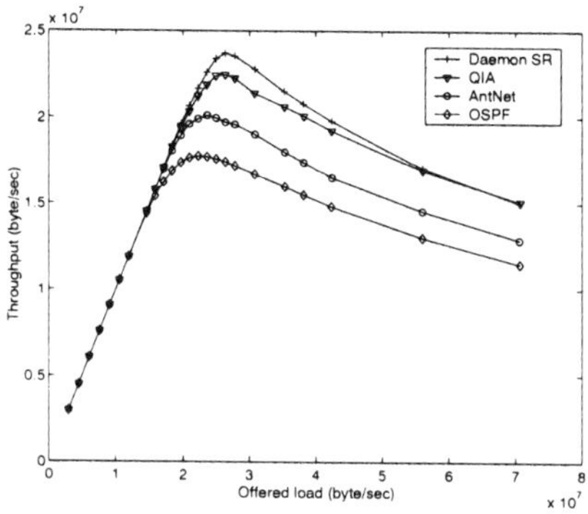

Figure 8 Bucky ball throughput

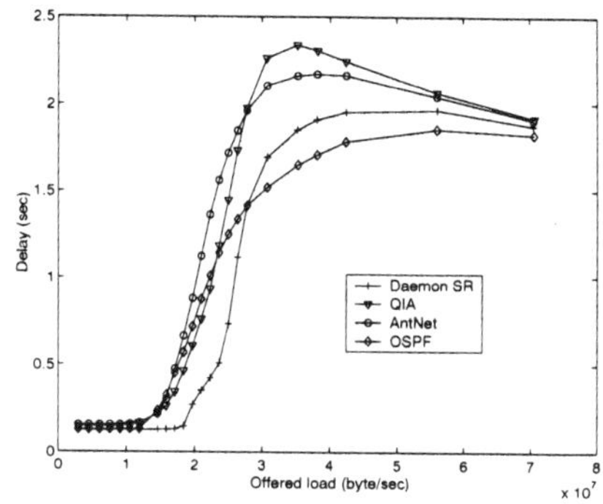

Figure 9 Bucky ball packet delay

From Figure 9 we see, as expected, that the daemon has the lowest delay until congestion arises. For higher loads, OSPF has the lowest delay, but we saw in Figure 8 that OSPF has at the same time the lowest throughput for high loads. The QIA system has a lower mean delay than AntNet until the network becomes congested.

Random network: The throughput graph for the random network (Fig. 10) is similar to the throughput graph of the bucky ball. Except for very high loads, QIA lies between AntNet and the daemon. The delay graph (Fig. 11) shows the excellent performance of QIA: for network loads below 2Mbyte/sec the curve is very close to the daemon, at a load of 1.62 Mbyte/sec AntNet has a mean delay three times higher than QIA. For loads above 2Mbyte/sec there is no significant difference between AntNet and QIA. OSPF is not shown for better readability, but performed bad due to the fact that the network is not well balanced and as a result some links were extremely overloaded.

To give an example of the differences in convergence time between AntNet and QIA, we ran the following experiment. We used the same random network as before, but initialized the routing tables in such a way that each link had the same probability of being used (i.e., random routing). We started the simulation for each of the following routing methods, with a relative low offered load (1.37 Mbyte/sec):

1 AntNet

2 AntNet Modified: We modified AntNet at two points: the agents now also use high priority on their forward path, and inspect the

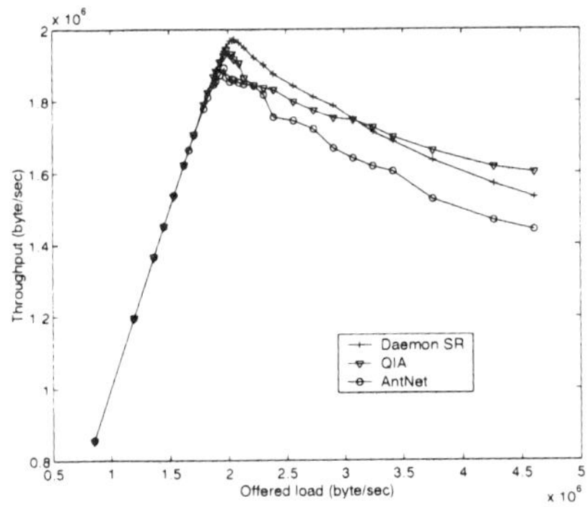

Figure 10 Random network throughput

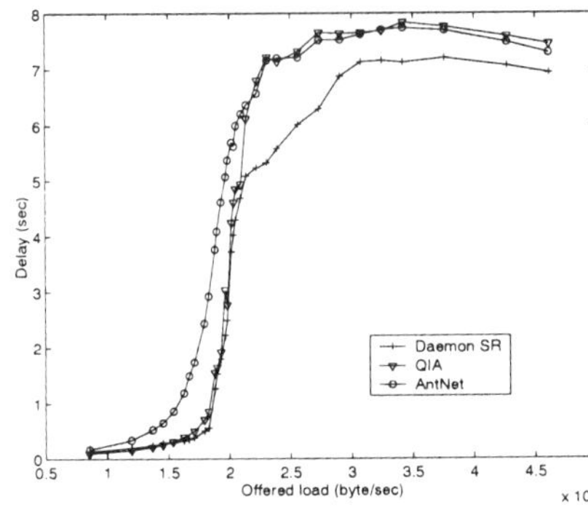

Figure 11 Random network packet delay

neighboring queues as well. Except for the inforcement rule, the modified AntNet system equals QIA.

3 QIA: Our system with high priority agents, using the neighbor queue information and the inforcement rule we found using genetic programming.

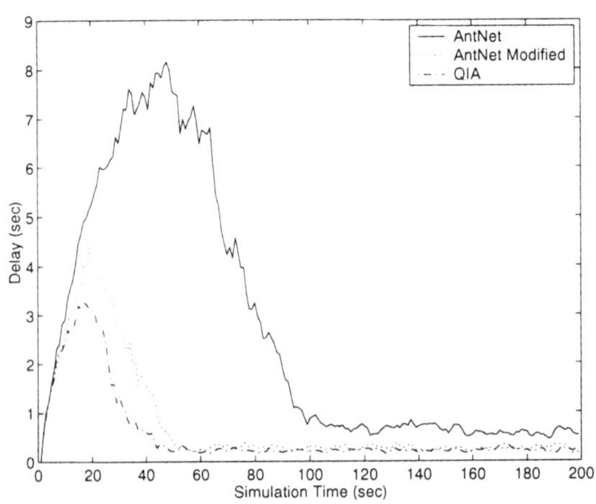

Figure 12 Effect on the convergence time.

Figure 12 shows the results of the experiment; AntNet takes about 100 seconds to reach a stable state, applying the modifications to AntNet reduces this to only 50 seconds. As a direct result, the maximum mean packet delay is about 50% smaller. The modifications make AntNet adapt faster, without destabilizing the system. The QIA system displays an even better performance.

6. CONCLUSIONS

This paper introduced new techniques for agent routing systems, and made a comparison with other routing methods. We have shown that genetic programming can be successfully applied to the non-trivial problem of agent based routing. Furthermore, we conclude that our other improvements on the AntNet system, which were: a) high agent priority, and b) using neighbor queue information, make it much more reactive on changes in the traffic load without destabilizing the system.

References

[1] G. Di Caro and M. Dorigo, "AntNet: Distributed Stigmergetic Control for Communications Networks", *Journal of Artificial Intelligence Research (JAIR)*, Vol. 9, pp. 317-365, 1998.

[2] T. Magedanz, K. Rothermel and S. Krause, "Intelligent Agents: An emerging technology for next generation telecommunications?", *INFOCOM 1996*, Vol. 2, pp. 464-472, 1996.

[3] S. Appleby and S. Steward, "Mobile software agents for control in telecommunications networks", *British Telecom Technol. Journal 12*, pp. 104-113, 1994.

[4] S. Goss, S. Aron, J.L. Deneubourg and J.M. Pasteels, "Self-organized shortcuts in the Argentine ant", *Naturwissenschaften*, 76, pp. 579-581, 1989.

[5] R. Beckers, J.L. Deneubourg and S. Goss, "Trails and U-turns in the selection of the shortest path by the ant Lasius Niger", *Journal of Theoretical Biology*, 159, pp. 397-415, 1992.

[6] J.R. Koza, F.H. Bennett III, D. Andre and M.A. Keane, "Genetic Programming: Biologically Inspired Computation that Creatively Solves Non-Trivial Problems", *Proceedings of DIMACS Workshop on Evolution as Computation*, 1999.

[7] R. Schoonderwoerd, O. Holland, J.Bruten and L.Rothkrantz, "Ant-based load balancing in telecommunication networks", *Adaptive Behaviour*, Vol. 5:2, pp. 169-207, 1996.

[8] M. Heusse, D. Snyers, S. Gurin and P. Kuntz, "Adaptive Agent-Driven Routing and Load Balancing in Communication Networks", *Ants'98*, Brussels, Belgium, October 1998.

[9] W. Lau, A. Erramilli, J.L. Wang and W. Willinger, "Self-Similar traffic Generation: The Random Midpoint Displacement Algorithm and its Properties", *ICC 1995*, Vol. 1, pp. 466-472, 1995.

[10] V.Paxson and S. Floyd, "Wide-Area Traffic: The failure of Poisson Modeling", *Proc. ACM Sigcomm*, London, UK, pp. 257-268, 1994.

Part Eight

Network Security

A Multi-agents System for Network Security Management

K. Boudaoud and Zahia Guessoum
[1]*Institut Eurécom*
2229, route des crêtes - 06904 Sophia Antipolis France
Phone: 33 (0) 4 93 00 26 38 / Fax: 33 (0) 4 93 00 26 27

[2]*LIP6-OASIS*
Case 169, 4 place Jussieu, 75252 PARIS Cedex 05 - France
Phone: (33) 1 44 27 87 43 / Fax: (33) 1 44 27 70 00

Abstract: The openness of business toward telecommunication networks in general and Internet in particular is performed at the prize of high security risks. Every professional knows that the only way to secure completely a private network is to make it unreachable. However, even if this solution was undertaken for many years, nowadays it is not possible to close private network especially for business purpose. Thus, security management becomes an important issue that must be considered carefully. The focus of our work concerns one critical security management issue that is intrusion detection. Some drawbacks of existing systems reveal the necessity of designing a new generation of self-adaptive systems. In fact, self-control, flexibility, adaptability, autonomy and distribution are the main features to be addressed in a suitable architecture that fulfils these requirements. The introduction of multi-agents system (MAS) in a network seems so promising to enable network entities to perform adaptive and "intelligent" behavior. "Intelligence" means that network entities provide reasoning capabilities, exhibit behavior autonomy, adaptability, interaction, communication and co-operation in order to reach specified goals. In this context, we propose a new approach for security management using intelligent agent (IA) technology. This approach provides a flexible integration of multi-agent technique in a classical network to enhance its protection level against inherent attacks.

Key words: Intelligent agents, multi-agents system, network security management, intrusion detection, distributed network management.

1. INTRODUCTION

During these last years, computer systems and networks have not ceased evolving, particularly in terms of number of users and offered services that are continuously increasing in complexity. Systems and networks become more complex (number of machines, number of users, number of connections...), making them more vulnerable to various kinds of complex security attacks. Therefore, security management of these systems and networks, particularly intrusion detection, requires more sophisticated models.

To deal with these requirements, multi-agents systems (MAS) are well adapted. They provide a powerful paradigm for the modeling and the development of complex systems. They are based on the decomposition of systems into several interacting and autonomous entities called agents. An agent refers to an entity that functions continuously and autonomously in an environment in which other processes take place and other agents exist. Recent applications show the growing interest of this paradigm in the network management domain [1].

This paper deals with the use of multi-agents system to detect security attacks. We first discuss some requirements for an efficient security management. In section 3 and 4, we describe the organizational model of our MAS and the functional model of a security agent. We give an overview on our implemented security management system in section 5. Finally, we conclude with some remarks and future work.

2. SOME REQUIREMENTS FOR AN EFFICIENT SECURITYMANAGEMENT

Security management aims to maintain the integrity, confidentiality and availability of systems and services. Securing a network involves protecting it against all possible attacks. But, in practice it is not possible to have a completely secure network. So, applying security management is a two-fold activity: 1) the security architecture is to be deployed to protect networks by detecting attacks; and 2) when attacks are detected the security architecture deals with these attacks in real time by taking security measures.

An attack can be defined as any non-standard activity, which compromise the information *confidentiality*, data *integrity* and *availability* of a resource. There are various kind attacks that can be classified in *Network attacks, System attacks and Web attacks* [2]. In this paper, we are interested in *Network attacks*, such as: *ICMP flooding*, *doorknob* and *Ping sweep* [2]. The aim of IDSs is to detect security attacks, especially in real-time fashion.

Among the existing systems, we can cite DIDS (Distributed Intrusion System Detection) [3] and CSM (Co-operating Security Manger) [4]. DIDS was designed to supervise a local area network LAN. Its centralized control represents a major disadvantage. In the case of WAN networks, where the communications with the entity manager, it can congest the network. CSM was developed for a distributed environment. However, it cannot be easily adapted to new environment. To enhance the security of computer systems and networks, we must deal with their distributed nature and dynamics, which are two important characteristics. The dynamics of networks is due to their increasing evolution in terms of offered services, used resources and number of users. In fact, users, known or not, have complex behaviors that vary considerably. Particularly a recent aspect, viz. the mobility of users, enhances this complexity. Once a user, known in a static network, moves, the knowledge related to his behaviors becomes different. This complexity and dynamics of networks make them more vulnerable to various kinds of security attacks. Therefore, the security policies may change over time. Moreover, some requirements are important for detecting attacks efficiently:

- **Distribution**: Many network attacks are characterized by abnormal behaviors at different network elements [7]. Detecting them by a single system, running on a single component, is too complicated. So it is easier to distribute monitoring and processing tasks among a number of entities at different points. This important aspect is provided by most existing IDSs [3]. For example, in DIDS, data collection is assured by several entities but a centralized director performs the analysis.
- **Autonomy:** Excessive data traffic between distributed entities can cause network congestion problems. So, it is more judicious to let the entity monitoring a network element perform local analysis and detect intrusive behaviors. Thus, distributed entities must be autonomous. The CSM approach has shown the necessity to use ***autonomous*** entities [4]. In CSM, there is no established central director but individual managers that are responsible of making local intrusion detection.
- **Delegation:** The high level of dynamics in networks requires modifying, at any time, security management functions to adapt them to changes occurring in the monitored network. The model of delegation among various management entities allows fulfilling this requirement. The delegated tasks are sent to the autonomous entities. Each one has to execute its own task. When new tasks must be added or existing one must be modified, this is done dynamically. The *Delegatio*n feature is not found in existing IDSs.
- **Communication and cooperation**: Coordinated attacks cannot be detected easily by an individual entity, which has a restricted view of the network. It is therefore necessary to correlate various analyses made by the autonomous entities. So communication and cooperation between

entities are needed to detect coordinated intrusive behaviors. In the CSM system, each security manager detects local intrusions and cooperates with other CSMs by exchanging information in order to detect cooperatively intrusive activities [4].

- **Reactivity:** The aim of efficient intrusion detection is to react against an attack before serious damages can be caused.
- **Adaptability** and **Flexibility**: When a new security policy is added or modified, intrusion detection and monitoring tasks must be adapted. Moreover, when new services and resources are added, the IDSs must consider these variations dynamically. IDSs must also detect new attacks when they happen. So it is important to learn new patterns of attacks. These two features are not provided by existing systems, which can not be *upgraded easily* and cannot *easily adapt* their intrusion detection tasks to changes in networks and user behaviors. In addition, they do not have the ability to learn new attacks.

The above-described features are considered as the main requirements for detecting attacks efficiently. They show that multi-agents systems are a suitable solution. Multi-agent system properties (distribution, cooperation...) and agent properties (adaptability, autonomy, pro-activity...) [1][13] match the whole requirements.
The adopted approach for designing our MAS is based on two levels [8]:

1) A **Macro** level, which describes the organizational and functional structure of the MAS;
2) a **Micro** level, which describes the architecture of a security agent.

The two levels are described below.

3. THE ORGANIZATIONAL MODEL OF THE PROPOSED MULTI-AGENTS SYSTEM

The *Macro* level defines the MAS organization. The latter defines a set of roles and relations between them [9]. A role can be defined as a set of tasks that an agent must do in order to make the organization reaching its objectives. To describe the different roles, it is necessary to identify the tasks that must be done by the MAS.

3.1 Identification of roles

We distinguish two types of tasks: **monitoring and management.** The monitoring tasks are identified correspondingly to the kind of activities to monitor. We identify five types of *monitoring*:

- *External monitoring* for the monitoring of external activities;
- *Exranet monitoring* for the monitoring of *extranet* activities;
- *Intranet monitoring* for the monitoring of *intranet* activities;
- *Internal monitoring* for the monitoring of internal activities;
- *Local monitoring* for the monitoring of local activities.

And two type of *management* tasks:

- *Policies management* for security policies management of the network;
- *Security management* for security management of a distributed or local area network.

According to the previous tasks, we identify the following roles:

- **Security policy manager** manages the security policies and communicates with the security officer.
- **Extranet manager** describes the security management functions of a distributed network. These functions concern the detection of complex attacks happening in a high level. Thus, an agent having this role, will have a global view of the network and will detect coordinated attacks. It also will specify monitoring and detection tasks to the low-level agents. This role manages the security of the distributed network with external networks and between LANs of the distributed network.
- **Intranet manager** manages the security of the LAN constituted of several domains. It concerns activities monitoring and detection of coordinated attacks within a LAN and between its various domains.
- **Local extranet monitor** includes *external* and *extranet monitoring* functions within the LAN. This role is associated to the detection of attacks originating or in direction to an external or extranet network.
- **Local intranet monitor** represents *internal and* in*tranet monitoring* functions within the LAN. It concerns also detection of attacks directed or originating from other LANs of the same distributed network.
- **Local internal monitor** that defines the *local monitoring* functions. It concerns the detection of attacks, which are local to a domain.

3.2 Organizational structure

The proposed security management architecture (MANSMA)[1] consists of several agents structured hierarchically. Agents, which have various roles, are located at specific network entities and distributed at different points of the network. The hierarchical organization of agents enables local as well as global intrusion analysis and detection. Each agent has its own perception of the network, which is limited by the domain to monitor. In MANSMA, we

[1] Multi-agents System-based Network Security Management Architecture

distinguish, accordingly to agent roles identified in the previous section, two functional layers: a *manager layer* and a *local layer* .

- The **manager layer** manages the global security of a network. In this layer we identify three levels of *manager agents*: a *security policy manager agent*, an *extranet manager agent* and several *Intranet manager agents.* The *extranet manager agent* controls *intranet manager agents,* which report pertinent analysis. It performs then another analysis to confirm the detection of an attack. It can also ask for more data processing and delegate new monitoring tasks to the *intranet manager agents.* The *extranet manager agent* is also responsible for distributing a set of *local agents* to each *intranet manager agent.* The *intranet manager agent* controls *local agents* and analyses the monitored events reported by these agents.
- The **local layer** manages the security of a domain. It is composed of a group of *local agents*, which have specific monitoring roles. We distinguish three kinds of *local agents*: *extranet local agent*, *intranet local agent* and *internal local agent*.

Domains are defined by the agents of the *manager layer*. In the *local layer*, a domain represents a group of sub-hosts, which are gathered either according to the organization chart of the company in terms of departments or according to security levels specified by the security policies of the company. In the *manager layer*, a domain represents either a distributed network or a LAN of this same distributed network.

In this hierarchical multi-agent model, each *manager agent* has the ability to control specified agents and to analyze data, whereas the *local agents* monitor specified activities. In each level, agents communicate and exchange their knowledge and analysis for detecting intrusive activities in a cooperative manner. The interaction between these two layers allows the detection of global attacks by correlating the various analyses of the *local layer*.

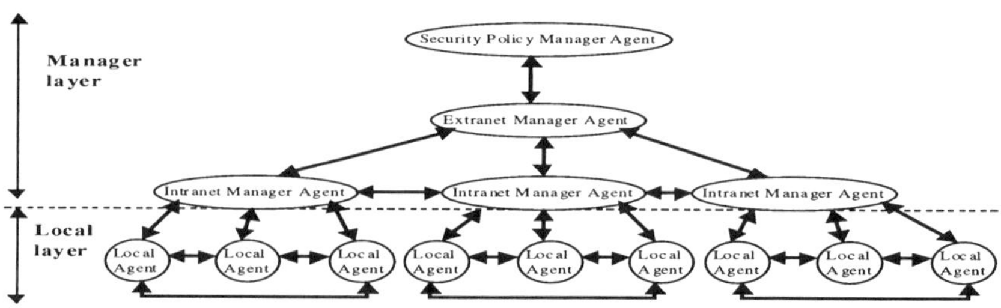

Figure 1: MANSMA Functional Architecture

4. THE AGENT MODEL

To model intrusion detection, agents must combine cognitive abilities cognitive (knowledge-based) to reason about complex attacks with reactive capacities (stimulus-response) to react rapidly to the environments changes. So, an agent has three functions: 1) a *filtering* function that filters security events, 2) an *interaction* function that manages its interactions with its environment and other agents and 2) a *deliberation* function that enable it to analyze new data and detect attacks. These functions are described in the following paragraphs.

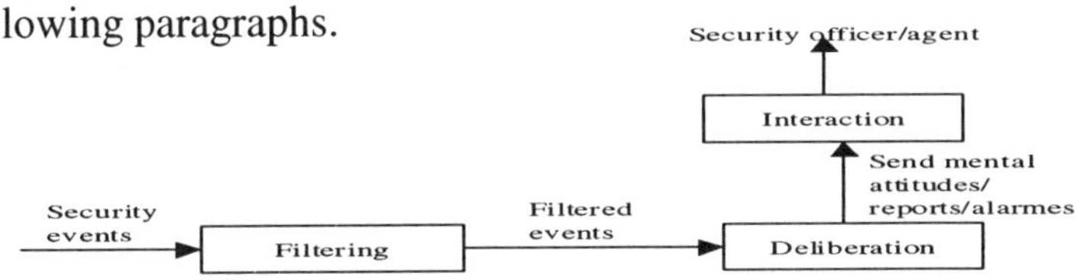

Figure 2: Interactions between agent functions

4.1 Event filtering function

A *security event* is characterized by its type, its observation point, a temporal attribute (representing the event occurring moment), and a set of non-temporal attributes. According to the event type and its observation point, we identify various event classes (see diagram below).

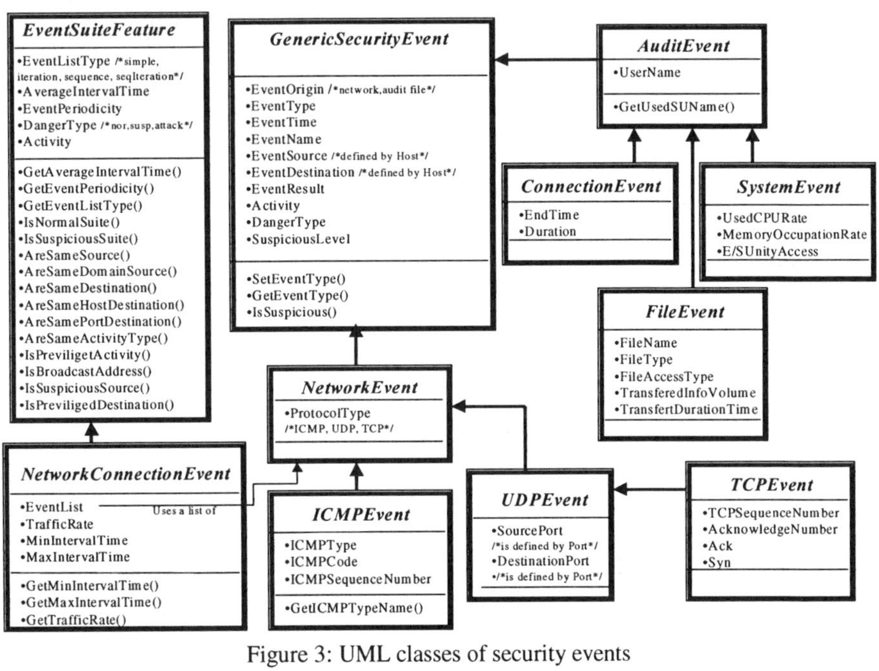

Figure 3: UML classes of security events

The *event filtering* function filters security events produced in the network, according to event classes specified in a detection goal. Indeed, the events occurring in network are not all collected. In fact, when a detection goal is sent to an agent, a set of event classes to observe is specified to it. Thus, when an event occurs in the network, the agent tests if it matches the event classes specified in the goal. If it matches, it is collected. The filtered events are then stored, waiting to be treated by the deliberation function.

4.2 Interaction function

This function describes interactions between the above-described agents. It allows them to communicate their analyses and knowledge and mental attitudes (beliefs, suspicions...). In fact, m*anager agents* interact with *local agents* by:

- sending goals, derived from security policies;
- delegating specific functions of monitoring/detection and specifying the various domains to monitor;
- asking particular information: the suspicion level of a specific user, the list of events generated by a user, etc.;
- and receiving the relevant reports or analyses results and alarms.

Interaction function also permits interactions between the security officer and *security policy manager agent/ extranet manager agent.* It ensures the reception of specifications and requests from the security officer such as security policies to apply. It allows the delivery of security reports and alarms when an attack is detected. The security officer can also ask for additional information (asking for the current security state of the network, the list of suspicious users...).

4.3 Deliberation function

As it has been outlined in section 2, security management must deals with significant network characteristics such as: 1) its continuous variation, particularly in terms of users and offered services; 2) and variation of its security problems such as new vulnerabilities and increasingly complex attacks. Considering the unpredictable character of the agent environment behavior (network), we adopted a BDI solution [10][11] for modeling the security management system. Thanks to the *deliberation* function the agent is able to reason and extrapolate by relying on its mental attitudes, built knowledge and experience, in a rational way, to find the adapted answers. The agent uses its beliefs resulting from the filtered events and beliefs of the neighboring agents for reaching its specified goals. When a goal is reached (an attack is detected), it executes appropriate actions.

In this section, we will start by describing the knowledge base of the agent and then the BDI-based information model.

4.3.1 Knowledge base

The knowledge base of the agent contains two types of knowledge:

- **Immediate character knowledge** that represent the observations made by the agent (events produced in the network) on its environment. This knowledge has a limited validity lifetime;
- **Permanent character knowledge**, which represent the necessary knowledge for managing security of the network (such as list of known user/user groups, list of administrators, list of known hosts, list of known addresses, prohibited addresses, reserved addresses...).

4.3.2 BDI-based information model

This model represents the mental attitudes of the security agent: *beliefs*, *goals*, *intentions*, su*spicions* and *policies*.

4.3.2.1 Beliefs

Beliefs represent the perception that the agent has on the network behavior and its security state. They indicate also knowledge that it has on other agents and itself. We distinguish three types of beliefs:

- **Personal beliefs** express its knowledge on its own state (information relating to it, in particular the domain, which it must monitor).
- **Relational beliefs** represent what the agent knows on other agents with which it communicates. These are all information (role, competencies...) that it needs to communicate with them
- **Environmental beliefs** include *local environmental beliefs* and *environmental beliefs of the others*. *Local beliefs* indicate what the agent believes on the behavior and the security state of the network whereas the *beliefs of the others* represent perceptions which have the other agents on the network. Within the framework of this project, we distinguish two types of *environmental beliefs*:
  - **Schema beliefs**, which are a description of attack scenarios to detect. These beliefs will not be instantiated until a goal is sent;
  - **Scenario beliefs,** which represent the sequences of security events. A *scenario belief* is associated to one or *several schema beliefs*. *Scenario beliefs* have a temporal validity, which depends on the temporal validity of the events constituting the event sequence

4.3.2.2 Goals

Goals represent the state that must be reached by the agent viz. its objectives. We distinguish three types of goals:

- **Monitoring goals**, which ask for monitoring specific activities (activities of a certain user, going-in activities...).
- **Informational goals** that request specific information on the security state of the network (detected attacks during a certain period of time, suspected users, current external connections...);
- **Detection goals**, which specify the attacks to detect and the measures to take if an attack is detected. They are the most significant goals within the framework of intrusion detection. A detection goal allows the instantiation of a *schema belief.*

4.3.2.3 Intentions

Intentions represent the list of actions that must be executed by the agent when it achieves its goal. These actions can be: sending alarms to the security officer or manager agent, closing a connection established by an attacker, reconfiguring a firewall...

4.3.2.4 Suspicions

This mental attitude, introduced within the framework of intrusion detection, expresses the suspicion that has an agent on a *scenario belief.* When an agent observes a sequence of events which corresponds neither to a normal sequence, nor to a known attack, then it identify it as suspicious sequence. To confirm that this *suspicion* is an attack, the agent needs further information or confirmations from other agents. The agent will then say to other agents: *"I suspect that this sequence of events is an attack"*. A *suspicion* is associated to a *schema belief* and is the result of the analysis of a *scenario belief* compared to a *schema belief.*

4.3.2.5 Policies

Policies represent the guiding mental attitude of the MAS behavior to manage the security of the company. Starting from the specified security policies, a set of *goals* are created and derived in order to maintain a certain security state of the network.

5. IMPLEMENTATION

The presented agent model has been implemented with the multi-agent platform DIMA [13]. The latter is mainly characterized by a modular agent architecture. DIMA proposes the extension of the single behavior of an active object into a set of behaviors. In our implementation, each agent has three behaviors:

- The *filtering* behavior filters security events. When an event occur in the network, it is collected only if it matches the event classes specified in the detection goal.

  EventFilter {
  Repeat
  security-event := get(security-event-to-filter);
  <u>*If*</u> is-in-list-of-event-types-to-filter(security-event)
  Update-list-of-filtered-event(security-event);
  end repeat }
- The *interaction* behavior manages the interaction between the agent and the other agents. It defines the mailbox of the agent and the way the messages are received and enqueued for later interpretation. An agent may need some others information to refine its analysis. In this case, it asks other agents to give it the necessary information.
- The *deliberation* behavior represents beliefs, goals, intentions and knowledge of the agent. It is responsible 1) for generating adequate responses to the messages received from the other agents and 2) for achieving the agent goal(s).

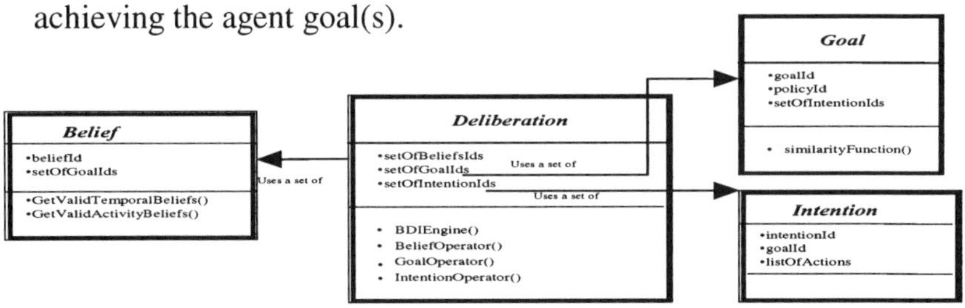

Figure 4: UML classes used by the deliberation function

When an agent receives a *detection goal*, it updates a set of event classes to filter. Then, when an event occurs it is filtered by the filtering module and sent to the deliberation module. This one, updates/creates agent *scenario beliefs* and then test if this belief matches a *schema belief.* If it matches, then a *detection goal* is reached and a list of *intentions* are sent to the interaction module for being executed (see figure 4 and 5).

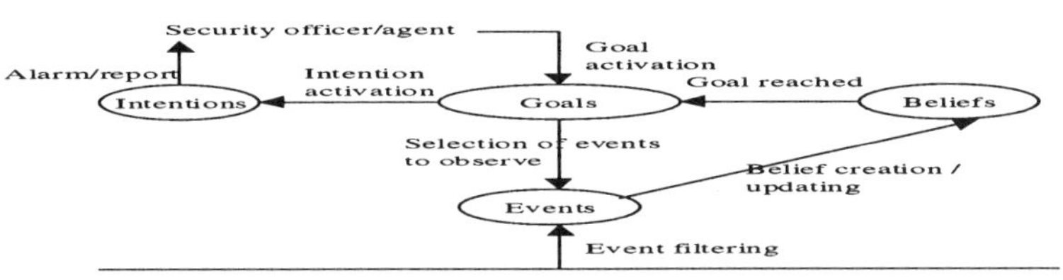

Figure 5: Interactions between mental attitudes

Our implemented system detects well-known attacks such as doorknob rattling, ping sweep and ICMP flooding attacks.

6. CONCLUSION

In this paper, we presented a multi-agents system, which aims to detect intrusions in a complex network. To deal with this complexity, we argued that multi-agents systems provide a suitable solution. So, we applied a well-known multi-agent methodology and showed thus that this methodology is useful for real-life application. Moreover, to model agent knowledge, we used the BDI theoretical model. This model required a hard work to deduce a practical implementation. The implemented system detects well-known attacks. We are now working on a new adaptive version, which deals with learning new attacks and react to non well-specified attacks.

References

[1] R. F. Teixeira de Oliveira, "Gestion des Réseaux avec Connaissance des Besoins: Utilisation des Agents Logiciel", PhD thesis, Eurécom institute, France, 1998.

[2] K. Boudaoud, H. Labiod, Z. Guessoum and R. Boutaba, "Network Security Management with Intelligent Agents", In proceedings of 2000 IEEE/IFIP Network Operations and Management Symposium (NOMS'2000), Honolulu, Hawaii, 10-14 April 2000.

[3] L.T. Heberlein, B.Mukherjee, and K.N.Levitt, "Network Intrusion Detection", IEEE Network Journal, pp. 26-41, May/June 1994.

[4] Maj.Gregory B. White, Eric A. Fisch and Udo W. Pooch, "Cooperating Security Managers: A Peer-Based Intrusion Detection System", IEEE Network journal, pp. 20-23, January/February 1996.

[5] M. Crosbie, E. H. Spafford, "Defending a Computer System using Autonomous Agents". Technical report CSD-TR-95-022, Computer Sciences Department, Purdue University.

[6] S.Corley and al, "The Application of Intelligent Agent Technologies to Network and Service Management", 5th IS&N Conference, Antwerpen, Belgium, 25-28 May 1998.

[7] C. Ko, D. A. Frincke, T. Goan, L. T. Heberlein, K. Levitt, B. Mukherjee, C. Wee, "Analysis of an Algorithm for Distributed Recognition and Accountability", 1st ACM Conference on Computer and Communication Security, pp. 154-164, 1993.

[8] J. Ferber and O. Gutknecht , "A meta-model for the analysis and design of organizations in multi-agen systems", ICMAS, 1998.

[9] J. Ferber,"Les Systèmes Multi-Agents, Vers une intelligence collective", InterEd. 1995.

[10] A. S. Rao and M. P. Georgeff, "Modeling Rational Agents within a BDI-Architecture", Technical Note 14, 1991.

[11] A. Rao and M. Georgeff, "BDI Agents:From Theory to Practice", Tech. Note 56, 1995.

[12] M. Wooldridge and N. R. Jennings. "Intelligent Agents : Theory and Practice ". Knowledge Engineering Review, 10(2):115 -152, 1995.

[13] Z. Guessoum and J.-P. Briot. "From Active Object to Autonomous Agents", *IEEE Concurrency,* volume 7 N° 3, pp. 68-78, July/September, 1999.

Policy and Trust in Open Multi-Operator Networks

Pekka Nikander, Lasse Metso
Helsinki University of Technology

Abstract: In the future telecommunications network, more and more services are based on open protocols and architectures. In such an environment, there is a clear need for controlling the access of users and other operators to the network services. If the network is based on internetworked facilities, traditional address based access control may not be sufficient due to the possibility of address spoofing attacks. Thus, the usage of strong cryptography is often the only possibility for providing authenticity and integrity. However, in such a setting both key management and trust management become challenging problems.

In this paper we present an architecture, called TeSSA, for managing service level access control and other security aspects in open telecom networks, and our experiences with an early prototype version of the architecture. The architecture is based on digitally signed certificates and explicitly presented permissions. The architecture itself is policy free, thereby allowing any reasonable security policy to be applied. Furthermore, the architecture fully supports both centralized and decentralized administrative models, thereby being especially suitable for internets where the networks of several operators are interconnected. In addition to security policies, the architecture can be extended to represent any types of policies. Using the TeSSA facilities, the operators do not need fully trust each other, but can explicitly vary the level and structure of trust.

Keywords: Security Management, Access Control, Security Policy, Decentralized Authorization

1. INTRODUCTION

Managing large networks is a complex and challenging task. Through a series of development steps, quite a lot of effort in the current management systems is circulating around the concept of *policy*. However, it is by far no clear what different authors exactly mean with policy; in this paper, we start with the IETF Policy Working Group definition [1]. Thus, according to this definition, a policy is a collection of rules, consisting of conditions and actions, that are used as a basis of configuring various network elements and controlling of their functionality. However, managing a collection of inter–

connected, more or less independent networks, is even more complex than managing a single network. In this case, managing and co-ordinating individual policies does not suffice, but the agreements and relationships between the inter-operating companies should also be handled. This effect becomes especially evident when considering security policies, where the individual protection needs and the desire to co-operate often conflict. In this paper, we extend the IETF policy definition with these security aspects.

Whenever two organizations agree to co-operate, the common understanding of the terms are recorded in an agreement. Based on the agreement and the backing legislation, the organizations can now expect each other to act according the terms. On the other hand, the actual nature of the common operations usually mirror the actual *trust* between the managers and employees of the organizations. Thus, the organizational behaviour in such a setting is characterized by both the terms of the agreement and the actual personal and social level trust relationships. From this point of view, even the terms of the agreement can be considered as a measure or expression of trust, as their fulfilment usually requires assumptions, or beliefs about the other party. Such assumptions or beliefs may be considered to involve trust, and all these trust aspects must be considered when defining policy.

Now, when we transfer this organization level relationship to the cyberspace, a number of difficulties arises. Based on the common agreement, the individual networks of the organizations should be configured in such a manner that they remain secure while controlled co-operation becomes possible. For example, one of the organizations may provide bandwidth to the other within a specified Service Level Agreement (SLA). Typically, such a configuration is handled through policy definitions.

In this paper, we present a framework for such policy definitions. The framework provides facilities for automated handling of the policy definitions, thereby making it possible to use policy-based network management in inter-organizational networks, e.g. in extranets or multi-operator networks. Furthermore, we consider the implicit security and trust requirements involved in the usage of such policies, and consider how services may be controlled between more than two organizations. The basic method for achieving this goal is to explicitly express the trust assumptions and stated agreements involved in the operations.

1.1 Trust

All human operation involves trust. Most of this trust is so inherent to the social nature of us human beings that we seldom think about it; consequently, inability to trust is considered to be abnormal. Usually, trust involves social or legal control to some degree. If the social control fails, or the legal system collapses, our basic security is fractured. The purpose of the social and legal system is to provide a solid foundation for trust. In a computerized system, as opposed to a physical system, trust relationships should be represented explicitly, and preferably in a way that their legal binding can be later non-repudiably proven in a court, if needed. However, in this paper we do not directly cover non-repudiation; instead, our framework is based on certificates and the assumption that the certificate's validity can be proven on need.

Thus, in the architecture that we later describe, *trust* in a service is a belief that the service, when asked to perform an action, will act according to a predefined description. In particular, this belief implies the belief that the service, or the operator behind, will not attempt to harm the requester independently of the way it fulfils the request. Thus, trust is always expressed in relation to a service provider *and* to an action assumed to be performed. Furthermore, trust is not necessarily transitive. Trusting someone for recommendation is different from trusting someone for direct action.

This definition limits the concept of trust within the distributed system itself. It does not, however, limit us from discussing other aspects of trust when needed. For example, some people consider trust only as inherently human behaviour. On the other hand, it is quite natural to speak about trust relationships between services, and therefore we keep speaking about trust.

Earlier theoretical studies of trust in a distributed setting have been conducted by Raphael Yahalom et al. [2] and later, independently, by Audun Jøsang [3]. In these studies, the goal has been to develop a calculus for trust. That is, the aim has been to create a generic calculus that shows how new trust relationships may be based on existing trust relationships and recommendations. In our work, we are more interested in what are the necessary trust relationships when applying security policies in the usage of services in open distributed networks. Thus, instead of presenting a metric, language, or calculus of trust, we present a software framework that allows trust to be expressed in a machine readable form. Indeed, we believe that the types and levels of trust are application specific, and should not be limited by the management framework. [4]

1.2 Organization of this paper

The rest of this paper is organized as follows. In Sect. 2. we describe the basic security functions in open distributed networks. Next, in Sect. 3. we describe the basics of policy based management. Sect. 4. describes the *Telecommunications Software Security Architecture*, or TeSSA, the top of which our present work is based on. The next section, Sect. 5., shows how the ideas of the TeSSA architecture can be applied to inter-organizational policy-based network management. Finally, Sect. 6. contains our conclusions of this work.

2. SECURITY AND ACCESS CONTROL IN OPEN NETWORKS

While confidentiality, integrity and availability are usually defined as the primary goals of any security system, *authentication, access control* and *authorization* are the usual means used to achieve those goals. In this section, we describe the traditional view to authentication and access control, and why the underlying implicit assumptions behind these terms may not be the right ones for open distributed systems.

2.1 Authentication

In most literature the term *authentication* is used as a synonym for *identity authentication.* However, a thorough analysis reveals that there is a number of fundamental difficulties in this definition. First, the concept of *identity* is very problematic when considering a large, multi-organizational distributed system, and second, limiting authentication to be used in connection with identity reveals to be too restricting.

Let us consider the problem with identity first. The term *identity*, stemming from Latin *identidem*, originally means sameness or oneness. For example, when we meet a previously unknown person for the first time, we cannot really *identify* that person, since there is nothing to identify with. This problems gets really evident in large networks, where we meet people without knowing anything about them, not even their face.

Taking a slightly different point of view, it has been argued that in a distributed digital system the only real "identity", with which anything can be later related to, is a private cryptographic key [4]. That is, when we for the first time meet someone in the digital world, we may be able to learn a public cryptographic key that corresponds to the private key possessed by our new acquaintance. Later on, we can really *identify* a future communicating peer with a known one by being able to convince us that the new peer possesses the same private key as the old one. Thus, we may say that only cryptographic keys should be considered suitable items to function as indications of identity.

Now, it becomes evident that the concept of identity authentication, as usually understood, is at least too limited if not outright misleading. Therefore, it is better to enlarge the term authentication to denote the act of proving the authenticity of any object or piece of information [5] instead of restricting it to denote the act of proving the authenticity of, e.g., the identity of a communicating peer or message originator, as it has been traditionally understood in the literature. Thus, we may speak about authentication of authorization information, or authentication of the possession of a cryptographic key.

2.1.1 Authentication protocols

An *authentication protocol* is a communication protocol whose purpose is to enhance the collection of evidence available to the communicating parties so that one or more of them can *believe* that a given piece of information is *authentic*. That is, it is a common practice to use cryptographic means in establishing new evidence. Typical belief goals include the belief that a (symmetric) cryptographic key is held by a communication peer, the belief that a cryptographic key or other random number has been generated during the protocol run, and the belief that the communication peer holds either of the former beliefs.

2.2 Access Control

Access control includes the means and methods with which the users and other active entities, such as processes and threads, are limited in their ability to manipulate objects within a computer system. In a communications system, it is usually used to denote facilities that limit the possibilities for opening connections or receiving messages. However, even in a communication system the term can be extended to control access to services in a more specific way. The purpose of an access control system is to maintain confidentiality, integrity and availability by making it impossible (or impractically hard) for unauthorized parties to read, modify or consume information or resources.

Capabilities are one possibility to represent access control information. A *capability* is a security token associated with a subject that lists a number of permissions. Each permission defines one or more objects, and an action or a set of actions that the subject may perform on the object(s). In this study, our interest is in explicitly *signed capabilities*, sometimes also called *credentials*, which are capabilities that are cryptographically bound to a specific subject. In the system to be presented, these signed capabilities are represented as *authorization certificates*. Thus, we define that a *signed capability*, or *authorization certificate*, is a digitally signed piece of information that assigns a subject, usually represented in the form of a cryptographic public key, one or more *permissions*, which allow the subject to perform specified actions on one or more specified objects in a target system.

What is probably interesting in this definition is the inclusion of a *target system*. By including this, we want to emphasize the local nature of capabilities in a distributed system. That is, a single capability should be valid only at a specific single system, the target system, or possibly at a (small) number of interrelated systems, e.g. a clustered server, which as a group can be considered to form a single target system.

2.3 Authorization and Delegation

Authorization, in general, denotes sanctioning or empowering someone, i.e., to make it valid, legal, binding, or official for a person to perform certain actions in the future. Delegation, on the other hand, denotes appointing someone to act as a representative, e.g., by means of a legal proxy.

In the context of a (distributed) computer system, both authorization and delegation can be basically defined as acts that change access control rights. In most current systems, local operating system usually assigns a number of rights to a process when the process creates a new object. More generally, any principal can create objects on the behalf of other objects, but this also means that the creating principal will be responsible for controlling the access to the created object [6]. This is important since it allows us to understand the fundamental issues that the concept of a *certificate loop* (see Sect. 5.1) is based on.

Thus, by definition, when a computer or communications node is installed, the node operating system is given implicit access permissions to all the physical and logical objects that are stored in the node. When a software based service is created, on the other hand, it is only given access to the objects that are parts of that service. In fact, these are the only implicit access rights; all other rights are delegated. Thus, we can say that a principal that has a permission to control another principal or access an object may *delegate*, on its will, this permission to a third principal, unless explicitly prohibited [6]. When delegating, some permissions assigned to the delegating principal are copied to the delegate.

Examples of existing distributed authorization and delegation systems include the Digital Systems Security Architecture (DSSA) [7], Kerberos [8], our TeSSA [9] (see Sect. 4). All of these are pretty similar in many respects. However, there are a number of differences as well. The most important differences between TeSSA and the other two can be summarized as follows.

- First, our system explicitly models more types of trust than either DSSA or Kerberos; especially, we model types that are not related to access control but to generic security conditions that must be met.
- The DSSA architecture is based on names and local access control lists, while our architecture uses signed credentials and thereby allows anonymous operation.
- The Kerberos architecture is based on symmetric cryptography and centralized key distribution centres. Our architecture is based on public key cryptography and is fully decentralized.
- Our architecture explicitly contains the concept of policy, and make it possible to represent various kinds of policies in addition to security policies.

Other existing prototype authorization systems that have had influence on our system include the PolicyMaker [10] and the SDSI/SPKI proposal [4][11][12]. Our system is based on the same (but independently developed) ideas. [6]

3. POLICY BASED MANAGEMENT

Policy-Based Networking (PBN) is gaining a wider acceptance in the management of open, IP based network, mainly due to it makes possible more unified control and management in complex IP networks [13]. In this section, we describe the related concepts and terminology, as defined by the IETF [1]. Later, in Sect. 5., we briefly describe how these definitions must be extended in order to take care of the various security aspects involved when applying policies in multi-organizational setting.

3.1 Policies, Rules, Conditions, and Actions

A *policy* is the combination of *rules* and *services*, where rules define the criteria for resource access and usage. Each policy rule is comprised of a set of *conditions* and a corresponding set of *actions*. The conditions define when the policy rule is applicable. Once a policy rule is so activated, one or more actions contained by that policy rule may then be executed.

Policies can contain other policies; this allows one to build complex policies from a set of simpler policies, so they are easier to manage. The inclusion possibility also enables the reuse previously built policy blocks.

3.1.1 Policy rules

Policy rules can be classified by their purpose; the following classification has been proposed by Moore et al. [14].

- *Service policies* describe services available in the network. For example, QoS classes are made by using service policies.
- *Usage policies* describe how to allocate the services defined by service policies. Usage policies control the selection and configuration of entities based on specific usage data.
- *Security policies* identify client, permit or deny access to resources, select and apply appropriate authentication mechanisms, and perform accounting and audit of resources.
- *Motivational policies* describe how a policy's goal is accomplished. For example the scheduling of file backup based on activity of writing onto disk is a kind of motivational policies.
- *Configuration policies* define the default setup of a managed entity, e.g., the setup of a network forwarding service or a network-hosted print queue.
- *Installation policies* define what can be put on the system, as well as the configuration of the mechanisms that perform the installation.
- *Error and event policies,* for example, ask the user to call the system administrator, if a device fails between 8am and 5pm.

3.1.2 Policy conditions and actions

A policy condition consist of two parts, which are policy condition *type* and policy condition *element*. A policy condition type is a set of predefined conditions, which all network vendors can implement. They can be attached to a policy rule for evaluation. A policy condition element is a policy condition type instance that is being evaluated. Policy condition elements are related together to form a Boolean expression.

A policy action is the changing of the configuration of one or more network elements in order to achieve a desired state; the state provides one or more behaviours. Like a policy condition, a policy action is comprised of two parts, a policy action type and a policy action element.

3.2 Policy Decision, Behaviour and State

A policy *decision* is the abstraction of activating and evaluating one or more policy rules. Each policy rule is interpreted in the context of a specific request for accessing and using resources. It connotes taking pre-determined actions based on whether or not a given set of policy conditions were satisfied.

A policy *behaviour* controls how traffic is treated, what network resources must be utilized, and what network services are provided. A policy *mechanism* is a set of vendor-specific commands that configures a network element to put a policy rule into effect. Policy behaviours define one or more mechanisms that are used to implement the policy. Therefore, different devices can carry out the same policy using different behaviours.

A policy *state* is a description of the settings of one or more network elements. These settings correspond to providing the set of services to be provided by the network. For example, a Service Level Agreement (SLA) can describe services contracted for by subscribers, this corresponds to a state that various network elements must be put into in order to provide those services.

3.3 Policy Functions

Policy is comprised of three functions: *decision-making (evaluation), enforcement* and *monitoring*. Policy *evaluation* is the determination of whether or not the network is in a desired policy state. This is usually determined by processing static or dynamic data against policy rules, the key point being that the decision is made by comparing definitional data stored in the policy repository with current data from the network that does not have to be stored in the policy repository. If it is found that the network elements are not in the desired policy state, then one or more policy actions will be taken to move the network elements from their current state to the desired state. This is called policy enforcement.

Policy *enforcement* is the action of placing the network in a desired state using a set of management commands. When this definition is applied to network elements, these management commands change the configuration of the devices using one or more mechanisms. Enforcement is carried out in the context of a policy rule.

Policy *monitoring* is an on-going active or passive examination of the network and its constituent devices for checking network health, whether policies are being satisfied, and whether clients are taking unfair advantage of network services. This is done for following reasons: to ensure that clients are receiving the services that they have contracted for, to monitor network statistics as part of checking network health, to monitor network statistics as part of checking whether policies that are currently in use are being satisfied, or to ensure that clients of a given set of policies are not abusing their privileges.

3.4 Security Policies

In strict contrast to the formal policy model presented above, security policy is *usually* defined to be the set of more or less informal rules that specify how the integrity, confidentiality and availability of the stored information and available resources are to be protected within an information processing system. The rules often involve manual operations and management conventions that are expected to be followed by humans.

Our point of view in this work, however, is somewhat different from the usual one. As the focus is on how to make such distributed system secure that span several organizations, our interest lies on how to express a security policy in a form that allows network nodes and services to enforce it. Furthermore, we want to apply security policies to all forms of trust involved in a distributed system, not just the access control type of type of trust usually considered. Based on this, we now attempt to define security policy, trying to collect the most important aspects together.

A (formal) *security policy* is a collection of policy rules (as defined above) that describe how a network node shall modify its behaviour when considering statements made by trusted third parties (TTPs), recommendations given by the TTPs, and statements considering claimed authority. The actions defined in a policy rule may involve access control, usage of cryptographic keys, or some other aspect of total system security. The purpose of the policy is to minimize the risk of losing the integrity, confidentiality, or availability of the protected information and resources.

In an open distributed system, the ability to publish a policy, and to base actions on the published policy of other principals, may prove to be an important feature. For example, it would allow software applications that work for a user to make policy decisions even when the user is off-line. [6]

4. THE TELECOMMUNICATIONS SOFTWARE SECURITY ARCHITECTURE

In this Section, we describe the *Telecommunication Software Security Architecture* (TeSSA) that we have developed in our research group. First, a conceptual view is presented, followed by architectural, functional, and implementation level views.

4.1 Conceptual overview

The basic layered structure of the TeSSA architecture, on the conceptual level, is shown in Fig. 1. The top component is the *trust and policy management infrastructure,* sometimes also called the public key infrastructure (PKI). It allows the users and administrators to explicitly define, for example, which network nodes are trusted for which operations, how users are allowed to access resources, and what kind of protection is required from the connections between different applications. In practice, the data present in the management infrastructure is represented in the form of certificates. The certificates are stored in a distributed *certificate repository,* which allows convenient storage and easy access to the certificates.

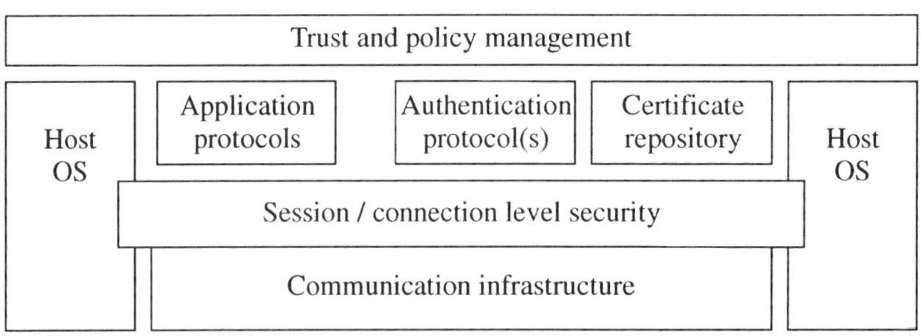

Fig. 1. The conceptual building blocks of the TeSSA security architecture

The certificates are used, among other things, to control and facilitate the operation of the authentication protocols. That is, the authentication protocols need the public keys of nodes and applications, along with semantic information about what the keys are good for. This information is available from the trust and policy management infrastructure.

An *authentication protocol* is used to create, modify and delete dynamic security associations and contexts between any two principals. The PKI layer above provides the initial security contexts on the base of which new contexts may be created. In practice, the protocol allows creation of session level keying material combined with authentication of the public key of the peer, the negotiation parameters, and the keying material itself.

Finally, the session or connection level security components take care of protecting the user data in transit. They allow user data to be protected from eavesdropping, modification, replay, and other active and passive attacks. Moreover, it is important to notice that each secure connection has a policy associated with it. That is, each secure connection is annotated with knowledge about the permissions of the peer object. This allows the local access control system to determine when a remotely initiated operation is authorized and when not.

4.2 Definition of architectural concepts

To give a more precise meaning to the various components of the conceptual architecture, the following definitions are given.

- A *security context* is a collection of security related variables, such as asymmetric or symmetric keys, policy rules and credentials, shared by two or more parties. When creating a new security context, authenticity and integrity of the information must be ensured. Typically, also some or all of the information is confidential as well. Technically, new security contexts may only be created on existing trusted security contexts. In the architecture, the trust and policy management infrastructure provides the initial security contexts, using which new contexts may be created. The initial security contexts are represented in the form of certificates. Creation of such contexts is based on management decisions external to the presented architecture.
- The *session / connection level security components* take care of the actual (cryptographic) protection of user data in transit. Typically, the layer consists of a number of cryptographic protocols and associated policy management functionality. The keying material and policy rules pertaining to a particular secure session are defined in a corresponding security context.
- The *authentication protocol* component of the architecture denotes that or those implementation level protocols that are capable of *creating, modifying and deleting* new security contexts, based on existing security contexts. Thus, the main function of the authentication protocol is the establishment and management of security contexts. A notable issue here is that, in addition to the key authentication and negotiation, the protocol takes care of managing the associated policy information as well.
- The *certificate repository* is a distributed database that facilitates on-line storing and retrieving of certificates. Ideally, the repository should be efficient, fault tolerant and transparent to applications.

4.3 Functional concepts

Changing the point of view from the layered one to a more functional one, the architecture can be depicted as in Fig. 2. First, there is a *reference monitor*, which is a functional concept within the local operating system that takes care of protecting local resources. Whenever a piece of software (e.g. a client program) attempts to access a protected resource, including services provided by other applications, the request is routed through the reference monitor. The reference monitor decides, using the local policy rules and the permissions the application has, whether the request is authorized or not. An unauthorized request is aborted.

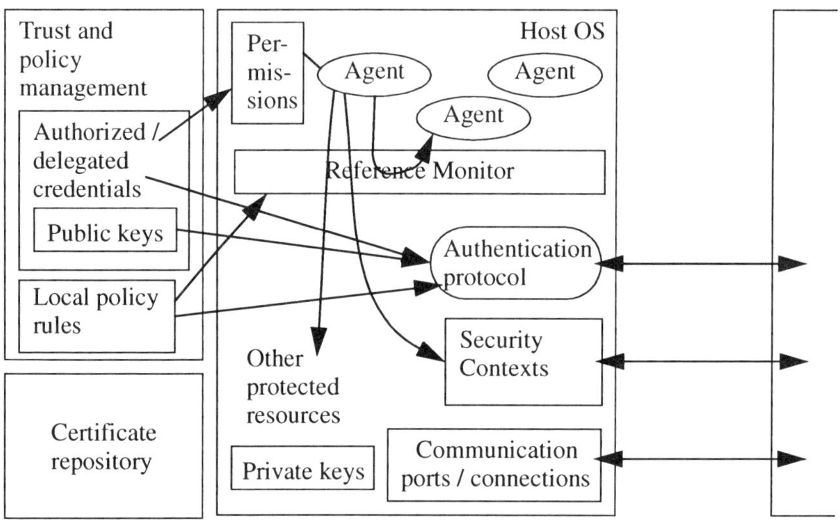

Fig. 2. A functional view to the TeSSA architecture

The protected resources include, among other things, *security contexts* and communication ports or connections. In order to have secure connections, and to enforce local security policy with respect to opening of connections and to having quality of protection on connections, both of these resources must be locally protected. When a software module wants to have a secure connection with a remote server, a security association and a connection (or a number of connections) that use the association must be established.

When a new security context is needed, an *authentication protocol* is used to negotiate one. The authentication protocol gets any needed public keys, local policy rules and credentials from the trust and policy management infrastructure. These allow it to perform the negotiation towards its peer, and to establish the local policy conditions that apply to the newly created association.

Thus, the *trust and policy management infrastructure*, or PKI, can be seen to provide control and management information to the various functions, including creation of security context, access control decisions made by the reference monitor, and control of quality of protection on connections. The PKI manages these resources by representing credentials of various parties.

A *credential* is a signed statement of authority, claiming that a principal has been authorized (possibly via delegation) to access a protected resource. In our architecture, the credentials are represented in the form of SPKI certificates. When a credential is accepted by a local execution environment, it is usually transformed into a local representation denoting permission to access some concrete resource. Credentials are almost always acquired through delegation. This means that a final credential does not alone qualify for creating permissions (or directly authorizing operations). Instead, a chain of authorization certificates is typically needed. Acceptance of certificate chains is controlled by the local policy rules.

A *local policy rule* is a locally understood instruction to the reference monitor that allows it to make decisions. Specifically, local policy rules may exclude credentials created by certain principals, limit the lengths of certificate chains, specify that stronger than default credentials are needed for certain operations, etc.

The local policy rules, similarly to the credentials, are represented as certificates in our architecture. This has a number of benefits. First, since they are signed, their management can be distributed. Only rules having a locally meaningful structure and a trusted issuer will be adhered to. Second, they need not necessarily be stored locally, but can be retrieved from the certificate repository only when needed.

4.4 Implementation

In addition to the conceptual model, we have also built a partial prototype of the architecture in our project. Taking the actual implementation components, the layered structure of the architecture is shown in Fig. 3. As one can see, the trust and policy management infrastructure is implemented using the IETF Simple Public Key Infrastructure (SPKI). Furthermore, the authentication protocol(s) are based on the Internet Security Association and Key Management Protocol (ISAKMP) infrastructure, the certificate repository is implemented within the Domain Name System (DNS), and connection security is taken care of with the IPSEC protocols.

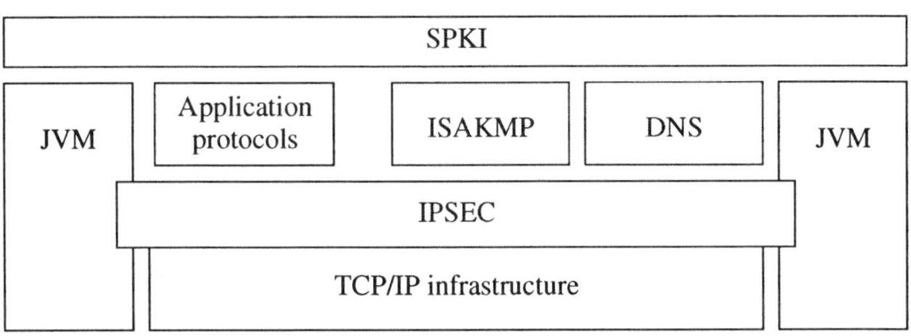

Fig. 3. Implementation architecture

All of the implementation is based on the TCP/IP protocol suite and the Java Virtual Machine version 1.2. The TCP/IP infrastructure provides ubiquitous communications and a multitude of application protocols. The JVM 1.2 provides an Object-Oriented operating environment with fine grained access control. The details of the implementation are beyond the scope of this paper, and described elsewhere, see e.g. [6][9][15][16][17][18].

5. APPLYING POLICIES BETWEEN ORGANIZATIONS

A crucial issue in Policy-Based Networking is the question of how to apply policies in a network with several operators that may have conflicting goals. If each of the operators just use PBN to manage their own network, much of the benefits of the policies are lost. That is, the operators would need to carefully co-ordinate their policies, and this co-ordination would be manual.

In this section, we show how the ideas of TeSSA based distributed security management can be applied to generic policy based networking in an inter-organizational setting.

5.1 Trust requirements at a node level

To start, let us consider a simple situation of two network nodes, one providing services and the other using the services provided by the first one. We assume that the nodes are managed by different organizations, and that the organizations do not fully trust each other. Furthermore, we don't assume anything about the security of the interconnecting link (other than that it transfers packets more or less reliably); there may be malevolent parties intercepting the link. In this situation, both nodes have a number of trust requirements.

- First, the node requesting services must determine that the target node is trusted to provide the services. That is, the requesting node must consider the target node known to run the requested services, and trustworthy enough to provide the services for the purposes of the requesting. This definition includes the idea that some nodes may be considered trustworthy enough only for some services.
- Second, the target node must be able to determine that the requesting node is indeed allowed to request the services if asks for. This includes, among other things, the permission that the new service may use resources such as network connections.

Both of these trust requirements may be represented with certificates [18]. In the first case, when the requesting node been started, it has been configured to trust a number of authoritative parties for determining trustworthy server

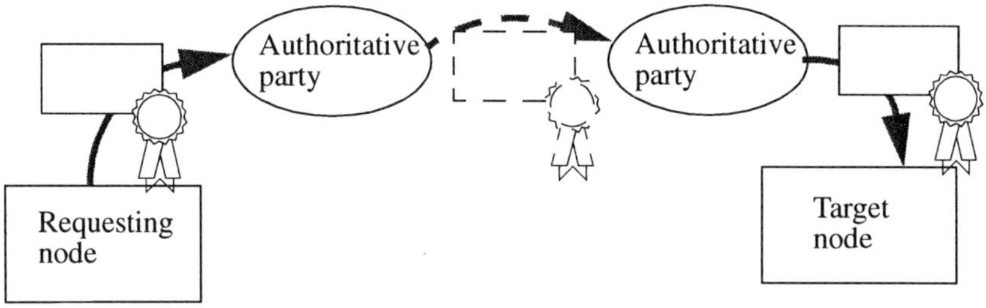

Fig. 4. Chain of certificates from the requesting node to the target node

nodes. One or more of these authoritative parties, on their behalf, have then certified the trustworthiness of the target node. This is illustrated in Fig. 4.

Similarly, there must exists a chain of certificates from the target node, through the security administrator of the node, to the requesting node. This chain must prove that the node is authorized to request the service, and thereby use the resource of the target node. This chain is illustrated in Fig. 5.

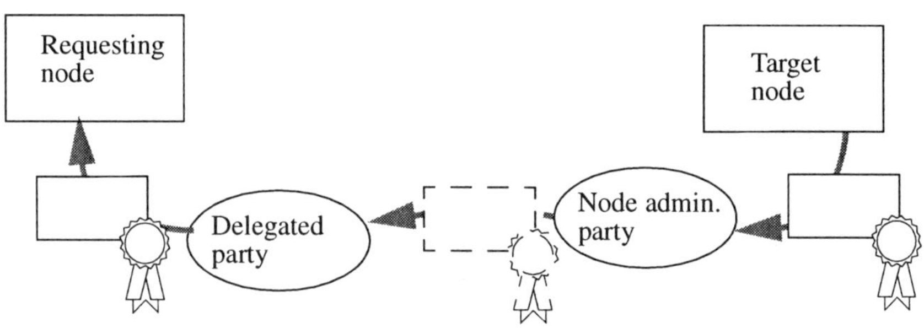

Fig. 5. Chain of certificates from the target node to the requesting party

When the requesting node contacts the target node, an authentication protocol is run. During the course of the protocol, the target node proves that it has its own private key in its possession. Similarly, the requesting node proves the possession of its key. These authenticating demonstrations may be considered to function as "virtual certificates" between the requesting node and the target node. In a way, these "virtual certificates" delegate back to the verifier the authority that the authenticating party was received through the chain of actual certificates. Using the local policy rules, the verifier can now determine if the peer actually does have the claimed authority by checking all of the certificates and seeing if the certificates really imply the claimed authority. These checks close the certificate chain into a loop, as illustrated in Fig. 6. [18]

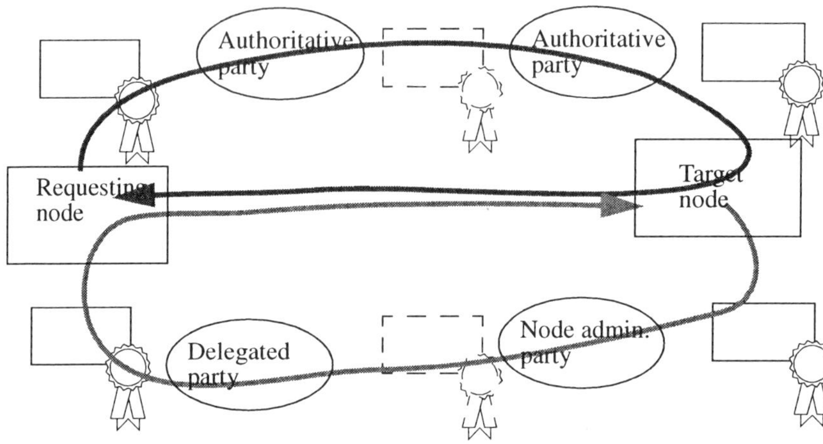

Fig. 6. Certificate chains are closed into loops by an authentication protocol

5.2 Organization level security policies

In Fig. 6, the authoritative parties, the node administrative party, and the delegated party are examples of Trusted Third Parties (TTP). Thus, architecturally, they belong to the trust and policy management layer of the architecture (see Fig. 1). Since these parties typically belong to different organizations, they, in fact, do represent organization level security policies.

So, on the simple example of Sect. 5.1., the *node administrative party* actually controls the behaviour of the target node in respect to who it allows to access the service. Accordingly, it *represents* the security policy regarding the target node as a set of certificates that control the access to the target node. More specifically, that particular certificate (or set of certificates) may apply simultaneously to several servers, thereby actually representing *organizational* level policy instead of a node specific policy.

In the same way, if we consider the trust mediated by the *authoritative parties*, the certificates provided by the parties control which servers the requesting node may use and which it must not use. Thus, they represent policies that control the security restrictions on the behaviour of the requesting node. Again, as the authoritative parties may be hosted by several organizations, they represent the security policies of their hosting organizations.

Possible conflicts in the expressed policies of the various companies are resolved in two distinct levels. First, at the certificate level, policies are represented in the certificates as sets of permissions. These sets may be easily intersected. Second, the actual node resolving the certificate chains may use additional local policy rules in order to resolve possible conflicts. We have discussed both of these aspects in more detail in [6].

5.3 Extending security policy representation to other kinds of policies

In the SPKI certificate standard [4][11], the so called tag field of the SPKI certificates are used to represent *sets of permissions.* However, the SPKI architecture allows the tag field to have local semantics that depend on the actual node interpreting the tag. Thus, it is possible to specify different kinds of semantics for the tag.

One approach is to use program fragments to represent policies, as suggested in [10]. However, that approach has the drawback that there must be an interpreter for the programming language. Additionally, such an approach is in direct opposition with the IETF policy definition, where the goal is to define policies in a descriptive way instead of using imperative program fragments.

Thus, we propose an approach where the policy rules are included in the tag field of the certificates. Furthermore, if the certificates are represented in XML, as suggested in [19], the policies can also easily be represented in XML, e.g., according to the approach suggested in [20]. This is very much in line with one of the original motivations for XML, namely the ability to use XML to represent configuration information. For example, the forthcoming Apple Mac OS X will use XML internally to represent configuration information.

Thus, the usage of XML allows any policies to be represented instead of security policies, and the usage of certificates allows the usage of the policies to cross organizational boundaries in a secure way. The certificates, and especially certificate chains, model the trust relationships between the organizations, and allow both mutual agreements and other trust conditions to be represented in the digital format.

6. CONCLUSIONS

In the field of Policy-Based Networking, the ongoing research has largely focused on modelling various concrete policies and applying them within a single administrative domain, or at most between domains that fully trust each other. On the other hand, various kinds of authorization certificate approaches have provided a model of how to express access control information in a fully distributed setting involving several organizations. In this paper, we have shown how mixing these two approached together with an XML based representation format can be used to represent all kinds of policies in a secure way. The certificate based security allows these policy expressions to be enforced even in a situation where there are several co-operating organizations that do not fully trust each other.

REFERENCES

1 John Strassner and Ed Ellesson, *Terminology for describing policy and services,* work in progress, IETF Internet draft, 1999, `http://www.ietf.org/internet-drafts/draft-ietf-policy-terms-00.txt`

2 Raphael Yahalom, Birgit Klein, and Thomas Beth, "Trust Relationships in Secure Systems - A Distributed Authentication Perspective", In *Proceedings of the IEEE Conference on Research in Security and Privacy,* 1993.

3 Audun Jøsang, A Model for Trust in Security Systems, in *Proceedings of the Second Nordic Workshop on Secure Computer Systems,* 1997.

4 Carl Ellison, *SPKI Certificate Theory,* RFC2693, September 1999.

5 Pekka Nikander, *Modelling of Cryptographic Protocols -- A Concurrency Perspective,* Licentiates Thesis, Helsinki University of Technology, December 1997

6 Pekka Nikander, *An Architecture for Authorization and Delegation in Distributed Object-Oriented Agent Systems,* Helsinki University of Technology, Doctoral Dissertation, 1999.

7 M. Gasser, A. Goldstein, C. Kaufman, and B. Lampson, "The Digital Distributed System Security Architecture," In *Proceedings of 1989 National Computer Security Conference.*

8 J. Kohl and C. Neuman, *The Kerberos Network Authentication Service (V5),* RFC1510, Internet Engineering Task Force, 1993.

9 Sanna Liimatainen et al., *Telecommunications Software Security Architecture,* Helsinki University of Technology, `http://www.tcm.hut.fi/Research/TeSSA`.

10 M. Blaze, J. Feigenbaum, and J. Lacy, *Decentralized Trust Management,* In Proceedings of the 1996 IEEE Computer Society Symposium on Research in Security and Privacy, Oakland, CA, May 1996.

11 Carl Ellison, *SPKI Requirements,* RFC2692, September 1999.

12 Ronald Rivest and Butler Lampson, "SDSI - A Simple Distributed Security Infrastructure", Proceedings of the *1996 Usenix Security Symposium,* 1996.

13 David C. Blight, Takeo Hamada, Policy-Based Networking Architecture for QoS Interworking in IP management. *Proceedings of Integrated network management VI, Distributed Management for the Networked Millennium 1999, IM 98,* IEEE 1999, pp 811-826.

14 B. Moore, E. Ellesson, and J. Strassner, *Policy Framework Core Information Model,* work in progress, IETF Internet draft, 1999, `http://www.ietf.org/internet-drafts/draft-ietf-policy-core-info-model-01.txt`

15 Jonna Partanen, *Using SPKI certificates for access control in Java 1.2,* Master's Thesis, Helsinki University of Technology, August 1998.

16 Sanna Suoranta, *An Object-Oriented Implementation of an Authentication Protocol,* Master's Thesis, Helsinki University of Technology, December 1998.

17 Tero Hasu, *Storage and retrieval of SPKI certificates using the DNS,* Master's Thesis, Helsinki University of Technology, April 1999.

18 Ilari Lehti, Pekka Nikander, *Certifying Trust,* Practice and Theory in Public Key Cryptography (PKC'98), Yokohama, Japan, January 1998.

19 Juha Pääjärvi, "XML Encoding of SPKI Certificates", work in progress, Internet draft `draft-paajarvi-xml-spki-cert-00.txt`, March 2000.

20 Manfred Hauswirth, Clemens Kerer, and Roman Kurmanowytsch, *A flexible and extensible security framework for Java code,* Technical Report TUV-1841-99-14, Technical University of Vienna, March 2000.

Enabling secure e-Commerce on Mobile phones

[1] Do van Thanh & [2] Jan A. Audestad
[1] Ericsson Norway Applied Research Center, P.O Box 34 N1361 Billingstad, Norway - etodvt@eto.ericsson.se - Tel: +47 66 84 12 00
[2] Telenor AS P.O Box 6701 St. Olavs plass N-0130 Oslo Norway – jan-arild.audestad@telenor.com – Tel: +47 22 77 99 52r

Key words: Mobile e-commerce, Mobile Internet Applications, Wireless Application Protocol (WAP), Mobile security, Wireless Public Key Cryptographic

Abstract: Mobile e-commerce enables the mobile user to buy and pay for things, to pay his bill or to make a bet via his mobile phone when on the move, anywhere and at any time. It will bring convenience and contribute to improve life quality of the users. However, in order to be successful, security measures must be strong enough to protect the user from illegal abuses and to get confidence from him. Unfortunately, current security measures for mobile phones are not yet sufficient. This paper describes the mobile e-commerce activities at Ericsson, which aim at making mobile e-commerce applications secure and enabling a full-scale development and deployment of them. The paper starts with a definition of mobile e-commerce. Next are a summary of the Wireless Application Protocol (WAP) and its achievements. The Web e-commerce is briefly explained. The problems related to security in mobile e-commerce are then described. Thereafter, the solution to the problems is presented. The paper concludes with a look on the future and discussions on what can be done.

1. INTRODUCTION

The convergence of mobile communications network and Internet has paved the way for a range of brand-new applications called wireless Internet applications. Which one of them will be the killer application is still unclear. However, there is one type of wireless Internet applications that are getting more and more popular and may even surpass their counterpart in the fixed Internet. They are called mobile electronic commerce applications. They enable the user to buy small things such as soft drinks, cinema tickets, train tickets, etc. or to pay his bills via mobile devices, i.e. mobile phones, PDAs (Personal Data Assistant), palmtops, etc. In a time when people are much on the move and focus is on life quality improvement, mobile e-commerce applications will bring both convenience and save a lot of time for the mobile user. However, in order to be successful, security measures must be strong enough to protect the user from illegal abuses and to get confidence from him. Unfortunately, current security measures for mobile phones are not sufficient. This paper describes the R&D activities in mobile e-commerce at Ericsson, which aim at making mobile e-commerce applications secure and enabling a full scale development and deployment of them. The paper starts with a presentation of mobile e-commerce. Next is a summary of the Wireless Application Protocol (WAP) and its achievements. The Web e-commerce is briefly explained. The problems related to security in mobile e-commerce are then described. Therafter, the solutions to the problems are presented. The paper concludes with a look on the future and discussions on what can be done.

2. WHAT IS MOBILE E-COMMERCE?

Mobile e-commerce is e-commerce brought to mobile users via mobile devices such as palmtops, PDAs or most dominantly mobile phones. With an ever-increasing number of devices in the market, mobile phones will undoubtedly play a crucial role in promoting mobile e-commerce. Mobile e-commerce allows users to conduct e-commerce on their mobile devices: obtain marketing and sales information, receive ordering information, make a purchase decision, pay for it, obtain the service or product and finally, receive customer support required.

Mobile e-commerce is more than a mobile and wireless extension of the Web-based e-commerce. It is an entirely new sales and promotion channel, and is the enabler for a whole range of new services such as buy a Coke, pay for parking, buy train ticket, etc. via mobile phone. Most importantly it is tailored to the users in many aspects. It follows the user and is available anytime and anywhere. Although mobility is a valuable characteristic to the user in general, it is especially precious for e-commerce because it enables a key factor, which is missing in other e-commerce forms, namely the ability to adapt to the user, his humor and his demands. In fact, the essence of commerce is to be able to satisfy the demands of the users. It is important not only to be able to offer whatever the user wants but also whenever he wants. Mobile e-commerce can also be customised such it fits the preferences of the user in combination with time and location.

Another important aspect of mobile e-commerce is the ability to mix electronic media with other media such as newspaper, TV, radio, natural communication in any of the commerce phases i.e. presentation, selection, ordering, payment, delivery and customer care. For example, the mobile user can browse on his mobile phone and obtain the location of the closest shop. He goes there and buys a Coke. In this case, the presentation and selection are done electronically via the mobile phone while the rest is done in a traditional way via natural communication. In another situation, the user buys groceries and pays via his mobile phone. The presentation, selection, ordering, delivery and customer care phases are carried out in traditional way and only the payment phase is done electronically.

3. MOBILE E-COMMERCE AND WAP

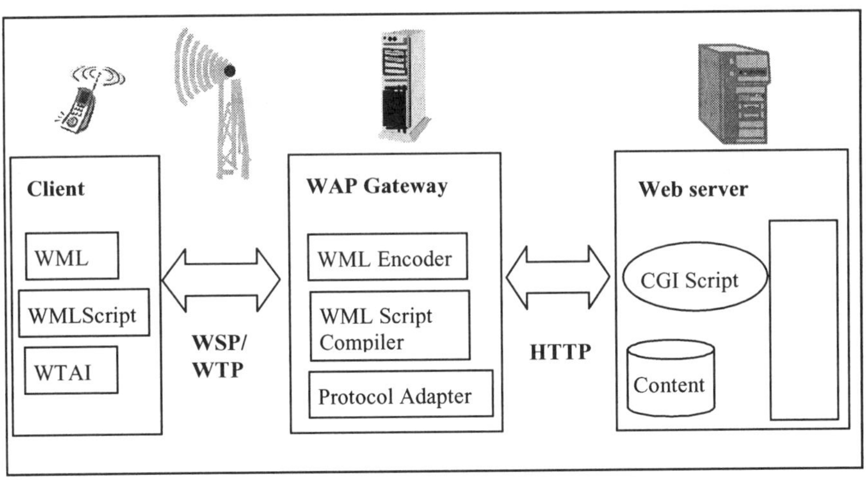

Figure 1. The Wireless Application Protocol Architecture

The *Wireless Application Protocol (WAP)* promoted by the WAP forum enables the access to the Internet for mobile devices. Taken into account the limited bandwidth of the wireless link, the limitation of mobile devices concerning processing, storage, battery life, size and weight, WAP is optimised for the wireless environment. The architecture of WAP is shown in *Figure 1*. Of course, WAP will contribute to the success of mobile e-commerce but it is worth noting that mobile e-commerce exists also without

WAP. For example, the first mobile e-commerce application in Norway,"The cinema ticket" that was jointly developed by Ericsson and Telenor Mobile is not based on WAP. It is based on SIM application toolkit where the commerce application is implemented on the SIM (Subscriber Identity Module) of the mobile phone. It is worth mentioning that WAP contains security specifications but they are not sufficient because they do not provide end-to-end security. In the future, mobile e-commerce can be extended further through the adoption of newer technology such as Bluetooth, which allows local communications between devices without the need of an on-line connection with the network.

4. SECURITY REQUIREMENTS IN E-COMMERCE

In e-commerce where the consumer and the merchant communicate indirectly via software entities and the Internet, trust must be somehow established between the two parties. In order to achieve trust the following security functions must be performed:

- **Authentication:** Each party needs to be able to authenticate its counterpart, i.e. to make sure that the counterpart is the one he claimed to be.
- **Integrity:** Each party needs to make sure that the received messages are not altered or fabricated by other than their counterpart.
- **Confidentiality:** Each party wants to keep the content of their communication secret.
- **Message authentication:** Each party wants to make sure that the received messages do really come from his counterpart.
- **Non-repudiation:** Each party wants to prevent that the counterpart later on denies the agreements that he has approved earlier.

Usually, the two parties do not and must neither know each other in order to do trading. In such a case, the *asymmetric cryptographic algorithm*, also called the *Public key algorithm* is more appropriate than the symmetric cryptographic algorithm.

Briefly, the public key algorithm uses a key pair, one private and one public for encryption and decryption. What encrypted by one key can only be decrypted by the corresponding one. It should also be practically impossible to derive one key from the other one. Confidentiality and integrity are prevailed when the sending party encrypts the message with the recipient's public key since only the later has the corresponding private key to decrypt the message. Authentication and non-repudiation are achieved when the sender encrypts the message or part of it with his private key. The receiver decrypts the message with the sender's public key and can be sure that it comes from the sender because only he is the only to have the private key. This later encryption scheme is known as *digital signature*, which usually consists also of a message digest (hash function) to reduce the size of the message to be encrypted and to optimize the signing process. There are currently several public key algorithms such as RSA [1], Elliptic Curve [8].

The issue now is to be certain who owns what key pair. A certificate issued by a trusted authority also called *Certificate Authority* (CA) attests that a public key belongs to an entity or individual with a certain name and attributes. Both certificates and keys need to be managed, i.e. generated, revoked, updated, recovered, etc. and a *Public Key Infrastructure* (PKI) is necessary for that. Unfortunately, no such global PKI does exist yet and ad-hoc solutions as we will explain on later sections, have been adopted on web e-commerce.

5. E-COMMERCE ON THE WEB

Since our intention is not to give a deep presentation about Web e-commerce but only an elucidation necessary for the explanation of mobile e-commerce later on, only simplified views of Web shopping and Web banking are described.

5.1 Web shopping

Web shopping is getting more and more popular, especially for books, music, films, etc. The procedure varies slightly depending on the visited web site but can be summarised as follows:

1. A user visits a web site of a merchant. He browses among the offers. Up to this point, no security measure is needed since everything is public.

2. He wants to order goods or services.

3. The web server asserts its site identity by signing its server certificate and sending it together with the unsigned certificate to the browser. In this case the server must be a secure server, i.e. having a server certificate and enabled for security. The browser uses the server's public key (from the server's certificate) to verify that the owner of the certificate is the same one who signed it.

4. The browser checks if the issuing CA is one that it accepts. The trusted CAs is specified in the list of so-called trusted root certificates. Such a list is embedded in the browser. Some browser like Microsoft's Internet Explorer allows the import of new trusted root certificates. If unknown, the browser informs the user that this server certificate was issued by an unknown CA.

5. The user manually (visually) authenticates that the site's certificate was issued by a trusted third party for the exact site the user is visiting.

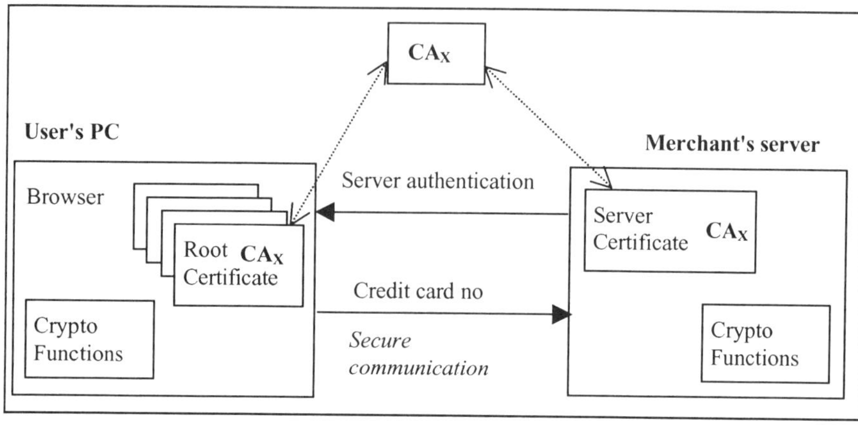

Figure 2. Web shopping

6. The browser generates a session key, encrypts this key with the server public key and sends it securely back to the server.

7. A secure channel is established, with the session key generated by the browser.

8. The user will be asked to enter his personal data, i.e. name, address, and email.

9. The user will be asked to enter his credit card number that will be charged for the purchase.

10. The server issues a receipt to the user or sends it back via email.

11. The merchant validates the credit card number and if valid ships the purchased goods to the user.

12. The transaction can be closed at this stage

The procedure to establish the secure channel described above is in accordance to the Secure Socket Layer (SSL) protocol.

The requirements on the user's side are as follows:

1. His PC must have a browser
2. The browser must be equipped with root certificates used in the authentication of the server
3. It must have access to cryptographic functions that are capable of validating server certificates and capable of encrypting and decrypting for the secure channel.

The channel is secure in the sense that confidentiality and integrity are prevailed. However, it is not a trusted channel. Neither merchant nor the user can be sure that he is dealing with the right counterpart. On the merchant side, only the web server authentication is executed but not the merchant authentication. On the user side, no user authentication is done. It is worth noting that only the validation of the credit card number is done i.e. the credit card number is valid and can be charged for the purchase. Nothing is said about whether the user is owner of the credit card and hence is entitled to use it.

The described web shopping scheme is used widely because it is simple and does not require much infrastructure and investment. However, it has the following limitations:

- The user has to trust the merchant's site. For well-known sites with good reputation, he can do that but for unknown site he faces a lot of risks. The site may be a fake shop that collects and abuses his credit card number.
- The merchant may deal with impostors that use credit card numbers from stolen cards or valid card numbers that are generated by an illegal process. In such cases, the validation of the card number is successful and the fraud can only be discovered long after the delivery of goods. The financial institutions refuse to cover losses for such cases because the merchant has not verified that the user has a valid credit card and the signature is identical to the one on the credit card.
- The financial institutions are not very satisfied because the authentication of the user and the authentication of the merchant are skipped. The risks for frauds and the number of disputes are higher.

Visa and MasterCard have jointly developed the SET Secure Electronic Transaction protocol [1] as a method to secure payment card transactions over open networks. SET requires however investments both on the merchant and the consumer side, and is not widely used.

5.2 Web banking

Many banks in Europe have realized that by providing banking services such as paying bills, money transfer, balance check, etc. on the Web they can reduce costs at the same time as better services can be offered to customers. However, they are very concerned about security and do not find the procedure used in web shopping secure enough since no client authentication is performed. In order to remedy the situation, the banks have adopted different authentication schemes.

Authentication using a set of numerated passcodes: The user receives from the bank by post a plastic card where a series of numbered passcodes are printed on. The number of passcodes varies depending on the bank. The user is supposed to keep this card in a secure manner. When the user visits the Bank's site, a secure communication is first established between the user's computer and the bank's server. Then, the user is asked to enter his username. The server will then ask him to enter for example passcode number n. The user consults his plastic card and enters the value of the passcode number n. If the passcode is correct, the user is authenticated.

Authentication using a passcode calculator: The user receives from the bank by post a calculator, which is capable of generating a one-time code. The calculator is secured by a PIN code chosen by the user at initialisation. When the user visits the Bank's site, a secure communication is first established between the user's computer and the bank's server. Then, the user is asked to enter his username. The server will then ask him to enter the passcode. The user enters the passcode generated by the calculator. The server has similar code generation function and does the comparison. If the passcode is correct, the user is authenticated. This method requires a synchronisation between the two calculators.

Authentication using software: Instead of a physical calculator the calculation function is delivered to the user as software in diskette or CD-ROM. The user installs it in his PC. Alternatively, the calculation function can be provided in a smart card but in this case the user must have a card reader and associated software. When the user visits the Bank's site, a secure communication is first established between the user's computer and the bank's server. Then, the user is asked to enter his username. The authentication is then carried out by the user's client program (browser) and the merchant's server without intervention of the user. The client software generates the passcode and sends it to the server. The server compares with the code it has generated. If they match, the user is authenticated.

All the three schemes described above although accepted by the banks because they provide sufficiently strong authentication still have weaknesses as follows:

- The two first schemes are not very user friendly since the user has to really concentrate in order to enter the numbers correctly.
- The user cannot be sure that the bank is performing the correct transaction that he wants.
- The bank on its side cannot prove that the user has requested a transaction and the latter one can deny it later on.

6. COMMERCE FOR THE MOBILE USER

6.1 Ideal mobile e-commerce system

At first glance, mobile e-commerce may appear to be identical to "fixed" e-commerce extended with mobile wireless access and the solutions used in

Web commerce, e.g. Web shopping, Web banking can be applied directly to mobile e-commerce. However, mobile e-commerce differs to "fixed" e-commerce in the following respects:

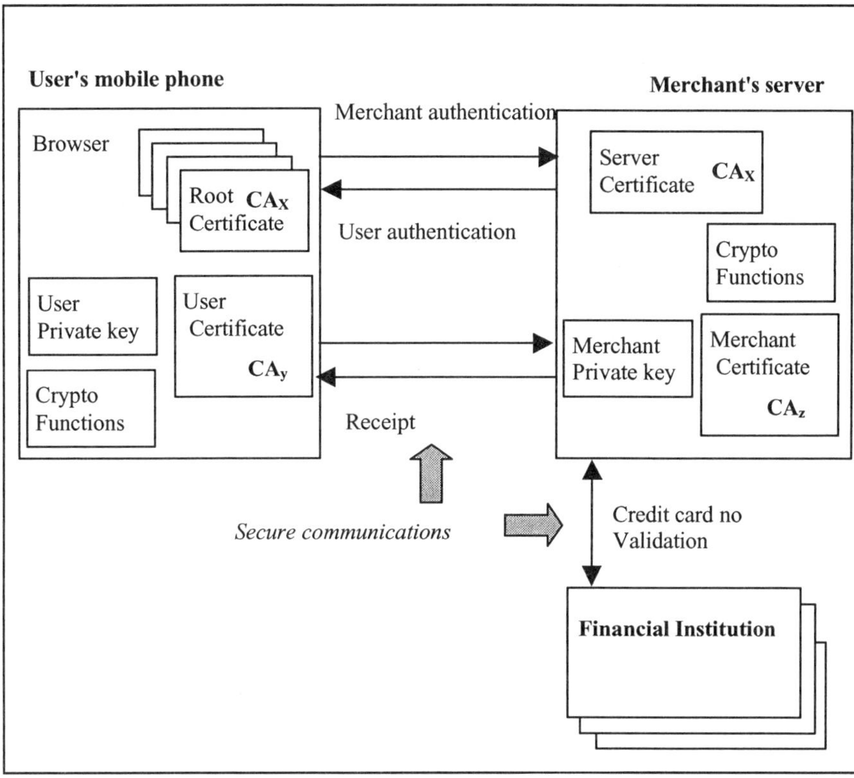

Figure 3. An ideal mobile e-commerce system

Instantaneous delivery: The mobile user is of course interested in having service like web shopping where the delivery of non-electronic goods is carried out later. But, in addition he may want to have the goods delivered to him immediately or in a short delay. For example, after paying for a Coke via his mobile phone he expects the can to run out from the Coke automate. When paying for a cinema ticket he expects to be able to collect the ticket within the same day. It is therefore necessary to have user authentication and also receipt delivery.

Micro payment: For mobile users it is also to be able to buy small things and to pay small amount of money. The fees for such payments must be small compared to the payments.

Mobile context: The mobile user in many situations must be able to operate the services with only one hand. The user may be in environments that are distracting, e.g. crowded, noisy and interactions with the e-commerce services must both simple and small in numbers. The payment scheme of Web shopping described earlier where the user has to enter his personal data and his credit card number is hence not appropriate for the mobile user. A user-friendly payment scheme is required.

An ideal mobile e-commerce as shown *Figure 3* should support the following:

- user authentication
- merchant authentication
- secure channel i.e., encrypted channel
- user friendly payment scheme supporting micro payment
- receipt delivery
- simple user interface

6.2 Limitations of the mobile phones

An ideal e-commerce system puts severe requirements that are difficult to be met by the mobile phone itself as follows:

1. It must also be equipped with a browser that has interface to the cryptographic functions.
2. It must be capable of digitally signing a message using the user private key in order to participate to the user authentication. For that, it must have public key cryptographic functions such as RSA. It must have a tamper-proof storage for storing the user's private key. It must also have enough storage for the user's certificate.
3. It must be capable of authenticating the merchant. For that, it needs to have enough storage for root certificates. It must have public key cryptographic functions.
4. It must also have symmetric cryptographic functions for the establishment of the secure channel between the mobile phone and the merchant' server.

Let us consider successively different type of mobile phones and see what capabilities they have and how to enable them to participate in mobile e-commerce.

Standard GSM phones

A GSM (Global System for Mobile communication) phone [4] [5] comprises of:

- an *ME (Mobile Equipment)* which is actually the "empty" phone with the display, keypad, microphone, speaker.

- and a *SIM (Subscriber Identity Module)* which is a removable smart card. The SIM contains *the International Mobile Subscriber Identity (IMSI)* which unambiguously identifies the subscriber. Without a valid IMSI, GSM service is not accessible. The SIM contains also the security features for subscriber authentication such as authentication algorithm (A3), subscriber authentication key (Ki), cipher key generation algorithm (A8), cipher key (Kc)

The ME is the master and initiates commands to the SIM and there is no mechanism for the SIM to initiate a communication with the ME. A standard GSM phone does not meet nay of the requirements mentioned above and is not capable to engage in mobile e-commerce.

GSM SAT enabled phones

The *SIM Application Toolkit (SAT)* provides mechanisms, which allows applications, existing in the SIM, to interact and operate with any ME supporting the specific mechanisms required by the application. A browser, the public key cryptographic functions and a user private key can be installed in the SIM. However, the SIM does not have enough storage capacity for all the certificates needed and is hence not capable of generating complete digital signature. In addition, in order to communicate with merchant's web server, the SAT phone needs assistance from an intermediary server that has similar functionality as the WAP gateway. We will not consider pure SAT phones since more powerful WAP phones have emerged.

WAP phones

The WAP phone is a mobile phone that has a WML browser and a WAP protocol stack on the ME. It is hence capable of communicating with any Web servers via the WAP gateway. The connection with the WAP gateway can be based on different bearers such as GSM circuit-switched connection, GPRS, SMS, USSD, etc.

The first version of WAP phones, called WAP 1.1 phones do not have public key cryptographic functions for digital signature. However, a combined WAP-SAT phones will both have a WML browser in the ME and public key functionality in the SIM. The only problem is the lack of the interface between the browser and the cryptographic functions on the SIM. The browser is hence not able to invoke the cryptographic functions necessary for user authentication.

In the WAP 1.2 phone, there will be a Wireless Identity Module (WIM), which incorporates both the SIM and also local memory in the ME. Public key cryptographic functions and also the user private key can both be stored in the WIM. There will also be implemented an interface, which allows the browser to communicate with the cryptographic functions. WAP 1.2 phones will be capable of generating digital signature according to the PKCS#1 standard [6], but they will not able to generate an electronic signature according to the PKCS#7 that are required in the validation process of the signature. It is possible to say that even WAP phones are not capable to participate in mobile e-commerce by themselves but they need assistance from the system.

6.3 The Mobile ePay

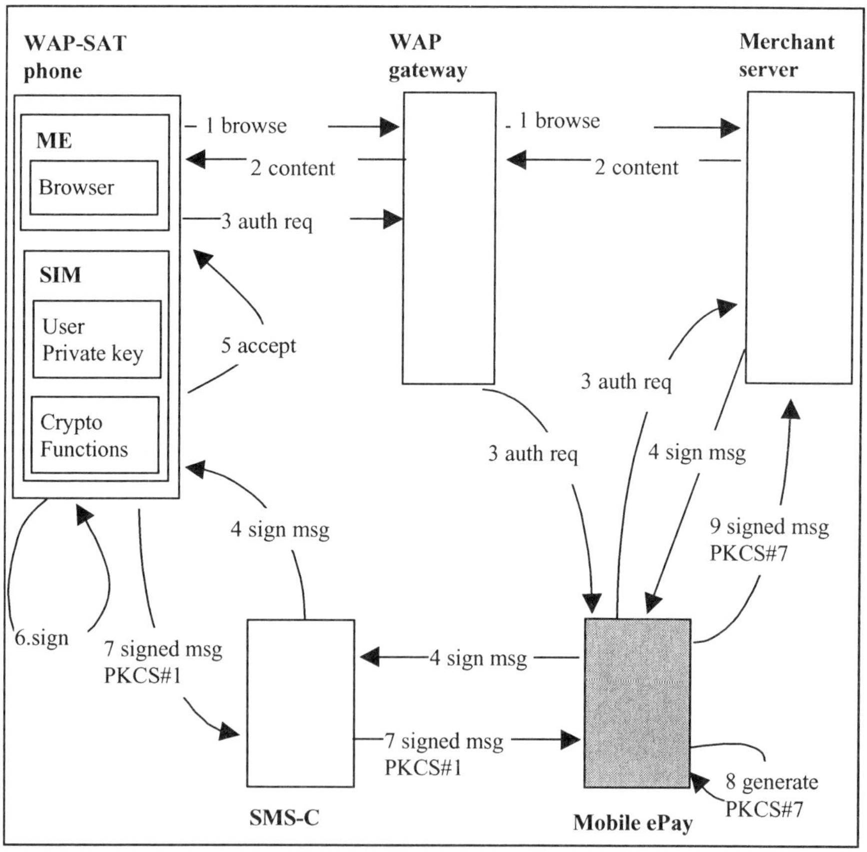

Figure 4. Mobile ePay role in the user authentication

To allow mobile phones to perform digital signature, we introduce a proxy server, called *Mobile ePay*. The Mobile ePay is responsible to perform on behalf of the mobile phones the tasks that the latter are not capable such as:

- Storing the user's certificates
- Generating electronic signature, e.g. PKCS#7 message format from digital signature, e.g. PKCS#1 format, generated by mobile phones.
- Validating of the merchant's servers

In addition to the security functions the Mobile ePay has also payment functions such as:

- Prepaid account supporting micro payment
- Interfacing with the systems of the financial institutions

To illustrate the role of the Mobile ePay in our payment system two operations namely user authentication for WAP 1.1 phones and payment from WAP 1.1 phones are described.

User authentication

The user authentication as depicted in *Figure 4* comprises of the following steps:

1. The user visits a merchant site.

2. The merchant server sends the content to the mobile phone via the WAP gateway.

3. The user wants to authenticate himself toward the merchant. The authentication request is sent to the WAP gateway, which sends to the Mobile ePay . The Mobile ePay sends it to the merchant server.

4. The merchant server generates an authentication message, e.g. a random number and sends it to the Mobile ePay, which sends to the SMS-C (Short Message Center). The SMC-C delivers it to the SIM on the mobile phone.

5. The SIM asks for permission to sign.

6. If the user accepts the SIM performs the signing, i.e. generating a digital signature in PKCI#1 format.

7. The SIM sends it back to the SMS-C, which sends it to the Mobile ePay.

8. The Mobile ePay generates an electronic signature in PKCS#7 format by using the received digital signature in PKCS#1 format.

9. The Mobile ePay sends the complete electronic signature to the merchant server.

Payment from WAP 1.1 phones

1. The user visits a merchant site.

2. The merchant server sends the content to the mobile phone via the WAP gateway.

3. The user wants to buy. The request is sent to the WAP gateway, which forwards it to the Mobile ePay. The Mobile ePay delivers it to the merchant server.

4. The merchant server sends an offer to the Mobile ePay.

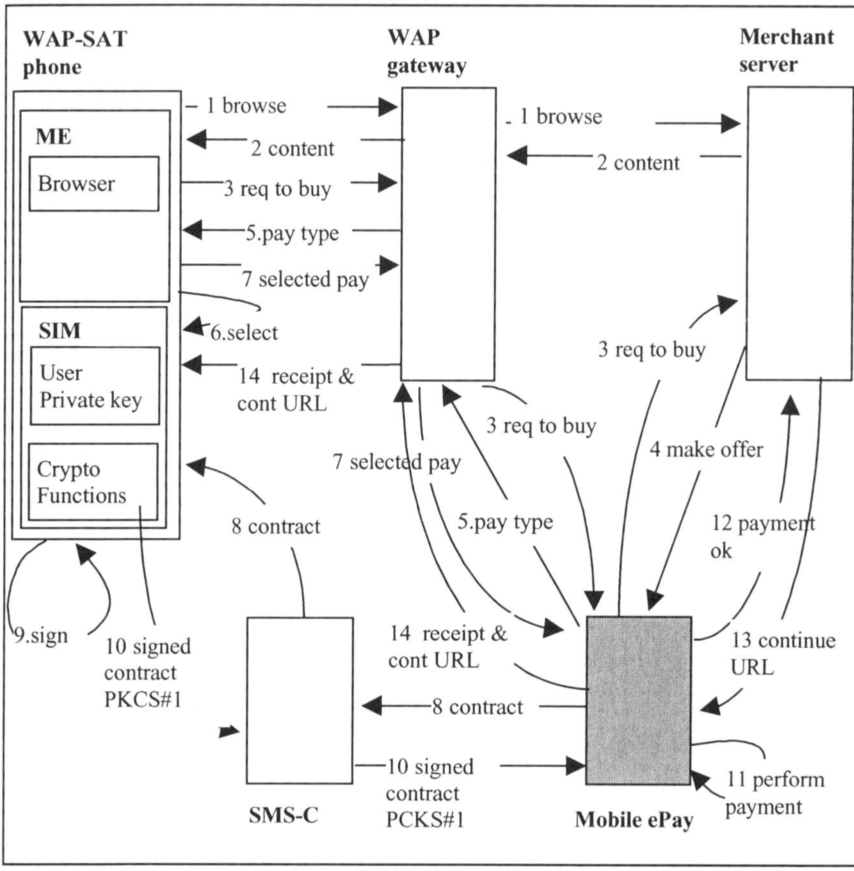

Figure 5. Payment from WAP 1.1 phones

5. The Mobile ePay sends a request for payment type to the browser via the WAP gateway

6. The user selects the payment type, e.g. prepaid account, credit cards, etc. and

7. The payment type is sent to the Mobile ePay via the WAP gateway.

8. The Mobile ePay sends the contract to the SIM via the SMS-C.

9. After asking for confirmation from the user, the SIM performs the signing

10. The SIM sends the digital signature back to the Mobile ePay via the SMS-C.

11. The Mobile ePay executes the necessary transactions according to the payment type. This may include transactions towards financial institutions in case of payment by credit card.

12. The Mobile ePay sends a confirmation to the merchant server.
13. The merchant server returns a URL for the continuation of browsing.
14. The mobile ePay generates a receipt and sends it together with the URL for continuation to the browser via the WAP gateway.

The browser can then continue with the browsing from the received URL. The shopping is hence completed.

7. CONCLUSION

In this paper a mobile e-commerce system is presented. Taking into account the physical and functional limitations that prevent mobile phones from participating to mobile e-commerce, the system introduces a proxy server that offers the necessary assistance to mobile phones. In addition to the security functions, the Mobile ePay also have payment functions such as prepaid account, interface towards financial systems. With Mobile ePay, the user can perform in a secure way any mobile e-commerce service such as doing bank transaction, buy goods or services, from mobile phones. The proposed solution is far from being perfect and quite a lot of issues remain to be done such as time stamping for electronic signature, the relation between the private public key pair and the user, i.e. how many key pair should the user have and the relation between key pair and certificates, i.e. how many certificates can be associated to a key pair

REFERENCES

[1] Visa & Master Card: SET Secure Electronic Transaction Specification - Book One: Business Description, version 1.0, May 31, 1997, http://www.setco.org/download.html/#spec

[2].Visa & Master Card: SET Secure Electronic Transaction Specification - Book Two: Programmer's Guide, version 1.0, May 31, 1997, http://www.setco.org/download.html/#spec

[3]. Visa & Master Card: SET Secure Electronic Transaction Specification - Book Three: Formal Protocol Definition, version 1.0, May 31, 1997, http://www.setco.org/download.html/#spec

[4]. ETSI: GSM 02.17 V8.0.0 Digital cellular telecommunications system (Phase 2+); Subscriber Identity Modules (SIM); Functional characteristic

[5]. ETSI: GSM 11.14 Digital cellular telecommunications system (Phase 2+); Specification of the SIM Application Toolkit for the Subscriber Identity Module - Mobile Equipment (SIM - ME) Interface

[6]. RSA Laboratories. PKCS #1: RSA Encryption Standard. Version 1.5, Nov 1993

[7]. RSA Laboratories. PKCS #7: Cryptographic Message Syntax Standard. Version 1.5, Nov 1993

[8] IEEE P1363: Standard Specifications for Public-Key Cryptography.

COMBINING AUTHENTICATION AND LIGHT-WEIGHT PAYMENT FOR ACTIVE NETWORKS

Rüdiger Weis
cryptolabs Amsterdam
convergence integrated media GmbH, Berlin, San Francisco, Amsterdam
ruedi@cryptolabs.org

Wolfgang Effelsberg
Praktische Informatik IV
University of Mannheim, 68131 Mannheim, Germany
effelsberg@pi4.informatik.uni-mannheim.de

Stefan Lucks*
Theoretische Informatik
University of Mannheim, 68131 Mannheim, Germany
lucks@th.informatik.uni-mannheim.de

Keywords: Active Networks, Payment, Authentification, Combined Schemes

Abstract Security functions are of critical importance for the acceptance of Active Networks in practice: network nodes must be protected from malicious code, and they should account for the cost of executing code; this also helps to prevent denial-of-service attacks. For the payment function code packets must carry some form of light-weight electronic cash. Cryptographic schemes can be used to solve both the security and the payment/resource management problem. In this paper we propose to *combine* cryptographic algorithms in order to solve both problems in an integrated way. Our scheme is secure, light-weight and efficient: It saves space in the packet headers, and the security is higher than that of separate algorithms for authentication and cost accounting.

*Supported by the Deutsche Forschungsgemeinschaft (DFG) grant KR 1521/3-1.

1. INTRODUCTION

Active Networks are considered to be especially useful for the rapid development of new services in a network. It is a key characteristic of Active Networks that packets carry not only data but also code. This code is written to be executed on internal nodes of the network. The internal nodes are thus expected to execute "foreign" code, i.e. code originating at remote sites. They must have means to execute authorized code only in order to prevent damage to local resources and to other network users.

Another problem with Active Networks is the fact that the execution of "foreign" code consumes local resources in the internal nodes. If there is no accounting in the internal nodes one packet stream could waste considerable processing power within the network; fairness between streams could not be enforced. Also denial-of-service attacks could not be prevented, i.e. an arriving packet could run into a code loop and block a node[1].

We propose to use a micro-payment protocol, initially developed for electronic commerce between humans, for this purpose. It turns out that the same cryptographic building blocks can be used for code authentication and for micro-payment[2]. We therefore propose to *combine the two functions* for Active Networks.

Since there are many different models for Active Networks in the literature we define the model on which our work is based in Section 2. In Section 3 we give a short overview of light-weight payment protocols. Section 4 presents a scheme to combine code error detection with the payment function. In Section 5 we introduce a slightly more complex scheme that combines code authentication with payment. Section 6 formally proves the security of our combined schemes. Section 7 discusses some implementation aspects, and Sections 8 concludes the paper.

2. OUR ACTIVE NETWORK MODEL

There are two fundamental ways of combining code and data in Active Networks:

- In the *immediate one-time execution* architecture each packet carries its own code; the code is extracted in each internal network node and executed locally. This architecture is very similar to Mobile Agent systems.

- In the *demand loading* architecture the code is initially loaded into an internal node when it is requested by a packet; it can remain there and be executed many times, typically when processing the subsequent packets of the same data stream.

In this paper we only consider the immediate one-time execution architecture where the code and data of each packet are self-contained. We call such a packet an *ANpacket*.

New ANpackets are generated by *client nodes*. Each client node is attached to a node of the active network, an *internal node*. The internal nodes are interconnected by secure links. All internal nodes are *trusted nodes*; our scheme does not provide protection of one internal node from malicious other internal nodes.

The purpose of our scheme is to

- authenticate ANpackets in the internal nodes (i.e., verify the client node they come from),
- provide a payment scheme by which the ANPackets pay for the services of the internal nodes. This payment scheme also helps to prevent denial-of-service attacks.

The ANpackets travel along an arbitrary path of internal nodes which provide code execution services. In order to distribute cryptographic keys and payment information efficiently we also assume that an efficient and reliable broadcast mechanism exists within the internal network. Our architectural model is illustrated in Figure 1.

A real-world example corresponding to our model could be a Virtual Private Network (VPN) where the internal network runs on trusted hosts interconnected by secure links. Another example would be an Internet Service Provider (ISP) where the inner network is a set of routers that use ANpackets for signalling or resource reservation purposes (but probably not in the main data path). It is typically assumed that one IP router can trust another IP router within the network of an ISP.

Without a payment function each ANpacket could execute as much code as it wishes on an internal node. This can lead to very unfair behaviour: a source node can inject a valid packet into the network that uses up most of the processing power of an internal node. Also, denial-of-service attacks are easy: a source node generates a packet with a code loop, to be executed on an internal node, and that node will become unavailable to other packets streams.

We therefore propose a resource management scheme based on micropayment. Each packet carries a chain of electronic coins. These coins pay for local services on internal nodes. When the money is used up the ANpacket is discarded by the internal node. The generation and

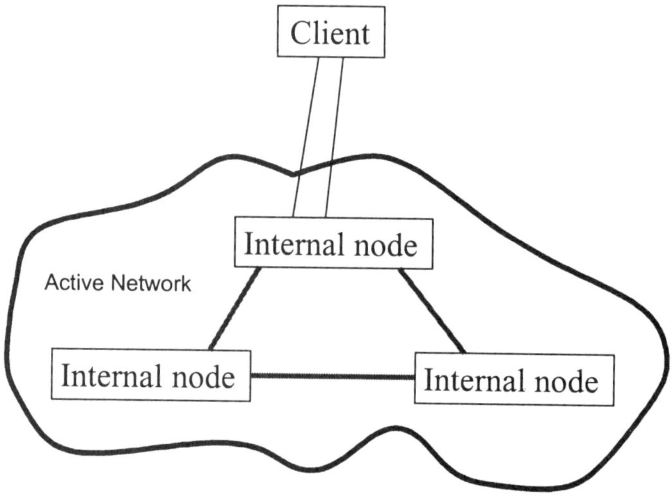

Figure 1 Architectural Model of the Active Network

consumption of coins is done with an efficient, light-weight payment protocol which we introduce in the next section.

3. A LIGHT-WEIGHT PAYMENT PROTOCOL

In the last years several light-weight payment schemes have been developed. Most of them are based on cryptographic hash functions. The main reason for this approach is the reduction of public key operations to gain better performance. As Rivest and Shamir have pointed out hash functions are roughly 1,000 times faster than RSA signature verification (with a small public exponent), and about 10,000 times faster than RSA signature generation [RiSh96].

3.1. THE PAYWORD SCHEME

An especially interesting scheme is the Payword scheme. It is fast, easy to implement and provides a high degree of security because of the use of simple operations. It is a payment scheme based on chains of hash values. It was developed by Ron Rivest and Adi Shamir [RiSh96]. Similar chains have been previously proposed by Lamport [Lamp81] and Haller in S/Key [Hall94], for access control (see also [OPIE]). The idea for such micropayments has also been independently discovered by Anderson et al. [AMS95] and Pedersen [Pede95].

With $\mathcal{H}$ we denote a cryptographically strong hash function, such as SHA-1 [FIPS180] or RIPE-MD160 [DBP96]. The important property of $\mathcal{H}$ is its *one-wayness* and *collision-resistance.* Note that the "main security" of the scheme is based on the non-invertibility of a cryptographic hash function, a weak cryptographic assumption.

Generating Paywords. In the first step the user U has to establish an account with the broker B. U creates a payword chain in reverse order by picking the last chain link w_n at random, and then computing

$$w_i = \mathcal{H}(w_{i+1}) \text{ for } i = n-1, \ldots, 1$$

We designate w_0 as the root of the payword chain. The broker B signs a digital certificate containing the broker's name, the user's name and IP-address, the user's public key, the expiration date, the root w_0 of the payword chain, the length of the payword chain and other information. This is illustrated in Figure 2.

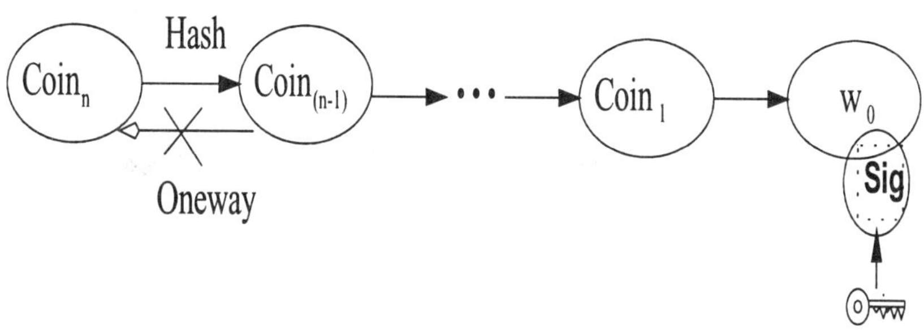

Figure 2 Hash chain

Performing a Payment. The i-th coin (for $i = 1, 2, \ldots$) from U to a vendor V consists of the pair (w_i, i). The vendor can verify the coin by checking

$$\mathcal{H}(w_i) \stackrel{?}{=} w_{i-1}.$$

So each payment requires no additional calculations by user U, and only a single hash operation by payment receiver V.

3.2. THE PAYWORD SCHEME IN THE ACTIVE NETWORK CONTEXT

The client is playing the role of the user U and the internal nodes play the role of the vendors V. The network owner might be the coin broker B. Before processing begins the payment root w_0 is broadcast to all internal nodes. The initial; distribution of "money" to the client nodes is beyond the scope of this paper.

An additional feature of our approach is that we can use the payment scheme for internal accounting in a very simple and natural manner. The Payword scheme supports variable amount payments without increasing the coin size. We can simply skip chain links to pay a higher amount, in multiples of the coin value.

Suppose that each coin is worth one cent. If U is given w_4 instead of w_1, the payment has the value of 4 cents. Now the first internal node has to perform four hash calculations to check the validity of the w_4 coin. He can "consume" one coin by using the "intermediate value" w_3 as the start of a new payment chain for the next internal node along the route (see Figure 3).

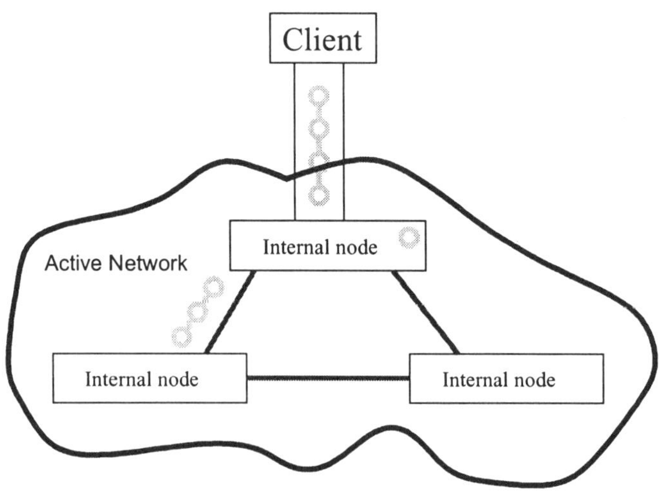

Figure 3 Payword Scheme for Active Networks

4. COMBINING ERROR DETECTION AND PAYMENT

We can check the *integrity* of an ANpacket by using a cryptographic checksum. Let $\mathcal{H}$ be a collision-resistant hash function and *coin* a coin of the payment scheme.

The payment scheme uses the boolean function $ValidateCoin$ to check the validity of a coin.

We define a function $\mathcal{HC}$ which combines both with a bitwise XOR

$$\mathcal{HC}(\cdot) \mapsto \mathcal{H}(\cdot) \oplus coin.$$

Now we can compute a "hash&coin" check field by

$$HCfield := \mathcal{HC}(ANpacket).$$

This is illustrated in Figure 4. The client node concatenates the contents

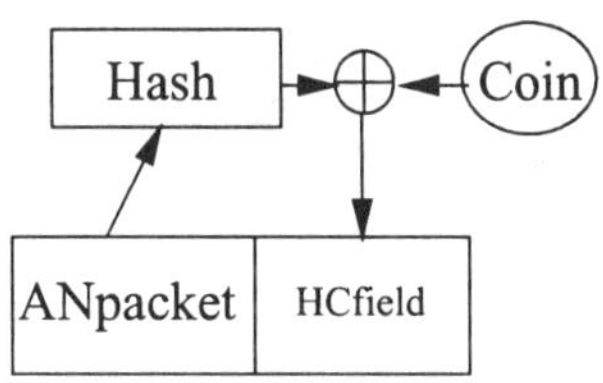

Figure 4 The Error Detection and Payment Scheme

of the packet with the HCfield

$$Send(ANpacket||HCfield)$$

An internal network node can check the validity of a packet by the $Check$ function using the $ValidateCoin$ function of the payment scheme. If we use the Payword protocol this implies only one other hash computation:

$$Check(ANpacket||HCfield) := ValidateCoin(HCfield \oplus \mathcal{H}(ANpacket))$$

If there was a modification of the ANpacket in transit, or the coin is invalid, the packet will be marked as "invalid" and discarded.

The possibility that two "false" values compensate each other can be neglected if we choose sufficient lengths for the hash code and the coins. We recommend to use $b := 160$ bit, which is the output-length of SHA-1 and RIPEMD-160.

Man-in-the-Middle Attack

Because of the fact that in our first scheme no secret information is used to calculate the checksum, the scheme can only be secure against a *passive attacker*. A passive attacker can read every protocol message, but is not able to modify, generate or destroy messages.

An active attacker can intercept an $HCfield$ and calculate the coin by using the public hash function $\mathcal{H}$:

$$HCfield \oplus \mathcal{H}(ANpacket) = Coin$$

Now he can use the coin for his own packet. This motivates us to provide a modified scheme for source authentification.

5. COMBINING AUTHENTICATION AND PAYMENT

We can check the *authenticity* of an ANpacket by using a Message Authentication Code (MAC) instead of a public hash function. This provides security against an active attacker. Note that there must be a secure channel for distributing the secret MAC key.

The Authentification and Payment Scheme is very similar to the scheme of the previous section. Let $\mathcal{MAC}$ be the Message Authentication Code and *coin* a coin of the payment scheme.

We define a function $\mathcal{AC}$ which combines the MAC and the Coin with a bitwise XOR

$$\mathcal{AC}(\cdot) := \mathcal{MAC}_K(\cdot) \oplus Coin$$

We then calculate the "Authentication&Coin" $ACfield$

$$ACfield := \mathcal{AC}(ANpacket).$$

The client node concatenates the contents of the packet with the AC field

$$Send(ANpacket||ACfield)$$

The scheme is illustrated in Figure 5. An internal node can check the validity of a packet by calculating the MAC and calling the $ValidateCoin$ function of the payment scheme:

$$Check(ACfield) := ValidateCoin(ACfield \oplus \mathrm{MAC}_K(ANpacket))$$

If there was a modification of the ANpacket in transit or the coin is invalid, the packet will be marked as "invalid" and discarded.

The possibility that two "false" values compensate each other can again be neglected if we choose a sufficient length for the MAC and the coins. We recommend to use $b := 160$ bit.

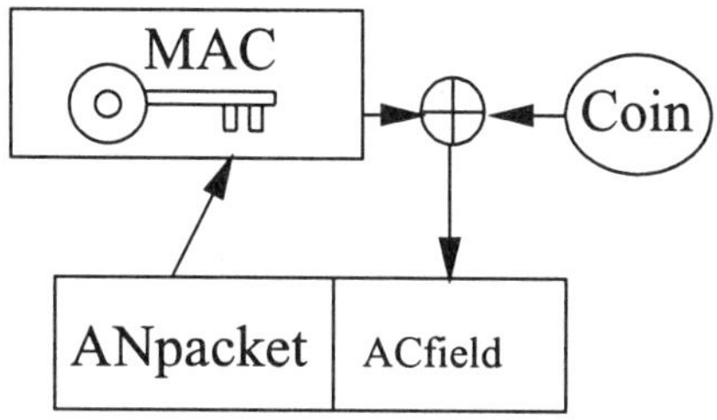

Figure 5 The Combined Authentication and Payment Scheme

6. SECURITY ANALYSIS

Our fundamental idea is to combine two cryptographic outputs using an algebraic group operation – such as the bit-wise XOR "$\oplus$" – in order to save space in the packet header. At a first look, the security of the combined scheme appears questionable. Intuitively, one would expect the combined scheme to be insecure if either of its components is insecure. In this section we give evidence that this intuition is wrong, and the combined scheme is secure if at least one of its components is secure.

In a formal model, we provide a proof of security for the combined scheme. For the sake of space, we concentrate on the combination of authentication and payment, as described in Section 5.

By "proof of security" we mean a formal proof that the combined scheme can only be broken, if *both* the authentication *and* the payment scheme can be broken. As a first step, we must define what it means if a scheme "can be broken".

Similar ideas to ours have been used in the context of cascading ciphers [MaMa93] and in the design of the RIPEMD-160 hash function [DBP96].

6.1. DEFINITIONS

Our scheme combines an authentication value, a "Message Authentication Code" (MAC), and a payment value, a "coin". We first define the security requirements for MACs and coins, then we define the security requirement for our scheme.

Adaptive Chosen Message Attack

A MAC depends on a secret key: everyone knowing the secret key K can compute the MAC

$$m = \mathcal{MAC}_K(x)$$

for the input x. On the other hand, it must be infeasible for any adversary to forge a MAC, i.e., to compute m without knowing K. While the key K is used by the legitimate users, the adversary may learn some pairs

$$(x_1, \mathcal{MAC}_K(x_1)), (x_2, \mathcal{MAC}_K(x_2)), \ldots,$$

and, in the worst case, may even be able to choose the values x_i. This is a so-called *Adaptive Chosen Message Attack*. Being quite pessimistic, we consider this type of attack to be possible. For example the adversary might modify the traffic on the link from the client to the first internal node.

In our formal model, we say that the adversary chooses x_i and learns the value $m_i = \mathcal{MAC}_K(x_i)$ from a "MAC-oracle". Then, the adversary chooses x_{i+1} and learns $m_{i+1} = \mathcal{MAC}_K(x_{i+1})$ from the oracle.

q-forgery-secure MAC Scheme

A MAC is *q-forgery-secure* if after choosing q values $x_1, \ldots, x_q$ and learning the MACs

$$m_1, \ldots, m_q \text{ with } m_i := \mathcal{MAC}_K(x_i)$$

from the oracle, it is infeasible to forge a MAC, i.e., to find a pair (x_0, m_0) with

$$m_0 = \mathcal{MAC}_K(x_0) \text{ and } x_0 \notin \{x_1, \ldots, x_q\}$$

without asking the oracle a $(q+1)$-st time.

Our payment scheme depends on sending coins c_q, c_{q-1}, ... to the receiver. A payment scheme is secure, if the adversary cannot forge coins, except when these are created by herself. In contrast to MACs, there is nothing to choose for the adversary.

q-forgery-secure Payment Scheme

A payment scheme is *q-forgery-secure* if after having learned up to q valid coins c_s, c_{s-1}, ..., c_{s-q+1}, it is infeasible for the adversary to produce another valid coin c_{s-q}. For the sake of simplicity, we assume the coin c_{i-1} to be uniquely defined by the coin c_i.

q-forgery-secure Combined Scheme

Similar to the above, we now define the security of our combined scheme for authentication and payment. We allow the adversary to choose a packet p_i, to learn the corresponding value a_i from an "AC-oracle", then to choose a packet p_{i+1}, to learn the corresponding value

a_{i+1} ... with

$$a_i = \mathcal{AC}(p_i) = \mathcal{MAC}_K(p_i) \oplus c_{s-i}.$$

Here, c_s, c_{s-1}, ... are our coins.

The scheme is *q-forgery-secure*, if after having chosen q values x_i and learned q AC-values a_i, it is infeasible for the adversary to find another pair (p_0, a_0) with

$$p_0 \notin \{p_1, \ldots, p_q\} \text{ and } a_0 = \mathcal{AC}(p_0).$$

on her own.

6.2. PROOF OF SECURITY

In this section, we give the proof advertised above.

Lemma 1: If the MAC is q-forgery-secure, then the composed scheme is q-forgery-secure.

Sketch of proof. An outline of the proof is as follows: assume the existence of an algorithm A to break the composed scheme. We describe an algorithm A* to forge MACs, using A as a "subroutine". The running time for A* is the running time for A plus a moderate amount of extra work. Essentially, A* computes the responses to A's q oracle queries:

1 A* creates $q+1$ coins c_q, ..., c_0.

2 A* runs A.

   Every time A chooses p_i and asks the AC-oracle for $a_i = \mathcal{AC}(p_i)$, A* asks the MAC-oracle for $m_i = \mathcal{MAC}_K(p_i)$ and responds with $a_i = m_i \oplus c_{q-i+1}$.

3 Finally, A produces an output (p_0, a_0) with $p_0 \notin \{p_1, \ldots, p_q\}$ and $a_0 = \mathcal{AC}(p_0)$. Then A*'s output is $m_0 = a_0 \oplus c_0$.

The value m_0 is a successfully forged MAC, i.e., $p_0 \notin \{p_1, \ldots, p_q\}$ and $m_0 = \mathcal{MAC}_K(p_0)$.
q.e.d.

Lemma 2: If the payment scheme is q-forgery-secure, then the composed scheme is q-forgery-secure.

Sketch of proof. Similarly to the proof of Lemma 1, we assume the existence of an algorithm A to break the composed scheme and describe an Algorithm A** using A as a "subroutine" to break the Payment scheme:

1 A** chooses a key K for the MAC scheme.

2 A** runs A.

Every time, A chooses p_i and asks the AC-oracle for $a_i = \mathcal{AC}(p_i)$, A** asks for the coin c_{q-i+1}, computes $m_i = \mathcal{MAC}_K(p_i)$ and responds with $a_i = m_i \oplus c_{q-i+1}$.

3 Finally, A produces an output (p_0, a_0) with $p_0 \notin \{p_1, \ldots, p_q\}$ and $a_0 = \mathcal{AC}(p_0)$. Then A** computes $m_0 = \mathcal{MAC}_K(p_0)$ and outputs $c_0 = a_0 \oplus m_0$.

The value c_0 is a successfully forged coin.
q.e.d.

Theorem: If either the MAC scheme or the payment scheme is q-forgery-secure, then the composed scheme is q-forgery-secure.

Proof: This follows directly from Lemma 1 and Lemma 2. q.e.d.

In other words, our scheme is secure in the above sense if either of its two components is secure, even if the other component is insecure.

6.3. REMARKS

In cryptography, one often requires MACs to be even stronger than forgery secure: If the adversary produces $q+1$ MAC inputs $x_1, \ldots, x_q$, and x_0 and learns the responses $m_i = \mathcal{MAC}_K(x_i)$ for $i \in \{1, \ldots, q\}$ and then learns another value m_0 which is either (a) $m_0 = \mathcal{MAC}_K(x_0)$ or a (b) random value m_0, then the MAC scheme is *q-distinguishing-secure* if it is infeasible for the adversary to distinguish between options (a) and (b).

If a MAC is distinguishing-secure, there is no way for the adversary to derive any useful information from the MAC. Similarly, we might define the distinguishing-security of payment schemes and composed authentication/payment schemes.

In this paper, we concentrate on forgery security because we believe that it meets most practical security demands for Active Networks. Note that hash-chain based payment schemes, as suggested in this paper, actually are distinguishing-*in*secure: given the coin c_i, everyone can verify $c_{i+1} = \mathcal{H}(c_i)$ and hence check a coin's authenticity.

Based on the famous work of Maurer and Massey [MaMa93], one can provide a formal framework similar to ours to prove the security of our composed schemes: if either of the two components behaves like a random function, then so does the composes scheme. But note that being random is a much stronger requirement than just being forgery secure. The formal treatment presented above depends on the much weaker assumption of forgery security and hence indicates an improved margin of safety, compared to a formal treatment based on requiring pseudorandomness.

We stress that the proof of security is still applicable if "$\oplus$" is replaced by any group operation, even a non-commutative one.

7. IMPLEMENTATION

In this section we discuss some implementation aspects. Based on the security proof, we suggest to use different realisations of the MAC scheme and the Payword scheme. First we discuss the HMAC construction. Then we take a short look at hash functions not based on MD4. Finally we suggest to use "salting" to improve the security of hash chains.

7.1. THE HMAC CONSTRUCTION

To save code and to be able to use standard cryptographic libraries, we suggest to use a Message Authentification Code (MAC) based on the same $\mathcal{H}$ as the payword chain. HMAC [BCK96] uses a cryptographic hash function as a black box:

$$\mathrm{HMAC}_K(x) = \mathcal{H}(\bar{K} \oplus \mathsf{opad}||\mathcal{H}(\bar{K} \oplus \mathsf{ipad}||x))$$

with ipad := $\mathsf{0x36}$ repeated 64 times and opad := $\mathsf{0x5C}$ repeated 64 times, $\bar{K}$ is generated by appending zeros to the end of K to create a 512 bit string.

This approach has several advantages. Cryptographic hash functions have been well studied, and are usually faster than encryption algorithms. In many countries, it is easier to export or import an authentication tool, such as a signature system, than to export or import a secure encrypting system.

Security of HMAC. In [BCK96] it was proven that HMAC provides security against collision and forging attacks with only weak assumptions on the underlying hash function.

This leaves an additional margin of security: even if some weakness in the hash function $\mathcal{H}$ (e.g. MD5) is found, the MAC based on $\mathcal{H}$ may still be secure. For example a collision of hash function means finding a collision with a fixed Initial Vector (IV) and known output. If an attacker wants to find a collisions in HMAC, she must find a collision in the underlying hash function even when the IV is random and secret, and the hash value is not explicitly known.

HMAC based on SHA-1 or RIPEMD–160 provides a 160-bit output. So even birthday attacks which need 2^{80} operations seem to be infeasible.

7.2. HASH FUNCTIONS WITH DIFFERENT DESIGN

It follows directly from the security proof in Section 6 that our scheme is secure if *one* building block is secure. So it might be a good idea to use building blocks based on *different* design. If a weakness is found in one design, we still have a security reserve. An attacker will have to break *both*. For example we can use RIPE-MD based HMAC and a hash chain for payment based on Tiger.

Tiger. Tiger [AnBi96] is a new fast iterative Hashfunction designed by Ross Anderson and Eli Biham. In contrast to MD5, RIPE-MD, SHA and HAVAL it is *not* based on MD4. Tiger uses a big 8×64-S-Box to gain a fast avalanche. Because of the extensive use of 64-bit operations Tiger is as fast as SHA-1 on 32-bit processors, and about three times faster on modern 64-bit processors. The algorithm produces a 192 bit output, which can be easily reduced to 160 or 128 bit by taking the first 160 respectively 128 bit from the 192 bit output.

7.3. SALTING HASH CHAINS

Salting is an inexpensive way to make dictionary attacks much more difficult. Simply replace $\mathcal{H}(\cdot)$ by

$$\mathcal{H}_s(\cdot) = \mathcal{H}(s||\cdot),$$

where s is a "salt" (random value) which can be specified in the commitment [RiSh96].

If we use a standard iterative hash function, the input (160 bit in our case) is padded to a longer bit string (typically $n \cdot 512$ bit). So we have no relevant performance disadvantage.

8. CONCLUSION

In Active Networks the security of packets and of internal network nodes is a vital necessity. Also, managing the active nodes' resources in the presence of unpredictable incoming code packets is quite difficult. Micropayment schemes help to manage node resources effectively and in a fair manner.

In our schemes, the same header field is used for both authentication and payment purposes. This saves space in the packet header. We have presented a scheme which combined bit error detection with micropayment and a scheme which combined authentication with micropayment.

We also gave a formal justification that our schemes are sound, i.e., secure if the underlying building blocks are secure. Our proof is very strong: even if one of our two components is broken, the other one only has to satisfy rather weak security requirements.

9. ACKNOWLEDGEMENTS

The authors want to thank the anonymous reviewers and Martin Mauve for useful hints to improve the paper.

Notes

1. Several authors have proposed the use of light-weight accounting schemes to control resource consumption; see for example [Tsch97].

2. Researchers at [IBM] have proposed a protocol family called KryptoKnight that is also based on heavy use of hash functions in order to avoid dedicated encryption functions for authentification and key distribution [BGH+92, MTVZ92, BGH+95].

References

[AnBi96] Anderson, R.,Biham, E., "Tiger: A Fast New Hash Function", Fast Software Encryption 3, LNCS 1039, 1996, pp. 89-98.

[AMS95] Anderson, R., Manifavas, H., Sutherland, C., "A practical electronic cash system", 1995.

[BCK96] Bellare, M., Canetti, R., Krawczyk, H, "Keying hash functions for message authentication", Advances in Cryptology - Crypto 96 Proceedings, Springer, 1996.

[BETT98] Banchs, A., Effelsberg, W., Tschudin, C., Turau, V.: "Active Multicasting of Multimedia Streams", Proc. IEEE Local Computer Networks Conference LCN'98, Lowell, MA, October 1998, pp. 150-159

[BGH+92] Bird, R., Gopal, I., Herzberg, A., Janson, P., Kutten, S., Molva, R., Yung, M., "Systematic Design of a Family of Two-Party Athentification Protocols", CRYPTO'91, Springer LNCS, 1992, pp. 44–61.

[BGH+95] Bird, R., Gopal, I., Herzberg, A., Janson, P., Kutten, S., Molva, R., Yung, M., "The KryptoKnight family of light-weight protocols for authentication and key distribution", IEEE/ACM Trans. Networking 3, 1, Feb. 1995, pp. 31 - 41.

[DBP96] Dobbertin, H., Bosselaers, A., Preneel, B., "RIPEMD–160, a strengthened version of RIPEMD", Proc. Fast Software Encryption (ed. D. Gollmann), LNCS 1039, Springer, 1996, pp. 71–82.

[FIPS180] NIST, "Secure Hash Standard", Washington D.C., April 1995.

[Hall94] Haller, N., "The S/KEY One–Time Password System", Proc. of the ISOC Symposium on Network and Distributed System Security, San Diego, CA, February 1994.

[HSW95] Hauser, R., Steiner, M., Waidner, M., "Micro-Payments based on iKP", December 17, 1995.

[IBM] IBM Applied Computer Science, KryptoKnight http://www.zurich.ibm.com/Technology/Security/extern/kryptoknight/

[Lamp81] Lamport, L., "Password Authentication with Insecure Communication", Communications of the ACM 24(11), November 1981.

[MaMa93] Maurer, U., Massey, J., "Cascade ciphers: the importance of being first" Journal of Cryptology. Vol. 6. Nr. 1., 1993, pp 89-105.

[MTVZ92] Molva, R., Tsudik, G., Van Herrweghen, E., Zati, S., "KryptoKnight Authentification and Key Distribution System", Europeam Symposium on Research in Computer Security (ESORICS'92), Toulouse, 1992.

[OPIE] One-time Password in everything, US Naval Research Laboratory, ftp://ftp.nrl.navy.mil/pub/security/opie

[Pede95] Pedersen, T., "Electronic payments of small amounts", Technical Report DAIMI PB-495, Aarhus University, Computer Science Department, August 1995.

[RiSh96] Rivest, R., Shamir, A., "Payword and Micromint", Security Protocols, Springer LNCS 1189, 1997, pp. 1-18. http://theory.lcs.mit.edu/~rivest/RivestShamir-mpay.ps

[Tsch97] Tschudin, C., "Funny Money Arbitrage for Mobile Code", Dartmouth Workshop on Transportable Agents, Extended Abstract, August 1997. http://www.icsi.berkeley.edu/~tschudin/dart97.txt

[WGT98] Wetherall, D., Guttag, J., Tennenhouse, D., "ANTS: A Toolkit for Building and Dynamically Developing Network Protocols", Proc. OpenArch 98, San Francisco, April 1998, pp 117-129. http://www.tns.lcs.mit.edu/publications/openarch98.html

Part Nine

Network Intelligence

Quality of Services for ISP Networks

Qi Guan
SIEMENS AG Austria
Siemensstrasse 88-92
A-1210, Vienna, Austria

Key words: Quality of Service, QoS, ISP Internet, Backbone, VoIP

Abstract: Quality of Service (QoS) is a basic functionality that the Internet needs to be able to cope with when the applications increased in number and varied in use. Internet QoS issues concerning the Internet involves many different areas: end user QoS, ISP network QoS, backbone QoS. Backbone QoS is the problem concerning traffic engineering and bandwidth. ISP network QoS is one of the major problems in the Internet, In this paper, we would like to present an architecture model for QoS enabled ISP networks. This model is based on differentiated services and is connected to QoS or non-QoS enabled backbones for IP upstream and downstream QoS traffic. At the end of paper, an example of VoIP application is introduced.

1. INTRODUCTION

Until now, two separate worlds have existed for telecommunication services and networks – one for voice and one for data.

The voice world is that of switched PSTN / ISDN networks, which generally provide real-time services and features such as call diversions, conference and display services. Switched PSTN / ISDN networks guarantee quality and reliability in the form of constant availability of voice services for their customers. As a result thereof, this quality of service has made voice networks extremely profitable.

The other world is that of low-cost and efficient data services such as e-mail, file transfer and access to 'information' resources of the World Wide Web (WWW) provided by packet-based networks, mainly the Internet. These data networks are based on commercial servers and widespread programming environments. They offer a degree of flexibility which allows the short term creation of applications.

Using slogans like "The convergence of the network" and "Everything over IP", more and more traditional telco services (e.g.: telephony, video conference, radio/TV broadcasting, etc.) are being adapted for Internet usage. This has caused an increase in Internet traffic as the number of users and applications that are constantly being introduced to the Internet.

In addition and due to the multitudes of new applications the character of Internet traffic has also been and still is in the process of changing to meet demands. Some of the new breed of Internet applications, for example IP-telephony, have very strict timing requirements. The end-to-end latency (phone-to-phone) determines whether the conversation sounds interactive or not. A delay of 100 msec is the threshold for phone companies. A delay larger the 200 msec is quite noticeable and will usually cause a switch to half-duplex conversation.

The latency of an Internet transmission depends on:
- network congestion,
- number of routers involved,
- type of prioritisation,
- delay of voice encoding, and
- packetisation.

In order to be able to ensure acceptable IP telephone services the Internet must provide enough bandwidth, low delay. Generally speaking, the Internet must offer the same QoS as conventional phone companies. Here QoS means that a network element (e.g. an application, host or router) to have some levels of assurance that its traffic and service requirements can be satisfied.

Satisfying demands for different types of applications requires that different more flexible methods be used for handling network traffic. Internet applications generate traffic at varying rates and generally require that the network is able to carry traffic at the rate they generated. In addition, applications are more or less tolerant of traffic delays in the network and of variations in traffic delays. Certain applications can tolerate some degree of traffic loss while others can not. For example, telephone service requires no significant delay and no jitter, data service requires no loss of data.

To be able to guarantee quality in Internet telephony it is absolutely necessary to achieve a so-called end-to-end QoS. This is, however, not always the case. The Internet is composed of many different types of networks (logical or physical) and operated by many independent carriers. For example, telcos provide twisted-pair copper lines to end-users as a means of accessing the network. Here the user can either dial-up a modem for low-band connections or use xDSL for broadband connections to its ISP. The Internet Service Providers (ISP) then provide the Internet services to their users, as well as the connection to the Internet backbones. The Internet backbones are again operated by many different carriers or organisations as wholesalers in different countries.

End-to-end QoS in Internet can be divided into various topics, for example:

- End users QoS
- ISP network QoS
- Internet backbone QoS

QoS problems have their own characteristics depending on the demands they have to meet (e.g.: end user, ISP and backbones). Each of these demands must be solved by special technologies. For example, QoS for Internet backbones is more or less a problem of bandwidth and traffic engineering. If we can provide enough bandwidth in the backbone, QoS is no longer a problem. However, in the ISP network, the connections of end users are concentrated and therefore traffic congestion takes place here. This is the main problems in the ISP network concerning QoS.

In this paper we present an architecture model for QoS for ISP networks. In the following section some QoS technologies for the Internet are described. Section 3 describes QoS applied to ISP networks. An example is shown in section 4.

2. QUALITY OF SERVICE TECHNOLOGIES

This section describes some technologies concerning QoS for IP services. The *Internet Engineering Task Force* (IETF) has proposed many service models and mechanisms designed to meet the demands for Internet QoS.

Among them are the *Integrated Services/RSVP* model, the *Differentiated Services* (DiffServ) model and the *Multi-Protocol Label Switching* (MPLS).

The ATM forum has also proposed ATM QoS services, for example, CBR (Constant Bit Rate), UBR (Unspecified Bit Rate),VBR (Variable Bit Rate) and ABR (Available Bit Rate) etc. ATM can be used in access networks, (e.g. ADSL). However, the applications involved in ATM network are mainly IP rather than ATM. ATM QoS can only be partially used as an Internet QoS. ATM is more suitable for traffic engineering on Internet backbones.

2.1 Integrated Services/RSVP

The integrated services model is characterised by resource reservation. Real-time applications must first set up paths and reserve resources, before data can be transmitted; RSVP (Resource Reservation Setup Protocol) is used for this purpose.

The sender sends a PATH message to the receiver specifying the characteristics of the traffic. Each intermediate router along the path then forwards the PATH message to the next hop determined by the routing algorithm. Upon receiving a PATH message, the receiver responds with a RESV message to request resources for the flow. Each intermediate router along the path can reject or accept the request carried in the RESV message. If a request is rejected, the router will send an error message to the receiver and the signaling process will be terminated. If the request is accepted, resources will be reserved for the flow and the related flow state information will be installed in the router.

RSVP is intended to provide the closest thing to circuit emulation on IP networks. RSVP is the most complex of all the QoS technologies for applications (hosts) and for network elements (routers and switches). As a result, it also represents the biggest departure from standard "best-effort" IP service and provides the highest level of QoS in terms of service guarantees, granularity of resource allocation and detail of feedback to QoS-enabled applications and users.

2.2 Differentiated Services

The differentiated services approach of providing quality of service in networks employs a small, well-defined set of building blocks from which a variety of aggregate behaviors may be built.

In DiffServ, packets are classified and marked in DSCP (DiffServ code poin, a field in the header of IP packets) at the network ingress routers to create several packet classes. DSCP is placed in the IPv4 TOS octet or the IPv6 traffic class octet. Packets having different classes receive different services, allowing delay indication and packet drop preferences. Thus, the IP packets can be mapped to several QoS classes which in turn will be handled in different queues within the network elements such as routers and switches.

Using different *classification*, *policing*, *shaping* rules and *scheduling* algorithms in the routers, many services can be provided, for example,

1) *Premium Service* (PS) with its scheduling priority for applications requiring low delay and low jitter,
2) *Assured Service (AS)* with high buffer priority for applications requiring reliable but not fixed delay bounds;
3) *Best Effort Service*, which is the same as today's internet service.

Note that the differentiated services model only defines the DS fields. It is the ISPs' responsibility to decide what services must be provided.

The main advantages of differentiated services are the simplicity and flexibility. The simplicity makes it possible for DiffServ networks to interwork easily with other QoS enabled networks. This is very important in a open environment like the Internet.

2.3 MPLS

MPLS (Multi-Protocol Label Switching) is a packet-forwarding scheme. Packets are assigned a label at the ingress of a MPLS-capable domain. Subsequent classification, forwarding, and services for the packets are based on the labels.

MPLS is similar to DiffServ in some respects, as it also marks traffic at ingress boundaries in a network, and unmarks it at the egress points. But unlike DiffServ, which uses the marking to determine priority within a router, MPLS markings (20-bit labels) are primarily designed to determine the next router hop. MPLS is protocol-independent (i.e., "multi-protocol"), so it can be used with network protocols other than IP (like IPX,ATM, PPP or Frame-Relay) or directly via data-link layers as well.

MPLS is more a "traffic engineering" protocol than a QoS protocol. MPLS routing is used to establish "fixed bandwidth pipes" analogous to ATM or frame relay virtual circuits. The difference is arguable since the end-result is service improvement and increased service diversity with more flexible, policy-based network management control, all of which the other QoS protocols also provide.

3. QUALITY OF SERVICE APPLIED FOR ISP NETWORK

As mentioned in section 1, QoS problems may occur at the end user, on ISP networks or on the Internet backbones. In this section, we will take a more detailed look at the QoS in ISP networks.

The ISP network, on the one side, can be seen as being connected to the end users. This interface is called UNI (User Network Interface). On the other side, it is connected to the Internet backbones and is called SNI (Service Node Interface).

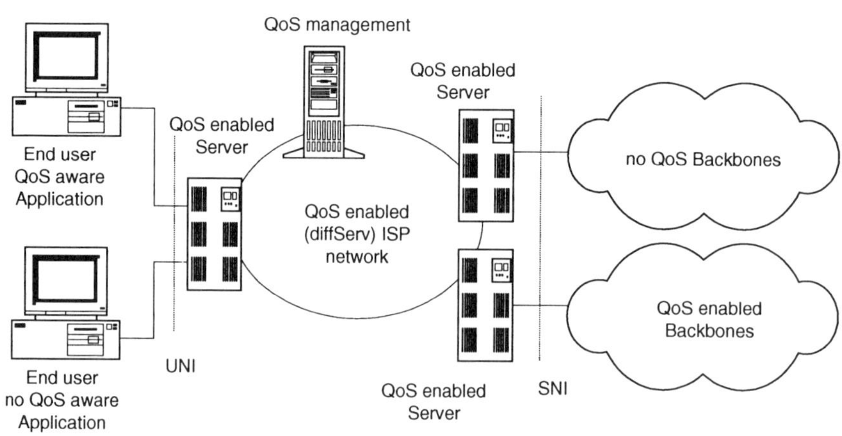

Figure 1. A QoS enabled ISP network

UNI can either be a low-speed access such as a dial-up or ISDN or a high-speed access such as xDSL and coaxial cables. SNI is usually a high-speed connection such as SDH, ATM, etc. Figure 1 shows our configuration for QoS enabled ISP networks.

The QoS enabled ISP is based on DiffServ. The ISP network is defined as a DiffServ domain. It has DiffServ boundary nodes on the UNI and SNI sides. These boundary nodes act as both ingress and egress nodes.

3.1 QoS management

QoS management is used to control traffic within the QoS enabled ISP network. It administer users profile, the load status of each ingress/egress node and applications flow. Based on these information QoS management may cluster (group) users into different service level, assign the application flow different priorities. In order to work with QoS management, the ISP network has some kind of "awareness" functions like

- User awareness,
- Traffic awareness, and
- Application awareness

User awareness means that the ISP network should cluster different user groups. A business user may, for example, be provided with a high level of service, regardless of ISP traffic load.

Traffic awareness is based on the analysis of IP flow. A IP flow can be checked by its IP header – IP address, its TCP/UDP header – port number and SYN/FIN bit. QoS enabled servers on UNI/SNI side may than make intelligent decision which kind service the IP flow can observe: with low delay or no loss of data.

Application awareness is the way to classify applications. A HTTP application may involve real-time or non-real-time applications. The traffic awareness cannot differentiate these applications. A higher layer knowledge is necessary here to optimise the usage of network resource.

3.2 Traffic control on UNI side

For low-speed access via dialup modems or ISDN connections, QoS is necessary for both bandwidth and delay control.

For high-speed access via xDSL or coaxial cables, it seems the QoS would not be necessary if there is enough bandwidth. But as usual there are congestion points, for example, the connections to backbones, which require QoS to be able to decide what traffic gets first resources at these points.

We use DiffServ servers on the UNI side and provide QoS for the end user.

According to SLA (Service Level Agreement), the end user sends application traffic into the ISP DiffServ network through the ingress node on UNI side. Each transmitted packet will be marked with its appropriate DSCP. Interior nodes within the DiffServ domain use the DSCP to classify packets and apply specific queuing or scheduling behavior (known as a *per-hop behavior* or *PHB*) based on the results of the classification.

A real-time application may, for example, apply the *expedited-forwarding* (or EF) PHB. EF PHB is defined to assure that packets are transmitted from the ingress node on the UNI side to the egress node on SNI side (at some limited rate) with very low latency.

The QoS enabled application must also register itself to the QoS management according to SLA together with other required information.

3.3 Traffic control on SNI side

An ISP network can be connected to QoS or to non-QoS enabled Internet backbones. A QoS enabled backbone with a QoS enabled ISP network provides end-to-end QoS. A non-QoS enabled backbone may provide a certain kind of QoS if there is enough bandwidth and traffic engineering in the backbone.

3.3.1 Connected to QoS enabled backbone

A QoS enabled ISP network connected to a QoS enabled backbone may act as two DiffServ domains. The DiffServ egress node may perform traffic conditioning functions on traffic forwarded to a directly connected peering domain, depending on the details of the TCA (Traffic Conditioning Agreement) between the ISP and backbone DiffServ domains.

The ISP and backbone DiffServ domains may support different PHB groups internally and different code points to the PHB mappings. However, to be able to permit services which span across the domains, the peering DiffServ domains must each establish a peering SLA which defines (either explicitly or implicitly) a TCA which specifies how transit traffic moving from one DiffServ domain to another is conditioned at the boundary between the two DiffServ domains.

Traffic conditioning performs metering, shaping, policing and/or re-marking to ensure that the traffic entering the DiffServ domain conforms to the rules specified in the TCA, in accordance with the domain's service provisioning policy. The extent of traffic conditioning required is dependent on the specifics of the service, and may range from simple code point re-marking to complex policing and shaping operations.

When packets exit the traffic conditioner of a DiffServ boundary node must set the DS code point of each packet to an appropriate value.

3.3.2 Connected to non QoS Backbones

Packets in a non DiffServ backbone have no settings in their DSCP. When the packets come to the ingress node of an ISP at the SNI, these ingress nodes must perform not only all of the necessary traffic conditioning functions on the incoming traffic but also all of the necessary traffic classifications.

Since there is no default setting in the DSCP of incoming packets, traffic classifications are done by the " awareness" functions in ingress nodes and QoS management. Does the traffic belong to a business user ? Is the traffic a http access or large data download? Based on the these analysis and services status on UNI side, traffic conditioning is performed on incoming packets. The DSCP of packets are marked with correspondent service classes.

4. AN EXAMPLE

This section shows an example how to solve the QoS problem related to Voice over IP (VoIP). Our VoIP application provides telephony service between a PC in ISP and a PSTN user. A VoIP gateway is responsible for converting PSTN voice signals into voice IP packets and then sends them to the diverted VoIP terminal (PC) or vice versa. Our VoIP gateway also provides call signalling for PSTN such as SS7 and for the Internet such as H.323 or SIP.

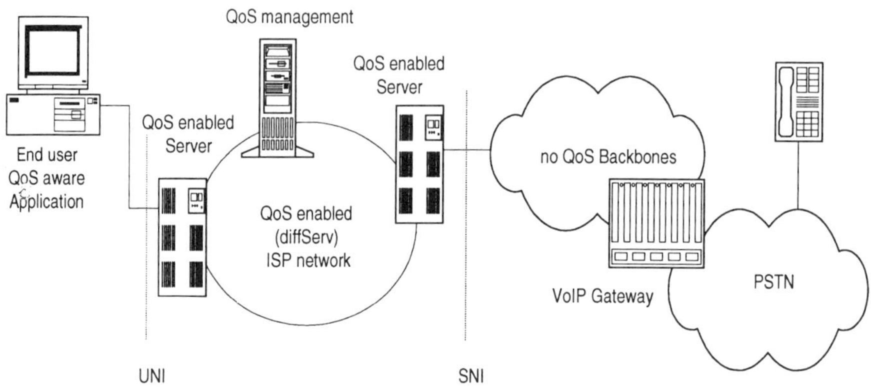

Figure 2. QoS enabled ISP and VoIP Applications

Internet users having a VoIP terminal on their PC can dial an E.164 (a usual telephone number system) code to call a PSTN user by means of Internet telephony signaling H.323 / SIP. The signaling process is like building a VoIP call connection to the nearest VoIP gateway from a called PSTN user. The gateway then sends SS7 signaling to the PSTN user. The PSTN user may pick up the phone and the call connection is established between the caller (the Internet user), and called party (the PSTN user). Up to this time no real-time packet was needed. After the connection is built-up the Internet user may register himself to the QoS management with information such as the IP-address of the VoIP terminal, the port number of the UDP packets (voice packets) and which service class applied to the real-time application.

The VoIP terminal sends the marked voice packet to the ingress node of the ISP. These packets apply the *expedited-forwarding* (or EF) PHB and are transmitted from the ingress node on the UNI side to the egress node on the SNI side with a very low latency. There may be many IP packets at egress node and that is where the congestion usually takes place. Only VoIP packets with a high service class are first transmitted with a high priority through the egress node. When the packets exit the ISP to a non-QoS backbone the DS code point of each packet must be reset.

The PSTN user then sends the voice signal to the VoIP gateway. The gateway converts the voice signal into voice UDP packets. The UDP packets have an IP-address and port number in their header. When the packets arrive at the ingress node of the ISP on SNI side, the node may control the UDP header and consult the QoS management. If these packets are registered as being part of a real-time application in QoS management, the *expedited-forwarding* (or EF) PHB is applied to them. The packets having a higher service class as compared to other lower service classes get the resource of ingress node first and are transmitted from SNI side to the egress node on the UNI side with a high priority and a very low latency.

5. RELATED WORKS

As mentioned above, IETF has proposed many service models and mechanisms designed to meet the demands for Internet QoS, e.g. RSVP, DiffServ and MPLS etc. Usually RSVP is used in ISP networks, while DiffServ and MPLS are used in Internet backbones. While our QoS enabled ISP network use DiffServ and can be connected to either QoS or non-QoS backbones.

Qbone of Internet2 is an interdomain testbed for DiffServ that seeks to provide the higher-education community with end-to-end services in support of emerging advanced networked applications. In Qbone each network participating is considered a "DS domain". The union of these networks - the QBone itself – is considered a "DS region". QBone participants cooperate to provide one or more interdomain services besides the default, traditional best effort IP service model. Virtual Leased Line Service in Qbone is such kind services as defined in DiffServ: for example "Premium Service". Every QBone DS domain supports the expedited forwarding (EF) per-hop behavior (PHB) [EF] and configure its traffic classifiers and conditioners (meters, markers, shapers, and droppers) to provide a VLL service to EF aggregates. As far as connection to QoS backbone concerned, our QoS enabled ISP network is the same as a DS domain in Qbone. But in Qbone it does not address any problem of connection to non-QoS backbone.

In the 46th IETF Conference an new architecture called DCS (Distributed Call Signalling) was introduced. DCS analyse the use of SIP in a cable environment and is a major push for IP-Telephony. QoS topic in DCS is RSVP and is tight coupling with call signalling. All the problems addressed in DCS are in cable environment. It is than easy to implement End-to-End QoS for VoIP in the domains where RSVP involved. The problems which we discussed are the interdomain VoIP applications. In such situation we must provide QoS for IP-Telephony even when our ISP connected to non-QoS backbones.

6. CONCLUSIONS

The Internet needs QoS for various applications. Internet QoS comprises various topics, i.e. end user QoS, ISP network QoS, backbone QoS. Backbone QoS is more or less a problem of traffic engineering and bandwidth. MLPS and ATM are more suitable for backbone QoS. ISP network QoS is one of major problems concerning to Internet. We presents an architecture model for QoS enabled ISP networks. This model is based on DiffServ and is connected to QoS or non-QoS enabled backbones for upstream and downstream QoS traffic.

7. ACKNOWLEDGMENTS

The author would like to acknowledge Josef Schmitt, Dr.Reda Reda, Wolfgang Aichberger, Albert Nill, Jean Kremlicka, and Josef Hosp for their useful comments, suggestions and improvements to this paper.

8. ABBREVIATIONS

PSTN	Public Switched Telephone Network
SIP	Session Initiation Protocol
VoIP	Voice over IP
QoS	Quality of Service
ISP	Internet Service Provider
CBR	Constant Bit Rate
UBR	Unspecified Bit Rate
VBR	Variable Bit Rate
ABR	Available Bit Rate
ADSL	Digital Subscriber Line
RSVP	Resource Reservation Setup Protocol
DiffServ	Differentiated Services
MPLS	Multi-Protocol Label Switching
SDCP	DiffServ codepoint
PHB	Per-Hop Behaviour
SLA	Service Level Agreement
TCA	Traffic Conditioning Agreement
DCS	Distributed Call Signalling

9. REFERENCES

[1] Qi Guan, "Voice Services for PSTN and IP Networks", Intelligence in Networks, SmartNet'99, November, 1999.

[2] IETF "Integrated Services" working group.
See http://www.ietf.org/html.charters/intserv-charter.html

[3] IETF "Resource Reservation Setup Protocol " working group.
See http:// www.ietf.org/ietf/html.charters/rsvp-charter.html

[4] R. Braden, L. Zhang, S. Berson, S. Herzog, S. Jamin, "Resource ReSerVation Protocol (RSVP) – Version 1 Functional Specification", RFC 2205, September 1997.

[5] IETF "Differentiated Services".
See http:// www.ietf.org/html.charters/diffserv-charter.html

[6] S. Blake, D. Black, M. Carlson, E. Davies, Z. Wang, W. Weiss, "An Architecture for Differentiated Services", RFC 2475, December 1998

[7] IETF "Multiprotocol Label Switching" working group.
See http://www.ietf.org/html.charters/mpls-chaeter.html

[8] QoS-Forum, "White Paper –QoS protocols & architectures", July 1999.
See http://www.qosforum.com

[9] QoS-Forum, "White Paper – The Need for QoS". July 1999.
See http://www.qosforum.com

[10] Microsoft Corporation. "White Paper – An Overview of QoS", 1999

[11] Xipeng Xiao and Lionel M. Ni, "Internet QoS: the Big Picture", Department of Computer Science, Michigan State University, 1999.

[12] Ben Teitelbaum, "Qbone Architecture(v.10)", 1999.
See http://www.internet2.edu/qos/wg/papers/qbArch/1.0/draft-i2-qbone-arch-1.0.html

[13] DCS - Distributed Call Signalling, 1999.
See http://www.packetcable.com/packetcable_specs.html

An Intelligent Network Architecture for Internet-PSTN Service Interworking

Menelaos K. Perdikeas, Iakovos S. Venieris
National Technical University of Athens, Athens, Greece.

Key words: Intelligent Network, Service Interworking, Distributed Object Technology

Abstract: This paper examines Intelligent Network (IN) approaches to interworking between the PSTN and Internet. It notes that until now, convergence between Internet and PSTN has taken place mainly at the transport and (less so) signalling layers. Even there however, the integration is not seamless enough to guarantee the ability to use the same IN superstructure over both constituents. This is an important concern given the investments that have been made in IN technology and its ability to introduce new services expeditiously. Further, the Internet world is more service-rich than PSTN and has a completely different service model. Telephony services are but a small fraction of the services that Internet can offer. Can IN retain its validity when examining provision of services other than telephony? How can IN concepts be put to use to allow interworking between, for instance, the web and the phone? In this paper we are describing an architecture that makes use of distributed object technology and IN principles in order to provide a comprehensive framework for service interworking.

1. INTRODUCTION

To date, convergence between the Internet and switched networks has for the most part taken place at the transport layer. This if for instance the case in dial-up connections when the local telephony access network is used to provide the missing last mile to connect the Internet Service Provider with the subscriber homes. This would also be the case when telephony service providers are routing calls through an IP network to reduce costs. In both situations only the lowermost layers are replaced and so convergence can be viewed less portentously as an instance of the encapsulating ability of layered protocol stacks. Where facilities of the one network are used by the other this is always on a rudimentary level. There are also some instances of convergence in the signalling layer for the particular class of telephony-like services. However, with the exception of the PINT protocol [1] for 'click-to' services there is no instance of service-level interworking between these two networks.

That the lower layers are interchangeable and accommodating their usage for the purposes of a different protocol stack should be attributed to the fact that these layers are always more generic and have more in common. Not so in layers above the data link one where differences between IP's connectionless mode and the (virtual) circuit switching Public Switched Telephone Network (PSTN) are irreconcilable. Differences in purposes and orientation are largely to blame for this disparity.

The PSTN (a term which in this paper will be used as encompassing also the technologies that have enhanced it like ISDN or ATM) is oriented towards the provision of the telephony service and other related services (like, more recently, video-telephony or video-on-demand). All these services share the concept of a call after which a (uni- or bi-directional) bit stream can be established between the parties participating in the call. The class of telephony-like services can be defined more narrowly as one comprising those services who have a call state model compatible to that of the Basic Call State Model (BCSM) as defined in the Intelligent Network (IN) specifications. Since IN exerted much effort to make the underlying network architecture service independent we can rationally expect to find in the BCSM a quite parsimonious service model that abstract as much as possible and identifies only what is intrinsic to the services provided by switched networks. So for the remainder of this paper we will use the term 'telephony class of services' to mean all services whose behaviour can be described in terms of a BCSM.

The Internet on the other hand, before the standardisation of popular application level protocols like the Hyper Text Transfer Protocol or the Simple Mail Transfer Protocol, was just a generic packet switching network used to interconnect mainframe computers. These services do not naturally lend themselves to description using a BCSM. Nevertheless once these services became available and commanded a critical mass of market attention, it was inevitable that demand would start to appear for two distinct abilities: (a) to access a network's service using the facilities of another and (b) to having discrete services coming from different networks work together for the needs of some envisaged solution to a real problem.

This paper is structured as follows. In the next section we will examine signalling interworking between PSTN and the Internet. Signalling interworking satisfies demand (a) and is shown to be narrower in scope and a sort of preamble for the support of a more thorough interworking at the service level which satisfies demand (b) and is covered next. We next introduce a layered service architecture that accommodates under its umbrella both signalling and service-level integration and interworking between PSTN and Internet. The architecture draws heavily from IN principles and can be used in a complementary manner with the specifications produced by the Parlay group [2].

2. SIGNALLING INTERWORKING BETWEEN PSTN AND INTERNET

The basic concern when seeking to integrate telephony in PSTN and Internet is that signalling emitted from PSTN terminals should transparently find its way to Internet nodes hosting Voice over IP (VoIP) applications and vice versa

The approach is usually to terminate some of the lower signalling layers and convey the upper layers transparently. Therefore the key point of comparison between different strategies is the level at which they terminate PSTN signalling. The decision of what types of Internet signalling components should be supported and of what the overall signalling architecture would look like largely depend on this subtle point. As an example work done in the IETF SIGTRAN group aims to relay SS7 signalling over the Internet (in effect, using it as a trunk line). There are at least two internet drafts [3] proposing to adapt the SS7 stack onto the Internet. The first provides an adaptation layer at Message Transfer Part 2 while the other at Message Transfer Part 3. Figure 1 depicts in a generalised manner these two approaches.

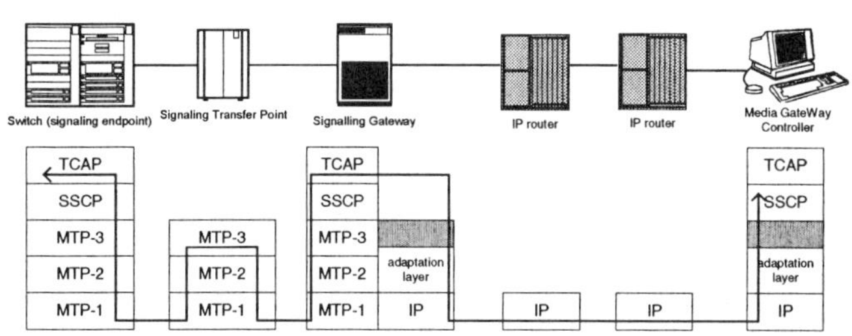

Figure 1. A number of signalling layers terminated, the others being relayed in the Internet side

There is another more clear-cut solution which can be seen as an extreme variation of the interworking approach. Namely, to terminate all signalling layers and completely encapsulate the signalling, conveying it transparently over the Internet. This however can lead to severe deficiencies mainly in terms of quality of service as no signalling information is made available to entities that could provide some sort of control over established connections at a local (e.g. enterprise) level. In general the main advantage of the more balanced approach to let some signalling layers pass on to the Internet side is that as signalling information becomes available at Internet nodes (these nodes hosting Internet Telephony components), it is possible to provide both (a) for the applications specific aspects of call control without needing to establish transactions all the way down to the actual terminals and (b) to impose some heuristic algorithm for QoS (e.g. by restricting the number of allowed connections within a domain). Naturally, the trade-off in this solution is that in addition to the terminals some Internet nodes are require to be augmented and able to play the role of their PSTN counterparts. For instance the H.323 architecture for VoIP foresees entities like Gatekeepers and Media Gateway Controllers which are equivalent in functionality and state information maintained with the PSTN switches.

Another equally important concern is the ability to allow the same IN superstructure to be used over both constituents. This is important in preserving the investments that have been made so far in IN and to take advantage of its efficiency in expeditious service creation and deployment. Signalling interworking as defined in the previous paragraph is an obvious prerequisite to the extension of the IN's realm of control to the Internet. But to support seamless IN operation more than that is required and some – to an extend artificial – constraints need be imposed.

The ability of IN to apply its services rests on the fact that state information on active calls is maintained by the network (at the Service Switching Points - SSPs). By examining this state and operating on it, IN services can be provided requiring only that a protocol is defined to communicate that state and IN-incurred operations that may modify it (a typical listener-controller pattern). Signalling as it is applied in the PSTN can be described as a protocol pattern where the uppermost layer protocol is not transparently conveyed end-to-end but is explicitly relayed and observed by intermediate nodes. We will not get into the discussion of why this particular protocol pattern is so prevalent in telecommunications while nearly non-existent in the Internet. One reason is that the progress of call and call control signalling is inextricably intertwined with the reservation of resources along a network path. One other is that it was necessary to keep the partial image of the state of a call in synch among the switches participating in a call and that therefore protocols that had a bearing on the call state had to be terminated and re-instated at each switch so that the partial images were kept synchronised. This protocol pattern which is not familiar in the Internet needs to be applied even if only for reasons of allowing IN to be used in the Internet as well. In order to do so it is necessary that i) at least call and call control signalling is not encapsulated but is relayed over to Internet and ii) that that signalling is not conveyed end-to-end but passes through Internet nodes which are designed to terminate it, observe it, operate on it and then re-instate it. Such nodes in H.323 terminology are known as Gatekeepers and this process is known as Gatekeeper-routed signalling.

In all cases and disregarding the Internet Telephony architecture that is used (H.323 or SIP) the key issue is of performing the appropriate mappings between the various protocols that can be used in IP telephony and making sure that a common BCSM can be suitable for all of them. Since the call control of H.323 is strongly influenced by Q.291 both Q.2931 BCSMs and (more easily) Q.931 BCSMs can be mapped to H.323's call model. The SIP model although much simpler can also be mapped to the previous call models. Refer to [4] for concrete examples.

Integrating the two networks at this level allows the same set of services to operate over both types of networks with no need to differentiate between the two. The notion of a service layer is thus given rise to, in which services invariably engineered are executed perceiving only a set of similar SSP-like entities and being completely oblivious as to whether the SSPs they interact with will effect their actions through the issuance of ISUP, H.323 or SIP messages.

3. SERVICE INTERWORKING

The second class of problems as has been mentioned earlier has to do with the ability to define interworking schemes between different kinds of services. The Pstn-Internet iNTerworking (PINT) specifications provide one such example in the click-to services. However the approach in PINT is too narrow in scope and qualifies for little more than an ad-hoc arrangement. In the more general case, service interworking aims for a framework that can enable services to find each other, communicate and share their capabilities. A protocol-based infrastructure is inhibiting the realization of such a framework for a variety of reasons. The lack of a common data model, common wire format and common procedures for directory and lifecycle operations are some of them. In a distributed processing environment the use of distributed object technologies offers a way out of most of these problems.

In particular, the exposure of the functionality of a service as a set of distributed, stateless APIs offers a neat way to export functionality in a self-explanatory manner bundled in coarse or fine-grained bundles as seen fit. That these APIs are stateless and usually synchronous further contributes to the robustness of the system as it minimises the number of states that the system may found itself into and it makes it harder for even an erroneously or maliciously operating client to throw the system out of synch. Another important advantage is that by using a distributed infrastructure we are replacing a fully-meshed graph of inter-service relationships, protocols and data models with a hub structure. All distributed objects plug into the same bus and share the same wire format, data model and protocol conventions. As it turns out service interworking comes naturally once participant services export their self-explanatory interfaces on a common scope allowing others to find their facilities, interact with them or employ them for their own purposes. A flat plane where remote objects (the service's proxies in the distributed world) interact is all that is needed for architectural support of service interworking to be in place. What flexibility is then allowed or how arbitrary or articulate interworking schemes can be defined is something that is determined solely by the nature of the interfaces that the services export into the distributed plane and by their expressiveness. This is just as it should be as it is impossible to abstract from arbitrary services. It is true that the telephony-like services that bear an underlying resemblance at the BCSM level have indeed be successfully abstracted and have lent themselves uniform treatment under IN but until now no other class of services has emerged with the same ubiquity, degree of penetration and level of coherence that would enable the appropriation of control models similar to the ones used in IN. The Parlay group considers this flat, API-based approach in order to export network-centric functionality to edge-of-network developers and to allow the development of cross-network services.

Once it has been decided that no higher level abstractions can be pursued for services other than telephony and that interworking between these services will be based on distributed APIs the focus shifts to architecture-related, service-independent considerations. These have to do with the ability of the services to find each other at runtime, with the binding semantics that the distributed processing infrastructure supports, with the copy and pass-by-value semantics and with mobile code. For the remainder of this paper we will assume that the distributed infrastructure is based on CORBA and that service code is in Java as the combination of these two technologies provides in our opinion the better solutions to the above problems (CORBA in the specifications of common object services and facilities and Java in code mobility).

4. A SERVICE INTERWORKING FRAMEWORK

At this section we describe a service architecture aimed at meeting the following requirements:

- Telephony-related services are integrated at the signalling and IN-control levels enabling existing IN architectures to control telephony in the Internet.
- Services expose their functionality to a control layer through open APIs allowing arbitrary interworking schemes between them to be defined
- Cross-network services can be built and operate seamlessly across diverse underlying network technologies
- Services can roam the network for load balancing or personal mobility reasons.

For a unified cross-network service framework there is no reason to reserve different treatment for the telephony services. These services like all the others will expose a set of primitives (a distributed API) in the form of CORBA objects on the distributed service layer. The only provision specific to the telephony services should be the accommodation of signalling interworking between PSTN and Internet and the use of IN to control and provide services in both segments. Even this however is not manifested to the service layer which only observes a set of service primitives exported in an API form (using the Interface Definition Language). Telephony IN seen from the global scope of our architecture is reduced to an implementation strategy for telephony services. The listener-controller pattern however by which IN services operate is adopted at the service layer where services register as listeners to each of the services proxy CORBA objects allowing them to be notified of events, changes in their status and offering them handles through which to effect some behaviour first on the CORBA objects and ultimately to the network resources which they represent.

Figure 2 shows the architecture of such a unified service provisioning environment.

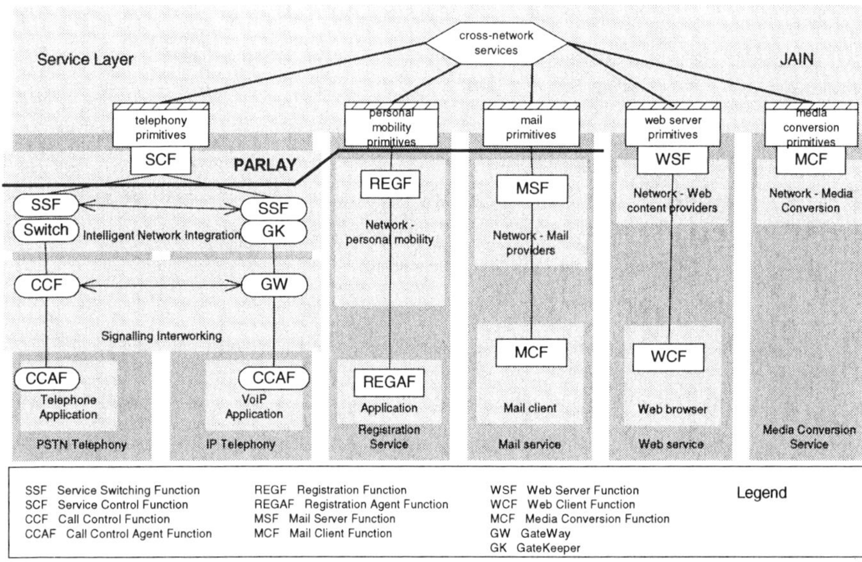

Figure 2. A Unified Service provisioning environment

According to Figure 2, telephony services seen from an architectural point of view are indeed one of the many vertical services that are available and that expose part of their functionality to a horizontal service layer above. It is also noted that the Intelligent Network as it is used for the purposes of the telephony services has little impact on the architecture at large. Provided that the same 'telephony primitives' or APIs could be exposed to the interworking service layer, it would have made no difference if services were switch-based.

Cross-network services exist as distributed objects in the service layer registering as event listeners to the CORBA objects representing the services and using the APIs the latter export to interact with the underlying networks. Figure 2 also depicts the interface line where Parlay standardisation efforts are focused and the area where ideas coming from JAIN [5] (for Java-based services) can be put to use.

Compared with telephony, services on the Internet have the disadvantage that they are ad-hoc and usually destined to spawn a shorter life-cycle. Few Internet services will reach the maturity of the telephone service and certainly none's will the standardisation be so carefully scrutinised. It is therefore necessary that any service interworking architecture takes that matter into account. Appropriate white or yellow pages mechanisms are necessary for service discovery and for allowing services to connect and use other services. The CORBA Naming and Trader services fall into that category. Resilience, fault-tolerance and transactions capabilities are all necessary components for a robust service infrastructure and fortunately standard interfaces to such services have been made available by the OMG (the consortium that is responsible for CORBA standardisation).

So the vertical services that appear in Figure 2 should be interpreted as pluggable components that do not form an integral part of the architecture. More dynamically yet the CORBA objects that represent these services in the service layer should be assumed to be of indeterminate uptimes and possibly replicated in different localities.

We will explore this issue as we delve more deeply into the structure of the service layer.

Figure 3 is a more detailed version of Figure 2.

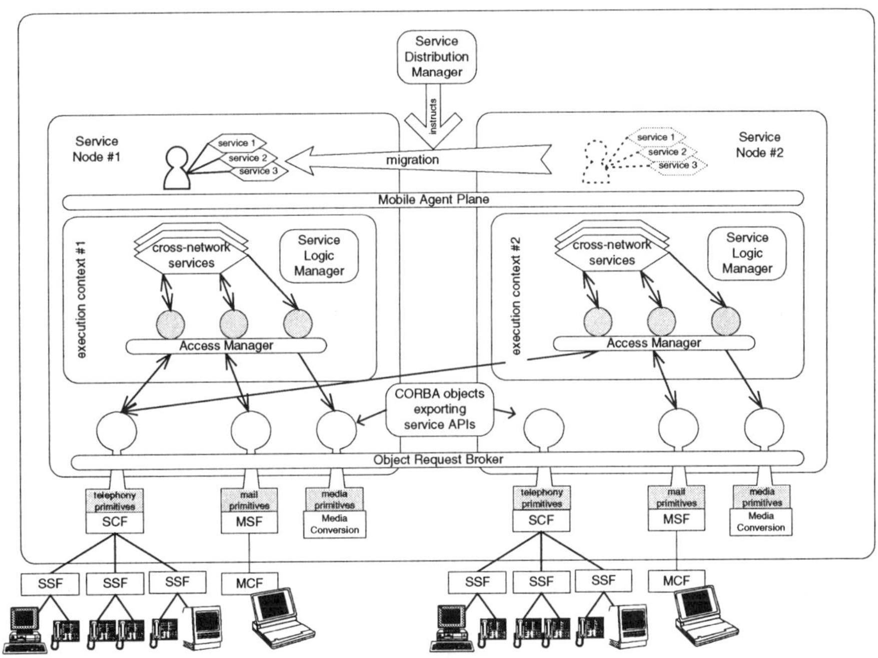

Figure 3. Service Layer for cross-network service provisioning.

It is seen that services are executed inside an execution context that is the same irrespectively of the service node (i.e. physical entity) where they are located. Programmatically the services are Java objects and thus their communication with their execution context is Java-based. Each cross-network service can register with Java objects that simply relay commands and events between the service logic programs and the underlying CORBA objects that represent the actual facilities. This is important as in a given service node a particular service might not be supported. For instance, in Figure 3 the SCF-based facilities are not available in the second execution context. The second service node's execution context (the access manager entity) transparently to the executing service substitutes the locally unavailable SCF facilities with a remotely located one.

Execution contexts hosts the service logic programs and are responsible for resolving such things as intermittent CORBA objects, local unavailability of a given facility, directory searching etc. So even if a given facility is not available at a particular physical location so long as a proxy CORBA objects exists somewhere in the distributed bus that provides access to the service capabilities, the execution context of the said physical node will be able to contact it transparently and still provide that facility to the service logic program that it hosts.

The alternative would be that services executing in service node 2 directly access the remotely located SCF object through CORBA. We want however to insulate services from the distributed platform we use and so this immediate step is necessary. With it we keep our services Java-based eliminating the need to upgrade them in case the distributed infrastructure we use changes. Services are simple Java objects that do not observer libraries other than the standard libraries that are available with the Java virtual machine. We also get to keep the location transparency offered by CORBA (or any other distributed processing technology used for that matter) as it is up to the access manager to connect the Java proxy objects with the real distributed objects to which they relay their commands or from which the receive event notifications.

Java nicely complements CORBA's implementation transparency due to its write-once-run-everywhere characteristics. Thus the same Java code can be executed in any service node disregarding its operating system or hardware architecture and that Java code will in the process ultimately (i.e. through the access managers and the proxy CORBA objects) interact using distributed APIs with proprietary code running in the SCFs, the mail servers or the media conversion centre without the latter pieces of code being able to distinguish the type of code they interact with. While the implementation transparency offered by CORBA is lost on the one side of the interface as service code will need to be written in Java, the uniform execution environment provided by the Java VM is more essential to any telecommunication architecture than the ability to have service code written in any language. After all, it is the diversity of equipment and operating systems that plagues telecommunication networks and could hinder an attempt of cross-network interworking. Furthermore, we get to keep implementation transparency at the side of the interface where it matters more: below the service layer where we are interfacing with legacy equipment.

By (a) using CORBA at the lower level in order to wrap native components and export their interfaces in a standard manner into the service's execution context and (b) using a Java-based execution context in our architecture's service nodes, we are exploiting the best of both worlds. With (a) comes a common data and object model, an API-based approach (with distinct advantaged over protocol-based ones) and seamless encapsulation of underlying network technologies. With (b) comes the power of Java in terms of mobility of code, safety (a simple language with automated garbage collection and fewer pitfalls for the developer to be aware of) and a standardised object oriented development and runtime environment that has entered the mainstream.

Another benefit accruing from the separation of the service logic programs from the actual distributed processing technology used for conveying event notifications and commands between the Java objects and the CORBA objects is that it is much easier to provide fault tolerance mechanisms at the access manager level than at the service logic programs level. Putting the code handling all possible eventualities into a service logic program would bloat the service code out of proportions and would hamper the mobility procedures described later in this paper. Note that we have already made considerable savings in terms of service logic size by allowing the services to remain 'pure' Java objects and putting stub and skeleton code in the access manager entity (which is responsible for CORBA to Java mapping and vice versa). Technically speaking our services are not CORBA objects (they do not extend a pre-compiled skeleton) nor do they need to be compiled with the stubs of other CORBA objects which they contact. All this is done at the access manager level.

We envisage a situation where the geographical dispersion of service nodes is indeterminate and not known beforehand. Service nodes may be added or removed from the network. It is clear that some of the cross-network services that can be supported by our architecture have topological associations with specific physical elements. For instance a service that has registered with the mail primitives provided by a mail server, expecting to receive notifications when new mails arrive would preferably be located somewhere 'close' to that mail server. Similarly it would be optimal for terminating or originating IN services to be 'close' to the terminating or originating switch. Personal mobility adds a further degree of freedom to this problem as users may out of their own accord move from one location to another rendering a previously optimal situation, sub-optimal. Services operating in such a fluid environment would have to be able to move between service nodes either to follow the user to whichever network he is roaming or to respond to varying network link conditions or even service node overloading or temporal unavailability. As service nodes should be able to be introduced or removed dynamically with no need of configuration scripts or pre-allocated distribution loads service could use all the mobility capabilities that could be afforded to them. That the service logic is implemented in Java and thus is in a semi-interpreted form ready to be interpreted on the fly helps promote mobility of service logic.

Java offers code mobility. It is however sometimes necessary to transport state as well as code. In these cases, a mobile agent platform should be used. Mobile agent platforms [6] transport state by serialising the objects and transferring their serialised form to their destination along with the code. In our system it is not necessary for service logic programs to be mobile agents themselves. Instead they can be simply transported by means of a migrating agent that holds pointers to them. Indeed the specifications for the Java serialisation describe that when an object is serialised the entire graph of objects to which it holds pointers (and, recursively, the objects to which the pointed objects hold pointers) are serialised. So a pattern can be easily derived by which service logic programs are being freighted by mobile agent objects implementing a simple 'freighter' interface. This has two distinct advantages:

- we are disentangling the properties of being a mobile agent and a service logic program. In narrower terms we are keeping the two class libraries separated and have removed from our service logic code any dependence on a particular mobile agent middleware. More than that, we have clearly delineated the role of mobile agent technology to that of an enabling technology for dynamic distribution of code for load balancing or other purposes – we have not made it an integral part of our approach. Note in this context that the same can also be said of the CORBA technology. From a functional point of view it does not matter if the access managers communicate with the entities they control through CORBA or some message-based protocol.
- Seen from a performance point of view, we are making migration operations less costly. Since services will not extend the – usually bulky – base mobile agent class their code size is reduced accordingly. Secondly, a single mobile agent can carry with it more than one service logic programs yielding a reduced overall bytelength.

For the architecture to be able to load-balance itself in that manner it will be necessary that a 'service distribution manager' takes into account two categories of events: (i) asynchronous events happening in the service node network. For instance, by the addition, removal, temporal suspension of a service node or by signalling overloading across a link or computational overloading on a service node. (ii) personal mobility events. A mobile agent is then instructed to migrate to the location the manager estimates as optimal and the service logic programs are attached to it so that they may follow it on the migration.

The architecture is in this way self-balancing without imposing any special requirements on the service logic programs.

5. CONCLUSIONS

In this paper we presented a service architecture for the purpose of providing an overall framework for service-level interworking between Internet and PSTN in general and for facilitating the creation and execution of cross-network services. We argued that IN is limited in scope and that in a environment where PSTN has converged with the more service rich Internet, it should be viewed simply as an implementation strategy for telephony services. Its listener-controller pattern can however be put into use by allowing services to expose their functionality to a service layer through open APIs allowing arbitrary interworking schemes between them to be defined. Using distributed object technologies it is possible to provide uniformity in data and object models and wire-formats that is indispensable in interworking between different services. By making use of distributed processing technologies the architecture can expose network-centric functionality to edge-of-network service providers. In that sense it is oriented towards the same objectives that have motivated the Parlay specifications. In doing this and by exploiting mobile code capabilities, it enables nomadic computing and personalised services that roam the network to follow their registered user.

REFERENCES

[1] Internet Draft: "The PINT Service Protocol:Extensions to SIP and SDP for IP Access to Telephone Call Services" (IETF PSTN and Internet Interworking Working Group).

[2] Parlay API Specification 2.0, available from http://www.parlay.org

[3] See the IETF's Signaling Transport working group at www.ietf.org

[4] Ackermann, D. and Chapron, J. "Is the IN Call Model still valid for new Network Technologies?", International Conference on Intelligence in Networks 2000, Bordeaux, France.

[5] JAIN: Integrated Network APIs for the Java platform, available from www.sun.com

[6] M.K. Perdikeas et al., Mobile Agent Standards and Available Platforms, Computer Networks, Elsevier, Volume 31, pp. 1999-2016.

I-centric Services in the Area of Telecommunication 'The I-Talk Service'

Stefan Arbanowski, Sven van der Meer, Radu Popescu-Zeletin

Technical University Berlin, Franklinstrasse 28/29, 10587 Berlin, Germany

Email: arb@cs.tu-berlin.de

Tel: +49 30 3463 7197, Fax: Tel: +49 30 3463 8197

In the last years, a variety of concepts for service integration and corresponding systems have gained momentum. On the one hand, they aim for the interworking and integration of classical telecommunications and data communications services. On the other hand, they are focusing on universal service access from a variety of end user systems. Many of the technical problems resulting from the service integration and service personalization have been solved during the last years. However, all these systems are driven by the concept of providing several technologies to users by keeping the peculiarity of each service.

Looking at humans' communication behavior, it is obvious, that human beings interact with a set of objects in their environment habitually. The individual preferences and needs, persons to interact with, and the set of devices to be controlled by each individual, defines his personal communication space. Following this view, a new approach is to build communication systems not on the basis of specific technologies, but on the analysis of the individual communication spaces. The result is a communication system adapted to the individual demands of each individual (I-centric). The communication system will act on behalf of users demands, reflecting recent actions to enable profiling and self-adaptation to contexts and situations. I-Centric services adapt themselves to individual communication spaces and individual environments and situations. In this context "I" means I, or individual, "Centric" means adaptable to I requirements and a certain user environment.

This paper introduces I-centric service as a new approach to integrate different services, and it illustrates how this approach can be applied to the area of telecommunication systems. The scenario proposed is derived from numerous development projects in the areas of: unified messaging systems, object and person tracking, mobile guides, smart homes and cars which have been implemented with industrial partners by our group in the last four years

Keywords: Service Personalization, Human Communication Behavior, Virtual Home Environment, Context Awareness, I-centric

1 I-centric Communications

In the former days, the communication space of human beings has been limited to the actual surrounding physical environment (like a home or a village) in general based on the geographical reachability of human senses. With the introduction of telephony, the spatial dimension of this space has been expanded. Now, it became possible to talk with people regardless of their actual location. With upcoming asynchronous services like email and fax, the dimension of time has been expanded. People can send emails and need not to care, if the addressee is ready to receive the message or not. That is, technology has eliminated distances in time and space or at least made these boundaries almost invisible. [1]

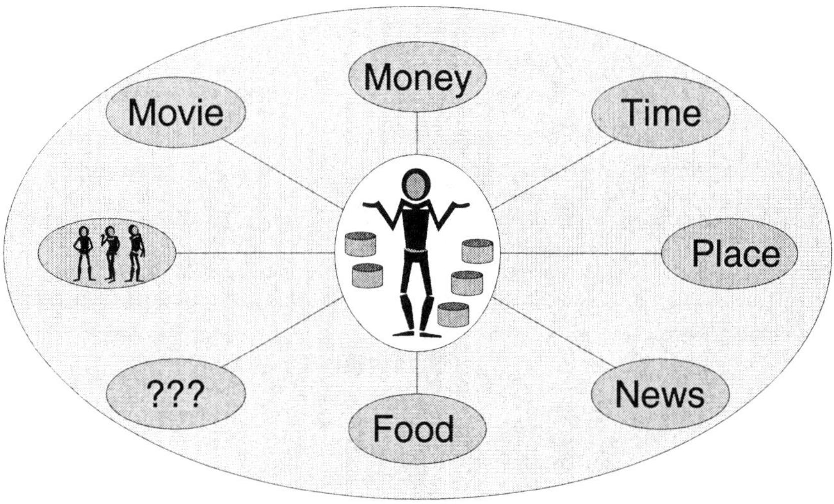

Figure 1: Objects in the Communication Space of Human Beings

I-centric means to take a bottomless look at human behavior and to adapt the activities of communication systems to it. Human beings communicate with their environment in different ways. They meet with other people to talk, to celebrate, they read and travel, they are listening to news or to music, they make decisions, just to give a small range of examples. People interact with several 'things of interest' to solve the problems of daily life: money and bank accounts need to be managed, food has to be bought and to be prepared for eating, movies can be watched for entertainment, places are visited and news are consumed to improve the education, other people are met for discussions. All these contexts and the related objects define the communication space of a human being. A context represents a certain "universe of discourse" in an individual communication space. In general, human beings communicate with 'objects' in their environment in a certain context. Note that same objects may pertain to different contexts. Objects pertaining to a certain context can be active or passive at certain moment in time depending on the situation the user is into. They can be activated or deactivated by the user or environmental condition. They can be directly addressable or represent a set of physical entities performing a certain service as a whole. [2][3][4][5]

A user might have different preferences in different situations. Sitting alone in a silent room might indicate that the user is willing to receive incoming communication requests. However, the same user can understand this as a disturbance being involved in a conversation with other people. [1] To be I-centric requires knowledge of the actual situation (context) of a user. A context defines a certain relationship of a human being to a particular number of objects of its communication space at a fixed moment in time. An I-centric service has to be aware of the context a user is in and has to adapt itself to that very context. [6]

The multitude of devices, wearables, different telecommunication technologies, positioning and sensing technologies, location-aware or context-aware applications etc. can be seen as enabling technologies for I-centric communications. Universal information access (including service interworking, media conversion), flexible control of equipment at facilities (e.g. smart homes), and personal communications (supporting personal mobility and terminal mobility) form the fundamentals of such systems. [7]

I-centric services describe the ability to define and to manage contexts that are tailored to the preferences of single users, facing individual ways to interact with the communication system. Based on the evaluation of 'profiles' that describe user preferences, service capabilities, and on sensing information about its actual environment, the user can be provided with individualized services for his current demands (the description of user demands opens a huge area of research, which has to cover both, technical parameters and abstract descriptions). Self-learning capabilities are employed to profile the behavior of users, numerous services or several features of different services are combined on-demand, and appropriate service facilities are evaluated.

Context influencing parameters with temporal and spatial characteristics like any user input, temperature, noise level, light intensity, presence of other people and objects in the vicinity are sensed by special sensing facilities like thermometers or microphones. Based on the sensed information, the I-centric service adjusts to the current situation and the user's needs. It chooses what equipment have to be controlled (presentation terminals, handhelds, microelectronic controlled devices), their quality of service (volume, brightness etc) to be finally connected via heterogeneous access networks (Infranets, Intranets, Internet, Extranets, telecommunication networks) to create an I-virtual private network. [8]

An I-centric Service Platform is responsible for shaping the communication system, based on identified contexts, which are adjusted, by sensed environment information. It activates the objects (things of interest) involved in the context, identifies causalities between them based on sensed environmental data, controls the services offered by these objects, and converts data structures and operations for interworking between different services. The underlying terminals and distributed computational objects are supervised dynamically by the platform to realize the physical service.

The aim of I-centric services is to model the entire communication behavior of human beings. This has to lead to an expandable system that is almost invisible to the user, that requires no time-consuming configuration, and provides customized interfaces to each single user based on its own preferences and situations in time. This paper shows that combining off-the-shelf technologies can provide I-centric services to human beings in an easy, for them understandable and therefore acceptable way.

2 Telecommunication systems and I-centric Services

The motivation for I-centric services is to enhance already implemented systems as well as developments, which are currently in standardization processes, by changing the way in which users have to interact with them. To give an example, in the following the application of I-centric services to the area of telecommunication systems, especially UMTS/VHE, is discussed. [8][9]

The idea of the Virtual Home Environment (VHE) has been introduced in the last years. It was proposed as the solution for the evolution from several different 2nd generation mobile systems to a single world standard for IMT-2000. This solution was thought to let user/service provider create and run consistent service and have the same 'look-and-feel' even when roaming to a dissimilar network.

Today, the VHE is a key concept within 3rd generation telecommunication systems that aims to provide access to services customized to the needs of end user irrespective of the serving network and terminal. Fundamentally, the vision of I-centric systems orientates to this. Because of its origin, the VHE concept is more related to mobile telecommunication networks (also considering data services). However, the idea behind can be adopted to all kind of communication systems - the user should be provided with his personalized usual environment that realizes a common look and feel of interface and service experience regardless of location, network and terminal type. Exactly this characteristic is also targeted by the I-centric concept.

I-centric services provide a high degree of personalization. The separation of I-centric services into the service logic itself and the user-interaction part enables the flexible adaptation to different kinds of end devices. That means the user can access services by using the communication equipment that he has on his disposal at the moment regardless of the particular service. The I-centric services are realized with multiple components that are distributed in the network in contrast to the classic approach of Intelligent Networks to have centralized service logic. At all possible edge devices (mobile or fixed devices) that are connected to the core network, users can access the services realized inside the network. [10]

3 I-centric Service Framework

Figure 2 depicts the main functional components of the I-centric service framework. I-centric services, Interactions in Smart Environments, and the I-centric Service Platform are decomposed to describe their functionality in more detail. Together, they are responsible for the mapping of abstract information about the user's communication space down to specific device instructions that realize a service physically. The user himself employs I-centric Services and interacts with Smart Environments using a human-machine-interface his current location provides.

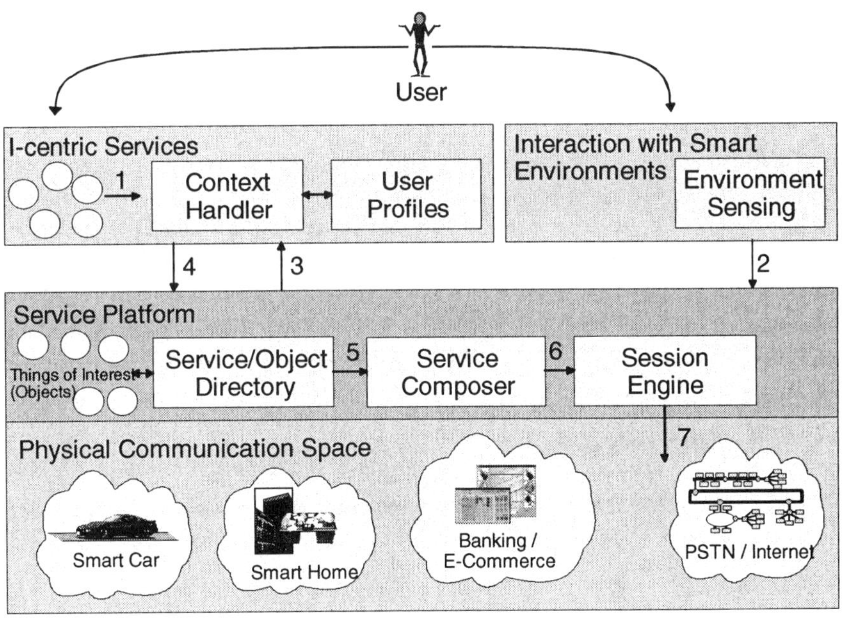

Figure 2: Architectural Framework for an I-centric Communication System

I-centric Services and components, able to sense the environment, are settled on top of the Service Platform. I-centric services process their tasks supported by the Context Handler, which determines what Objects (things of interest) have to be considered in a certain context (1). The user can benefit from I-centric services via the interaction with his environment. That interaction can be a dialog with an interactive user interface (touch-screen application, telephone, microphone/audio speaker combination) or a daily action, which is sensed by the Smart Environment automatically (e.g. the user enters a room and is located instantly). The knowledge about the interaction is transmitted via the Service Platform to the Objects (2), where the Context Handler can recognize the updated information for a certain context. The component User Profiles contains information about user preferences as well as information about recent service usage. This component has to be updated, for instance when habitual behavior is monitored (3). Furthermore, the Context Handler deals with the problem of interpreting/converting any kind of user input, sensed information, and occurred events. It maps resulting information or user instruction into information the I-centric service can 'understand'[1]. Finally, the I-centric Service interacts with the determined Objects, facing User Profiles. The results in a formal description of the requested service, the configuration of involved Objects, and information needed to perform the I-centric services.

The Service Platform has two main objectives, the conversion and exchange of information between all components of the framework, and the mapping of causalities of Objects to causalities of legacy services. Each 'thing of interest' is realized as an Object inside the platform and can be managed by an I-centric Service and updated by 'Interaction with Smart Environments' (real world objects which are not realized as objects inside the platform cannot be used or accessed at least by the platform). The Service Composer receives the request from the Objects. The Service & Object Directory contains necessary information about services, service features, service building blocks, and policies for the mapping of causalities. The Service Composer can combine legacy services to fulfill the request or it can create new services on-demand out of a pool of service building blocks (5). Additionally, it selects appropriate terminals, devices, and/or applications (6). The Session Engine represents the final component in the processing of a service request (7). It controls legacy services and configures selected hardware and software.

The mechanism how the system combines different legacy services to provide one I-centric service is described in [10]. We used our recent experiences from the area of Unified Messaging, Media Conversion, and Job & Stream Control to develop the I-centric service platform. [1][10]

4 I-Talk

Based on the proposed framework several I-centric services have already been implemented at the department for Open Communication Systems of

---

[1] Note: In this paper we use the Context Handler as a container for several functionalities. For the physical architecture we have decomposed the Context Handler in several functional components to separate tasks (e.g. one component is responsible to handle user preferences in an extendible XML-based profile database).

the Technical University Berlin. They cover areas including telecommunication, home automation, mobile guidance, and electronic commerce. In this paper, the I-centric service "I-Talk" from the area of telecommunication is introduced.

4.1 I-Talk – the service

I-Talk realizes an I-centric telephony-service, where a user has no longer to type in or to memorizes any telephone-numbers (Figure 3). He only has to say his personal nickname of the person he wants to talk to, or he can request a person that has never been contacted before by saying the whole name. After that, I-Talk analyzes the user input and contacts the requested person via telephone. In case that I-Talk cannot determine automatically which phone number belongs to the requested person, the user can dictate a phone number and assign a name (alias) to it for later usage.

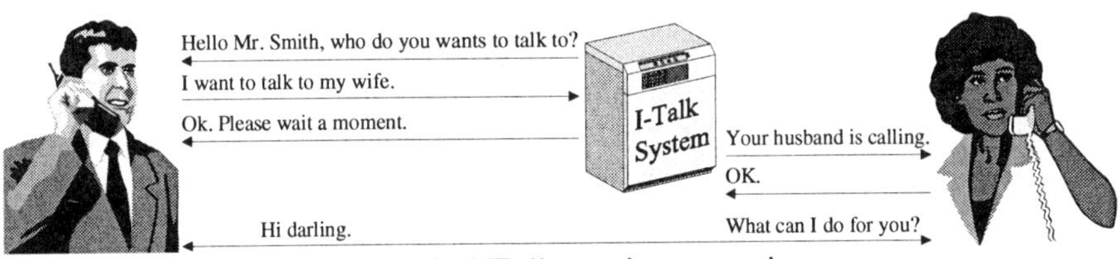

Figure 3: I-Talk service scenario

The I-Talk service features the functionality of an electronic secretary, which hides the peculiarities of different kinds of telephony services (IP-T, PSTN, ISDN) to the user. It does not matter, what kind of physical service underlies the I-Talk service – users just contact and talk.

4.2 I-Talk – the realization

Following the framework, now it is discussed what kind of functional components are used for I-Talk.

The Interaction in Smart Environments component is responsible for the interaction with the user. In this case, it provides a telephony based speaker independent and speech driven human-machine-interface. The interface is realized within an Interactive Voice Response (IVR) server that is based on specialized hardware. The front-end for users can be each kind of telephone able to dial the I-Talk service number. This gives users the freedom to use their personal I-Talk service from any telephone they want.

The IVR application deals with receiving user input and responding to it. The Context Handler, which is part of the I-centric services component, evaluates the user input. Within this component, we provide on one hand Automatic Speech Recognition (ASR) to recognize which person should be contacted, and on the other hand Text-to-Speech (TTS) conversion to generate voice-prompts dynamically for negotiating with the user. The separation of user interface (Interaction in Smart Environments) and the evaluation of user input (Context Handler) lead to a flexible system architecture. Both, the user interface and the logic behind the interface can be exchanged without affecting the other.

The Context Handler analyzes each user input by applying an ASR to it. The result is evaluated considering the User Profile, which contains a personal phone book for each I-Talk user. The phonebook entries are used by the Context Handler to map user input (spoken words) to persons, which has to be contacted. Each entry consists of a list of words, representing personal aliases for other persons, and the name of a person, which is mapped later to a phone number (e.g. saying ‘Cynthia’, ‘Ms. Smith’, or ‘boss’ means: ‘talk to user Cynthia Smith’).

A very important component in our system is the ASR. The more reliable the automatic recognition of user input will be processed, the more the I-Talk users will be satisfied. Therefore, we separated the ASR component for being exchanged quickly, if a new and better technology is available.

I-Talk (as an instance of an I-centric service) receives the result of the context evaluation (e.g. ‘talk to user Cynthia Smith’) from the Context Handler and queries the Service/Object Directory to find a service ‘talk to’ and an Thing of Interest (object) ‘Cynthia Smith’. The Service Composer establishes the service ‘talk to’ using the legacy service ISDN. The phone number, which has to be called to talk to ‘Cynthia Smith’, is taken from the corresponding object. Each person who could be contacted via an I-centric service is represented inside the Service Platform by an object. Information necessary for I-centric services (e.g. the whole name of a person) as well as technical parameters (e.g. corresponding phone number) are stored in such objects. Finally, the Session Engine connects the IVR application via ISDN to the phone number that belongs to the request person.

The implementation is based on Euro-ISDN on top of a Dialogic™ BRI80SC-ISDN board and a supporting Dialogic™ Antares™ ASR/TTS board. The I-Talk system can handle 8 simultaneous talk-requests, providing real-time ASR and TTS on each line. The host computer is a state-of-the-art Pentium III with 128 MB running Windows NT 4.0.

The functional component, which have been mentioned above, are realized as CORBA-objects using the C++ CORBA 2.0 implementation Orbix™ from IONA Technologies. Therefore, the system can be set up on a single host system or it can be distributed for scalabilities reasons.

5 Conclusion

This paper has introduced the vision of I-centric services and how these services relate to today’s and upcoming communication systems. After that, a general framework for the realization of I-centric service has been shown and finally, a first realization (I-Talk) has been illustrated.

I-Talk shows how an I-centric service can be applied to the area of telecommunication. Facing the service personalization requirements of 3rd generation mobile telecommunication systems, I-Talk can be seen as partial realization of the VHE concept. Therefore, we are going to integrate I-Talk in two UMTS/VHE trial implementations (Eurescom / IST). The main advantage of services like I-Talk is that the service logic and processing is realized inside the network, instead of having it on the terminals. This let users access their services from different terminal not caring about what kind of terminal it is and what person the terminal owns.

Further developments: We want to enhance our Interactive Voice Response System with automatic speaker recognition to enable automatic authentication procedures for security reasons. The separation of user interface and analyzing the user input offers the possibility to exchange the telephony-based interface with a microphone speaker combination. The idea is to equip a room or a car with this kind of interface to enable hands free I-Talking.

The targeted implementation should realize the functionality of today's immobile phones inside the network. That comprises many features like phone book, recent call lists, and the handling of incoming calls especially. This functionality will be available within the next half year.

6 References

[1] Arbanowski, St.; van der Meer, S.: Service Personalization for Unified Messaging Systems. Proc. of the fourth IEEE Symposium on Computers and Communications, ISCC'99, Red Sea, Egypt, July 6-8, 1999, 156-163

[2] Greenberg, S.;Roseman, M.: Using a Room Metaphor to Ease Transitions in Groupware, Research report 98/611/02, University of Calgary, Alberta, Canada, 1998

[3] Abowd, G. D.; Dey, A. K. et. al.: Towards a Better Understanding of context and Context-Awareness. First International Symposium on Handheld and Ubiquitous Computing, HUC'99, Karlsruhe, Germany, 27-29 Sept., 1999, 305-307

[4] Popescu-Zeletin, R.; Pfeifer, T.; Magedanz, T.: Applying Location-Aware Computing for Electronic Commerce: Mobile Guide. Proc. of the 5th Conference on Computer Communications, AFRICOM-CCDC'98, Tunis, October 20-22-1998, 233-246

[5] Pentland, A.: Perceptual Intelligence. First International Symposium on Handheld and Ubiquitous Computing, HUC'99, Karlsruhe, Germany, 27-29 Sept., 1999, 74-87

[6] Ljungstrand, P.: Context-awareness in distributed communication systems, CHI Work-shop #11: The Who, What, Where, When, Why and How of Context-Awareness, The Hague, April 2000

[7] van der Meer, S.; Arbanowski, St.; Popescu-Zeletin, R.: A Platform for Environment-Aware Applications. Proc. of the First International Symposium on Handheld and Ubiquitous Computing, HUC'99, Karlsruhe, Germany, 27-29 Sept., 1999, 368-370

[8] Eckardt, T.; Magedanz, T.; Ulbricht, C.; Popescu-Zeletin, R.: Generic Personal Communications Support for Open Service Environments. Proc. IFIP World Conference on Mo-bile Communications, Canberra, Australia, Sep. 1996

[8] ETSI Technical specification TS 22.01v.3.1.0: Universal Mobile Telecommunication Systems (UMTS). Aspects, Principles. - Sophia Antapolis, France, 1997

[9] ETSI Draft 22.70v.0.0.3: Virtual Home Environments. - Sophia Antapolis, France, 1997

[10] van der Meer, S.; Arbanowski, St.: An Approach for a 4th Generation Messaging System. Proc. of the 4th International Symposium on Autonomous Decentralized Systems, ISADS'99, Tokyo, Japan, 1999, 158-167

7 Acronyms

ASR	Automatic Speech Recognition
CORBA	Common Object Request Broker Architecture
IMT	International Mobile Telecommunication
ISDN	Integrated Service Digital Network
IVR	Interactive Voice Response
PSTN	Public Switched Telephony Networks
TTS	Text-to-Speech
UMTS	Universal Mobile Telecommunication System
VHE	Virtual Home Environment

Centralized network intelligence vs. decentralized network intelligence

Josef Ferchenbauer
Siemens AG
E-mail: Josef.Ferchenbauer@icn.siemens.de

Key words: Network intelligence, Private line service, Permanent Virtual Connection, Soft Permanent Virtual Connection

Abstract: Often operators and vendors have to decide whether a new service shall be introduced into the network via centralized or decentralized service logic. This paper shows the points that have to be considered in such cases. It works out the advantages and disadvantages of a centralized and decentralized network intelligence approach on an example. As an example an existing service, the private line service offered via an ATM network, is taken. The example is used to picture and verify the results, but the results are universally applicable. In cases where they are not it is stated explicitly.

1. INTRODUCTION

This paper shows the points that have to be considered by operators and vendors to decide whether a service shall be introduced into the network via centralized or decentralized service logic.

The first part of the paper describes the service that has been taken as an example. The second part describes the advantages and disadvantages of a centralized and decentralized approach using the example.

2. PRIVATE LINE SERVICE

2.1 Private Line Service via PVCs

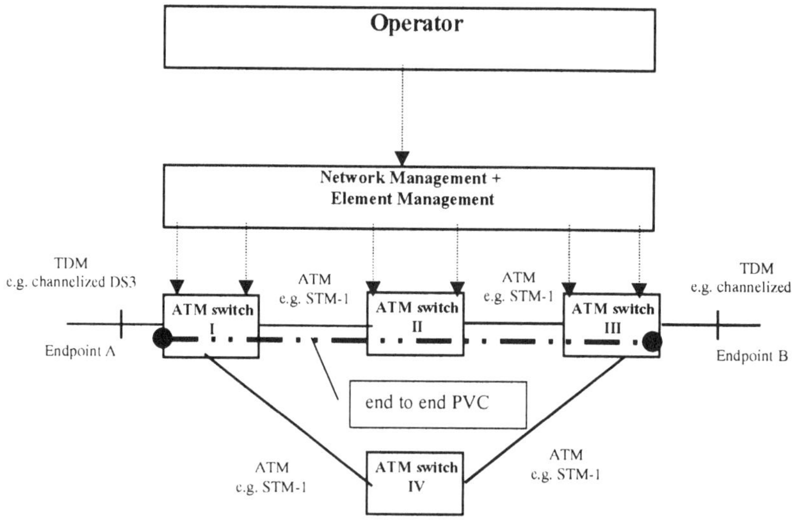

Figure 1. Private line service offering via PVCs (centralized network intelligence)

A private line is a permanent connection between two endpoints, in the given example endpoint A and endpoint B. The establishment of the permanent connection can be done by the operator via TMN, i.e. the operator initiates the through connection between Endpoint A and the ATM link in switch I. Then the operator connects both ATM links in switch II. Finally the through connection between the ATM link and Endpoint B has to be performed. Then a so-called PVC between Endpoint A and Endpoint B is established. To support the operator an intelligent application in the network management can do the routing, therefore the operator has to define only Endpoint A and Endpoint B. The rest is performed by the network management itself. This application is called end-to-end PVC establishment.

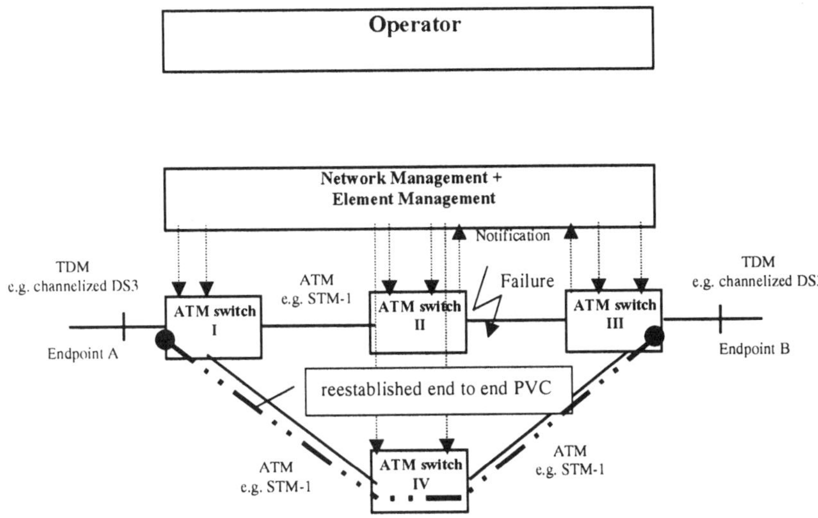

Figure 2. Private line reestablishment via PVCs

It is very important that these connections are highly reliable, i.e. the downtime of such connections has to be minimized as much as possible. Therefore in case of a link or node outage the network management is notified about the outage. This triggers the end-to-end PVC application in the network management to establish the connection via an alternative path, e.g. via switch IV instead of switch II.

End-to-end PVC establishment is an example for centralized network intelligence.

2.2 Private Line Service via SPVCs

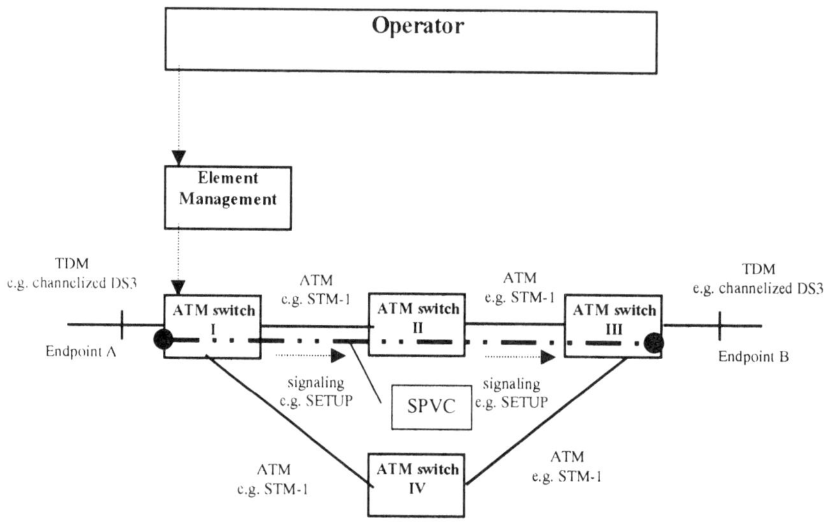

Figure 3. Private line service offering via SPVCs (decentralized network intelligence)

Another approach is to use soft Permanent Virtual Connections (SPVCs) to establish a private line between the endpoints. The operator defines Endpoint A (originating endpoint) and Endpoint B and sends the request to establish a connection between both endpoints to switch I. Endpoint B is identified by a unique address (called party number). Then switch I sets up a connection, using SVC functionality (signaling, routing etc.), towards endpoint B at switch III. This connection is a so-called Soft Permanent Connection (SPVC). Each switch itself has to perform routing and through connection, therefore this is an example for decentralized network intelligence. For this approach no network management is required; a simple element management is fully sufficient.

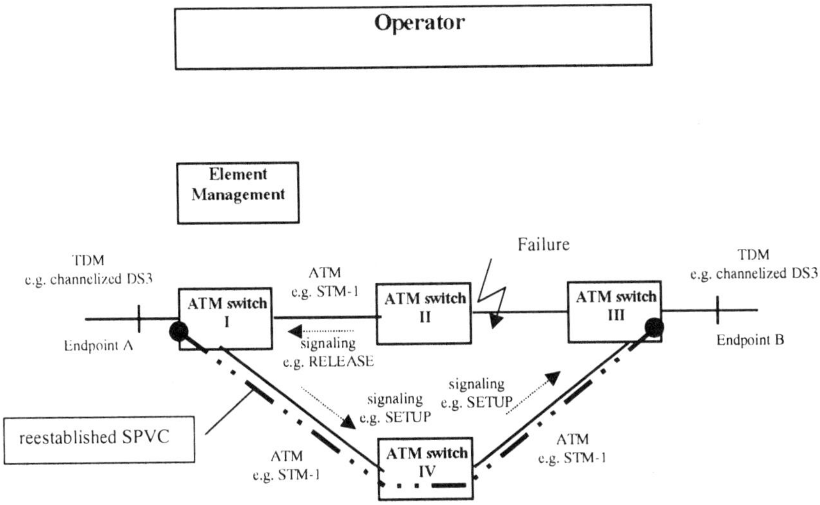

Figure 4. Private line reestablishment via SPVCs

The originating endpoint (Endpoint A at switch I) is responsible for maintaining the end-to-end connection by setting up a new switched connection automatically in case the existing switched connection fails. This is to minimize the downtime of such connections, which is very important for this service

One approach of failure handling is to tear down the connection, i.e. the neighbor node of a faulty entity issues a release of the connection via signaling. When Switch I receives the release message it tries to find another path towards Endpoint B through the network. This reestablishment will result in a connection via switch IV instead of switch II.

All this is achieved without any centralized network intelligence, i.e. without a centralized network management. Therefore SPVCs are an example of decentralized network intelligence.

To support SPVCs in a multi vendor environment the signaling enhancements for this service has been standardized by ATM-Forum /1/ /2/ and ITU /3/.

3. COMPARISON

3.1 Time to market

The PVC (centralized) approach has the advantage that only one network element, the network management, has to be upgraded. A precondition for this is that the ATM switch already offers an interface to establish PVCs inside the switch and to notify the network management of faults. These are basic functions, therefore most of the ATM switches offer such functions.

In case of the SPVC (decentralized) approach each switch has to be upgraded to support SPVCs. In addition the switches have to support SVCs because this is the basic function used for SPVC establishment. In addition the signaling protocols to support SPVCs has to be standardized. For SPVCs this has already been done but for new services this might take some time and defer new features. For a short term solution signaling might be enhanced proprietary, but for multi vendor networks standards are crucial.

Assessment: As long as only already available basic functions of the switch are used to create a new service the centralized approach has clear advantages.

3.2 Performance

The establishment and reestablishment of PVCs has to be performed by network management. Only one PVC after the other can be established respectively reestablished. Thus network management becomes a bottleneck especially in case of failure. When a failure occurs the network management has to search a new path through the network and create the new connection. This reestablishment is very time critical because the interruption of the connection, due to a failure, has to be very short.

SPVC are established and reestablished from the originating endpoint. Only the nodes, which are carrying the SPVC, are involved in establishing and reestablishing. The establishment and reestablishment of SPVCs will be done in parallel. Each switch only processes its SPVCs.

Assessment: From the performance point of view the decentralized solution has clear advantages. This is because the functions (in our example establishment and reestablishment) are performed in parallel.

3.3 Scalability

As the network grows the processing power which is usable by the service logic grows with it in case of the decentralized service support. In case of the centralized approach it does not.

Assessment: The decentralized approach has its advantages, but this is not a point in its own rights because it is already covered by the performance investigation.

3.4 Reliability

Central approach: In case that the central entity which hosts the service logic, in our example the network management, fails the service cannot be supported anymore. For instance no reestablishment can be performed.

Distributed approach: In this case when the entity, which hosts the service logic, in our example, an ATM switch fails, only a fraction of the service is influenced by the failure. In addition when the switch fails it has the same influence on the central approach. In our example when the switch which has the endpoints connected (switch I and switch III) fails no establishment or reestablishment can be performed. This is valid for both approaches.

Assessment: The influence of hardware failures of the controller that host the service logic is much smaller in the decentralized approach.

3.5 Self Organizing Network Support

In self-organizing networks, e.g. PNNI networks, there is no central entity required that has the knowledge of the whole network. But for certain services like the private line service this is necessary when the centralized approach is chosen.

Assessment: The decentralized approach does fit best into a plug and play environment. By softening the plug and play philosophy, e.g. by configuring the physical nodes, physical links and user addresses via a central management entity, also the centralized approach can be used. But then the question is whether it is still a plug and play network?

3.6 Multi Vendor Ability

Central approach: Network management has to be able to support the different management interfaces of the switches, but the interface can be negotiated between the network management vendor and each switch vendor separately.

Distributed approach: The interface has to be negotiated between all vendors of the network. A practical way to achieve this is to standardize the feature if it shall be supported in a multi vendor environment.

Assessment: Both approaches can be used in multi vendor networks. The distributed approach needs more coordination effort, but this becomes a time to market issue in the end.

3.7 Operational Costs

The operational cost for both approaches is the same, because the operator has to put in the endpoints and maybe some ATM traffic parameters for both solutions.

In case of SPVCs the signaling and routing infrastructure might have to be configured. This is only valid for networks where no SVC service is offered. If the SVC service is offered anyway the same signaling and routing infrastructure can be used.

Assessment: In certain cases (see above) the operational costs for the centralized solution are less.

3.8 Cost of Purchase

This point can only be answered on a case by case base. The only thing, which can be stated, is that a centralized solution needs a network management entity, which causes additional hardware costs.

3.9 Service limited to network

The end to end PVC management application, which is running on the network management entity, can only establish PVC via switches that are managed by it.

An SPVC need not to be limited to a network. This is because the terminating endpoint, i.e. Endpoint B in the example above, is identified by a unique address. The only precondition is that the other network supports the according standard /1/ /2/ /3/.

Assessment: This point is very special for this service; therefore it cannot be taken to derive a general statement for a centralized or decentralized approach.

3.10 Summary

The following table summarizes the above made assessments.

	Centralized Network Intelligence	Decentralized Network Intelligence
Time to market	+	
Performance		+
Scalability		+
Reliability		+
Self organizing network support	no	yes
Multi vendor ability	*	
Operational cost	*	
Cost of purchase		*
Service limited to network	yes	no

Table 1. Summary

Legend:
+ ... strength of the solution
* ... slightly better than the other solution

4. CONCLUSIONS

In this paper the advantages and disadvantages of a centralized and decentralized network intelligence approach have been investigated. This has been done for a specific service, the private line service. This does not mean that the result is only valid for this specific service, in the contrary the results are universally applicable.

The main advantage of a centralized approach is that services can be introduced in a network very fast (time to market). Therefore if a new service is introduced first on a small scale this approach is advantageous.

The main advantage of an decentralized approach is the performance and its scalability, i.e. when a service is commonly used in a network it has to be decentralized else the performance limits will be reached quite soon.

5. REFERENCES

/1/ ATM Forum Technical Committee: Private Network-Network Interface Specification
Version 1.0 (PNNI 1.0); af-pnni-0055.000; March 1996

/2/ ATM Forum Technical Committee: ATM Inter-Network Interface (AINI) Specification
AF-CS-0125.000; July, 1999

/3/ International Telecommunication Union:
ITU-T Recommendation Q.2767.1; Soft PVC capability

6. ABBREVIATIONS

PVC	Permanent Virtual Connection
SPVC	Soft Permanent Virtual Connection
SVC	Switched Virtual Connection
TMN	Telecommunications Management Network

Part Ten

Video Broadcasting over Packet-Networks

A NETWORK BASED REPLAY PORTAL

Service Scenarios, Architecture and Implementation

J.E. van der Merwe[1], C.J. Sreenan[2], A.N. Donnelly[3],
A. Basso[1], C.R. Kalmanek[1]

[1] *AT&T Labs - Research*
Florham Park, NJ, USA
{kobus,basso,crk}@research.att.com

[2] *University College Cork*
Cork, Ireland
cjs@cs.ucc.ie

[3] *Cambridge University*
Cambridge, UK
and1000@cl.cam.ac.uk

Abstract

Technologies based on cable modems currently use the capacity of a single TV channel to offer 25 - 30 Mb/s downstream for Internet access. With the advent of Digital TV and the significant bandwidth savings it gleans from video compression, it is expected that providers will increase the access capacity available for IP traffic, while retaining the bulk of the bandwidth for the primary service of broadcast TV. Motivated by the increasing popularity of on-demand streaming media, our work questions this IP-versus-broadcast distinction, and proposes a hybrid model which combines the familiar broadcast model with the conveniences of on-demand TV viewing over IP. In this paper we discuss the potential benefits of this approach, and some of the difficult technical challenges it raises. We propose an architecture of network-based portals and sketch the types of services we envisage it will enable. Further, we describe the design and implementation of a replay service that operates within this architecture to offer a new on-demand TV-viewing experience.

Keywords: streaming media, network services, on-demand TV

Introduction

Recent years have seen the rising popularity of streaming media applications on the Internet. While the amount of this traffic is still small relative to web traffic [van der Merwe et al., 1999], there is general agreement that it will grow significantly, driven by bandwidth-efficient compression techniques, a wider array of globally accessible content, and increased capacity of backbone and access networks. Our research is motivated by this expected growth in Internet streaming content and by developments in the Cable-TV industry, where there is a concerted effort to provide high-speed, full-duplex Internet access to residential customers. Current standards in the US allow up to 30 Mb/s for Internet access by allocating a single 6 MHz wide TV band. In just a few years it is expected that the broadcast TV industry will fully embrace and deploy digital TV based on MPEG-2. The benefits of digital compression will allow several TV channels to be carried in a single 6 MHz band, freeing up a significant amount of system capacity. It is expected that this will be mainly used to offer new premium TV channels, and new services such as movies on demand, while allocating a relatively much smaller amount of bandwidth for IP traffic, including streaming media. Our work questions this TV versus IP distinction and proposes instead an all-IP model. This approach has several potential benefits, including the ability to use multicast delivery to efficiently scale from large broadcast-style distribution to smaller, more select audiences. Also, the Internet provides access to content which is globally accessible, unlike the broadcast TV model, where access to content depends on the channels offered by your local TV operator. Integration with the web also becomes much more convenient and promotes interactive services.

We believe this approach has the potential to revolutionize the TV viewing experience by moving away from a broadcast model to a hybrid on-demand model. In that context, several difficult technical challenges arise, including the design of on-demand services, storage management, efficient use of multicast, caching models, and protocol interactions for live and on-demand viewing. In Section 1 we propose an architecture of network-based portals and sketch the types of services we envisage it will enable. In Section 2 we describe the design and implementation of a replay portal that operates within this architecture to offer a new on-demand TV-viewing experience. Section 3 provides a summary of related work, and Section 4 concludes the paper.

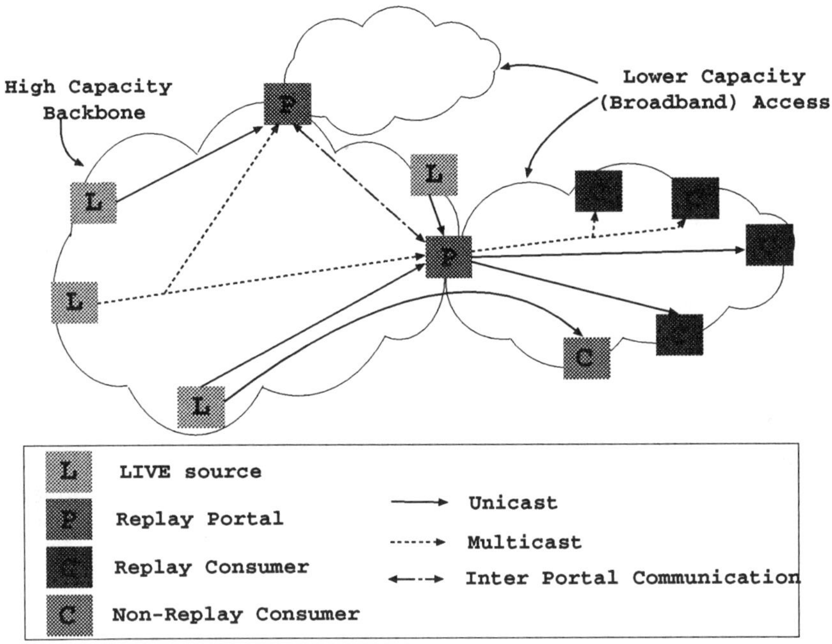

Figure 1 Network Based Replay Service

1. NETWORK BASED REPLAY SERVICES

Our basic assumption is that we are dealing with an environment where high quality live video is being distributed across an IP network. This live content might enter the IP network "locally" (e.g. from a satellite feed at the head-end of a local access plant) or might indeed be carried on IP from the remote live source. This world essentially duplicates or emulates the "pure broadcasting" (or more correctly for IP, multicasting) model of current TV networks. In our architecture we add the notion of a **Replay Portal** to this basic infrastructure to change the broadcasting model to a hybrid on-demand model. This arrangement is depicted in Figure 1 which shows a high level view of our architecture, its different components and variations.

A Replay Portal becomes the local video access point for customers and provides the following functions:

- Access to live content (subscribed to by portal on behalf of users or based on the content offering of the replay service).
- Moving window recording of recent (e.g. the last 24 hours worth of) live content, enabling on-demand viewing of such content.
- "Pause" and "Replay" functionality of live content.

- Personal recording facilities.
- Subscription to non-local live content, (obtained on subscription or on-demand basis from other Replay portals).
- Indexing and search functionality providing access to the video content of multiple cooperating portals.

As shown in Figure 1, content is delivered to the portal either by unicast or multicast, or indeed by a special content distribution mechanism. Similarly, from the portal downstream, content can be delivered by means of multicast, as would be the case when live content is watched through the portal, or unicast when previously stored content is watched on-demand. Customers that do not make use of this service, but are located on the same access network can connect directly to the original live sources (assuming that this is allowed by the live content provider) as they would do in the absence of the replay server. (Such customers will of course not be able to make use of the replay portal functionality.)

The high quality service we envisage will only be realizable on broadband access networks. While these access networks will increase the access capacity with an order of magnitude or more [Eldering et al., 1999], access capacity is expected to lag behind backbone capacity for a considerable time to come. We position the replay portal at this capacity discontinuity. We view our architecture as enabling the replacement of current TV offerings and as such schedule based streaming of content from the headend is still the basic service offering. However, in our architecture, video is only streamed downstream of the portal if there are actually consumers of a particular stream downstream of the portal. This can easily be achieved in an IP environment, where streams are delivered via multicast rather than broadcast, and is expected to result in bandwidth savings across the access network.

For example, in a cable based access network the portal is expected to be located in the cable headend. As access capacities increase, the capacity discontinuity, and the portal location, will move downstream towards the home. Eventually, in a fiber to the home scenario, the discontinuity will disappear or move into the home. When this happens, the bandwidth savings envisaged by our architecture will become less important but the service interaction and integration aspects discussed in Section 1.1 will be as important only at a much larger scale.

1.1. SERVICE SCENARIOS

In this section we motivate our approach by considering a variety of services or service features enabled by our IP-based architecture. We require the basic functionality to be equivalent to current TV and as such the basic service offering is still schedule driven access to a variety of live channels. These live channels are transmitted downstream by means of multicast delivery. The services and features considered below provide some enhancement to this basic service.

Moving window replay of subscribed channels: For a predetermined set of channels the portal stores a moving window of the most recent N hours worth of content for each channel. This stored content is made available to subscribers for on-demand viewing. Different ways of indexing can be provided to this stored content ranging from simple time based schemes to indexing that is content aware.

Library archive of popular programs: A complementary way of providing access to previously recorded content is to maintain a library of certain popular programs in the replay portal. For example, all programs in a certain series such as the X-Files or Star Trek can be archived for subscribers to the portal service.

Replay and pause of non-portal-subscribed channels: Customers of the portal service can also watch other live content, i.e. content not subscribed to by the portal, through the portal. In this case the portal will contact the actual live upstream server. Content is streamed to the downstream customer via the portal which stores a small moving window (say M minutes) of the stream. This small stored window allows customers to request a replay of a recent part of the stream, or pause the live stream (causing the window size to increase) and to then join the live stream again.

Network-based personal recording: A small extension of the above services allows customers to make their own personal recordings which are stored in a personal account on the portal. Reliable recordings can be initiated in a variety of ways and the content can be from either subscribed or non-subscribed channels. Note that combining this functionality with the moving window store allows a user to record content after it has been "aired". In this manner a user can "retro-actively record" something from the replay store thereby adding it to the user's personal collection.

Friends access to personal library: Since the portal is located in the network and on a high capacity backbone it is possible for users to allow access to their personal library of recordings to friends. A user might for example see something that he/she knows is of interest to a friend and start recording it. Having finished the recording the user can simply mail a pointer to his/her friend who can then stream (or transfer) the content from the portal where it was recorded.

A very powerful extension of the above service offerings, enabled by the fact that the portal is network based, is to allow interportal exchange of content. The novelty, over current video-on-demand offerings, is that the user has control over the source of the material.

Subscription based exchanges: The simplest form of interportal exchange will be interportal-subscription based. In this case content stored by a remote portal is transferred to the local portal for on-demand viewing by local customers. Certain non-local channels (stored by a remote portal) might be of sufficient interest to the local community to warrant it forming part of the local portals regular offerings rather than having customers access it from the remote portal directly. An example might be regular (or seasonal) European sporting events made available in the U.S.

Personalized content aware exchanges: A much more sophisticated means of content exchange between portals might involve users specifying a profile of interest, with portals exchanging content based on the profiles of its local users. In this way users are ensured of receiving up to date streaming content on the topics they find interesting.

In the above examples the assumption was that content produced by some live sources was stored, indexed and made available for retransmission by the portals. Such actions will clearly require some agreements between the portal operator and the producers of live content, and a number of business arrangements are possible between content providers, portal service providers, customers and advertisers. However, in that scenario the mode of operation of content providers remains unchanged from the current cable TV scenario. A logical extension of this involves negotiation between the portal operator (or a third party that use its infrastructure) to directly negotiate with the content provider for specific content:

Local target audience: In this case the portal becomes the access point for a particular mix of programs targeted at a specific audience. At one end of the spectrum this might resemble the service currently offered by a local TV station. The key point however is that basing this architecture on a packet network allows this type of service to scale to arbitrary small target audiences. For example the the target audience might be the local bird-watchers club that obtains (and adds to its library) programming information of interest provided by a variety of content providers.

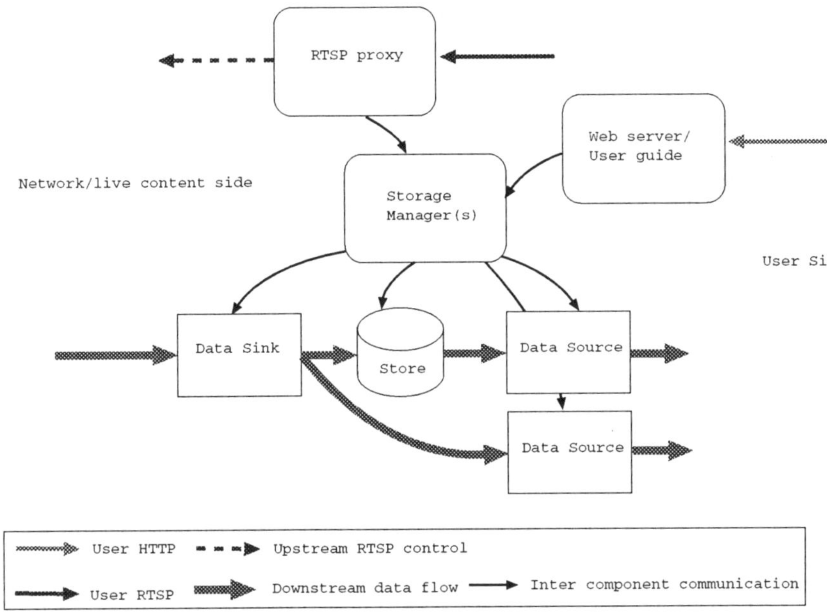

Figure 2 Replay Portal Architecture

2. REPLAY PORTAL DESIGN AND IMPLEMENTATION

2.1. PORTAL ARCHITECTURE

The main architectural components of a Replay Portal are shown in Figure 2. Our architecture is built around standard IETF protocols namely the Real Time Streaming Protocol - RTSP [Schulzrinne et al., 1998], the Real Time Transport Protocol - RTP [Schulzrinne et al., 1996] and the Hypertext Transfer Protocol - HTTP [Fielding et al., 1999]. A user typically starts interaction with the portal by means of accessing a portal **Web-server/User-guide**. This interface provides the user with personalized access to and control of the portal content. Personalized portal content includes the portal-subscribed content (either live or on-demand) as well as any content stored in the user's personal store. The Web interface offers a number of ways of indexing the content that is of interest to the user and allows the user to initiate streaming of any such content. In the case of a personal store the user can also perform management functions such as removing previously stored content or setting up the recording of a future streaming event. When a user initiates streaming through the portal Web interface, a helper file is downloaded to the user's browser. The Mime type of this file instructs the browser to start up a streaming client application on the user's PC or set-top box passing to the application the RTSP URL contained in the helper file.

A user might also make use of the portal for content that is not subscribed to by the portal. In this case the user would not typically make use of the portal web interface. Rather the user will go to a web interface associated with the content source and obtain an RTSP URL in similar fashion as described above. This also means that in this case the user's first interaction with the portal will be through the RTSP interface as described next.

The RTSP URL obtained by the streaming client through either of the above approaches is presented to the **RTSP proxy** on the replay portal. The proxy establishes whether the URL represents content currently stored in the proxy or whether it is necessary to establish a connection to an upstream server. If the requested content is available on the server, either live or stored, the proxy initiates delivery of the content to the client. (In these cases the proxy would have contacted the relevant upstream servers beforehand.) If the content is not locally available, the proxy will contact the upstream server and on success will initiate local handling of the content as well as delivery to the client.

The actual manipulation of content on the portal is performed by a set of **storage managers**. Each storage manager is in control of a specific physical data store and controls the way content is added to and removed from the store. The storage manager provides the Web interface with information about the contents of a particular store, for example to create an RTSP URL to pass back to the client. Similarly the storage manager can tell the RTSP proxy whether a particular URL is currently to be found in the local store. The storage manager(s) manipulate the store(s) under control of the RTSP proxy. For example, in the case of live content being viewed through the portal, the RTSP proxy will instruct the storage manager to create a **data sink** and a **data source** for the data path handling of the stream. The data sink receives the content from upstream and writes it to the store, while also making the content available to the data source for immediate delivery to the client.

The portal architecture lends itself to a number of implementation options depending on the required scalability. In the simplest case a small replay portal can have all the components executing on a single physical machine. This is the nature of the prototype implementation which is discussed in more detail in the remainder of this section. A more scalable realization could involve **frontend web and RTSP servers** which hand off processing of streaming content to a farm of **backend**

RTSP servers and storage managers. This arrangement is depicted in Figure 3. In this case access to the portal through the web interface will result in one of the backend servers being chosen based on server load, the content to be accessed or some other policy. Similarly direct accesses to the portal RTSP interface, handled by the RTSP frontend, will be handed off to one of the backend servers.

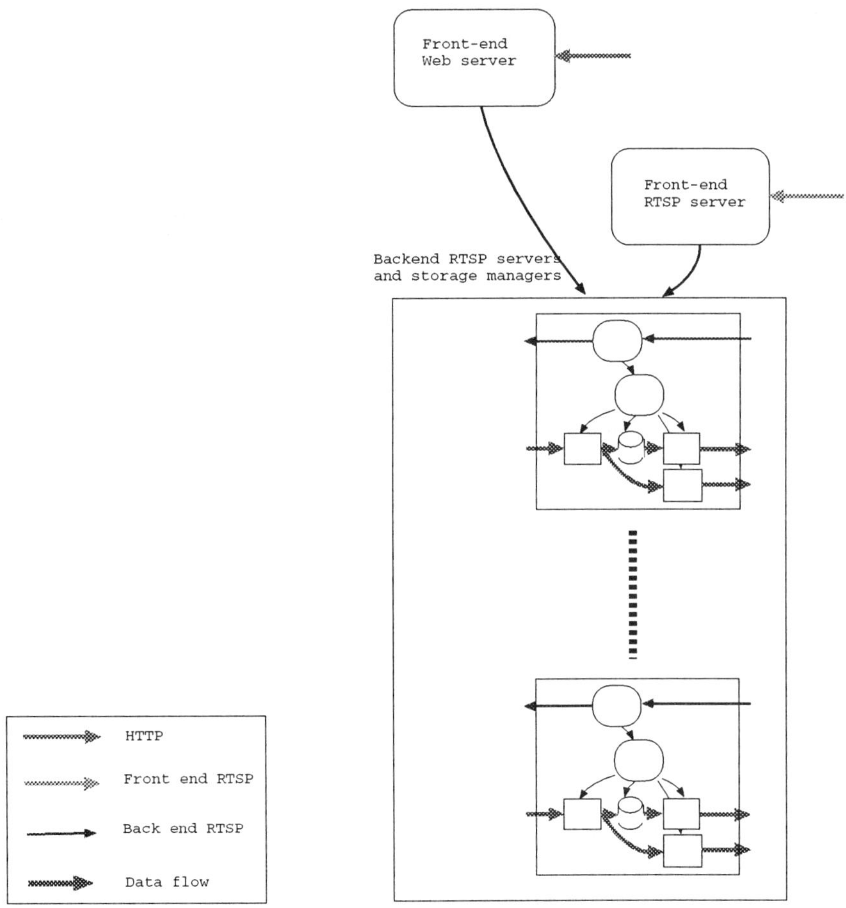

Figure 3 Replay portal with RTSP proxy/storage manager farm

The data path handling of streaming content can similarly be realized in a variety of implementations. Again in the simplest case an RTSP server, storage manager combination can simply execute on a single server machine potentially with two network interfaces. In such an implementation the server could however easily become a bottleneck, as

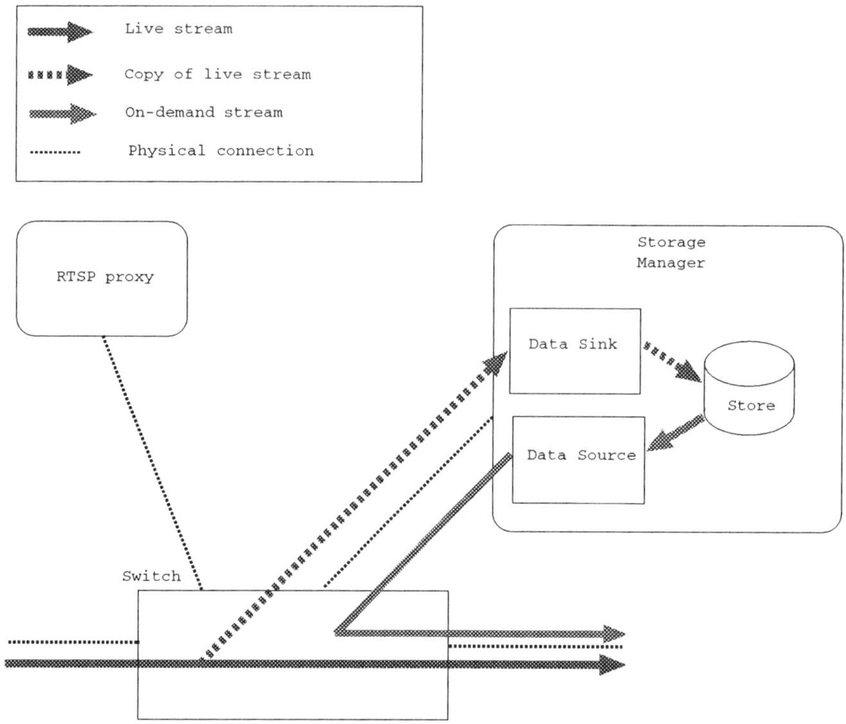

Figure 4 Scalable portal realization

it has to handle re-delivery of any live streams as well as any on-demand delivery of streams. An alternative realization is depicted in Figure 4. In this case an RTSP proxy and its associated storage manager is separated by means of a forwarding device such as a switch or a router. As before the storage manager is effectively controlled by the RTSP proxy based on user requests. The RTSP proxy also has some control over the forwarding device. In particular the RTSP proxy can instruct the switch to have a **copy** of a particular stream delivered on the switch interface connected to the storage manager. As before the RTSP proxy instructs the storage manager to expect and store this stream. In this case the storage manager does not handle the live stream at all and is only responsible for handling any on-demand requests.

While we do not expect any major obstacles in realizing the portal architecture in a scalable manor, many details need to be worked out and this is the subject of current and future work. In the remainder of this section we will concentrate the discussion on our current prototype implementation.

2.2. PROTOTYPE IMPLEMENTATION

In order to demonstrate the feasibility of our architecture we have developed a prototype system consisting of all the elements in our architecture:

- A live server
- A replay portal (consisting of web server, RTSP proxy and storage managers)
- A streaming client

Since we expect to provide a high quality service we use MPEG2 encoding for the video streams making use of hardware encoders and decoders we have used in earlier work [Basso et al., 1999]. We use hardware encoders from VisionTech [Vision Tech, 2000], while the decoders are from SigmaDesign using a Microsoft Windows environment [Sigma Designs, 2000]. RTSP is the control protocol that binds all our components together and we have developed an RTSP library (librtsp) which has been derived from an early public domain implementation from Real Networks [RealNetworks, 1999]. The portal was implemented on a Linux infrastructure and the web server is an unmodified Apache server [Apache Software Foundation, 2000]. Since we knew from the outset that we would be dealing with a diversity of platforms and operating systems, code portability was a major concern. We addressed this by developing a basic portability library (libcommon) that dealt with operating system specific issues and provided a common interface to other libraries and applications.

Each of these libraries and the applications built on them are discussed in more detail in the sections below.

2.2.1 Support libraries: libcommon and librtsp. The main functions provided by libcommon are an event scheduling mechanism and IO handling of both network and file systems across all supported platforms. The event scheduling mechanism allows specific functions to be called based on time, network or file events. This include the running of "background" tasks when the system is idle. Libcommon also contains a number of general mechanisms such as safe string handling and ring buffer and table manipulation.

Librtsp builds on libcommon and provides a simple way for either client or server applications to use RTSP. For example a client application simply calls "rtsp_connect" to initiate communication with an RTSP server. On success the client obtains a handle with which all further interaction with the server (i.e. describe, play etc) is conducted through a remote procedure call (RPC) like interface. The library deals with message formatting and parsing and presents the content of messages to the application in the form of well defined structures, or form RTSP messages out of structures provided by the application.

2.2.2 RTSP client and server. The structure of the client software is depicted in Figure 5. A graphical user interface (GUI) allows the user to specify the RTSP URL of interest and initiate streaming. (As explained earlier an alternative is for the URL to be supplied to the client software by means of a helper file downloaded by a web browser on the client device.) On successful RTSP interaction with the server, the client sets up a datasink and a ring buffer and initiate the MPEG hardware decoder. The datasink receives an RTP encapsulated MPEG stream from the network, strips off the RTP encapsulation and puts the MPEG packets in a ring buffer for asynchronous collection by the MPEG decoder hardware. The decoder driver performs an upcall into the application whenever its buffers are below a certain threshold at which point data is transferred from the ring buffer to the decoder.

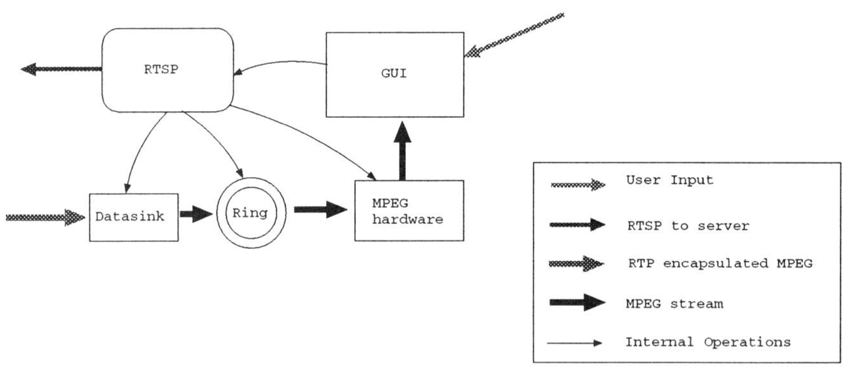

Figure 5 RTSP client

Figure 6 shows the main components of our RTSP server implementation. RTSP requests received by the RTSP library are passed to a **Media Manager** which determines if there is a media backend that can handle a request of this type. A number of media specific backends have been implemented namely backends for MPEG2 audio, MPEG2 transport and WAV streams. These backends deal with media specific issues such as the frame format of streams, the rate at which streams should be played out and how to encapsulate media frames in RTP. The content on which the backends operate can be either stored on disc or be supplied in real time from an encoder. For example, our live server is implemented as an encoding thread which supplies an MPEG stream to an MPEG transport stream backend.[1]

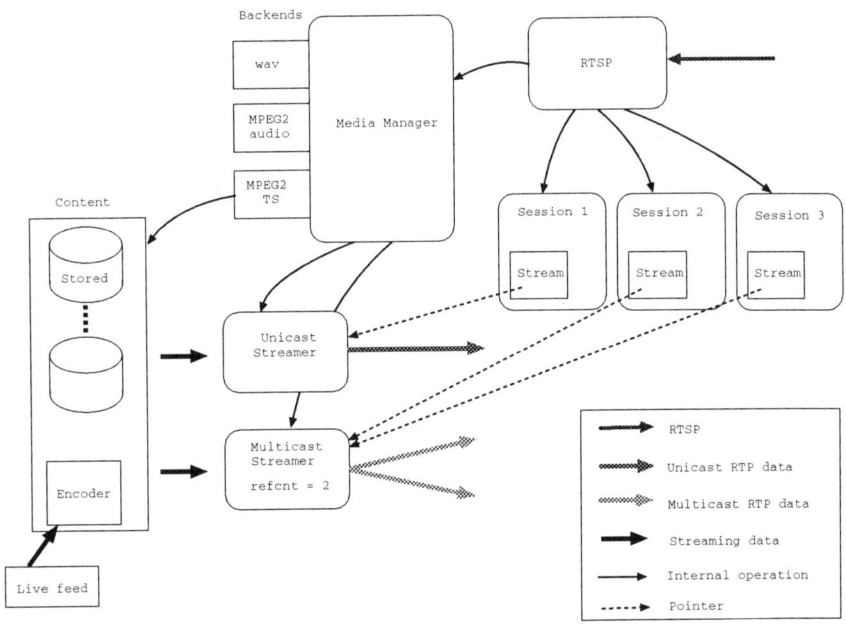

Figure 6 RTSP server

Once the Media Manager has determined that the requested content is available (i.e. a successful RTSP DESCRIBE interaction), the client application normally issues RTSP SETUP and PLAY requests. The SETUP request results in session state being created in the server and a **streamer** is initialized to deliver the stream. A PLAY request starts delivery of the stream. The session state contains stream information that is relevant **for this stream for this session** (e.g. the time a session joined a stream), whereas the streamer contains only session independent information about the stream. This separation is important in order to deal with multicast streams. The first client to request delivery of a multicast stream will result in a streamer being created. Subsequent sessions for the same stream will be served by the same streamer and

[1]Currently only the live server makes use of threads as the encoder hardware and SDK operates as a threaded application on the Solaris platform.

a reference count in the streamer ensures that the streamer does not disappear when the initial session is terminated.

2.2.3 RTSP proxy and Storage Managers. The RTSP proxy functionality required by our replay portal is realized by having the proxy as another media backend. As is the case with other backends, the proxy backend determines whether a request can be satisfied from its local stored content. However in the case of the proxy, the server address of the RTSP URL is not the proxy address and if the request can not be satisfied locally the proxy backend can issue an upstream RTSP request to the server specified in the URL. (In our current implementation the client has to be configured to establish an RTSP connection to the proxy server rather than the real content server.)

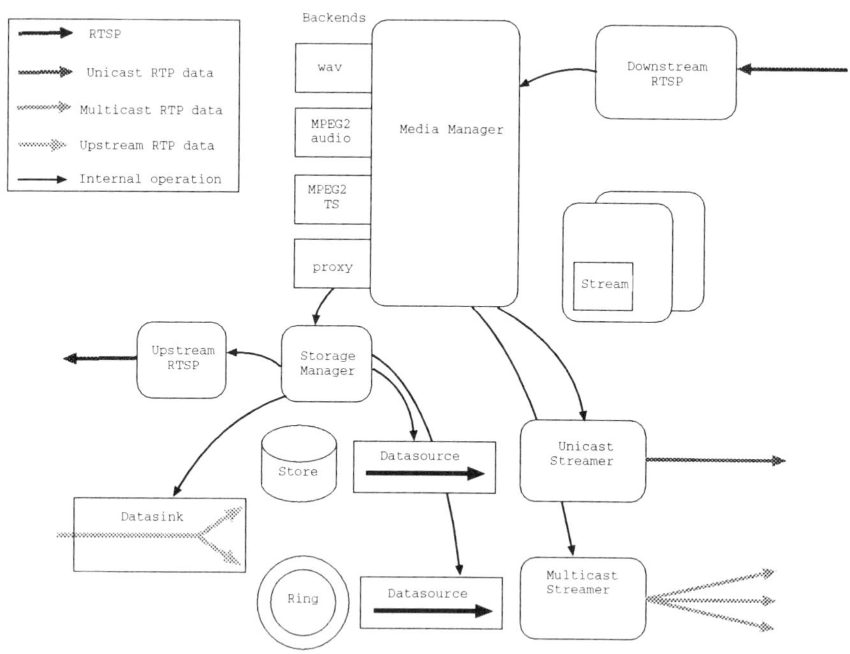

Figure 7 RTSP proxy and Storage manager

If the request can be satisfied from the local store, a streamer is set up as described above and the stream is delivered to the client. If content is received from upstream, a datasink will receive the packets writing them to disk and putting a copy in a ring buffer for delivery to live viewers of the stream (if any). In the case of the proxy backend, content is stored to disk with the RTP header it was received with intact. Subsequent playout of stored content is then based on the RTP timestamp of the stored packets while the RTP sequence numbers are replaced for retransmitted downstream delivery. Storing content in the proxy with RTP headers intact has the desirable property that our proxy is media independent so long as the stream is delivered using RTP and the clock frequency used for RTP timestamps is known from RTSP interaction.

The storage manager(s) handles the manipulation of stored content. This includes the eviction policy associated with a particular stream. In the case of portal subscribed content which is made available for on-demand viewing the policy is simply to keep the last N hours worth of content. This is currently implemented as a logical circular set of files where the oldest file gets overwritten after N hours with new content. In order to have the N hour window move forward in time with a small granularity and to ease time based indexing into the stored content each of these files holds a relatively small amount of data, in the order of one or two minutes worth of content.

In the case of a user watching non portal subscribed content through the portal, the store manager in the proxy still stores some amount of the content to disk. This is needed to facilitate replay of very recent content (i.e. in the order of the last few minutes). However in this case the eviction policy of the storage manager is much more aggressive.

A key aspect of the replay service is that it provides access to arbitrary time offsets into the past of live streams. This requires each client and the proxy to agree on a certain reference point in time in the live stream. We call this reference point the **fixedpoint** relative to which all time sensitive interaction is performed. We rely on a mapping between global time (UTC) and the RTP timestamps in the media stream for this purpose as is described below.

Consider the interaction between a live server and a proxy: The server picks an RTP packet (the first for a unicast stream) that it considers the start of the stream for a particular user and records the absolute time that corresponds to this timestamp as the server fixed point for the session. This information, i.e. fixed time and RTP timestamp is relayed to the proxy via the control channel, e.g. in the PLAY response message. Given this information, the proxy, when it receives an RTP packet can work out the absolute time that the server would have associated with this packet (the clock frequency for the RTP time stamp is known from the RTSP interaction). This absolute time is stored with each RTP packet on the portal and is used for obtaining stored previous live content from the portal. For example, a client might obtain from the portal web interface a URL of the form

`rtsp://pc-green:8554/live.m2t?pausepoint=utc:19991011T103400Z`

based on the selection made from the schedule on the web interface. The absolute time in the URL (the pausepoint), represents for example the time when a particular program was aired live, and is used by the proxy to serve the appropriate content by comparing it with the absolute times stored with each RTP packet. The main point here is that all time offsets into the media stream is effectively based on the RTP time stamps of the live source which allows the indexing based on the time the content was aired, which is crucial to our approach.

For per stream (or VCR-like) functions between the proxy and the client we employ a similar approach. Again the server (or more correctly the proxy in this case) gets the absolute time of the first RTP packet it is about to send (i.e. the proxy fixed point), to the client and sends this to the client in a control message. The absolute time for this packet is the absolute time the packet was first sent by the live source not the absolute time when the proxy is about to send it. When the client is about to **play out** this packet it takes a local time stamp which becomes the client fixed point. Now when the user performs a per stream operation, e.g., a PAUSE, the client works out the time difference between the time at which the operation was performed and the client fixed point and adds this to the known proxy fixed point which is specified in the request sent to the proxy as an absolute pausepoint as before.

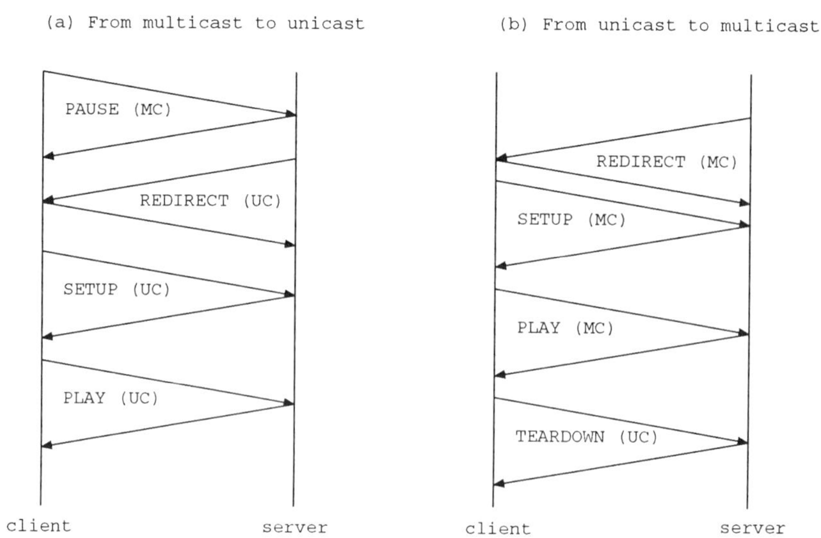

Figure 8 (a) Moving from multicast to unicast and (b) moving from unicast to multicast

A user watching a live event that performs a PAUSE, will have to be switched from (typically) a multicast stream to a unicast (per-user) stream on which these operations can be performed. This can be done by having the proxy send the client an REDIRECT request with the transport parameters for the new unicast stream after the per-stream operation, followed by the client doing a SETUP and PLAY with the new transport parameters as per normal. Figure 8 (a) shows this arrangement. At some later time the proxy might realize that the client's unicast playout point has moved close to the playout point for the live stream. This might for example happen as a result of the user fast-forwarding through the stream. The proxy might then sent another REDIRECT message to the client as an invitation to rejoin the live (multicast) stream. At its discretion the client might then rejoin the live stream by performing a SETUP and PLAY with the multicast transport parameters and eventually tearing down the unicast connection. This is depicted in Figure 8 (b). An alternative is for the client to explicitly rejoin the live (multicast) stream because of for example the user clicking on a "joinlive" button.

3. RELATED WORK

If we focus the discussion on a single replay portal then its functionality is similar to that of consumer electronic devices such as those provided by TiVo [TiVo, 2000] and ReplayTV [ReplayTV, 2000]. The products of these companies are very similar – both sell a combination of a hardware device and a TV listings service. These devices provide an in-home replay service, and do not allow the benefits of content sharing provided by a networked replay portal. A networked solution can also offer higher degrees of reliability, more sophisticated search and indexing, and relieves the consumer of the burden of having to keep pace with advances in technology. Also, as indicated in Section 1 the Replay Portal architecture avoids the blanket broadcasting of content across access networks which is not possible with regular consumer electronic devices, which only deals with the video signal once it has already been delivered to the home. Having a network based service also implies a managed service freeing users from the chore of managing their own content unless they so wish. Finally, in the Replay Portal architecture a user is not limited in the number of simultaneous recordings that can be performed by tuner limitations in a home device. Since all storing of content happens inside the network, on shared service provider infrastructure, a user might be simultaneously recording multiple streams (at the portal) while watching any one of these (or indeed any other stream) live. Prior work on network-based services for processing TV content includes Agora [Hyden and Sreenan, 1996] from Bell Labs. Agora allows users to have personalized access to TV newsfeeds, its main focus being techniques for efficient content extraction and event notification.

Addressing some of the same issues from a different angle are the activities of the Advanced Television Enhancement Forum [Advanced Television Enhancement Forum, 2000]. This type of interactive television aims to add HTML data as overlay information on TV signals. This approach does not change the fundamental broadcasting-everything model and is therefore unlikely to succeed in an environment where users are demanding personalized services.

Another important area of related work is Internet based content distribution. The replay portal architecture presented in this paper will be a value added service to a "basic" streaming content distribution network. Our architecture will make use of a content distribution network in order to get content to portals and to exchange content between portals and will then add the replay and related functions in a service offering. Current product and service offerings in this space mainly cater to Web traffic but support for streaming content is becoming available from both the vendor and research communities [Francis, 1999, Sight-Path, 2000, RealNetworks, 2000b, Fast Forward Networks, 2000]. One part of the problem solved by these offerings resolves around on-demand streaming of fairly short (low quality) clips where the objective and solution is very similar to that of Web content, i.e. to get content closer to users and to make intelligent choices as to what server will serve a particular request. The problem is generally addressed by creating an overlay network of cooperating content distribution servers which interact with each other and the actual content servers to offer load balancing, redundancy and reduced latency. In the domain of live streaming content the overlay network can also provide efficient application level distribution trees between the content distribution servers and offer retransmission facilities in the overlay network to compensate for the lack of such mechanisms in streaming protocols. Indeed one of the major problems with current streaming offerings [RealNetworks, 2000a, Microsoft, 2000] is the lack of standard protocols on which to transfer streaming content. This means that content distribution server vendors are required to support a number of proprietary protocols in order to realize their goals thus increasing the price and complexity of their products. More seriously though is the fact that these proprietary protocols are not subjected to the same amount of scrutiny TCP has undergone and its impact on the stability of the Internet is therefore unknown.

The final substantial area of related work is that of video-on-demand (VOD). The work presented here is not VOD in the "traditional" sense, where video content is somehow uploaded to a server and then made available for on-demand viewing. Rather in our architecture, live schedule-based content is made available for on-demand viewing as soon as it has been "aired". Nonetheless, as soon as content is viewed on-demand, we expect that many of the techniques and methods developed for VOD will be applicable in our architecture. For example, access to popular content might well benefit from batching [Dan et al., 1994] or patching [Sen et al., 1999] techniques. Batching involve slightly delaying a particular request for content in the hope that other requests for the same content will arrive soon so that all requests can be served with a single response and content delivery. Patching on the other hand tries to exploit the buffering capabilities of endpoints by allowing a client to receive (and buffer) part of a clip from an existing stream, and the server then only has to send the missing initial part of the stream.

4. CONCLUSION

We presented a hybrid IP-based architecture which explores the space between broadcasting and personalized on-demand access to streaming media. Our solution maintains the current schedule driven approach of present day TV, while making previously "aired" content available for on-demand viewing in a variety of ways. The architecture presents an attractive means for service providers to gradually introduce a variety of services, on a common IP transport infrastructure, which enables the possibility of rich interaction between different packet based service offerings.

References

[Advanced Television Enhancement Forum, 2000] Advanced Television Enhancement Forum (2000). `http://www.atvef.com`.

[Apache Software Foundation, 2000] Apache Software Foundation (2000). `http://www.apache.org`.

[Basso et al., 1999] Basso, A., Cash, G., and Civanlar, M. (1999). Implementation of a real-time MPEG-2 delivery system based on RTP. Packet Video 99, NY.

[Dan et al., 1994] Dan, A., Sitaram, D., and Shahabuddin, P. (1994). Scheduling policies for an on-demand video server with batching. Proceedings of the second ACM international conference on Multimedia.

[Eldering et al., 1999] Eldering, C. A., Sylla, M. L., and Eisenach, J. A. (1999). Is There a Moore's Law for Bandwidth ? *IEEE Communications Magazine*, 37(10):117–121.

[Fast Forward Networks, 2000] Fast Forward Networks (2000). http://www.ffnet.com.

[Fielding et al., 1999] Fielding, R., Gettys, J., Mogul, J., Frystyk, H., Masinter, L., Leach, P., and Berners-Lee, T. (1999). Hypertext Transfer Protocol – HTTP/1.1. RFC 2616.

[Francis, 1999] Francis, P. (1999). Yallcast: Extending the Internet Multicast Architecture. http://www.yallcast.com.

[Hyden and Sreenan, 1996] Hyden, E. and Sreenan, C. J. (1996). Agora - a personalized digital newsfeed. NOSSDAV.

[Microsoft, 2000] Microsoft (2000). http://www.microsoft.com.

[RealNetworks, 1999] RealNetworks (1999). RTSP: Reference Implementation. http://www.real.com/devzone/library/fireprot/rtsp/index.html.

[RealNetworks, 2000a] RealNetworks (2000a). http://www.real.com.

[RealNetworks, 2000b] RealNetworks (2000b). Real Broadcast Network. http://www.rbn.com.

[ReplayTV, 2000] ReplayTV (2000). http://www.replaytv.com.

[Schulzrinne et al., 1996] Schulzrinne, H., Casner, S., Frederick, R., and Jacobson, V. (1996). RTP: A Transport Protocol for Real-Time Applications. RFC 1889.

[Schulzrinne et al., 1998] Schulzrinne, H., Rao, A., and Lanphier, R. (1998). Real time streaming protocol (rtsp). RFC 2326.

[Sen et al., 1999] Sen, S., Gao, L., Rexford, J., and Towsley, D. (1999). Optimal patching schemes for efficient multimedia streaming. NOSSDAV'99, Basking Ridge, NJ. Available form:http://www.nossdav.org.

[SightPath, 2000] SightPath (2000). http://www.sightpath.com.

[Sigma Designs, 2000] Sigma Designs (2000). http://www.sigmadesigns.com/.

[TiVo, 2000] TiVo (2000). http://www.tivo.com.

[van der Merwe et al., 1999] van der Merwe, J., Chu, Y.-H., Caceres, R., and Sreenan, C. (1999). mmdump: A tool for monitoring internet multimedia traffic. AT&T TR 00.2.1, available from http:\\www.research.att.com/resources/trs.

[Vision Tech, 2000] Vision Tech (2000). http://www.visiontech-dml.com/.

A NEW MULTICAST TECHNIQUE FOR VIDEO TRANSMISSION OVER ABR SERVICES IN ATM NETWORKS

Santiago Felici
Andres Lopez-Penaranda
Juan Perez
Institut de Robotica. Universitat de Valencia
Poligono La Coma (Paterna) 46071 Valencia
Ph: +34 96 398 35 63 Fax: +34 96 398 35 50
Santiago.Felici@uv.es

Jorge Martinez
Dept. Comunicaciones. Universidad Politecnica Valencia
jmartinez@dcom.upv.es

Abstract Multicast techniques are the only way to simultaneously provide flows of information from one source to several destinations. The intention of this paper is to study and to evaluate different multicast techniques using a video coder based on an adaptive video compression algorithm with subband coding and a best effort network service like ATM with the Available Bit Rate (ABR) service. This video transmission can adapt faster and easily to changing network conditions. In this way, we present an evaluation process for a determined network configuration. Thereafter we discuss the results obtained by simulation and propose for this video transmission a trade-off between these multicast techniques, in order to obtain the best perceptual video quality.

Keywords: ATM-ABR services, multicast, QoS, adaptive video compression algorithm, multiresolution

1. INTRODUCTION

Different multicast techniques and different network technologies are being analyzed to determine which one offers better performance for multimedia traffic. This study is more relevant when the network offers best effort services because in this scenario it is more restrictive to maintain a certain Quality of Service (QoS) using the available network resources. We focuses the study on ATM with Available Bit Rate (ABR) service.

The ABR class of service was initially conceived to support data traffic. Its service model is based on the *best-effort* paradigm but enhanced by some specific characteristics: *fair sharing of the available resources among the contending ABR connections* and *a closed-loop feedback mechanism with Resource Management Cells (RM) with each destination.* Nevertheless in a multicast tree, when different connections over the same source are running simultaneously, this closed-loop feedback with each destination becomes a problem, because each destination is providing different information to the source. The switches within the *multicast connection* (or *multicast tree*) have to manage different RM cells to the same source, what is called a multicast congestion control.

The intention of this paper is to evaluate different multicast techniques for ATM-ABR, using a subband based video coder, and finally to propose a suitable multicast technique for this kind of information exchange. The rest of the paper is structured as follows: in Section 2 is explained the network configuration for the evaluation process of these multicast algorithms; Section 3 describes the operation of the adaptive video coder based on subband coding; in Section 4 different multicast algorithms over ATM-ABR are evaluated, trying to compare them; in Section 5, we propose a new algorithm and in Section 6 we evaluate the performance of the proposed technique, providing some numerical results. Finally, Section 7 presents the conclusions and ideas for future work.

2. DEFINITION OF A NETWORK CONFIGURATION FOR THE EVALUATION PROCESS

This section explains the basic assumption for the evaluation of the different multicast techniques. It is necessary to use a network configuration which let us stress the video sequence. The network configuration used in the experiment, is given in Figure 1. It has 4 switches which are multicast capable. Source A is a multicast one and source B is an unicast one working as a greedy source (what means that it uses as much bandwidth as it available for them). The *multicast connection* (or *multicast tree*) has three leaves, one in the second (A1), third (A2) and fourth (A3) switch (from left to right). The links L1 and L3 are 50 km long and the link L2 is 100 km long. The access link is 0.2 km long. The propagation delay is 5 μsec/km. The ABR sources are explained in [1], and are specified by the ABR source parameters: PCR (Peak Cell

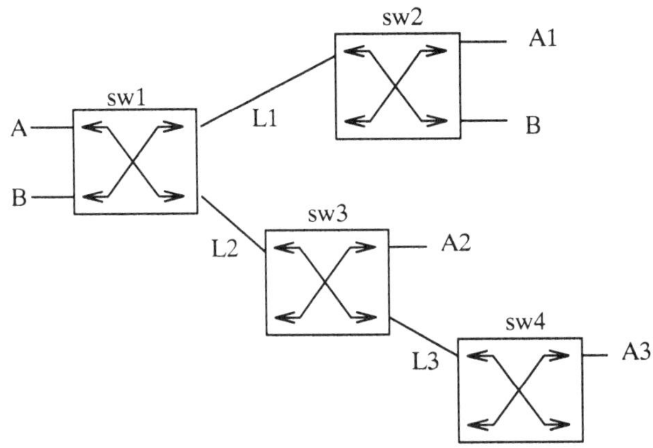

Figure 1 *Network configuration*

Rate)= 23.85 cells/msec, MCR (Minimum Cell Rate)= 0.8516 cells/msec and ICR (Initial Cell Rate)=2.24 cells/msec. The rest of parameters used for these ABR sources are: Trm=10 mseg, ADTF=10000 msec, RIF=0.0625, TCR=0.8516 cells/msec, CDF=0, RDF=0.0625, Mrm=2 cells, Nrm=32 cells, TBE=0, Crm=1000 cells and FRTT=0.

The links have 10 Mbps of bandwidth, apart from link L2 for which the bandwidth is changed. Because the intention of this paper is to evaluate different multicast congestion controls, we compare the ACR (Allowed Cell Rate) of the sources when different changes of the bandwidth are introduced in the configuration. The bandwidth change in link L2 from 10 to 3 Mbps at 150 msec and from 3 to 10 Mbps at 300 msec.

3. ADAPTIVE VIDEO COMPRESSION ALGORITHM AND SUBBAND CODING

Video-based applications that are rate adaptive can obtain substantial benefits over best effort network services, as can be seen in [2]. In ABR connections, these benefits can be summarized in the following three aspects. First, these applications typically require some guarantee on bandwidth, for example a minimum encoding rate for a video stream, but can take advantage of spare bandwidth. This can be supported by an ABR connection using a Minimum Cell Rate (MCR) at the connection set up. Second, when explicit rate feedback is used and the ABR connections supporting these applications are multiplexed on a dedicated queue at the switches, the cell transfer delay is more predictable because the congestion control mechanism keeps the queues almost empty.

And third, the feedback mechanism keeps each source informed of the available bandwidth it has at his disposal.

A video compression algorithm is based on three steps: a decomposition process, a quantization process and finally a entropy coding process. But if adaptive performance is required, each process requires a suitable design. Adaptability means multiresolution, which can be implemented using a subband decomposition or subband coding. A subband decomposition is a process where the information is decomposed in subbands at different levels of resolution; a video signal can be decomposed in a 3D domain, see reference [3].

Each subband has different resolution levels of the original video and if we add all subbands in a reverse decomposition process, we obtain the original video. Obviously, depending on the video information of each subband, not all subbands have the same importance in the human visual system (HVS), because the HVS has different perceptual responses to these subbands. This perceptual priority will determine the order in which subbands are going to be transmitted.

3.1. OPERATION OF THE VIDEO CODER

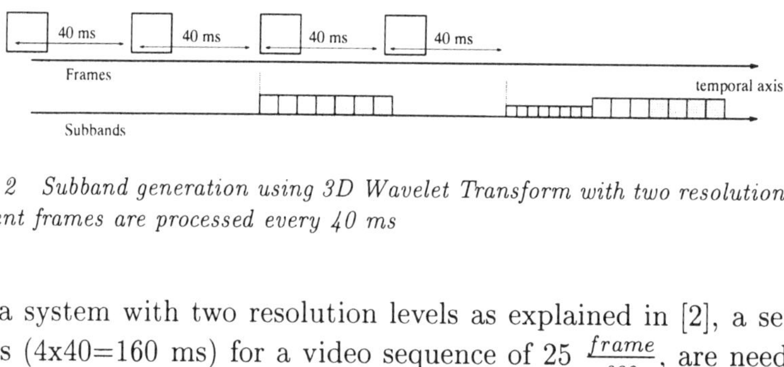

Figure 2 Subband generation using 3D Wavelet Transform with two resolution levels. Different frames are processed every 40 ms

In a system with two resolution levels as explained in [2], a set of 4 frames (4x40=160 ms) for a video sequence of 25 $\frac{frame}{sec}$, are needed to perform a complete 3D subband decomposition. This represents a trade off between the decorrelation ratio and the number of frames that need to be stored at the coder. The process can be observed in Figure 2.

For example, let assume that our system is going to perform the decomposition of 4 frames, that we label as frames 1, 2, 3 and 4. The system uses the pair of frames 1-2 and the pair 3-4 to obtain the first resolution level. This process generates 8 subbands from each pair of frames, but because we use the pair of subbands with the lowest resolution (fewer frame details) of the original pair of frames (1-2 and 3-4) to generate the second resolution level, then only 7 subbands remain at the first resolution level. By this process we obtain 8 additional subbands from the second resolution level. Therefore, at the end of the decomposition process we obtain 7+7+8 subbands. A full explanation of this subband based video coder can be found in [2].

Once each subband is available, the coder creates an information unit per subband, which contains the necessary information to reconstruct each individual subband. The information unit is called Packet Data Unit (PDU). The order in which the different PDUs are transferred, will define their priority. This order is determined by their perceptual weight as said below.

4. OVERVIEW OF DIFFERENT MULTICAST TECHNIQUES

In this section, we are going to describe the five multicast algorithms published in ATM Forum, which will be used for a multicast video transmission with the video coder of Section 3. Our purpose is to compare these algorithms by the ACR of the sources and how they distribute the available bandwidth between the different sources at the switches. In order to evaluate these parameters, we select the configuration of Figure 1

The multicast algorithms from the literature like [4] are:

- **Fast Indication (FI)**: this algorithm proposes that the source transmits at the minimum available bandwidth of all the c branches of the *multicast connection*. When a FRM (Forward RM) cell is received, the switch changes it to a BRM (Backward RM) cell and fills the ER (Explicit Rate), NI (No Increase ACR at source) and CI (Congestion Indication) fields with two classes of information: *external information*, which arrives to the switch by the BRM cells and *internal information*, like queue length and Fair Share calculated in the switch. The minimun between the ER calculated both with the *internal information* and with the *external information*, is written in the BRM cell which is sent. This algorithm has a very fast response but a big consolidation noise (fluctuations).

- **Wait For All (WFA)**: this algorithm eliminates the consolidation noise, because it wait one BRM cell for each branch of the *multicast connection* to send a BRM cell to the source. Now, the information is more reliable, because we wait for the complete information of the entire connection which is given feedback to the source. But we have to wait a period of time (called consolidation time) to feedback. Thus, this algorithm has a slow response, because we have to wait the BRM cell from the slowest leaf of the *multicast connection* to send a BRM cell to the source

- **Fast Overload Indication (FOI)**: it avoids the consolidation time of the WFA algorithm during the overload period. On the one hand, if there is overload, the switch sends a BRM cell with the information currently it has, like in the FI algorithm. On the other hand, during the steady state the switch works as the WFA algorithm. This algorithm has a problem: the $\frac{BRM}{FRM}$ ratio is greater than one, and the desirable $\frac{BRM}{FRM}$ ratio is 1 [4]. However, this algorithm has a faster response than the WFA algorithm and it has less consolidation noise than the FI algorithm

- **RM Ratio Control (RMRC)**: in the FOI algorithm the $\frac{BRM}{FRM}$ ratio is larger than one. To avoid this situation, the RMRC algorithm proposes to control this ratio. During the overload, BRM cells are sent like in the FI algorithm, but during the steady period we do not send BRM cells like in the WFA algorithm.By stead we recover the excess of BRM cells sent in the overload period. This mechanism does not guarantee a $\frac{BRM}{FRM}$ ratio equal to one, but it guarantees a $\frac{BRM}{FRM}$ ratio to be lower than in the FOI algorithm

- **Memory Enhanced (ME)**: it is not exactly an algorithm, but a new mechanism to improve the multicast algorithms in order to reduce the consolidation noise. With the ME mechanism, the switch has information of each branch in the *multicast connection*, not only for the whole connection. This mechanism avoid fluctuations around the operating point, because we have more information about the state of each branch of the *multicast connection*

The above algorithms are mechanisms which decide when a BRM cell has to be sent to the source. Nevertheless a unicast algorithm is also necessary to calculate the portion of the available bandwidth for each connection (ER). In this case, the unicast algorithms are CAPAC [5] and ERICA [6]. In summary, the unicast algorithm calculate the ER written in the BRM cells, but these cells are decided to be sent by the multicast algorithm.

The simulations results can be seen in the following figures. Notice that multicast source A is stressed by link L2 meanwhile, the unicast source B uses the excess of bandwidth. In Figure 3 the ACR of the sources are shown, with important oscillations at 150 and 300 msec (with bandwidth changes, see Section 2). It is remarkable that this algorithm has a fast response as can be observed by the slope of the ACR at these times. The unicast algorithm used in these simulations is CAPAC.

By comparison, we simulated the WFA algorithm. The results can be seen in Figure 4. The oscilations are smaller than in the FI algo-

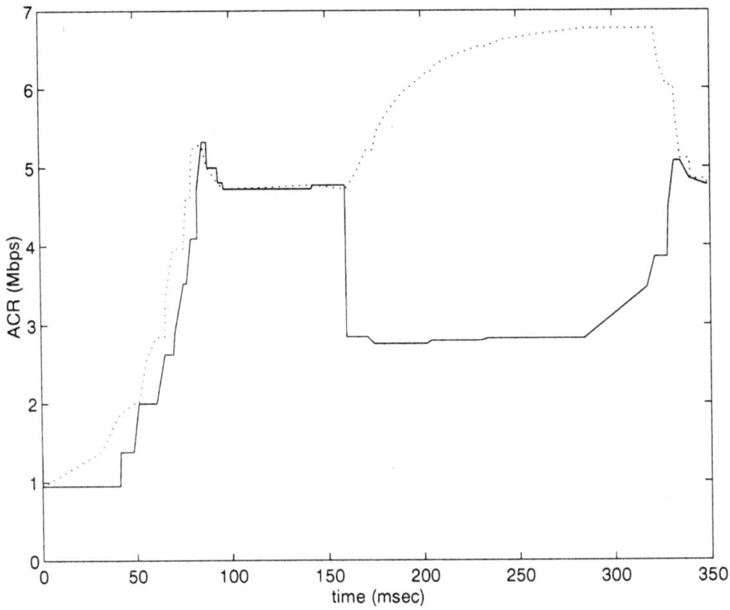

Figure 3 ACR of source A (continuous line) and source B (dashed line) with the Fast Indication algorithm using CAPAC

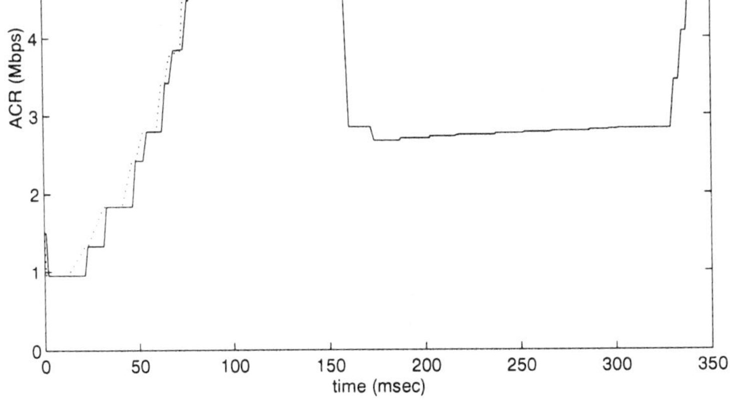

Figure 4 ACR of source A (continuous line) and source B (dashed line) with the Wait For All algorithm using CAPAC

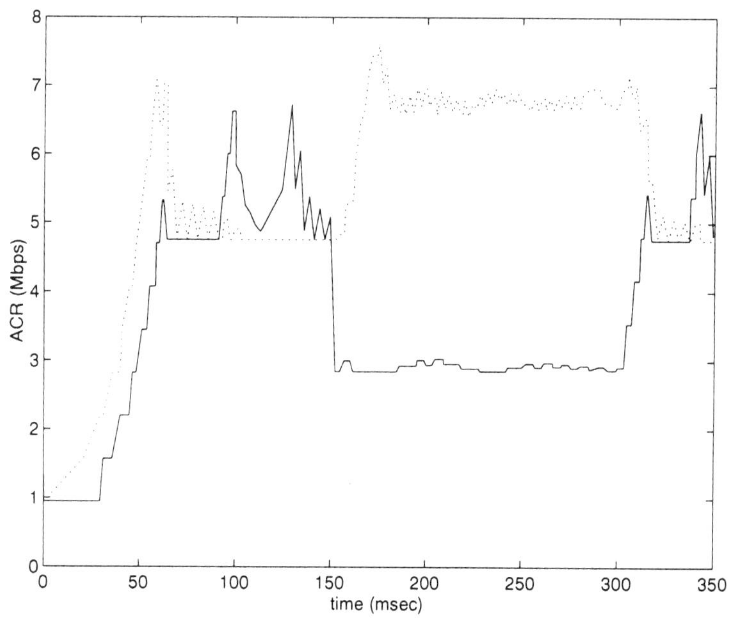

Figure 5 *ACR of source A (continuous line) and source B (dashed line) with the Fast Indication algorithm using ERICA*

rithm, but the response to these changes is slower. Finally, using the FI algorithm as a multicast algorithm with the ERICA as the unicast algorithm, it has a faster response, with steeper slope than in Figure 3, as can be seen in Figure 5. Notice that oscillations are not relevant for the video coder, because its time scale is greater than these oscillations. These oscillations are produced by the multicast connection.

In conclusion, we see that these multicast algorithms have been designed for data traffic. They need to adapt to the worst situation, the bottleneck of the *multicast connection*, transmitting at the minimum bandwidth. Because video application are rate adaptive and a multiresolution process can be used as explained in Section 3, a better choice than a minimum criteria is proposed as presented in next section, trying to trade-off between minimum and maximum values of ER.

5. PROPOSED MULTICAST ALGORITHM

Previous multicast algorithms try to fit the bottleneck bandwidth in the *multicast connection*. Operating in this way, we force the rest of the destinations to work at the video quality determined by this bottleneck. A maximum value would be unrealistic thus as always, a trade-off between minimun and maximun is the best option to assign the available bandwidth to the video source.

It is interesting to determine this value through a probability distribution function of the explicit rates, given by each leaf in the *multicast connection*, independently of the distance between source and each destination. This should be a complex but necessary task to implement, however the switch has to do that in an easy way. A valid approximation is presented in this paper.

Our proposed multicast algorithm, called *Trade-Off in a Fair Share* (TOFS) is based on the number of hops from each destination to the switch. The usage of the number of hops in the RM cells, suppose to declare a new field. This is not a real problem because RM cell data are always nearly empty.

When a particular switch within a *multicast connection* receives several BRM cells, for each branch, it calculates the FS (Fair Share) by the minimum value between MER (Medium Explicit Rate, containing *external information*) and the FS given by the unicast algorithm (*internal information*), like the FI algorithm. The calculated FS is the value which corresponds to the FS available for each branch, whereby each branch has a different number of destinations connected. For instance, it can have either one destination or another switch connected to more destinations. To take the number of destinations into account , we compute the number of hops (*nhops*) associated to each branch, weighing each FS with this values.

For the selection of an unicast algorithm to calculate the FS, we have observed in the Figures 3, 4 and 5 that the better response is given by the FI algorithm with the CAPAC congestion control. Nevertheless, because the ERICA algorithm is well known and more oftenly used, we will choose ERICA for the unicast algorithm in the TOFS algorithm.

The proposed TOFS multicast algorithm is resumed by the next expression:

$$TOFS = \sum_i FS_i \frac{nhops_i}{nhops_{total}} \tag{1}$$

where $FS_i = min(MER, FS_i)$ and $nhops_{total} = \sum_i nhops_i$.

Because the subband video coder is adaptive, it easily achieves a suitable working compression point with a properly bit allocation, where the perceptual distortion is improved.

The proposed multicast algorithm is based on the FI algorithm, but with the ER given by the TOFS algorithm and calculating the number of hops associated to each switch as $nhops = \frac{nhops_{total}}{i}$, being i the number of branches of that switch . These few modifications let us introduce a more realistic scenario, where the video source can take profit and adjust to a number of different destination, not only the worst case or bottleneck. Also, because this algorithm is a modification of FI, it keeps its properties, like the $\frac{BRM}{FRM}$ ratio nearly to one.

6. RESULTS AND DISCUSSION

First of all it is important to notice that the operation of the explained algorithm matches the ERICA behavior, as it should be expected, when there is one leaf at the multicast tree. On the other hand, as it had been said, the video source does not operate at the ACR given by the bottleneck but a tradeoff within the overall multicast tree.

In Figure 6 the ACR of different sources is shown in order to evaluate and to compare the behavior of TOFS algorithm against the previous multicast algorithms. The multicast source gets more bandwidth than in the previous multicast algorithms.

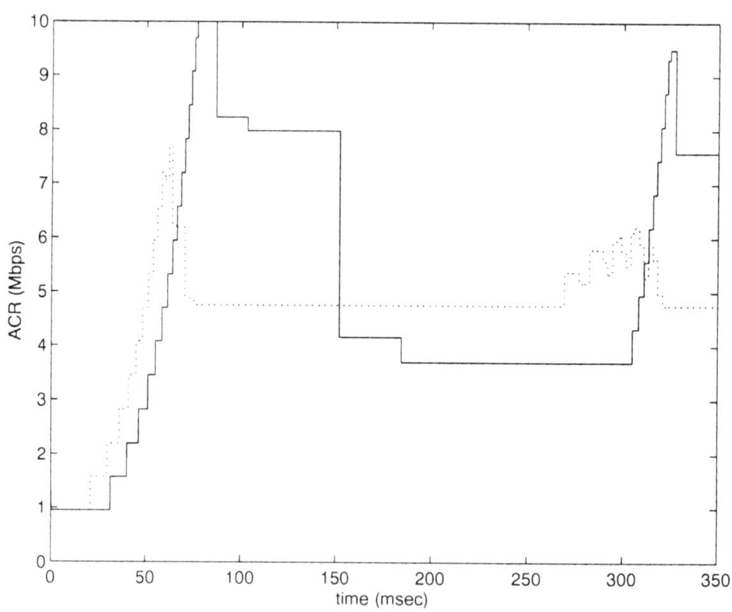

Figure 6 ACR of source A and source B with the TOFS algorithm

A more exhaustive evaluation with several video sequences and a new methodology to get more reliable measures in a subband video coder can be found in [7].

7. CONCLUSIONS AND FUTURE WORK

In conclusion, while ABR services over ATM have been designed for data traffic, a number of studies show that video transmission can take profit from these best effort services, even in a *multicast connection*. Furthermore, we have shown that for adaptive video transmissions, in this case based on subband coding, working at the bandwidth given by the bottleneck of the *multicast connection* is not the best choice and then an intermediate solution is better. The proposed solution in this paper is called TOFS algorithm. Further studies, will be carried out using CAPAC as unicast algorithm.

In future work, the same could be done for IP networks, because multicast techniques although the technology of ATM and IP differs show certain similarities. Over IP, we have similar mechanisms to RM cells using Real Time Protocol and Real Time Control Protocol (RFC1889 and RFC1990), but in this case, we should need to consider the QoS using Differentiated services(DS)(RFC2475).

References

[1] S.Fahmy R. Goyal R. Jain, S. Kalyanaraman and S. Kim. Source Behavior for ATM ABR Traffic Management: An Explanation. *IEEE Communications Magazine*, 34(11):50–57, November 1996.

[2] S. Felici and J. Martinez. A video Compression Algorithm for ATM Networks with ABR services, using visual criteria. *Lectures Notes in Computer Science: 6th Workshop Interactive Distributed Multimedia Systems and Telecommunications Services Spriger-Verlag*, 1718:245–258, October 1999.

[3] Gilbert Strang and T. Nguyen. *Wavelets and Filter Banks.* Wellesley-Cambridge Press, USA, 1996.

[4] Rohit Goyal Bobby Vandalore Shivkumar Kalyanaraman Sastri Kota Pradeep Samudra Raj Jain, Sonia Fahmy. Feedback consolidation algorithms for abr point-to-multipoint connections. *ATM Forum Document/97-0615*, Julio 1997.

[5] J.R. Vidal J. Martinez and L. Guijarro. A Low Complexity Congestion Control Algorithm for the ABR Class of Service. In *Internation Distributed Multimedia Systems IDMS98, Oslo (Norway)*, pages 219–230, September 1998.

[6] Shivkumar Kalyanaraman Rohit Goyal Ram Viswanathan Sonia Fahmy, Raj Jain. Erica switch algorithm: A complete description. *ATM Forum Document/96-1172*, Agosto 1996.

[7] S. Felici and J. Martinez. A statistical method to evaluate adaptive video appplications based on subband codign over best effort networks, particularly with abr services. *Proc. of Access to the Internet in the next centuary, EUNICE99*, pages 65–70, 1999.

Part Eleven

Network Monitoring

Flexible COM - based software solution for HFC Network Performance monitoring

Hung Nguyen Chan Ph.D. student, Judith Redoli Granados Ph.D., Rafael Mompó Gómez Ph.D., Belen Carro Martinez Ph.D. student.

Department of signal theory and Telematic Engineering – ESTI, University of Valladolid

47011 Valladolid - Spain.

Abstract: Emerging HFC (hybrid fiber coaxial) networks with activated return path for data communications requires careful performance monitoring and preventative maintenance over the entire network. Return path is especially problematic due to the noise funneling effects that can strongly affect digital services. Besides, the HFC networks which base on existing CATV infrastructures require many considerations. The special characteristics of HFC network lead to the needs of taking into account many issues related to software, hardware, measurements, etc. This article focuses on the designing issues of HFC network monitoring software to deal with the noise problem in the physical layer, in respect to the trends of measurement software, and describes a performance monitoring software framework, which applies the Component Object Model (COM) approach. The software has been developed in the CEDETEL Lab – ESTI, University of Valladolid, in collaboration with RETECAL – Spain. The article is organized as follows: the first section addresses the general problems of HFC performances. The second section presents the typical issues of monitoring the performance of HFC physical layer. In the third section, the key issues of designing software for HFC network monitoring, and the COM approach will be discussed. In the fourth section, an HFC performance monitoring application, which has been developed in respect to the issues mentioned on the previous sections will be described. Finally, the last section presents conclusions and future works.

Keywords: CATV networks, hybrid fiber coaxial, performance monitoring, physical layer, software, COM.

1. INTRODUCTION

An HFC network is a communication network in which a variety of telecommunication systems coexist. The most critical problem of the network is the noise and ingress in the return path. The noise is produced by many sources, such as TVs, Set-top Boxes, and other electrical domestic devices. Due to the noise funneling effect, the noise introduced by one subscriber can influence the network performance and service qualities through a wide region. In other words, a strong source of noise from one subscriber's home can affect other subscribers. That is why HFC networks change over time and need periodic measurements to maintain proper performance. Each time a new user gets into the network, the characteristics of the shared medium are modified. This changing medium depends mainly on the users themselves. That fact leads to the important role of the preventative maintenance program, which includes the status monitoring to focus maintenance efforts. The monitoring system should be able to automatically measure and collect the most relevant parameters of both downstream and upstream channels such as vision carrier level, sound carrier level, C/N (Carrier/Noise), CTB/CSO (Composite Triple Beat / Composite Second Order), for digital service: BER (Bit Error Rate), EVM (Error Vector Magnitude), etc. Technicians will consult these results to detect network problems. In the case that the problem already has occurred, measurement data also should be reviewed and compared with the previous data. Only the most critical places in the network will be visited and tuned. (BARCO [14]). Therefore, some monitoring equipment should be placed in the important points such as the headend and distribution centers. Different equipment should be dedicated for different kinds of measurements.

As the number of measuring points is high due to the structure of the network, automatic measurements and a properly defined data base are necessary to control an adequate performance of the HFC network.

So far, the HFC network is not strictly standardized. Currently, the DOCSIS, IEEE 802-14, DAVIC/DVB standards are coexisting. Several vendors even develop products of their own technical specifications. In most of the cases, upgrading for data service requires a complete solution from a specific vendor including headend, cable modem termination system (CMTS), cable modems and software [23]. There are several software solutions on the market to monitor noise and ingress on the HFC network ([11] and [24]). Their common drawback is that they only support a small set of hardware from specific vendors. These facts raise the need of developing a flexible solution for monitoring HFC access network, which can handle a large set of instruments, support hardware/software reusing, and easily be adapted to the future needs.

2. PHYSICAL LAYER PERFORMANCE ISSUES

So far, many trials and models have been developed to study the effects of noise and interferences to HFC network. Some laboratory models of Motorola, General Instruments, and CableLabs were described in the Joyce [6], Eldering [7] and Williams [16].

In the operating network, it is required to automatically monitor performance as well as quickly detect problems with minimum affect on normal service. The hardware requirements and measurement procedures were covered in detail in Thomas [1] [2], Bill [15], Feather [17], Parker [18] and NCTA [22].

In general, both models have the similar configuration, which is illustrated in Figure 1.

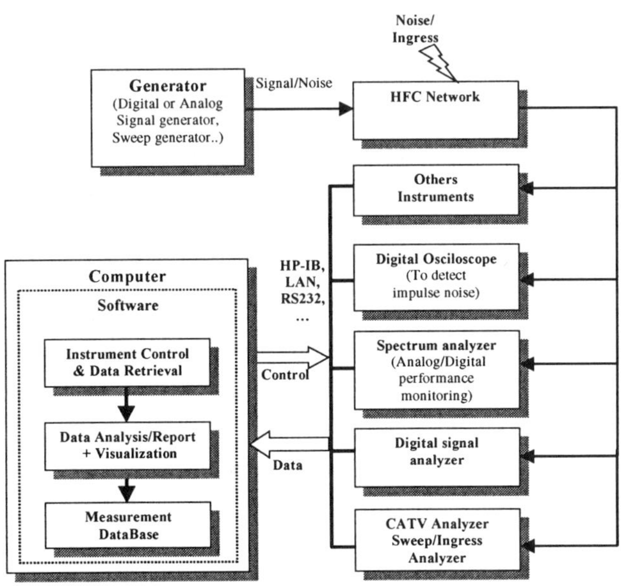

*Figure 1.*Typical configuration of monitoring performance of HFC physical layer.

As can be seen from the figure, the software should provide interoperability between various types of instruments (can be from various vendors). The second thing can be seen is that the laboratory model can be adapted to work in the real network because of their similarity.

3. DESIGNING ISSUES OF HFC PERFORMANCE MONITORING

3.1 Measurement issues

As mentioned in previous sections, measurement functions play the most important role in HFC performance monitoring because of its use to combat with noise/ingress problem and observe digital/analog performance. The designing issues of measurement software will be discussed thoroughly in this section.

3.1.1 Background

The measurement software stack is illustrated in Figure 2.

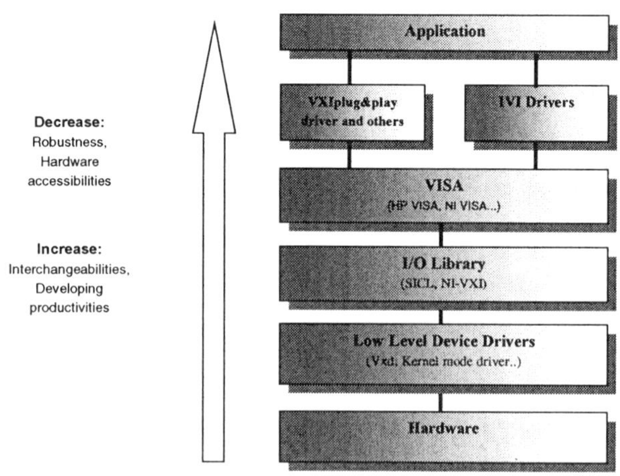

Figure 2. The measurement software stack

The goals of introducing several layers in the software model are:

- Providing compatibilities between various instrument vendors.

- Increasing the interchangeabilities between vendors and hardware/firmware versions.
- Reducing the programming efforts of measurement software developers by eliminating the needs of accessing low-level drivers/firmware.

More details can be found in the article of Rowe and Kerridge [19].

3.1.2 New trends of measurement software

Recently, measuring software is moving toward the interchangeabilities between vendors and hardware/firmware versions. In other words, it is desired that the specific measurement software not be changed when the hardware is replaced. To achieve that, one layer must be added to the legacy measurement software model. This new layer, known as Interchangeable Virtual Instrument (IVI) drivers, provides the interchangeabilities for the software model. If the hardware is replaced, only IVI drivers would be changed and the application should be kept unchanged.

So far, there are not yet standards for the IVI drivers. The IVI foundation [21] is still working on the drafts. There are many possibilities that IVI drivers will be based on Component Object Model (COM) technology of Microsoft due to the following reasons:

- COM is directly supported by MS Windows OS and the most popular developing tools like MS Visual C++, Java, MS Visual Basic, Borland Delphi, etc.
- The new emerging developing tools, known as Rapid Application Development (RAD), such as MS Visual Basic and Borland Delphi, are based on COM/ActiveX technology. Therefore, supporting COM can greatly reduce developing time by making use of these tools.
- COM is now available on most major platforms. Therefore it can provide interoperability between various types of computer hardware running on various operating systems ([25], [26]).
- COM has become an industrial standard.
- A great number of COM-based software components, known as ActiveX components, have been offered by most of software vendors. Therefore using COM will greatly take advantage of software reuse.
- COM provides the location and platform transparency. [20]

Oblad [3] and Fertita [4] studied deeply the benefits of implementing COM into Test& measurement industry. The previous efforts of VXIplug&play Systems Alliance led to the introduction of VXIplug&play C-based drivers (to support multi-platform hardware and software).

However, together with the rapid progress of Information Technology, this approach is recently found inappropriate.

Other competitive software component technology, such as CORBA (COM Object Request Broker), is not considered adequate due to the far less popularity in Test and Measurement filed. However, the IVI foundation is considering the possibility of allowing migration into CORBA and other technologies [27].

3.1.3 Location independence

To achieve location independence between hardware equipments, several possibilities can be implemented:

- GPIB LAN gateway: the measurement instruments will connect to a gateway (a computer or dedicated hardware), which has a LAN interface.
- Web Server: the measurement software has a built-in Web server that allows remote user access using popular Web navigators.
- Users can connect to measurement software on remote servers through communication software components.
- The instrument drivers based on DCOM (Distributed COM), which has remote capabilities, or in other words, location transparency. This option has been recommended for future IVI drivers.

3.2 Data analysis

The laboratory models require powerful data analysis and visualization tools such as MATLAB, while for real network reports, MS Office documents are usually used. In both cases, applying COM technology is very beneficial since MATLAB, MS Office and many other popular software tools support OLE automation (COM-based).

4. APPLICATION SOFTWARE FRAMEWORK FOR HFC PERFORMANCE MONITORING

4.1 Designing issues and their implementations

Taking into account the above-mentioned issues, a software framework has been developed to support HFC performance monitoring in the laboratory environment and easily adapted to the demand of real networks.

The application, named "Magic Monitor for CATV", runs in the MS Windows 9x/NT OSs, which are the most popular OSs at the moment.

Figure 3 is the software block diagram of the software framework. Designing issues and their implementations will be discussed next.

4.1.1 Developing tools

In order to become a high performance multi-threaded application, the framework was written by MS Visual C++ and implement MFC (Microsoft Foundation Classes).

Several software components, such as software agents (client and server side) were written in Visual Basic in order to reduce developing time.

The interface between the software components and the main application were written using ATL (Active Template Library), which simplifies COM programming.

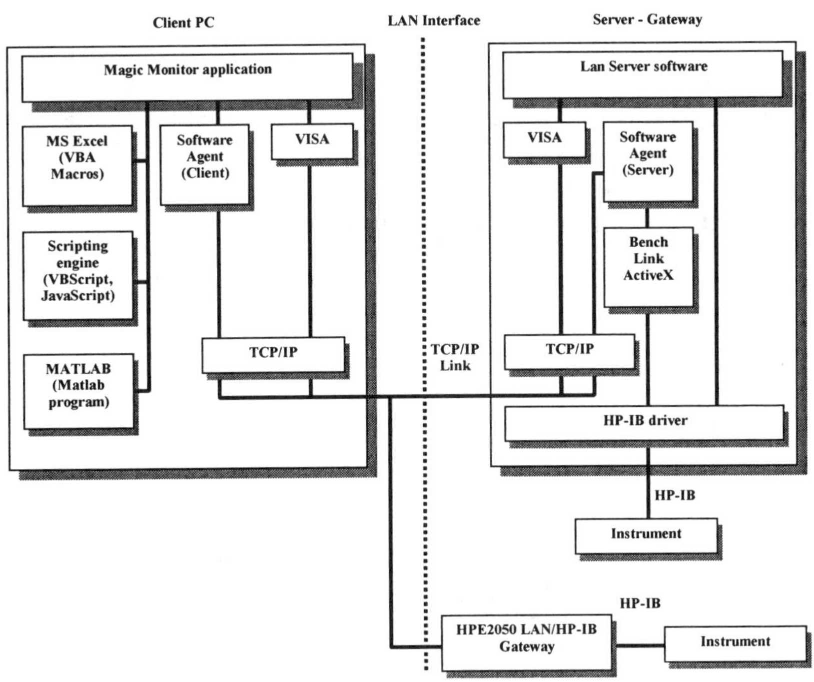

Figure 3. Software block diagram of the "Magic Monitor" framework

4.1.2 Software component based

Instead of building a fixed-function application, it has been broken down into many small components to increase flexibilities.

As Brockschmidt [20] indicated, one of the strongest features of COM is software component reuse. Therefore, instead of rewriting a feature for the application, it is better to connect to a software component (ActiveX), which already has this feature, through COM interfaces. This approach has been thoroughly implemented in "Magic Monitor".

4.1.3 Scripting and macro language support

Scripting & Macro languages such as Visual Basic for Application (VBA) and VBScript strongly support COM.

Therefore, Magic Monitor provides the simple Application Development Environment (ADE) for creating, editing, and debugging simple macros that can launch other scripts such as VBA program and VBScript in a new thread.

Writing macros and scripts is rather simple. In order to write a VBA macro, users can simply record a macro in MS Office, then optionally edit it in the VBA developing environment, which goes together with MS Office.

Supporting scripting language has a great advantage because of the introduction of the new MS Window Scripting Host technology [28], [29]. This new technology allows automating OS tasks by executing VBScript and JavaScript directly in Windows environment and controlling ActiveX-based software via COM interfaces. The power of scripting & macro language allows them to directly handle instruments if their drivers are COM-based.

4.1.4 Flexibilities

The application itself took the idea of macro language: in order to do a task, write a macro, assign it to a button or menu, and do the task with just one mouse click. This approach provides great flexibilities, since once the task needs change; one just has to modify the macro. Magic Monitor macros will have the power of VBA, VBScript, JavaScript and MATLAB programs, since they can be integrated into the macros.

On the measurement side, Magic Monitor controls the VISA layer directly, and therefore, has the flexibilities of hardware access and robustness, but will also be able to handle future IVI drivers, which will be COM-based and intended to use in scripting/macro languages.

To facilitate macro writing, some solutions have been implemented. For example, users can select a command from a drop down list using mouse, the whole command will be automatically completed by the software, and added to macro text. The help system is also improved, as described in the next section. A macro usually contains from 10 to 200 lines, including user's comments.

The procedures to make monitoring tasks using the application are illustrated in Figure 4.

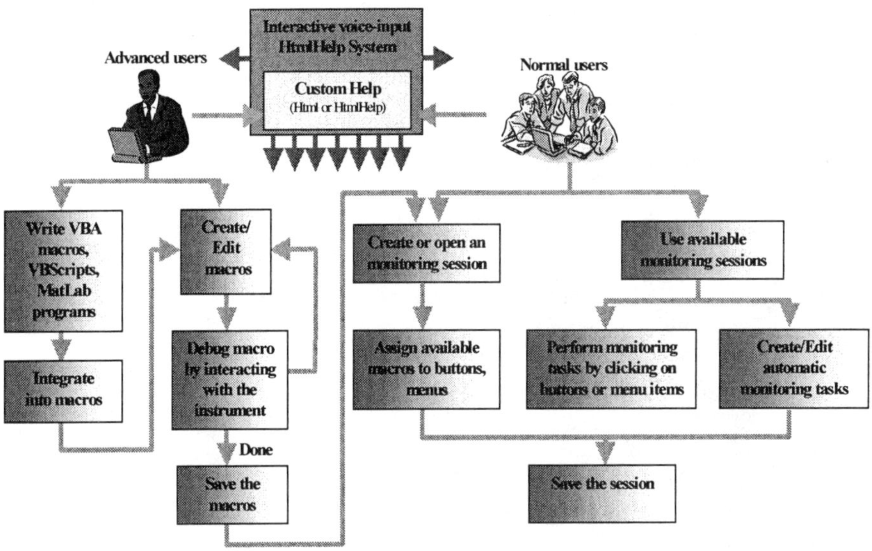

Figure 4. Creating monitoring tasks using Magic Monitor

4.1.5 Improvement in Help tools.

In our experiences, the roles of documentations and Help systems have increasing value as the amount of information expands, especially when facing programming and workgroup co-operating problems. Therefore, an advanced expandable HtmlHelp/Html system has been developed to assist group users/programmers. The new MS HtmlHelp system has the full-text Boolean search capability, very much like Internet search mechanism, can effectively facilitate programming and using software. Besides, the help system was designed to easily expand. Should a user write a new macro, he or she can also write its documentations and integrate it into the global help system. The new MS Agent and speech technologies (speech synthesis and recognition) are also made available as an option. So far, speech engines are available in all major languages. Therefore, the speech-enable interactive help system will have increased values.

4.1.6 Location independencies

The application can run locally as well as remotely. It has several remote options:

- Remotely control instruments via a small-size dedicated hardware HP- E2050 gateway, which is very convenient for remote operating network monitoring.
- Control instrument via a PC, which works as a measurement server. In this case, client PC can also view the real instrument image screen by controlling the BenchLink ActiveX component via two software agents. Figure 5 is the application's screenshot.

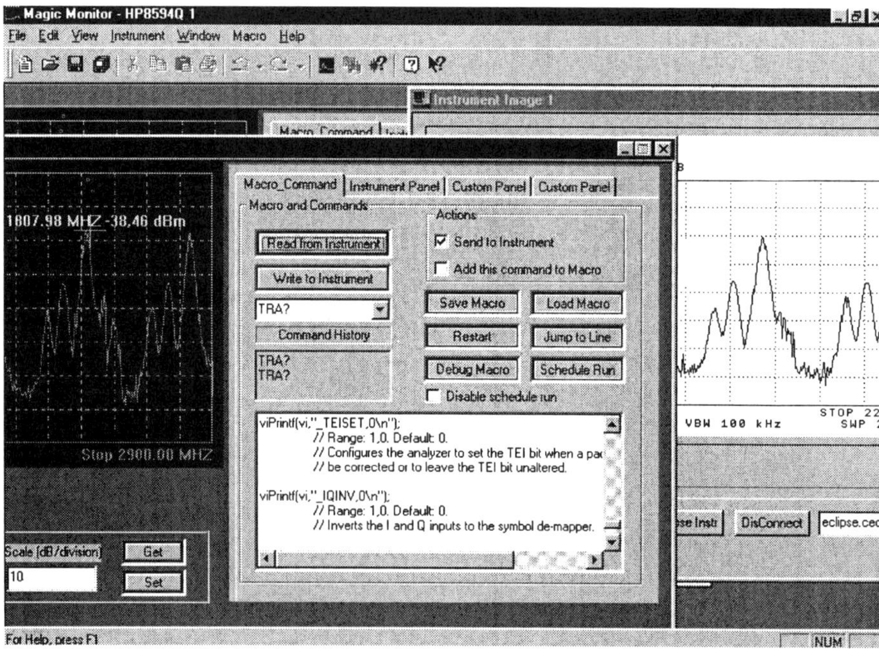

*Figure 5.*Magic Monitor Screenshot

4.2 Project Results

4.2.1 Software capabilities:

- Providing the capabilities to perform automatically scheduled remote measurement tasks on most of the programmable instruments such as spectrum analyzers, oscilloscopes, arbitrary/function generators, etc

through GPIB and RS232 interfaces and collect the measuring data into files.
- Remotely retrieving screen images of various instruments and saving them in compressed formats.
- Having the advanced expandable voice-interactive Help System to assist users to write macros and perform monitoring tasks.
- Performing data analysis automatically from the application environment by executing VBA macros.
- Executing VBScript, JavaScript and MATLAB programs from inside macro.
- Dynamic Graphical User Interface (GUI): users can create their new buttons and menu items to the GUI, link them to macros and perform tasks simply by mouse.
- Allowing automatic customizable CATV measurements with standard spectrum analyzers.
- Providing easy-to-use application development environment (ADE) to write and edit macros.

4.2.2 Applications

Based on the software framework, a research model has been constructed in the CEDETEL CATV Laboratory - University of Valladolid, in collaboration with RETECAL, a northern Spain Cable TV operator. The main target of the project is studying the digital service aspects in Spanish HFC network conditions and assisting CATV/HFC network OAM&P. The model is shown in Figure 6.

The goal of the model is comparing and analyzing measurement data from the CATV Lab trial (CEDETEL – Spain), the operating network of RETECAL, and the simulation model. The automatic data retrieving and analyzing are integrated and synchronized by using the software framework.

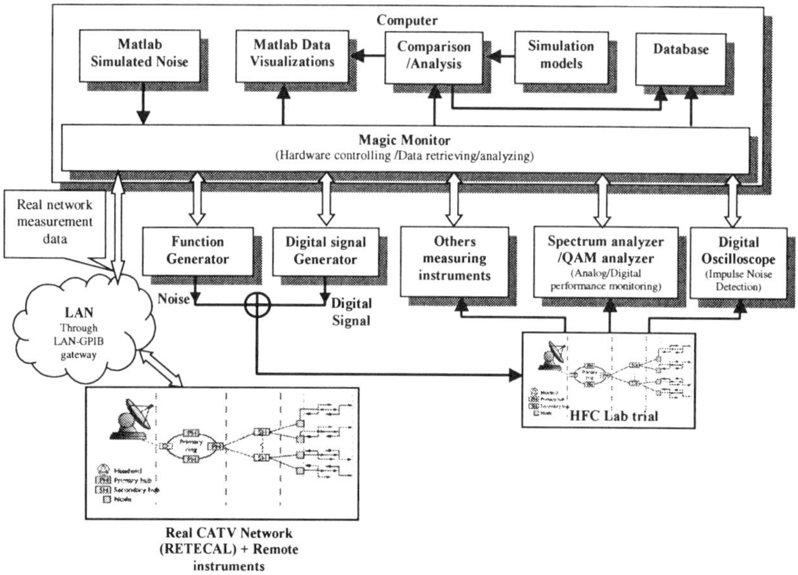

Figure 6. Hardware/software model to study HFC performance applying the "Magic Monitor" framework.

5. CONCLUSION AND FUTURE WORKS

This article has addressed a several issues of design and implementation of HFC performance monitoring software to deal with the noise problems in the physical layer, regarding the latest progresses in Information Technology and Test & Measurement trends. The measurement problems have been highly focused on.

An application that has been developed implementing the considering issues has been described as an example.

It can be concluded that COM technology is a promising candidate for the HFC performance monitoring software due to its power and popularity. This will cut down developing time as well as increase flexibilities and hardware/software interchangeabilities.

As HFC network is new and always subject to change, flexibility is one of the most important features that the monitoring software should have.

Supporting COM, and macro/scripting languages are good options to increase the flexibilities.

With the current framework, we will develop more software components to serve our researching demands. The application in the real network and the integration of hardware-software simulation will be further studied.

References:

[1] Thomas J., Edginton.M, Digital Basics for Cable Television Systems; Prentice Hall 1999.
[2] Thomas J., Cable Television Proof of Performance: A complete guide to Cable TV Compliance Measurement using a Spectrum Analyzer, Prentice Hall 1995.
[3] Oblad R., Achieving robust interchangeability of Test assets in ATE systems, Proceeding AutoTestCon 1999. www.hp.com
[4] Fertita K., Harvay .J. The Role of ActiveX and COM in ATE, Proceeding AutoTestCon 1999. www.hp.com
[5] Bahatti S., Knight .G, On Management of CATV Full Service Networks: A European Perspective, IEEE Network September/October 1998. p28-39.
[6] Joyce G., Moran .J, Henderson .S Gaetani .D, The Ultimate Information Capacity of HFC Return Path, Motorola 1997. p4-6.
[7] Eldering C., Himayat .N, Gardner .F, CATV Return Path characterization for Reliable Communications, IEEE Communication magazine August 1995. Vol 33 , p62-69
[8] Sucharczuk G., Advanced Network management, International Cable ,October 1997. p 40.
[9] Fox S., Five steps to Reduced Return Path Ingress, International Cable, May 1998. p 62.
[10] Walter C., James F., David L., Modern Cable Television Technology, Morgan Kaufman Publisher, Inc.1999.
[11] Hewlett-Packard http://www.hp.com.
[12] National Instrument http://www.ni.com
[13] Microsoft Developer Network (MSDN) http://msdn.microsoft.com
[14] BARCO Communication Systems, Achieving accurate network monitoring in the cable environment to improve the level of service and reliability, http://www.barco.com/communic/products/Monitoring_Management/network/accuratenet monitoring.htm.
[15] Bill M., Noise and ingress performance in the return path; CED Magazine March 1997.
[16] Williams T., Plant architectures for diagnosing and maintaining bi-directional HFC network, SPIE proceeding November 1996. SPIE Vol.2917, p12.
[17] Feather B., Understanding Software standard, International Cable December 1997. p48 - 51.
[18] Parker D., Network management: The key to system reliability, International Cable. May 1997. p10-14.
[19] Rowe M., Kerridge B., Software drivers: What are they exactly?, Test and measurement World, March 1999. http://tmworld.com/articles/03_1999_Drivers_cover.htm
[20] Brockschmidt K., What OLE is really about, Microsoft Developer Journal July 1996, p43-45.
[21] IVI Foundation. http://www.ivifoundation.org/Operations/charter.htm
[22] National Cable Television Association, NCTA Recommended practices for measurement on CATV systems, Supplement on upstream issues, Oct 1997.
[23] Kiniry J., Mertz C., Cable Modems: Cable TV delivers the Internet, IEEE computing May/June 1998. p13.
[24] Tektronix www.tek.com
[25] Microsoft COM technology, http://www.microsoft.com/com/

[26] SoftwareAG solutions, http://www.softwareag.com/entirex/product/EntireXengl.pdf
[27] IVI foundation meeting summary, Nov 1999
http://www.ivifoundation.org/Meetings/1999Nov/99NovMeetingSummary.doc, p.42.
[28] MS Window script Technology, http://msdn.microsoft.com/scripting/
[29] MS Windows Scripting Host programmer's reference.
http://msdn.microsoft.com/scripting/

Statistics Warehouse for Intelligent Networks

Applying data warehouse technology in the telecommunications domain

J. Schabauer, H. Portschy, R. Siebenhofer, A. Holzinger
SIEMENS AG AUSTRIA
Program and System Engineering
Telecommunications Management Networks - Intelligent Networks Department
Gudrunstrasse 11, 1100 Vienna, Austria
e-mail: {johann.schabauer, hans.portschy, rudolf.siebenhofer, andrea.holzinger}@siemens.at

Key words: Intelligent Network, Data Warehouse, Decision Support, Database

Abstract: Network operators and service providers are faced with technology and marketplace changes. Enhanced decision-supporting capabilities are needed to respond quickly to these changes. In this paper we present a system which applies a data warehouse approach for collecting and analysing data from an Intelligent Network to support business decision support systems. The system continuously gathers data from disparate systems in different formats, transforms the data into a consistent format and loads it into a database. Sophisticated and flexible analysis report facilities are provided, which are important for network planning, system operation and marketing.

1. INTRODUCTION

The telecommunications market is becoming increasingly competitive. Service providers and network operators are facing dramatic changes in the way they do business. The model of the past, one provider or just a handful of competitors in each country, is being replaced by a model shaped by hundreds of competitors competing for global presence. Under these conditions, the customer becomes the central focus of a provider's activities, and the ability to react quickly to market trends and to tailor products and services to individual customers is more critical than ever. Providers need to have a complete understanding not only of their technologies, networks and offered services but also of what their customer usage behavior and desires are for the network operation, system planning and marketing.

The system we have developed collects and analyzes information from an Intelligent Network and presents this information in an understandable and flexible way to end users through a user-friendly Web-interface. These analysis reports and evaluations may then support a variety of decision analysis functions as well as strategic operational functions.

2. BACKGROUND

This section gives a short overview of the Intelligent Network (IN) and the application domain.

The IN essentially separates service control functions from service switching functions. Typically, both types of functions are implemented in different physical equipment. Due to this separation, it is possible to introduce new services into the network without having to change the functionality of the switches in the network. *Figure 1* shows the IN architecture.

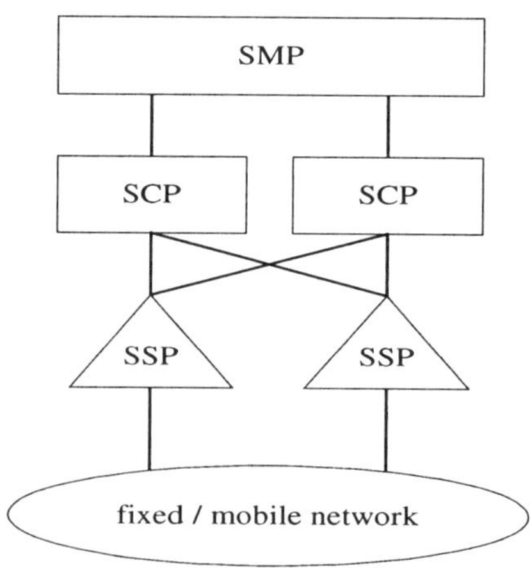

Figure 1. IN architecture

Service switching points (SSPs) are stored program control switches that interface to the SS7 signalling network. The SSP embodies the call control function (CCF) and service switching function (SSF) entities. The SSF recognizes IN service calls and routes the appropriate queries to the service control function (SCF) that resides in a service control point (SCP) via the

SS7 network. SCP functions are used by the SSP to process calls. The SCP is a fault-tolerant, high-capacity, transaction-processing entity that provides call-handling information in response to SSP queries. The service management point (SMP) provides operation, administration, and maintenance functions for the IN.

All of these basic elements form the infrastructure of IN, which supports the notion of separating service-control functions from service switching-functions to realize more rapid-services development and deployment.

For each service usage SCP and SMP collect and log a lot of information. The SCP creates call detail records (CDRs), which contain all call related data including service number, calling party, called party, date, time, duration, release cause and so on. The SMP produces administration tickets, which include service number, date, time and service parameters. All these records and tickets contain a gold mine of information about the network behavior, the service behavior and the user behavior. The analysis and evaluation of this information is essential for the service providers to improve their quality of service and to understand the customers' needs and behaviors. It is also useful for the service subscribers and end-users to control and optimize their service usage.

3. CHALLENGES AND REQUIREMENTS

One of the characteristic challenges of the telecommunications business is managing enormous volumes of data. Not only do telecom systems generate huge numbers of call detail records (and others), but the records themselves are very large. Thus one of the most important and also critical success factors is to handle tremendous insertion rates, as well as to manage databases that will likely grow into multiterabyte range. Building a system that will not scale efficiently as data volumes grow can be one of the most frustrating (and expensive) experiences a company can endure.

Over the last years a popular approach for collecting and analyzing data and information is the use of data warehouse technology. Therefore, we have adopted a data warehouse approach to collect and aggregate statistical data in a continuous way (24 hours 365 days a year) for millions of calls per hour. We have termed our system Statistics Warehouse - collection and analysis of statistical data form an Intelligent Network using a data warehouse approach.

To support the call rate of IN services a data throughput of up to 50 million call detail records per hour is necessary. Call detail records have to be archived over 90 days. 50 million call detail records per hour over 90 days means that a storage volume up to 50 Terra Byte has to be managed. *Figure 2* illustrates the data rates and the data volumes.

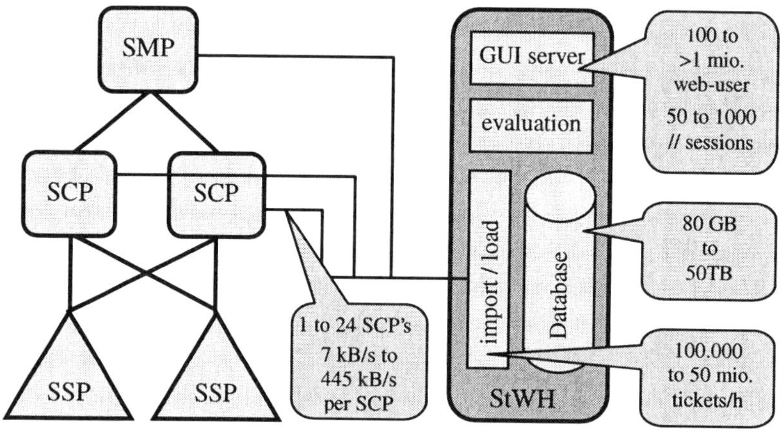

Figure 2. Data rates and volumes

4. STATISTICS WAREHOUSE

This section describes the overall architecture of the system, the data model, the database configuration and the process of importing, transforming, and loading the data.

The overall architecture consists of three main parts: the mediation component, the warehouse component and the user interface component (see *Figure 3*).

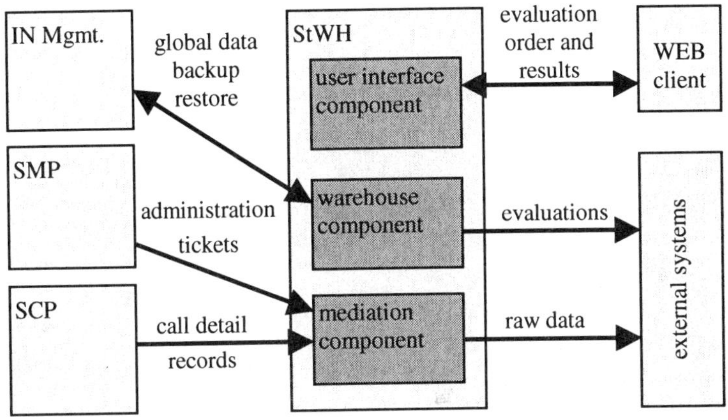

Figure 3. Overall architecture

The mediation component periodically imports call detail records from the SCPs and administration tickets from the SMPs via file transfer. This data is cleaned and transformed into an internal uniform format for loading into the database. In addition, an export to other external systems (e.g. a billing system or another data warehouse) is also possible. This concentrated data is then loaded into the database to populate the tables.

The warehouse component consists of the database, a global data import function, data analysis and evaluation functions and an export mechanism to export filtered data and reports to other external systems. The global data import function is used to import IN-specific global data, such as announcements, origin areas and other network related data, from the IN management system. Backup and restore of the database is integrated with the OAM concept of the IN system.

The user interface component, which provides a user-friendly Web-interface, performs access control, invokes the evaluation and analysis functions and sends the reports to the Web-browsers.

The system has been developed using relational database technology, CORBA technology and Web Technology. Java has been chosen as the programming language for client and server components.

Due to the enormous amount of data and the data rate, the critical core components are the data model and the import and load mechanism. The following subsections describe these components in more detail.

4.1 Data model

For the data model we have adopted the standard practice of the star schema (see *Figure 4*). The star schema is a concept that provides superior data retrieval power to the data warehouse. The name "star schema" is derived from the appearance of the data model, with a large central table surrounded in a star formation by subordinate tables. The central table, known as the fact table, is surrounded by many dimension tables. This type of modelling is known as dimensional data warehousing. Fact tables contain measures that are used to perform analysis as well as the keys that link the dimensions. The dimension tables contain the attributes that describe the data components and provide the information to do comparative analysis.

Our model consists of the main fact table raw_fact, several dimension tables and the table raw_rest. All call detail records contain large fields like service number, origin, etc. which are identical in numerous records. These fields are not unique primary keys for all pre-defined report queries. Most queries affect a range of such keys, for example all origins of a district. To optimize the data access and minimize the data volume, the most-queried fields are stored in a separate fact table and ranges are stored in dimension tables. The rest of the raw data is stored in a so called rest table.

The raw_fact table contains the most-queried fields - such as number of calls, number of dialogs, number of reroutings, etc. - and the keys to the other tables. The raw_rest table contains all remaining fields. The dimension tables include amongst others time, origin, and service. Time is a dimensional table that holds date and time information. The origin dimension contains geographical information including country, district and city. The service dimension contains information about the IN services and numbers.

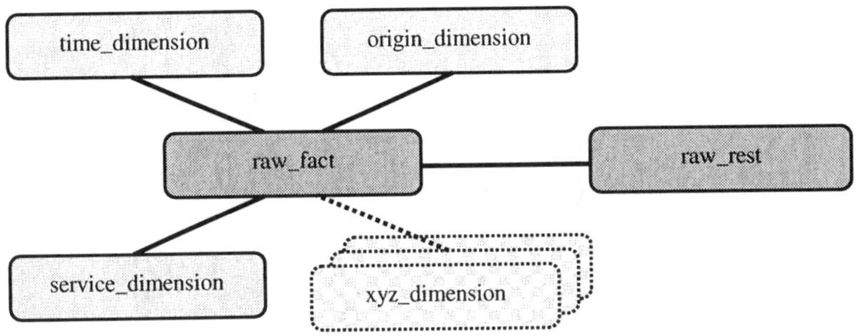

Figure 4. Data model

All raw data is located in raw_fact and raw_rest tables, which can contain up to 50 thousand million records. To manage such an amount of data, these logical tables are divided into smaller parts by partitioning of the tables by day. Partitioning by the day has the following advantages:

- Data for a specific day can be directly accessed.
- Loading of new data affects only one partition.
- Deletion of expired data is achieved by dropping the partition.
- Backup and restore can be done separately for each partition.

A daily partition can contain up to 500 million records, which is still a large number. Therefore each daily partition is divided into several sub partitions. This composite partitioning concept (see *Figure 5*), a combination of range (daily) for primary partitioning and hash for sub-partitioning divides one large table into manageable parts. These sub-partitions are distributed to several discs to enable good I/O balancing to achieve high performance data access.

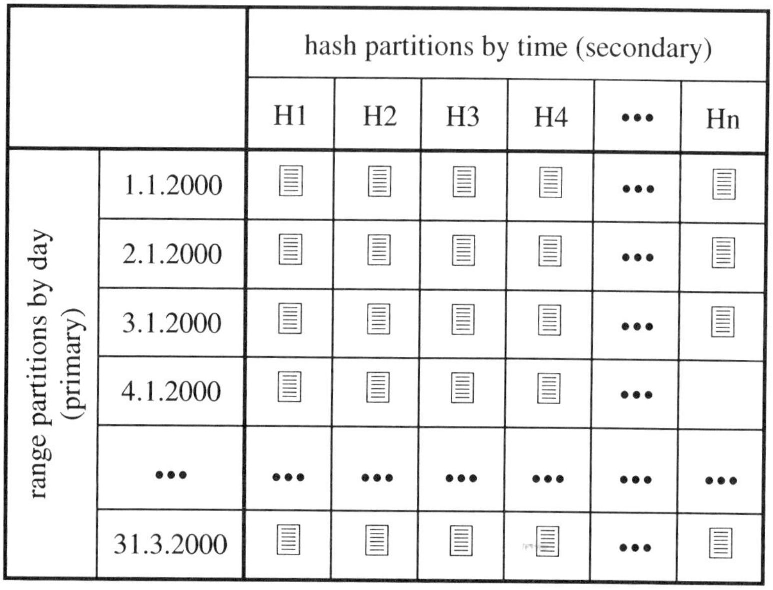

Figure 5. Composite partitioning

The data model and the partitioning concept enable to handle a database with 50 thousand million records and a data volume of up to 50 TerraByte.

4.2 Data Import and Transformation

Data is imported from several different data sources; call detail records from the SCPs and administration tickets from the SMPs. This data is periodically imported via file transfer and stored in files. Due to the fact that the data is imported from several systems it usually has a different format. For each data source with a distinct data structure a corresponding interface module is provided. These interface modules perform data cleaning and transformation of data into a uniform structure. The output of this process are loadable files ready to be loaded into the database. *Figure 6* shows the architecture and the data flow of the import and transformation process.

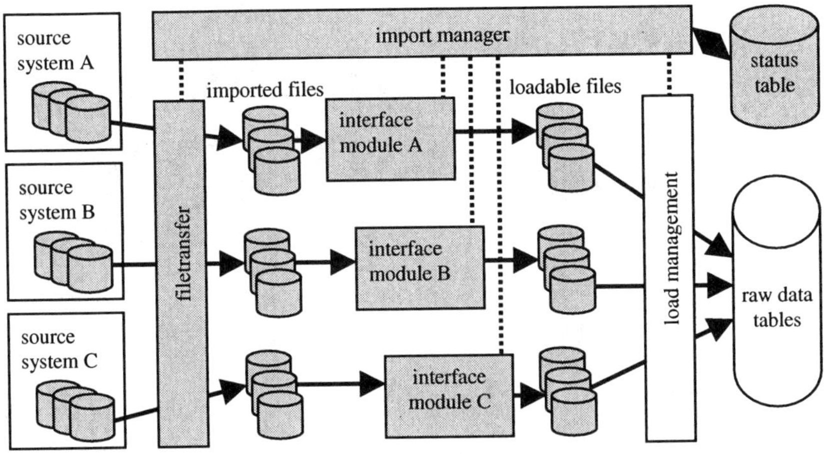

Figure 6. Import and transformation process

To achieve high data throughput and to prevent loss of data, an import manager synchronizes and controls the parallel processing of the file transfers, the cleaning and transformation of imported data into uniform loadable files, and the hand-over of loadable files to the data loading process. For synchronization and error recovery without corruption or loss of data, each step in this process is treated as a transaction. The current status of each step and file is logged into a status table.

4.3 Data Loading

The data loading process has to support a load rate of up to 50 million records per hour. Therefore a sophisticated load mechanism has been implemented, where the data loading is done in several steps (see *Figure 7*):

Step 1: Loading of the data into temporary tables to enable parallel processing of the bulk loader and to prevent I/O conflicts.

Step 2: Adding foreign keys, creating indexes and new dimensions.

Step 3: Insert-append of the temporary tables to the fact and rest_raw tables.

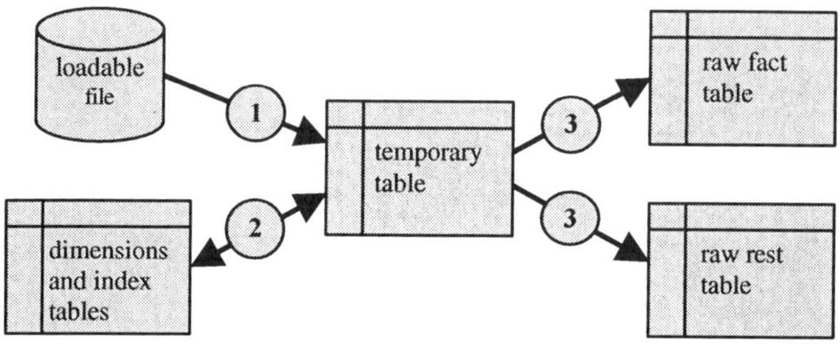

Figure 7. Data loading mechanism

This multi-step loading mechanism has the advantage that the large data tables are only affected in the short time of the insert-append and that all other processing steps are done with small, independent temporary tables.

5. ANALYSIS REPORTS

The system provides a set of pre-defined analysis reports. These reports can be customized to fulfil the different requirements and needs of the various users. This enables the user to perform any user-specific analysis and to access all of the data that he or she could possibly desire. Further evaluations can be done using Online Analysis Processing (OLAP) tools. The data model as well as the system architecture allow an easy integration with other systems and analysis programs. In addition raw data filtering and download of the filtered extract of raw data to the client workstation for further processing/analyzing is also supported.

The analysis reports are grouped into IN service-independent evaluations, which are based solely on call detail record fields available for all services, and IN service-specific evaluations, which are based on fields in call detail records, fraud records and confirmation records available only for certain services. All analysis reports are grouped and sorted by IN number, date and time.

The service-independent analysis allows the user to obtain the following reports:

- Traffic curve:
  total number of calls, average call and set-up duration, number of successful calls sorted by call origin, SCP number or SSP number
- Analysis of successful calls:
  number of calls sorted by set-up and call duration

- Analysis of rejected calls:
  number of calls sorted by reject cause
- Structural analysis:
  number of calls sorted by origin and/or destination location

The service-specific analysis allows the user to obtain the following reports:

- Analysis of voting result:
  number of calls sorted by service embedded counters
- Fraud analysis:
  number of events sorted by event type (e.g. wrong PIN, simultaneous call)
- Call charge analysis:
  number of calls sorted by amount of charge
- Recharge analysis:
  successful rechargings, unsuccessful rechargings, recharge amount, average recharge amount
- Outpayment analysis:
  successful outpayments, unsuccessful outpayments, average amount of outpayment

6. WEB-BASED USER INTERFACE

All user interactions such as analysis reports, raw data filtering, system administration and user administration are done with a graphical user interface running in a standard Web-browser.

Analysis reports and raw data filtering are performed on demand. The user initiates an evaluation. This order is stored and processed by the warehouse component. The result is written to a file, which can be displayed on the browser or exported to an external system. A user management provides privacy of information and prevents unauthorized access and user actions. Access rights, roles and restrictions can be defined for each user by the system administrator.

7. CONCLUSIONS

This paper describes the Statistics Warehouse for Intelligent Networks, a Web-based system for the analysis of the behavior of IN service users. To fulfil the performance and flexibility requirements a data warehouse approach has been used. The Statistics Warehouse provides different kinds of analysis reports and evaluations which are essential for decision support purposes in the area of network operation, system planning and marketing.

REFERENCES

[1] Harry S. Singh, *Data Warehousing Concepts, Technologies, Implementations and Management,* Prentice Hall, 1998

[2] Michael J. Corey, Michael Abbey, Ian Abramson, Ben Taub, *Oracle8 Data Warehousing*, Osborne/McGraw-Hill, 1998

[3] Ralph Kimball, *The Data Warehouse Toolkit*, Wiley, 1996

[4] Gregory F. Pfister, *In Search of Clusters*, Prentice Hall, 1998

[5] Thomas J. Mowbray, Raphael C. Malveau, *CORBA Design Patterns*, Wiley, 1997

[6] Andreas Vogel, Keith Duddy, *JAVA Programming with CORBA, Advanced Techniques for Building Distributed Applications*, Wiley, 1998

[7] *Oracle Express Server - Delivering OLAP to the Enterprise*, An Oracle White Paper, January 1997

Part Twelve

Advanced Network Services

Service Interoperability through advanced Media Gateways

Sven van der Meer, Stefan Arbanowski

Technical University Berlin, Franklinstrasse 28/29, 10587 Berlin, Germany

Email: vdmeer@cs.tu-berlin.de

Tel: +49 30 3463 7158, Fax: Tel: +49 30 3463 8158

The convergence of telecommunications systems and the Internet causes a variety of concepts for service integration. The focus of the recent research studies and the work of several standardization bodies lies, mostly, on the interworking of services and the universal service access from end-user systems including both, fixed and wireless terminals. All approaches are driven by the concept of providing several technologies to users by keeping the peculiarity of each service alive.

But, developments should not only concentrate on media adaptation between VoIP and PSTN, but also consider the adaptation among completely different types of applications as for example email, fax, or voice. Unified Messaging, which is an already accepted service on the market, provides solutions for conversions of different application protocols into each other. The functionality of converting one media into another is implemented here in so called Media Gateways.

This paper provides an overview of the current developments in the area of Media Gateways. Next, it introduces the basic architecture of an existing, CORBA-based unified messaging system and the implementation of Media Gateways for the adaptation of different services. Finally, both approaches are combined to Unified Messaging enabled Media Gateways that improve the user's reachability because of their capability to adapt services to different kinds of end devices.

Keywords: Media Gateways, Service Personalization, Service Adaptation, Unified Messaging, Media Conversion, IP, User Mobility, UMTS, VHE, VoIP

1 Introduction

Internet technologies have proven their applicability to the future telecommunications environment. Emerging architectures for Internet Telephony, such as the ITU-T H.323 standard [1] or the IETF drafts of MGCP [2] and Megaco [3], offer solutions for the integration of the telecommunication world and the Internet. However, these activities concentrate mainly on the provision of a basic voice communication service, leaving value-added services (like Intelligent Network (IN) services) out of scope. New developments try to combine VoIP services with the IN, resulting on so called soft switches able to process IN-like services over IP networks. In this context, a Gatekeeper or a Media Gateway Controller takes over the signaling and Media Gateways are responsible for the adaptation of transport protocols.

The concept of Unified Messaging has recently emerged from the research of Personal Communication Support (PCS). Reaching industrial relevance, it addresses the task of overcoming the multiple-mailbox approach of today's messaging systems, with separated facilities for e-mail, voice storage, fax reception, etc. This coincides with the vision for future communication, to deliver information any time, any place, in any form, as it is described in the Virtual Home Environment (VHE) concept within the Universal Mobile Telecommunication System (UMTS) standards. [4] [5]

Messaging systems take into account new application fields such as location-awareness, terminal-awareness, and context-awareness. This means, that the user is supported by the seamless integration of new network access technologies (e.g. UMTS), new terminal equipment (e.g. Personal Digital Assistants and enhanced mobile phones), and new multimedia applications. The more such technologies are in common use, the more the user's communication behavior will change. The main interest of the users is no longer the ubiquitous access to their messages but the control of their reachability. Messaging systems have to support the user to define exactly when, where, and for whom he is reachable. This means that people will define their own individual communication and services environment, specifying how they want to use services and in which way.

The restricted capabilities of a terminal limit users in being reachable for different kind of services. To overcome this restriction, messaging systems adapt one service to another employing conversion technologies. The system provides the user the capability to configure service-interworking scenarios for specific types of messages and invokes the appropriate conversion to deliver the received messages. The evaluation of an appropriate conversion strategy applies application based Quality of Service (QoS) preferences made by the user. But, users are not interested to define technical QoS parameters like jitter, delay, or bandwidth. Most users even do not understand the technological background of them. Users can describe their preferences by i.e. means of costs of delivery and intelligibility of converted information.

One prerequisite for the vision of information any time, any place, in any form is global connectivity, based on a fast developing web of interconnected communication networks, comprising fixed and wireless networks as well as the Internet (figure 1). In addition, the provision of a global service infrastructure, based on network-independent open service platforms is the other fundamental prerequisite, hiding the complexity of network diversity. This platform allows the fast and efficient creation, provision, and management of services.

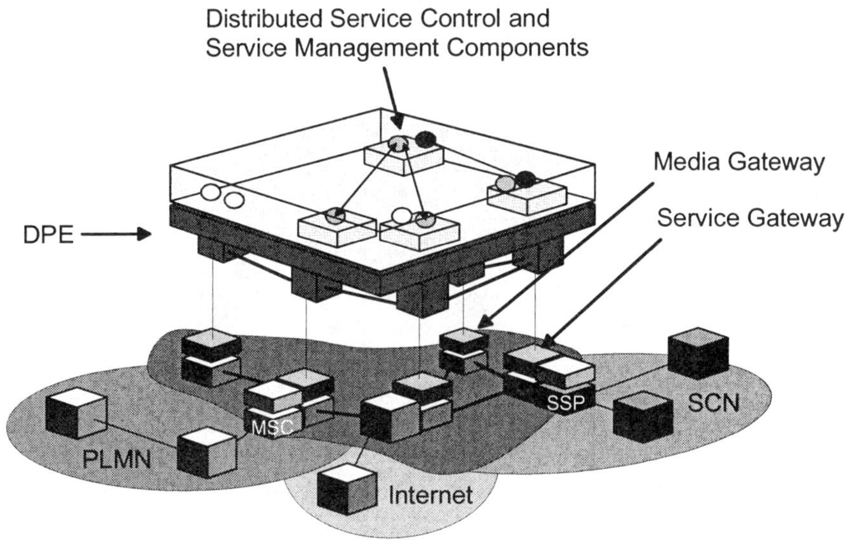

Figure 1: Future Telecommunications Environment

The spectrum of services ranges from simple communication services up to complex distributed applications. In particular, intelligent information and messaging services are commonplace. Additionally, services providing information retrieval (news on demand, brokerage information on demand), services emerging from the field of electronic commerce (hotel or flight reservation), and services like the intelligent house are gaining momentum.

With regard to such services and in addition to the understanding of different standards of Media Gateways, the developments should not only concentrate on media adaptation as for example between VoIP and PSTN, but also consider the adaptation among completely different types of applications as for example email, fax, or speech. To analyze such kind of adaptation, the concept of Unified Messaging, which is an already accepted service on the market, provides solutions for conversions of different application protocols into each other.

2 VoIP Media Gateways – Current Situation

Fundamentally, Gateways are used for connecting different networks. If these networks are of the same type, the gateway simply performs the routing of information between them almost without any adaptation. This relates for example to the Internet Protocol where the gateways are responsible for the forwarding of IP packets to other subnetworks. However, in order to enable the interconnection of different kinds of networks, gateways also have to provide an adaptation to the respective transport media and protocols.

The more interesting and today still topical aspects regard to the interworking of services that are provided in different networks. The IP telephony is a well-known example for this. The VoIP applications shall be able to establish connections to legacy telephone devices and vice versa. Here, there is on the one side the packet orientated Internet and on the other side a Switched Circuit Network (SCN) like PSTN/ISDN. The later is traditionally used for the telephony service. With the continual development of the Internet, the multimedia communication was given an important role. In 1995, the first Internet Telephony software was developed by Vocaltec corporation. The H.323 recommendation is the principal ITU-T standard for Internet telephony. It is an umbrella standard defining a protocol framework including several specification concerning the signaling, coding, and transmission of multimedia data over the Internet protocol.

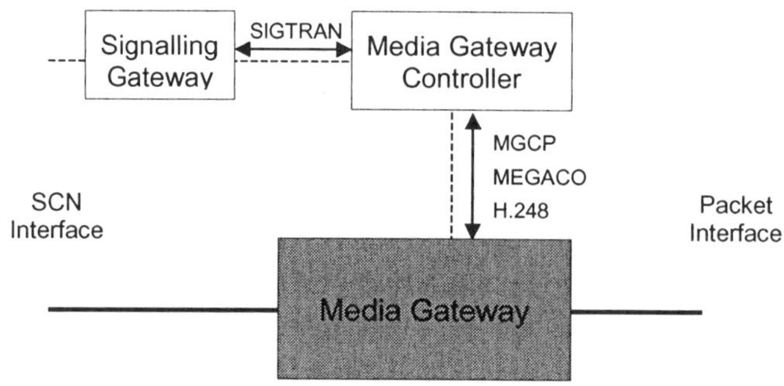

Figure 2: General Architecture of Media Gateways

Whereas PSTN is strongly established worldwide, IP telephony is to a certain point of degree still topic of research and development but it has huge potential to revolutionize the world of telecommunication and is a serious challenge to PSTN. There are still problems to be solved, like the paradigm 'always on' feasible for any common telephone or the immediate delivery of an emergency call. Other questions are security issues of telephony equipment with an IP address that is connected to the internet. Such hardware can easily be controlled remotely or can be interfered with denial of service attacks. To become successful finally, the interworking to traditional telephony must be made possible. This leaded to the introduction of Media Gateways.

TIPHON ("Telecommunications and Internet Protocol Harmonization Over Networks") is an ETSI project that has defined a distributed architecture for realization of such gateways between IP-based Internet and PSTN. This decomposed architecture is based on three components - Media Gateway, Media Gateway Controller, and Signaling Gateway (figure 2). Because of the distribution, the necessity for standardized interfaces and protocols arose. Several standardization forums such as ITU-T and IETF have been working in this. TIPHON considers exemplary a couple of Media Gateways for different kinds of applications in [9], including:

- **Trunk gateways** that interface between SCN networks and IP networks. Such gateways typically interface to SS7 or other NNI signaling on the SCN and manage a large number of digital circuits.
- **Access gateways** that interface UNI interfaces like ISDN (PRI and BRI) and traditional analogue interfaces to a Voice over IP network.
- **Residential gateways** are access gateways for a small number of voice terminals that can be co-located with a cable modem or set top box.
- **Network Access Servers**, which convert modem signals from an SCN network and provide data access to the packet network.
- **Interactive Voice Response** systems that provide automatic voice response and switching services in response to DTMF signals from the SCN.
- **IP Gateways** that are used to interface either between administrative domains which apply different policies (e.g. proxies), or to transform media streams formats (e.g. transcoding).

Originally developed for the interconnection of different types of networks, gateways in form of Media Gateways are today increasingly deployed to enable the interworking of services.

In this section, the concept of media gateways was introduced with media gateways as the intermediate components that enable the interconnection of different networks and the interworking of similar services. The next consequent step in this evolution will be the enhancement of media gateways to enable a very capable interworking of different kind of services.

3 Media Gateways & Unified Messaging

The idea of media gateways in unified messaging systems is to establish an intelligent control of these conversion capabilities by mapping user preferences to actual system behavior. This means, a media gateway converts a specific message according to user or systems demands. User demands are taken from profiles, whereas system demands result from network related aspects (e.g. the media gateway has to adapt an incoming ISDN call to the GSM world, because the callee is using a cellular phone).

The department for Open Communication Systems (OKS) at the Technical University Berlin has developed a 4th generation Unified Messaging System (UMS) [10]. The focus of the system lies on the interworking and integration of legacy and future communication services, such as telephony, voicemail, fax, email, paging, Tele-conferencing, and IP-Telephony. It offers a universal access from a variety of end user systems, including both fixed and mobile terminals. The common attribute is the notion of customer control / customer profile management enabling users to define rules for the handling of incoming calls and messages in accord to their individual preferences. [Vdm99a]

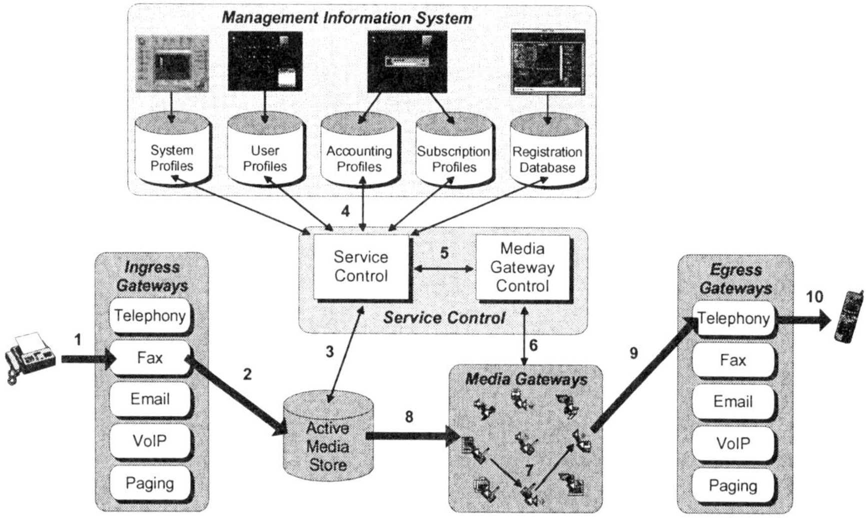

Figure 3: UMS – Functional Components

Typical scenarios featured by UMS are the support of mobile users with universal access to their services. A special capability is the selection of the most appropriate terminal or application for the delivery of an incoming message. Furthermore, the system can adapt any kind of services by conversion processes, e.g. to deliver an incoming e-mail via Short Messaging Service to a mobile phone. UMS provides features like:

- Call Screening/Scheduling/Forwarding,
- Customer Profile Management,
- Media Conversion, Content Screening/Handling,
- Multimedia Message Storage, and
- Automatically/Manually User Registration at locations or terminals.

3.1 Functional Components

The functional components of UMS and an example of their interworking based on the forwarding of a fax message to a telephone are given in Figure 3.

1) The Ingress Service Gateways connect the messaging system to SCNs and the Internet. They can be associated to synchronous and asynchronous services. Asynchronous services, also known as store and forward services, receive information in form of messages. After the complete reception of a message, it is handed over to the Active Media Store.

2) The Active Media Store provides persistent storage capabilities of multimedia information and universal access to them. After receiving the fax message from the gateway, it processes the content of the fax to collect all information for Service Control. E.g., to select an appropriate conversion strategy it is necessary to extract the text out of the fax image and to identify the language the text is written in. Because a conversion of a message always means a loss of information, the system is not going to remove this stored message, unless it is requested to do so by the user.

3) Service Control is the service logic of the messaging system. It queries other components (4) to get on-demand information on user/system preferences. It finds out where a user is located at, which communication capabilities he can use, which services he is permitted to use, and when, for whom, in which way the user wants to be reachable. This information is passed on to Media Gateway Control.

5,6,7,8,9) Media Gateway Control controls both, the conversion process and the stream binding between resources. This comprises the instantiation, configuration, start-up, suspension, reconfiguring, restart, termination, and cleanup of any kind of software/hardware realized within the pool of Media Gateways. We have defined a unified interface to access the different kinds of Media Gateway.

10) Finally, the fax message is read out to a telephone line.

3.2 Media Conversion

Conversion technology reached a state, where it can be used in communication systems to adapt services to limited terminal capabilities. The so called media converters may be defined as a system entity, which input is information I1, with the semantic S1, carried by a specific medium M1, using a specific format F1. The media converter will distribute information I2 as output in medium M2, format F2, carrying semantic S2. The quality of the conversion can be measured by comparing S1 and S2. Generally, two major classes of conversions can be identified. [11]

Media Format conversions change the format the information is coded in or modifies specific parameters of the media format, e.g. converting a GIF coded image to a JPG coded image or changing the scaling of a picture. Such converters are expected to be highly flexible, allowing most possible conversion combinations.

Media Type conversion alters the type of the medium (M1 to M2). The term medium is related to the human senses. E.g., these converters are able to transform an image containing characters (like the received fax) into spoken words. Because formats of images differ from formats of voice streams, they have to be converted, too. A media conversion is specialized to a specific task, e.g. Optical Character Recognition (OCR) or Text-To-Speech (TTS) conversion. The accepted input formats and the produced output formats are limited.

According to the fixed amount of supported input and output formats, media type converters have to be accompanied with media format converters. This leads to a large range of supported media formats for a single media type conversion. The result is a framework of media type converters and media format converters. This framework can be seen as a single and flexible new converter. Inside of this framework, messages have to be stored temporarily (storage dimensions depends on the supported services) and stream binding between the implemented converters has to be supervised (but can be done in a homogeneous environment using proprietary technologies).

For the conversion of messages or communication services, media converters have to be chained to so called converter chains. To illustrate this, the following example shows the steps necessary to read out the fax to a telephone device:

1) analyze the fax to gain information about the content. Result should be a decision, if the fax images contains text or images, and the language of the probably included text;

2) convert the image via an OCR to plain text;

3) preprocess the text to reformat it (for increase of the intelligibility), to remove unnecessary information, and to adapt it to a coding format understood by the TTS conversion;

4) alter the text to speech by a TTS conversion. Output is a stream of audio data;

5) change the format of the audio stream to a format, which can be read out to a telephone (e.g. WAV to μlaw);

This particular service adaptation needs five different converters. All of these converters have to be started, parameterized, activated, deactivated, and finally terminated. During the activation of the converters, the data has to be streamed from one converter to the next. Streams have to be bound between the input and output interfaces of pairs of converters and supervised to guarantee the correct transmission of data. In our implementation, we analyze the actual language the text is written in on a paragraph basis. So all conversion steps after the OCR can be done for single paragraph written in different languages.

For implementing such kinds of media gateways the industry provides off-the-shelves products, like automatic speech recognition, optical character recognition, interactive voice response, speech synthesis, text-to-speech conversion, language detection, text filtering, content analysis, etc. These products are combined to a powerful conversion environment, which is able to adapt any media type to any other.

4 Advanced Media Gateways for Interoperability of Services

Facing the functionality of Unified Messaging Systems, the demand for a variety of conversion technology is obvious. In general, such systems provide functionality to convert email, voicemail, facsimile, and short messages into each other. These conversion processes are executed on certain media converters that could be seen as media gateways. They are capable to convert a specific medium type or media format into another one.

The concept of Media Gateways in the understanding of different standardization bodies such as ITU and IETF concentrates actually on media adaptation between VoIP and PSTN. In the area of Unified Messaging, such gateways are also responsible for the conversion of media format and media type keeping the semantic of the message intact. Such conversion of service data depends not on the transport network. It is essential for the integration of asynchronous messaging services to differ between the actual signaling of a service and the representation of the streaming of the service data. For instance, a Text-to-Speech conversion can be handled in the same way for each kind of text, regardless the original service like fax or email. This provides a service independent handling of service data for the conversion between services like show in figure 4.

Multiple or chained conversions may result in several drawbacks. The first is the delay introduced through the conversion process depending on the computing power of the employed hardware platform. The second is the loss of information within the converted message. No conversion process can be done without the loss of information. The question is, how this information loss can be described, measured, and configured by the user. We have implemented an application based QoS evaluation mechanism that provides this functionality (for details see [11]).

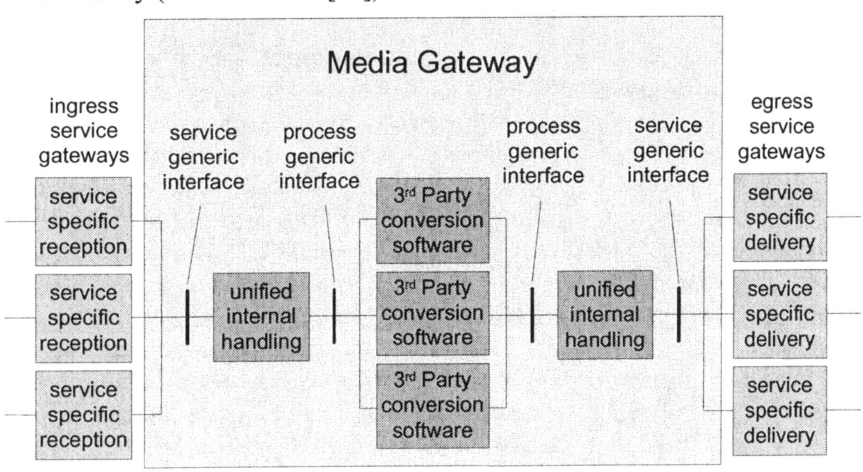

Figure 4: Media Gateways for Service Interoperability

The proposed solution is to distinguish between Service Gateways that provide the access to various transport networks and specific service signaling and Media Gateways that realize the adaptation of service data. A set of distributed components grants the evaluation of the user's preferences and the selection of appropriate conversion strategies on top of these two kinds of gateways.

4.1 Call Processing

The opportunities of Media Gateways impact the call processing of telecommunication systems and the information contained in user profiles:

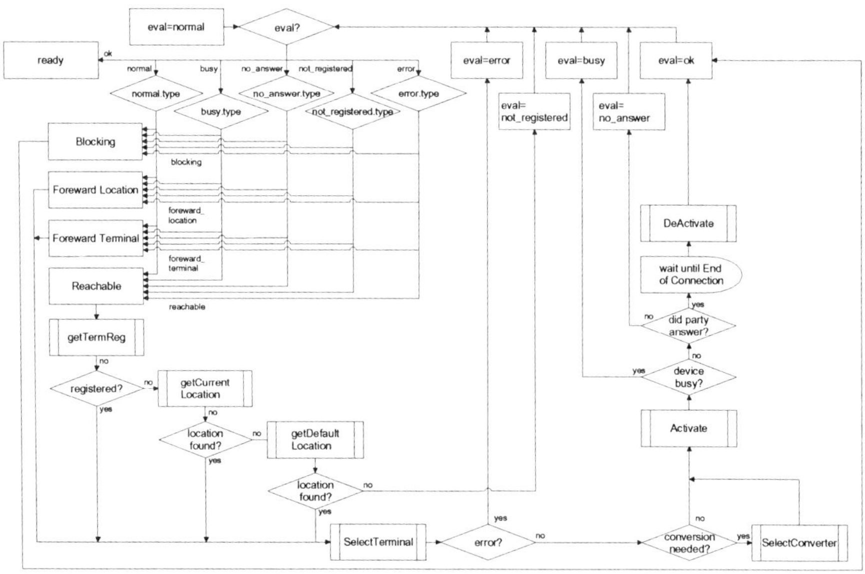

Figure 5: Call Processing Logic

All user preferences necessary for the call processing are stored in a rule based customer service profile. Rules define conditions on which several actions have to be processed (e.g. from whom the message is sent and at what time, the terminal a user is currently registered at, or the type of messaging service, the message is transported by). Possible actions for the call processing are blocking the call (blocking), to forward the call to a specific terminal (forward terminal), to forward the call to a specific location (forward location), or on-demand evaluation of the user's registration (reachable). Each action can be coupled with additional information, like forced/forbidden services. The execution of a rule can result in numerous states; normal delivery (normal), busy terminal (busy), user does not answer (no answer), user is actually not registered (not registered), or various cases of failures (error). The user can define the system behavior by specifying a rule for each state.

The resulting call processing (figure 5) is not trivial. The evaluation begins with the state normal. If a valid rule for this state can be found in the user profiles, the related action will be executed. The actions can be blocking, forwarding, or reachable. In case the action is set to reachable, the call processing has to request further information from the user's profile. If the user is registered at a specific terminal (e.g. a mobile phone), the call should be delivered to that terminal. If the user is registered at a specific location (in-house via active badge systems, via GSM cell detection, or via GPS), the terminals available at this location have to be identified and the most appropriate terminal is selected for the delivery. If the finally selected terminal is not able to receive the call directly (e.g. fax to phone), a suitable Media Gateway has to be chosen. Then, the call can be delivered with appropriate media conversion.

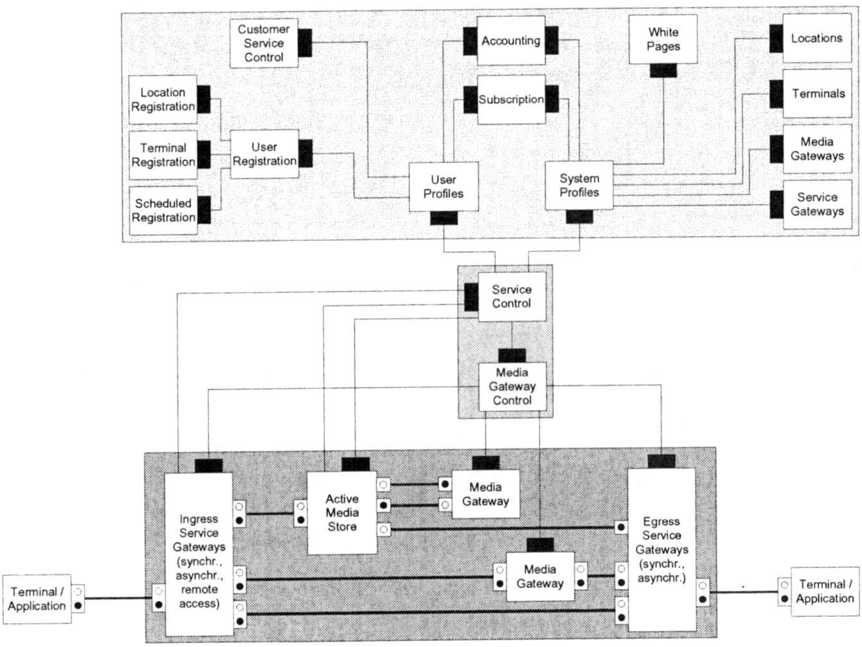

Figure 6: Distributed Components

During the delivery of the call, the system can run into other states. The terminal is busy or the user does not answer. Then, the call processing is invoked again and will now evaluate the rules specified for the new state. Therefore, mechanisms have to be provided to prevent internal loops and other failures.

Please note, that this call processing has been developed for the design and the implementation of our UMS system and the control of our Media Gateways. The evaluation of a possible adaptation to UMTS and 3GPP developments is still in progress.

4.2 System Architecture

The messaging system architecture comprises a certain number of distributed objects, representing different functional blocks of the system as introduced in section 3.1 and shown as computational objects in figure 6. Each of them has a well defined interface which offers its functionality to other objects. In the following these functional blocks are called components. A detailed description of each component can be found in [10].

The intention to subdivide the system into components results from the requirement of a freely scalable and distributed system. In a component based architecture it is possible to substitute a component by an extended one, without affecting the rest of the system. A future extension could be made easily. The only precondition are carefully designed interfaces between the object, which should be unchanged during the implementation process.

The components can be grouped in to three main blocks. The lowest block consists of all objects which are responsible for streaming and processing of service data. This includes the Service Gateways and the Media Gateways. The middle block represents service logic, service control and the media gateway control.

The implementation covers all components shown in figure 6. The system itself is up and running since November 1999. After a phase of intensive testing by our development team, a first real-life version is going to be installed at the end of May 2000.

For the realization of our system we use Orbix 2.3 (including Name Service), a CORBA 2.0 conform implementation. All components are realized on Windows NT using Microsoft's Visual C++ 6.0. The core implementation is based on Rogue Wave's Tools.h++ 8.0 Professional library package.

5 Summary

his paper described an approach, how Media Gateways currently used for Internet Telephony can be enhanced with the capability of media conversion. The basic concepts for this enhancement have been developed in the area of Unified Messaging, where the conversion of different medias is used for the interoperability of applications.

One central point of our approach is the distinction between a Service Gateway and a Media Gateway. Service Gateways are responsible for service specific signaling, whereas Media Gateways are responsible for the actual conversion of the content of the services. This provides an easy integration of new services, because the conversion process does not rely on the service signaling. On the other hand, new conversion technologies can be applied to the system without affecting the supported communication services.

The described approach and the already implemented system are currently analyzed in several European projects with regard to existing standards and working groups, namely the 3GPP, ITU-T, and IETF.

6 References

[1] ITU-T Rec. H.323, Packet-based multimedia communications systems, 1998

[2] Greene, N., Ramalho, M., Rosen, B.: Media Gateway control protocol architecture and requirements. IETF Internet draft, September 1999

[3] Cuervo, F., Huitema, C., Kelly, K., Rosen, B., Sijben, P., Zimmerer, E.: MEGACO Protocol, IETF Internet draft, July 1999

[4] ETSI Techn. specification TS 22.01v.3.1.0: Universal Mobile Telecommunication Systems (UMTS). Aspects, Principles. - Sophia Antapolis, France, 1997

[5] ETSI Draft 22.70v.0.0.3: Virtual Home Environments. - Sophia Antapolis, France, 1997

[6] Lennox, J., Schulzrinne, H., La Porta, T.: Implementing Intelligent Network Services with the Session Initiation Protocol. Tech-Report Number CUCS-002-99

[7] Magdanz, T., et.al.: Towards an Integrated Architecture for the Harmonisation of PSTN and Internet Services. International Conference on Intelligence in Networks, ICIN 2000, Bordeaux, France, January 18-20, 2000

[8] Zizza, F., et.al.: Towards a Distributed Intelligent Network. International Conference on Intelligence in Networks, ICIN 2000, Bordeaux, France, January 18-20, 2000

[9] ETSI TIPHON Homepage: http://www.etsi.org/tiphon

[10] van der Meer, S.; Arbanowski, St.: An Approach for a 4th Generation Messaging System. Proc. of the 4th International Symposium on Autonomous Decentralized Systems, ISADS'99, Tokyo, Japan, 1999, 158-167

[11] Pfeifer, T.; van der Meer, S.: The Active Store providing Quality Enhanced Unified Messaging. Proc. of the 5th Conference on Computer Communications, AFRICOM-CCDC'98, Tunis, October 20-22-1998

WEB ENABLED TELECOMMUNICATION SERVICE CONTROL USING VOXML

Bilel Guedhami, Cornel Klein, Wolfgang Kellerer
Siemens Information & Communication Networks,
Hofmannstr. 51, 81379 Munich, Germany
and
Munich University of Technology, Institute of Communication Networks
Arcisstr. 21, 80333 Munich, Germany

Email: Bilel.Guedhami@icn.siemens.de, Cornel.Klein@icn.siemens.de and Wolfgang.Kellerer@e-technik.tu-muenchen.de

Abstract: Techniques for enabling voice access to the World Wide Web have been extensively investigated both within the research community and industry. The voice markup language VoxML[1] has been proposed as a markup language enabling the HTML-like description of voice services by describing such services as "dialogues". By means of some sample services, we show that VoxML would highly benefit from additional language features for call control. We propose a set of tags that allow to access call control from within VoxML dialogues, and describe a prototype system that implements these tags.

Keywords: VoxML, Service Architecture, Web, Call Control

1. INTRODUCTION

Almost every day one can read a press article reporting about someone who got rich within some months simply by opening up a new ".com"-service, and almost every day several dozens of new such services are

[1] VoxML is a trademark of Motorola Inc.

deployed on the world wide web. Besides the commercial attractiveness of offering WWW services, from a technical point of view, the rapid pace of growth of the WWW is spurred by the fact that web programming is relatively simple: only a little know-how in HTML authoring and database programming are needed to develop a new internet service. For this reason, web-based services can be deployed by almost everyone, simply by registering a domain address and connecting a new server to the internet.

Compare this with approaches for deploying value-added services within the public switched telephone network (PSTN), such as the IN (intelligent network) architecture. Although offering clear advantages w.r.t. reliability, security and access to billing facilities, the required infrastructure is relatively expensive, controlled by few telecom operators, and the service development requires programming proprietary APIs. For this reason, an approach combining the best of both worlds would be highly desirable.

Access to the world wide web is, however, still limited mainly to PC-based clients, due to the nature of HTML and the web-pages expressed therein, which both assume that large, graphical displays and sufficient processing power on the clients are available. Several attempts have been made to overcome this deficiency. Most famous is WAP, an architecture and a suite of protocols which are tailored for bringing the web to mobile phones, which have both limited display capabilities as well as limited processing power. Less known, but interesting because of the immense amount of installed clients, are approaches that aim to provide voice-based internet access from ordinary telephone sets.

The objective of so-called "Voice Markup Languages" such as VoxML [VoxML] is to provide, similar to HTML, a simple approach for developing speech applications and making them accessible to a broad audience via any voice-enabled client (phones, mobiles). The idea behind voice markup languages is that a "voice surfer" dials into a "voice browser" via the public switched telephone network (PSTN). The voice browser accesses, via HTTP, web servers in the public internet or within corporate intranets. The content, which is fetched from these web servers, which is expressed in VoxML and which is structured into so-called "dialogues", is interpreted and transformed by the voice browser into speech signals. For output, speech synthesis or the play back of prerecorded audio-files is used. For input, technologies for speech recognition or for DTMF-decoding are used. VoxML does perfectly fit for information services such as

- voice-surfing the web (via dedicated VoxML pages or via HTML to VoxML gateways),
- inquiries of email (via VoxML gateways to email systems or via VoxML interfaces of existing webmail systems),

- access to information systems e.g. for travel information, weather information, stock quotes etc.

However, services which require access to network features such as call-control or billing can not be implemented conveniently this way, i.e. only by accessing platform-specific APIs. This makes VoxML inappropriate for services like

- call centers, which require redirecting calls from voice-surfers to other telephone lines or
- services which require connecting voice-surfers with other subscribers, like e.g.
    - establish a call to a subscriber after this subscriber has been found by a VoxML-based lookup of a telephone directory
    - establish a call to the originator of a voice mail after a VoxML based inquiry of the voice mail box.

In this paper, we will propose an extension of VoxML with features for call control. This will allow the portable and platform-independent description of the above-mentioned kinds of voice-based services. By integrating call control into VoxML, content authors – with little know-how in programming – will be enabled to develop web-based speech applications by separating the underlying call control from content. In particular, these VoxML based service descriptions will be independent of any underlying network APIs (TAPI, JTAPI, Parlay), signaling protocols (H.323, ISDN signaling) or service control platforms (e.g. SCP, CTI Servers).

The paper is structured as follows: In Section 2, we will review existing approaches for voice-based access to web-based information systems and discuss some services to show the need for an enhancement of VoxML. In Section 3 we will propose a set of language extensions ("tags") for VoxML. In Section 4 we will present the architecture of an enhanced VoxML enabled telecommunication system and describe our prototype implementation. In Section 5 we will discuss the benefits of our approach, and in Section 6 we will discuss related work. We will conclude in Section 7 with a summary and an outlook on future work.

2. VOICE-BROWSER BASED ACCESS TO WWW BASED INFORMATION SYSTEMS

The main idea of voice browsing is that a "voice surfer" dials into a "voice browser" via the public switched telephone network (PSTN) (Fig. 1). The voice browser accesses, via HTTP, web servers in the public internet or within corporate intranets. However, the proposed solutions differ in the form the content is stored on the web server – HTML or specialized markup languages like VoxML – and the intelligence which is built into the browsers.

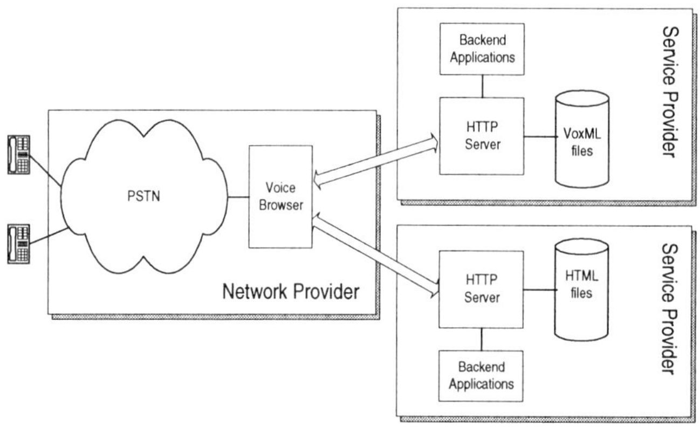

Figure 1. Voice Browsing Service Architecture

Voice browsers are browsers that allow people to access the web through voice interfaces. A combination of keyboard and speech recognition is usually used for input, and speech synthesis and pre-recorded sound are used for output. An extended concept of a voice-browser may also control small displays on telephone when these are available. Despite these limited capabilities for input- and output, voice browsers represent a new and attractive alternative to the classical browsers (that use screen and/or keypad). They make the need for the hands and the eyes unnecessary and reduce the required equipment for web browsing to a telephone.

Beside opening the web doors for people with telephone access, voice browsers also provide a field for new ideas for web-based applications and services. Call centers and hands-free mobile applications especially in automobiles can make use of this speech technology.

Today, almost all the information published in the web is authored in HTML. So it is attractive to offer voice access to HTML content. As HTML is designed with a visual interface in mind, it is difficult to create high-quality voice services using today's HTML standards. Some people suggest the extension and the adaptation of HTML to the new requirements. Others propose to create a new mark-up language that target the voice medium precisely and directly.

2.1 Extension of HTML

HTML (HyperText Mark-up Language) is a set of "mark-up" symbols or codes inserted in a file intended for display on a World Wide Web visual browser. The mark-up specifies for the browser how to present a document on a screen, including words, images etc.

HTML has been developed with visual rendering in mind. As a result, many of its elements would not be amenable to speech rendering (e.g. image maps). Other elements could be misinterpreted or even senseless for voice browsers (e.g. frames). In addition, lots of voice specific aspects cannot be supported since there are no corresponding HTML tags. For example, using a specific recognizer or synthesizer that associate specific recognition grammar with input elements and with the control over the synthesizer volume, speed, pitch, etc., and also the interaction for timeouts, errors etc.

Obviously, HTML in its current form will not be of great help when authoring web sites for voice applications. As they assume it to be unlikely that web designers will develop different pages for both visual and voice browsers, some researchers suggest the extension and reform of HTML to support voice browsing. These "pro-HTML" researchers underline that adding voice capabilities to the web should be regarded as a natural evolution of the web and not as an industry specific derivative, as this could be the case when resorting to a new mark-up language. They assume that a combination of the following techniques would provide a suitable solution for the voice extension of the web:

- *Extension of HTML:* Extensions of HTML such as Aural Style Sheets, e.g. [Fer98] (which are part of the Cascading Style Sheets, Level 2 specification [RHJ99, CSS]) provide a level of control in spoken text. With the use of an aural style sheet, authors are allowed to specify characteristics of the spoken text such as volume, pitch, speed and stress. It is also possible to indicate pauses, insert sound files (analogous to icons in HTML) and show how certain phrases, acronyms, punctuation, and numbers should be voiced.
- *Style guides for HTML authors [Gun98]:* Styleguides provide rules for writing HTML-files which can easily be translated into useful dialogues by voice browsers. An example is to provide useful text choices for links, where text descriptions instead of URL's make the links easier to speak. Also, names of persons are usually simpler than their e-mail-addresses. Furthermore, ambiguity could be reduced by avoiding same words for different links. Here are some examples:

This version: http://www.w3.org/WD-USERAGENT-19980814
Instead of:
This version : http://www.w3.org/WD-USERAGENT-19980814

Select local news or national news.
Instead of:
Click here for local news, or here for national news.

- *Intelligence in Voice Browsers [Gun98]:* When browsing into Web pages, it is easier to determine their structure and retrieve the relevant information by visually scanning them rather than by listening to all their content read by voice browsers. This could turn to be very uncomfortable especially with long pages. Voice browsers can apply intelligent abstraction techniques such as forming a structure based on the <H1>, <H2>, ... headers or listing or the links on a page, to provide a kind of a structured summary to the user. This suggests that the author uses the HTML elements properly and meaningfully. For example he should make useful choices for link text or makes use of the SUMMARY attributes for tables.

The main point is whether these proposed strategies together with appropriate HTML extensions would be sufficient to support voice access to the web, or whether new specialized formats need to be created. The question that has to be asked is whether people really want general telephone browsing, or do they just want access to particular important information in more than one way.

2.2 Voice Markup languages

Based on the fundamental requirements for a mark-up language that supports voice access to the web, we have found that the extension of HTML has some important limitations. HTML could not be easily extended in ways that would make voice browsing possible. In addition, generating a dialog system from a graphical representation can easily fail to create a usable or useful system. We also suppose that the majority of people wanting to access the web via voice interfaces are not interested to hear the content of some people's private sites "spoken" on the telephone but rather aim to make use of some important and helpful documents and services on the web. Such documents and services will have to be designed with a specific mark-up language for the interactive voice rendering. The ideal would be to develop a mechanism that allows both graphical web pages and telephone dialog systems.

Experience shows that the complexity of dialogs and interactions can overwhelm a programmer not skilled in both concurrent systems and human-computer interfaces. A structured mark-up language is a convenient way to hide hardware complexities, to abstract away the difficulties of concurrent programming and to codify proven principles of IVR (Interactive Voice Response) design. The mark-up language should allow the designer to concentrate on the essential matter: the contents, choice of pre-defined dialogs and associated help menus and prompts.

For this reason, several companies have proposed voice markup languages. In this paper, we will stick to VoxML [VoxML], which has been developed by Motorola and which is used within our prototype. A comparison with the upcoming industry standard VXML is given in Section 6.

It is important to note that several concepts and elements of HTML apply in the voice world. For example the LINK element in the header section that define the relationship to other documents, the FORM element that defines the name/value pairs to be returned to the server. The incorporation of HTML elements in the new mark-up language and the adoption of some of its basic concepts makes it easier to learn and to use.

Voice Markup languages will enable a greater flexibility in accessing internet resources, as everyone can launch a variety of information and communications applications from anywhere provided that one could access a telephone. They also offer new business opportunities for content developers and network operators and they allow rapid and easy application development for voice access. However, the wide spread of voice browsers would depend on the ease with which web designers can add speech support to their site.

2.3 Service Examples

VoxML enables a large spectrum of voice services to be created. The main characteristic of these services is that they provide an interactive dialog system for users, known also as IVR (Interactive Voice Response). Users would then have access to multiple sources of information that can be adapted to their personal needs. This adaptation occurs dynamically within the user-browser conversation, as the user can decide which information is relevant to him and he wants to hear.

Since the access device for the voice services is the ordinary phone, service providers can extend their function as an information source to provide telephony and networking applications, such as setting up and forwarding calls, arranging and managing voice or video conferences (depending on the user calling device) and sending mails.

What is currently missing in VoxML are language constructs for call control. The following two examples services indicate that an appropriate extension of VoxML would be highly beneficial:

Information Inquiry and Call Management Service: A company may offer a voice-based inquiry service for telephone and e-mail addresses of its employees. By dialing a special service number, users can look up the corporate's directories (which is e.g. stored on an LDAP server) by naming the person through voice recognition or by typing it on the phone keypad. Users can then ask the system to establish a connection to the employee.

Note: VB stands for Voice Browser. U means User.

VB: *Welcome to Siemens. You can use our service to get the co-ordinates of our employees such as their phone, fax or email addresses. We can as well transfer your call to any destination or set up conferences with whom you wish. Please say "search", transfer" or "conference".*

U: search

VB: Please say the surname of the person you wish to look for, or type it on the keypad of your phone

U: Mayer

VB: Please say the first name of the person you wish to look for, or type it on the keypad of your phone

U: Peter

VB: Are you looking for the phone, the fax number or the email address?

U: phone number

VB: The telephone number of Peter Mayer is 722-12345

If you wish to get connected to him now, please say connect. And if you want to have him as a part of conference please say conference.

U: connect

VB: I'll try to establish the connection now. Please don't hang up...

(Trying to call Peter Meyer)

VB: I am sorry, the line is busy. If you wish to leave him a message please say message. If you wish to send him an email, please say email

U: message

VB: You can speak after the sound signal for a maximum of 30 seconds

U: .....

(After longer breaks or 30 seconds)

VB: Thanks a lot for exploiting our service. Can we help you any further?

U: No.

VB: Bye.

Figure 2. Information inquiry and call management service

```
<?xml version="1.0?>

<DIALOG>
 <STEP NAME="init">
  <PROMPT> Welcome to Siemens. You can use our service to get the co-ordinates of
  our employees such as their phone, fax or email addresses. We can as well transfer
  your call to any destination or set up conferences with whom you wish.
  </PROMPT>

  <INPUT TYPE="NONE" NEXT="#main_menu"/>
 </STEP>

 <STEP NAME="main_menu">
  <PROMPT>  Please say search <BREAK/> transfer <BREAK/> or conference
  </PROMPT>

  <INPUT TYPE="hidden" NAME="inputn_first_name" VALUE="FIRST_NAME"/>
  <INPUT TYPE="hidden" NAME="inputn_last_name" VALUE="LAST_NAME"/>
  <INPUT TYPE="hidden" NAME="search_type" VALUE="NEW"/>

  <INPUT TYPE="optionlist" NAME="service">
   <OPTION NEXT="#search_service"> search </OPTION>
   <OPTION NEXT="#transfer_service"> transfer </OPTION>
   <OPTION NEXT="#conference_service"> conference </OPTION>
  </INPUT>
 </STEP>

 <STEP NAME="search_service">
  <PROMPT> Are you looking for the phone, the fax number or the email address?
  </PROMPT>

  <INPUT TYPE="optionlist" NAME="search_option">
   <OPTION NEXT="#firstname_request" VALUE="phone number">  phone </OPTION>
   <OPTION NEXT="#firstname_request" VALUE="fax number">   fax  </OPTION>
   <OPTION NEXT="#firstname_request" VALUE="email address"> email </OPTION>
  </INPUT>
 </STEP>
 ....
```

Figure 3. Sample VoxML Script

The convenience and quality of this service can of course be extended in many ways. For instance, to improve the recognition, the server can compare the recognized name to the entries of its database. In case of slight differences the server can correct some input faults. Moreover, if it is not possible to setup the connection, users are offered to send an email or voice mail alternatively.

Figure 3 shows a sample VoxML script, and Figure 2 the corresponding interaction between a user U and a voice browser VB of such a service.

Virtual Call Center: VoxML can provide an alternative to realize small virtual call centers with simple functionality. A virtual call center is a call center that is composed of multiple units that have not to be located geographically at the same place. These tiny units can represent call centers on their own or unique persons working for instance at home. Callers have to dial just one service number. A server would then distribute the incoming calls over the several associated units. Virtual call centers represent a considerable way to achieve great flexibility, increase productivity, and reduce costs when creating a call center.

3. EXTENSION OF VOXML WITH FEATURES FOR CALL CONTROL

As we have seen, VoxML in its current form enables already a large spectrum of voice services to be created, in particular access to web-based information systems where users have access to multiple sources of information that can be adapted to their personal needs. This adaptation takes place dynamically within the user-browser conversation, as the user can decide which information is relevant to him and he wants to hear.

However, to define more sophisticated telecommunication services like conference calls within a voice browsing session, VoxML has to be extended with call control commands. By analyzing several services, we have identified the following new tags that can form the basis for a future standardized extension of VoxML. In particular, we propose the following tags:

- TRANSFER: Transfer a call to another destination
- MCALLS: Transfer a call to some destination within a group of destinations (e.g. of call center agents)
- CONFERENCE: Set up a conference between more than two parties
- SEARCH : Search a data base
- EMAIL: Send an email.

In the sequel, we will describe these tags, which we have included in our prototype voice browsers system, in more detail. Since all functions are selected by user input the tags are defined as part of the INPUT tag family.

3.1 TRANSFER Tag

Syntax:

```
<INPUT  TYPE="TRANSFER"     NAME="name"
        DEST="phone number">
```

Attributes:

DEST: The telephone number the call is transferred to.

NAME: The name of a VoxML global variable, to be set when accomplishing the transfer action. Its value can be retrieved within further steps.

3.2 CONFERENCE Tag

Syntax:

```
<INPUT  TYPE="CONFERENCE"   NAME="name"
        PARTIES="party, party,..."
        CONFTYPE="type">
```

Attributes:

PARTIES: List of the conference participants

CONFTYPE: The type of the conference, e.g. "voice" or "video"

The CONFERENCE tag is quite different from the TRANSFER tag since here we aim setting up one or more completely new connections in addition to or beneath the existing voice browser session.

3.3 EMAIL Tag

Syntax:

```
<INPUT  TYPE="EMAIL"        NAME="name"
        FROM="name@host"
        TO="name@host"
        SUBJECT="subject"
        ACK="value"
        ATTACH="file.doc"
        TEXT="body">
```

Attributes:

FROM: email address of the sender (user)

TO: email address of the recipient

SUBJECT: subject of the email

ACK indicates if an acknowledgement message is to be sent to the user account

ATTACH: list of files to be attached
TEXT: The body of the email

The EMAIL tag goes beyond the purpose of providing telecommunication services. We have included this tag on the one hand to show the flexibility of the enhanced VoxML based telecommunication service control. On the other hand email services are an essential improvement for today's telecommunication services. Just have a look on the success of the SMS service in GSM.

3.4 MCALLS Tag

In order to create the service "virtual call center", it is necessary to be able to connect a caller to the next available call center agent. This requires that the voice browser should be able to try to simultaneously establish a connection with any agent within a specified group. This can be described using a new tag <MCALLS>. "mcalls" means "multiple calls". This is similar to the "fork" command within the IETF Session Initiation Protocol (SIP), which deals with the same scenario [SR99].

The MCALLS tag has got the following syntax:

```
<INPUT   TYPE="MCALLS"      NAME="name"
         PARTIES="number, number,..."
         SIMULT="4"
         LOOP="yes"
         TIMEOUT="time">
```

Attributes

PARTIES: The list of telephone numbers of the concerned units
SIMULT: Indicates how many numbers the browser tries to reach simultaneously.
LOOP: Can take the value "yes" or "no". "yes" means the voice browser should go through the telephone numbers list once again, if he fails to setup a connection during the first attempt.
TIMEOUT: Indicates the time interval, after which the connection setup procedure should be aborted.

3.5 SEARCH Tag

Database access represents today one of the main features of the services offered on the web. Search machines, online dictionaries, booking services and innumerable other applications live on the dynamic information resources provided by databases.

Voice-based web services and screen-based web services should be all the same except for their access interfaces. In this way it is obvious to integrate a

database functionality also in the voice based system. To support the developers with an easy way of handling data base requests we integrated a basic database functionality on the mark-up language level in contrast to HTML-based services, which use CGI-scripts whenever database queries have to be sent. Precisely, a click on an appropriate button (often labeled "search", "start" or "send"...) in the service page, launches a CGI-script, that collects and analyses the information filled in the entry boxes by the user, conducts the communication procedure with the database to finally generate new HTML pages containing the new acquired data.
Our approach is to enhance VoxML with a new tag, which provides to the service designer the following basic database functions:

- search a given database for a specific key value
- retrieve the matching entries successively
- refine search results

Based on these functions we have defined a new VoxML tag:

```
<INPUT TYPE="SEARCH"          NAME="name"
        DB="database"  QUERYNAME="search1"
        INPUTNAME="name"
        INPUTVALUE="value"
        OUTPUTNAME="name"
        OUTPUTVALUE="value"
        SEARCHTYPE="type">
```

Attributes:

NAME: The name of a global variable, which is set according to the result of a data base search, e.g. no_entries, one_entry.
DB: Name of the database the request goes to
QUERYNAME: Label of the search request for identification
INPUTNAME: List of the field names to be checked
INPUTVALUE: List of the field values
OUTPUTNAME: List of the searched field names
OUTPUTVALUE: List of the variable names where the searched values are to be stored
SEARCHTYPE: Type of the current search (NEW, REFINE,...)

In order to create a flexible voice system, the mentioned database functions are to be provided by an independent module, called the directory server. The communication between the VoxML browser and the directory server is message based. Thus, the role of the browser is limited to parsing the VoxML files, generating and sending the search messages to the directory server, and interpreting the returned messages. This logical architecture also implies a possible physical separation, as the two modules could be located on different servers (see Figure 4).

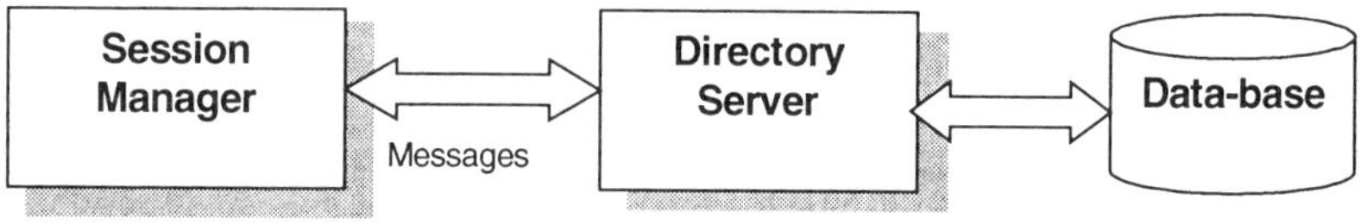

Figure 4. Principle Architecture of the database query enhancement

4. ARCHITECTURE OF A VOXML ENHANCED TELECOMMUNICATION SERVICE NODE

In this chapter, we present the architectural concept of a voice browsing platform, enabling the creation of voice-based telecommunications services. Furthermore we describe a prototype implementation that has been built according to the general concepts.

The platform is composed of several basic units (see Fig. 5). Their interaction and data flow is controlled and supervised by the central unit. The platform also includes software and hardware components for data (voice and tones) input and output such as Telephone API (TAPI) compliant hardware or speech synthesis and recognition tools.

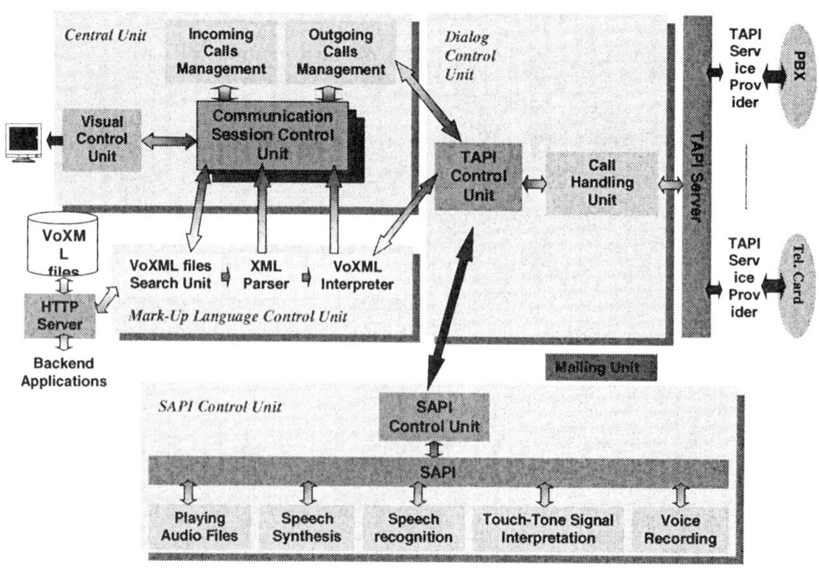

Figure 5. Architecture of an enhanced telecommunication service node

4.1 Microsoft TAPI

The Microsoft Telecommunication Application Interface (TAPI) is an open industry standard for the support of telephony service control. It provides network and device independence and supports a great variety of telephone features.

The MS TAPI model supports four different physical configuration alternatives. The first is phone based. That means the telephone is in between the application running on a PC and the switching system. The second alternative is the PC based configuration, in which the PC is the means to access the telephony lines. In the shared or unified line alternative the PC and the phone have the same access to the phone lines.

All of the above configuration types only allow the control of one telephone line per workstation. The last of the four, the multiline configuration, supports the control of many lines and is therefore suitable for the realization of PBX applications or for the discussed voice server.

For the programming of the controlled telephony resources the MS TAPI provides a number of functions, which support the development of services on telephone networks. For simple basic telephony services the TAPI basic functions are sufficient for the control of incoming or outgoing calls. Advanced services like call forwarding, suspend or conference control use the supplemental telephony functions. The class of extended telephony services is for device specific functions of hardware vendors.

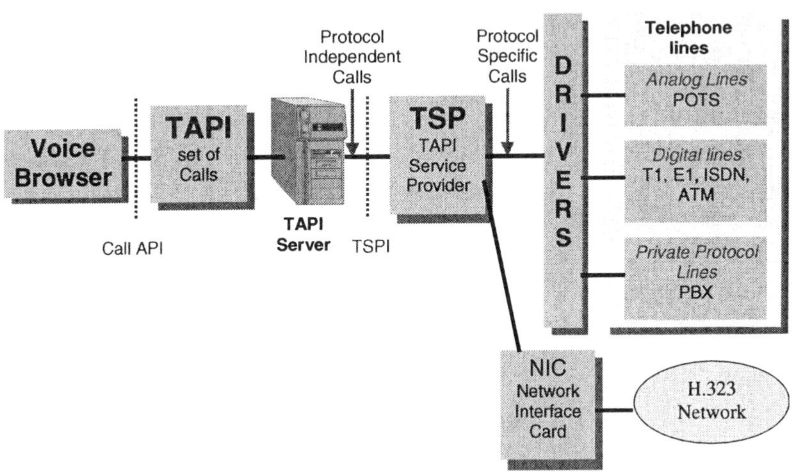

Figure 6. MS TAPI Architecture

The control of a telephony resource e.g. a telephone line or a PBX is illustrated in Fig. 6. An application, which may be a voice browser for example, uses the TAPI functions. The TAPI server transmit these function calls to the TAPI service provider (TSP). The TSP translates protocol-independent functions in specific functions of the underlying driver that controls the telephone resource. A telephone resource may be a simple analogous line or a digital PBX or even a H.323 network. MS TAPI is supported by a large number of resource providers. Most of the PBX products on the market provide a TAPI interface.

4.2 Components of the Telecommunication Service Node

Coming back to our voice browsing platform from Fig. 5, whenever a call request is detected by the TAPI control Unit, the central Unit starts a new session. It is then the role of the Mark-up Language Control Unit to get the VoxML files and their DTD (Document Type Definition) from the local memory or from a remote voice-service server via an HTTP-server. The browser has then to interpret the parsed VoxML resources and to deploy the TAPI and SAPI (Speech API) units that provide the call control and speech control mechanisms. The directory server assures database access functionality. The mailing unit helps to realize voice mail services. More precisely, the different units serve the following tasks:

Central Unit. Its main function is to manage the user sessions and to control the underlying units. This includes launching their mechanisms, checking their state of function and coordinating the communication between them.

Mark-up Language Control Unit. This Unit consists of three main modules: the VoxML-Files Search Module, the XML Parser and the VoxML Interpreter.

The VoxML-Files Search Module is responsible for seeking after VoxML files and their DTD. This may imply setting up connections to remote HTTP-servers. The VoxML files are retrieved when a user session is launched or during the interaction itself, if the browser steps on a link request. The VoxML files are to be handed over to the XML parser.

The XML parser checks the received VoxML source code for validity and conformance with the DTD. If the document is well-formed, it is translated into a parse tree, handed over to the VoxML interpreter and an indication is sent to the central unit. Otherwise, the central unit is informed of the parsing failure. The error handling procedure is then to be applied.

The VoxML Interpreter is the most important element in the voice browser. It processes the received DOM tree from the XML parser and starts the related applications. It holds the control on the SAPI and TAPI control units as well as the directory server and the mail services unit, by sending them commands and interpreting their responses and indications. It analyses user inputs delivered by the SAPI Control Unit and generates output commands.

TAPI Control Unit. The TAPI control unit is the intermediate unit between the VoxML Interpreter and the TAPI compliant hardware. Its tasks involve the execution of the network access commands received from the VoxML Interpreter by translating them to TAPI calls and sending back confirmation messages, state and failure indications. The TAPI control unit has to interact with the underlying TAPI hardware and handle all the events reported by the TAPI call-back mechanism. For instance, in case an incoming call request is detected, the central unit is informed, so that a new session can be started.

SAPI Control Unit. The SAPI control unit has the direct access to SAPI. So it transforms the received commands from the VoxML Interpreter into SAPI specific commands, that control the various speech technology engines. It detects user inputs and forwards them to the VoxML interpreter.

The main tasks of this unit include the speech synthesis, playing audio files on an established connection, voice recognition and recording.

Directory Server. The directory server performs database access and realizes basic functions, such as searching for specific field values in database records or retrieving and refining results. It locally stores the results of search queries, to use them for refinement purposes until it is overwritten by a new query or the session ends up.

Mail services Unit. This unit is contacted each time an email is to be sent.

Visual Control Unit. This unit represents a monitoring system for the voice browser and its running applications. It may also be used for debugging purposes. It shows the working units, the progress of the calling procedures and displays the received messages from TAPI and SAPI interfaces. It provides also a graphical overview on the actual proceeding connections, the number of the available and active (or inactive) line and phone devices, and the incoming and outgoing calls.

4.3 Realization

To show the feasibility of our approach we have realized the concepts described above in a prototype implementation. To support rapid prototyping we have based our implementation on existing components. It is two main components we need. First we have to take a system that allows control over telephone calls. This enables the development of services within the telephone network. Second we have to integrate a voice browser system. This system has to be enhanced by the new defined tags and has to be connected to the call control.

4.3.1 Siemens VoxPortal

VoxPortal is a commercially available product that has been developed by Siemens Information and Communication Networks [VoxPortal]. It represents a powerful and flexible IVR system basing its interaction rules on the VoxML language. Consequently, the VoxPortal could be considered as a voice-based access tool to any type of information, provided that it has been previously translated to VoxML.

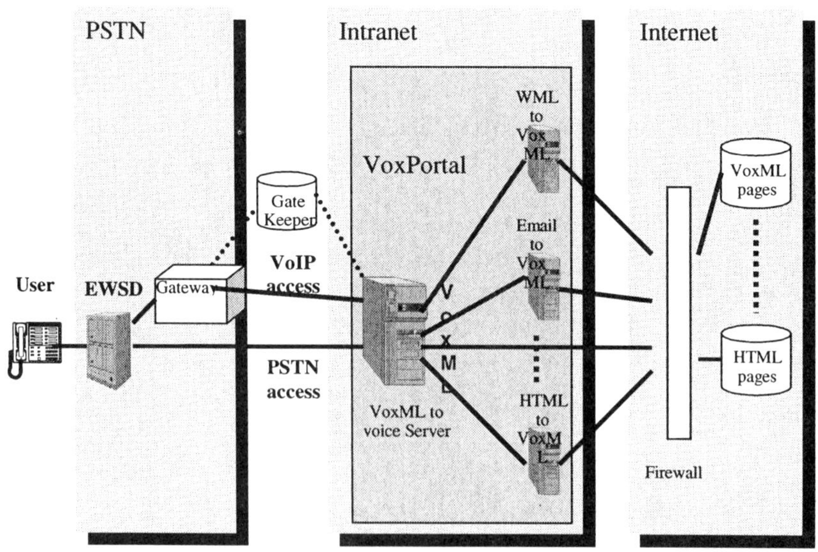

Figure 7. VoxPortal Architecture

The VoxPortal architecture is depicted in Figure 7:

- *PSTN:* The user dials into the VoxPortal via the public switched telephone network. Depending on the operator's infrastructure,

VoxPortal is connected to the PSTN either via E1 lines or via a VoIP Gatekeeper (H.323). To interact with the VoxPortal, the user can make use of DTMF dial tones (currently supported by nearly any phone) or voice commands (planned in future releases).

- *VoxPortal:* The VoxPortal, which runs a VoxML browser, represents a bridge between the voice world and the internet world (HTML, VoxML)
- *Add-on servers:* Several application-specific add-on servers enable a great variety of interactive voice services to be realized such as remote email access or simple retrieval of HTML pages.

Altogether, VoxPortal represents a perfectly suitable platform for the realization and the demonstration of our concepts and ideas.

4.3.2 Enhancement of the Vox Portal

The commercial implemantion of the existing VoxPortal product has been the basis for our prototype. For the realization of our prototype we had to integrate the telephone network via TAPI to this VoxPortal system. In addition we had to add and to supplement several modules. Fig. 8 shows the main points of work for our demonstration application.

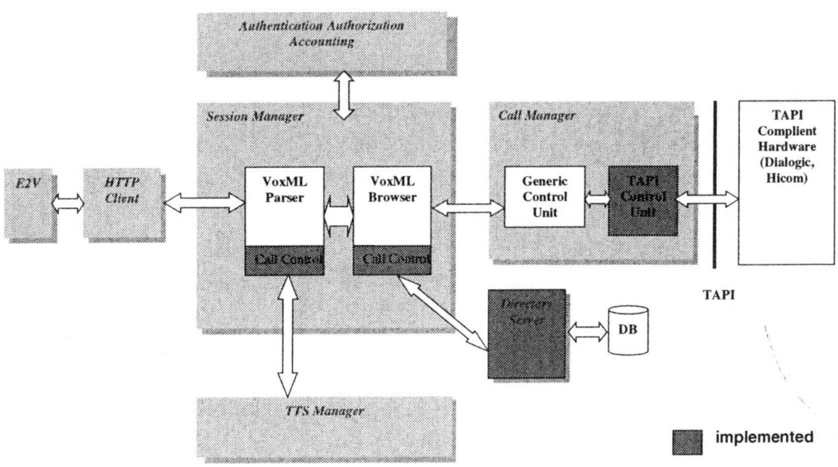

Figure 8. Integration of the VoxPortal and a TAPI based call control

The VoxML parser as well as the VoxML Browser had to be enhanced to include the new call control features. All actions resulting from the new tags had to be included in the session manager system.

For the interaction with the telephone network we have chosen the MS TAPI since this is supported by our switching hardware consisting of a dialogic board. A new TAPI module had to be developed acting as an application on the TAPI interface to generate and control calls from the voice browser. The TAPI module had to be integrated in the existing Call Manager module in addition to the existing module controlling the functions of the dialogic board for incoming calls.

For the demo system we decided to include the directory server and the data base in the voice browser system. The directory server interacts as a client with a database to handle information queries.

In order to be able to test and to demonstrate the system, we have implemented several applications, in particular the one depcited in Figure 3, as enhanced VoxML scripts.

5. OPPORTUNITIES AND BENEFITS OF USING ENHANCED XML VOICE INTERFACES

Beyond the new features introduced for call control, the presented system concept reveals opportunities that are also applicable in a broader context. The general ideas of our proposed approach can be applied in many other areas of telecommunication service development as well.

5.1 Paradigm Shift

The voice browser concept (as is, or applied to telecommunication service control) changes the traditional Web browser/server architecture. In the traditional HTML based world the information is displayed on a common Web browser like Netscape Communicator or Microsoft Explorer. The web browser is running as an application on the user terminal. The only way a information service provider can influence the user domain is to send cookies or to provide applets for download. As we can see in Fig. 9 this strict separation of domains between browser and server that means user domain and service provider domain is totally changed with the voice browser system.

Since the user terminal is no longer an intelligent end system but an ordinary telephone system, mechanisms in the network provider domain have to be provided to realize voice browsing. That means in other words that the voice browser is located in the service provider domain. This offers new opportunities for the service provider who offers a voice WWW portal. Examples are:

- New browser updates e.g. for new versions of VoxML can be introduced more rapidly, because only the browsers of the service providers have to be updated.
- This way, new tags for the creation of new services can be introduced more rapidly.
- The browser is located in a trusted area, enabling an easier introduction of securitiy-critical features such as billing (difficult in the internet).
- Furthermore, user can be bound to voice WWW portals by offering personalized services e.g. personal bookmarks access from everywhere.

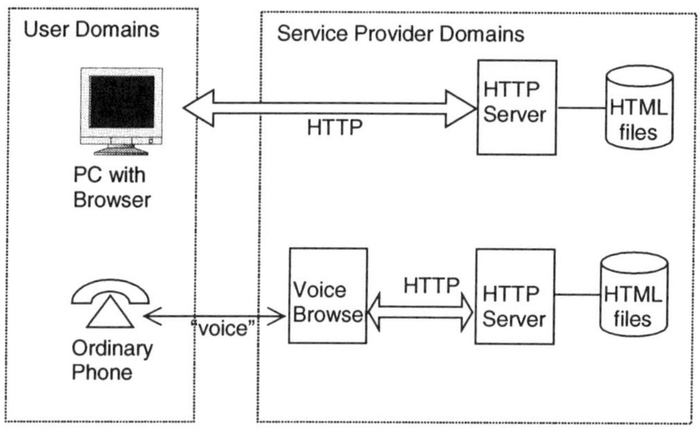

Figure 9. Voice browsing changes browser/server responsibility

When we look back on Fig. 1 we can see, that the voice portal service provider does not need to be the same as the content service provider (VoxML content) but also a network provider can earn revenue by this new role.

5.2 XML based service definition

As we have shown, the XML based VoxML language allows a very flexible definition of voice accessed information services. We have enhanced the VoxML by new tags to support some telecommunication services, additionally. Looking beyond voice based services the Extensible Markup language XML provides a basis for the definition of telecommunication and information services in general. For example the definition of personalized services like unified messaging in the IETF IP Telephony approach uses the Call Processing Language CPL [RLS99], which is based on XML.

In comparison with traditional service definition methods like the use of SIBs in the Intelligent Network the XML based service definition by scripting languages has a number of advantages:

- XML very well supports representation of structured data (e.g. service logic trees)
- XML allows definition of all necessary tags and is open for all kinds of extensions
- XML is easily readable also for human users
- XML allows easy verification against a Document Type Definition
- XML is fault tolerant by skipping unknown sequences by the parser

6. RELATED WORK

We have chosen VoxML as voice markup language and TAPI for call control as a basis. An alternative would have been to use the upcoming industry standard VXML and/or the PARLAY API.

6.1 VXML

Many companies recognized the importance of accessing the resources of the web by voice interfaces and developed new mark-up languages based on XML to enable easy creation of voice-enabled applications. Recently, AT&T, Lucent Technologies and Motorola decided to form the Voice eXtensible Mark-up Language Forum (VXML Forum) [VXML] to unify their efforts and contribute their technologies and experiences in this domain to the development and promotion of the open standard VXML specification.

AT&T, Lucent and Motorola worked long time on their own independent projects. While AT&T was building a mature phone mark-up language and launching applications, Lucent was continuing on a similar project known as TelePortal. Motorola's language, VoxML [VoxML], has focused on hands-free access for its mobile phone customers and uses speech recognition rather than touch-tones as an input device. With the VXML specification, the three companies hope to leverage the best of their approaches for the benefit of the entire industry. Since its launch in March 1999, the VXML Forum has more than tripled in size by adding 44 leading technology industry players to as supporters including: Conversa, Ericsson, France telecom, Siemens and Sun Microsystems.

Unlike VoxML, VXML contains a <TRANSFER> tag that can be used to forward calls to other subscribers. Not only is this transfer tag reserved for operator calls only, but also more sophisticated tags e.g. for setting up conferences are still not available in VXML. Moreover, in contrast to VoxML, VXML capable voice browsers are still not available. This has been the motivation for using VoxML instead for VXML as a basis for our work.

6.2 PARLAY

The Parlay working group [PARLAY] was formed in April 1998 by BT, Ulticom, Microsoft, Nortel Networks and Siemens. It aims the development of an API specification that enables applications to control a range of network capabilities and to access network information independently of network specific details. In December 1998, the first version (Version 1) of the specification was released together with an application for demonstration purposes.

The PARLAY architecture is similar to the TAPI architecture, but is targeted towards IN-like service control. Applications developed by a service provider make use of the PARLAY functions that are provided by the network provider on top of his network resources and are converted via resource interfaces in the network specific functions (see Fig. 8). The main difference compared to TAPI is the mechanism how the functions are provided for the application developer. The PARLAY API is therefore separated into two classes: The framework interface and several service interfaces.

The framework interface contains all necessary support functionality for the access control as well as for security, service discovery and maintenance. After successful authentication the service discovery mechanism supports the request of the application for a suitable service interface. The security is an essential feature to imply the adoption of the Parlay API by both service and network providers.

Service interfaces provide the network capabilities of the underlying resources including call control, messaging and user interaction. The Generic Call Control Service (GCCS) for example is based on a third party model, which allows calls to be instantiated from the network and routed through the network. It supports the same functionality like today's Intelligent Networks.

Due to the availability of the TAPI-based VoxPortal, we have based our call control features on TAPI. However, a PARLAY-based realization may be needed for additional features such as third-party call control or access to billing facilities by new tags.

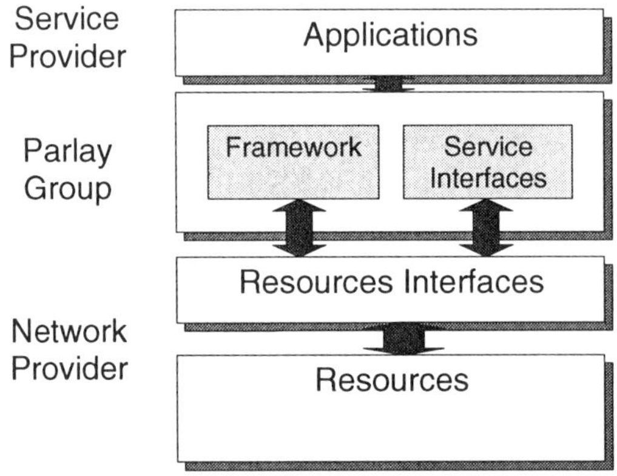

Figure 10. PARLAY Architecture

7. CONCLUSION AND OUTLOOK

We have introduced VoxML and VoxPortal as a web-based language and architecture for deploying voice services to users of voice-enabled clients such as ordinary telephone handsets or mobile phones. By means of some examples, such as call centers, we have seen that VoxML in its current form is not sufficient for the platform-independent description of such services. For this reason, we have proposed some tags as a minimal extension of VoxML which allows to access call control directly from within VoxML. The feasibility of this approach has been shown by extending the VoxPortal from Siemens ICN accordingly. We expect the following benefits from our approach:

- *Platform independency*: New services such as call centers or inquiry systems of telephone dictionaries with automatic call setup can be specified in a platform independent way. In particular, these services can be developed by anyone in a web-like fashion, without consideration of the technology of the underlying voice browsers accessing these services.
- *Separation of concerns*: Bringing new language features for call control into the VXML is a clear separation of concerns, which allows the use of such features by web authors with little or no know-how about programming.

Issues to be investigated further include tags for billing and approaches for personalized services (e.g. via cookies).

8. REFERENCES

[CSS] W3 Consortium. Cascading Style Sheets (CSS). http://www.w3.org/Style/CSS

[Fer99] J. Ferraolo. Scalable Vector Graphics (SVG) 1.0 Specification Appendix E.2 Aural Style Sheets. W3C Working Draft 3 December 1999. Http://www.w3.org/TR/1999/WD-SVG-19991203/access.html

[Gue00] B. Guedhami. Enhancing the VoxML with Language Features for Call Control. Dipl.-Ing. Thesis, Institute of Communication Networks, Munich University of Technology, 2000.

[Gun98] J. Gunderson. WAI Accessibility Guidelines: User Agent. W3C Working Draft, 14 August 1998. http://www.w3c.org/WAI/UA/WD-WAI-USERAGENT-19980814.html

[MSTAPI] Microsoft Windows NT Server TAPI White Paper, Microsoft, 1996.

[PARLAY] PARLAY, http://www.parlay.org, 1999.

[VoxPortal] SURPASS Application VoxPortal. Siemens ICN, product description A30828-X8010-S2-1-7618.

[SR99] H. Schulzrinne, J. Rosenberg. The IETF Internet Telephony Architecture and Protocols. IEEE Network, May/June 1999, pp. 18-23.

[RLS99] J. Rosenberg, J. Lennox, H. Schulzrinne. Programming Internet Telephony Service. IEEE Network, May/June 1999, pp. 42-48.

[RHJ99] HTML 4.01 Specification. W3C Recommendation 24 December 1999. Http://www.w3.org/TR/html4

[VXML] VoiceXML Forum. Voice Extensible Markup Language (VXML). Http://www.voxmlforum.com

[VoxML] Motorola. VoxML 1.2 Language Reference Version 1.0 Rev. 1.5. Http://www.motorola.com, January 2000. See also VoxML Developer Site http://www.voxml.com/voxml.html.

[XML] World Wide Web Consortium. Extensible Markup Language (XML). Http://www.w3.org/XML/

AUTHOR INDEX

KEYWORD INDEX